AF891047

The Yamuna River Basin

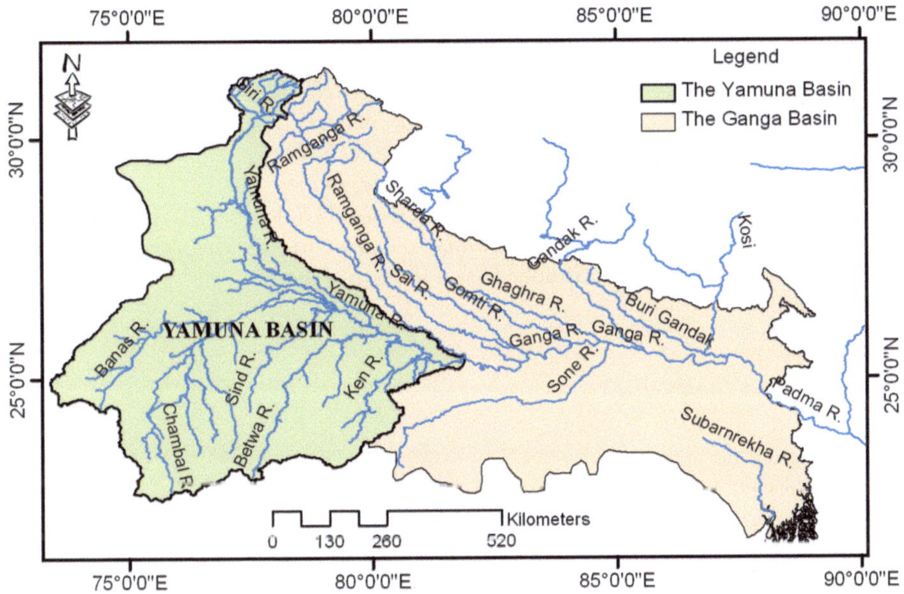

Water Science and Technology Library

VOLUME 66

For further volumes: http://www.springer.com/series/6689

The Yamuna River Basin

Water Resources and Environment

Raveendra Kumar Rai
DHI (India) Water & Environment Pvt Ltd,
New Delhi, India

Alka Upadhyay
DHI (India) Water & Environment Pvt Ltd,
New Delhi, India

C. Shekhar P. Ojha
Department of Civil Engineering, Indian Institute of Technology
Roorkee, Uttarakhand, India

and

Vijay P. Singh
Department of Biological & Agricultural Engineering,
Texas A and M University, College Station, TX, USA

Springer

Dr. Raveendra Kumar Rai
DHI (India) Water & Environment Pvt Ltd
III rd Floor, NSIC-STP Complex
Okhla Industrial Area
New Delhi-110 020
India
rai.raveendra@gmail.com

Alka Upadhyay
DHI (India) Water & Environment Pvt Ltd
III rd Floor, NSIC-STP Complex
Okhla Industrial Area
New Delhi-110 020
India
upadhyayalka1@gmail.com

Dr. C. Shekhar P. Ojha
Indian Institute of Technology Roorkee
Department of Civil Engineering
Roorkee, Uttarakhand
India
cspojha@rediffmail.com

Vijay P. Singh
Texas A and M University
Department of Biological & Agricultural Engineering
Scoates Hall, 2117 TAMU
77843-2117 College Station
Texas
USA
vsingh@tamu.edu

ISSN 0921-092X
ISBN 978-94-007-2000-8 e-ISBN 978-94-007-2001-5
DOI 10.1007/978-94-007-2001-5
Springer Dordrecht Heidelberg London New York

Library of Congress Control Number: 2011936862

Printed on acid-free paper

Springer is part of Springer Science+Business Media (www.springer.com)

Preface

Challenges faced by more and more countries in their struggle for economic and social development are increasingly related to water. Water shortages, quality deterioration and flood impacts are among the problems which require greater attention and action. Globally the good quality of water for various uses has been scarce. Therefore, its management and allocation to meet out the various demand such as domestic, agriculture, industrial and environmental become important. However, in many cases due to involvement of inter-state and inter-country boundaries, the water disputes have been become a common feature leading to poor management of water resources and trans-boundary issues in the basin. To overcome this issue, various river basin authorities have been come into picture worldwide. India has also constituted various river basin authorities for better integrated water resources managements (IWRM), such as Ganga River Basin Authority, Narmada Valley Development Authority, Bhagirathi River Valley Development Authority, Upper Yamuna River Board, etc under Ministry of Environment and Forests and Ministry of Water Resources of Government of India; and State Governments. The IWRM is a participatory planning and implementation process, based on sound science, which brings stakeholders together to determine how to meet the society's long term needs for water resources while maintaining the essential ecological services and economic benefits. IWRM helps to protect the world's environment, foster economic growth and sustainable agricultural development, promote democratic participation in governance, and improve the human health. In IWRM, evaluation of natural resources is one of the major components, which requires scientific approaches.

Further to this, in India several river basins exist. Among many others, the Ganga River Basin forms nearly twenty six percent of total geographical area of the country which supports forty three percent of its population; of which Yamuna basin contributes approximately 40.2% of the drainage area of the Ganga River basin. Beside this, Yamuna River is amongst the most polluted rivers of India, which carries huge magnitude of wastewater from various urban centers and industrial towns. More than 14 towns/cities are civilized along the river banks, of which National Capital Region of Delhi (NCR-Delhi) is one of them. Due to wastewater disposal in the river, water quality has been deteriorated and is further aggravated due to less water availability during the lean periods (i.e. November to June). Considering this situation, the importance of detailed study on Yamuna River Basin has been increased, which form the basis to write a book that follows the concept of IWRM. This book "The Yamuna River Basin: Water Resources & Environment" has been designed from our continuous efforts from last three years. This book not only provides the good understanding and insight into the Yamuna River basin but also can be used as guidelines for carrying out the integrated evaluation of water resources for other river basins of India.

This book has been organized into thirteen chapters starting from the introduction to the River Basins of India to the Crop Management Plan for Yamuna basin for water resources management. Chapter 2 provides brief idea of Yamuna River basin along with inter-state water sharing. The detailed analyses of climatic characteristics of the basin have been presented in Chapter 3. This chapter also includes the development of isopluvial map along with the spatial variation of climatic parameters important to the water resources and agricultural planning. In recent years, the climate change has been come out as a major issue, and therefore, the impact of climate change is included in Chapter 4 of this book. This chapter presents detailed statistical methods for investigating the changes in climatic pattern in the Yamuna basin. The spatio-temporal variability in the climatic parameters has been adequately described through mapping and tabular results.

In Chapter 5, topography, geomorphology and geology of basin have been addressed in details. This chapter also includes the detailed analysis of geomorphological characteristics of the basin and their catchments, which are further used to derive the catchment wise unit hydrographs in Chapter 7. The soil, landuses and agriculture of the basin have been covered in Chapter 6. Detailed analyses of spatial variability of these parameters are presented in this chapter. Chapter 7 include the detailed analysis of socio-economics of the Yamuna basin starting from the demographic variation to employment followed by the poverty index and human development index (HDI) for the basin.

Chapter 8 include the catchment wise water budgeting for analyzing the surplus and deficit catchments in terms of water resources, so that adequate attention can be made for integrated water resources management. A brief methodology of flood estimation based on the approach suggested by Central Water Commission and Geomorphological Unit Hydrograph has been presented for hydrological design in the catchments.

Water pollution is the major concern of the Yamuna River, and therefore, thoroughly covered in Chapters 9, 10 and 11. The Chapter 9 presents the critical issues of the Yamuna River as far as the quality is concern. This chapter also includes the water quality parameters and their role on human body and aquatic system with their standard limits given by Bureau of Indian Standard (BIS), World Health Organization (WHO) and Central Pollution Control Board (CPCB). However, the wastewater analysis including the available treatment capacities, trends and gap has been discussed in Chapter 10.

The Chapter 11 gives a better insight into the water quality indices (WQI) and their application for assessing the water quality. In this chapter, little mathematical expression has been added for fast computations. Beside the use of WQI, the water quality trend in the river has been analyzed using the actual data. The water quality analysis has been further carried out using the bio-mappings of dissolved oxygen through the river length.

In Chapter 12, a detailed methodology on the environmental flow estimation has been presented along with the applications. A methodology for polluted and over exploited rivers has been developed and applied for the Yamuna. The developed methodology is highly useful for such other rivers in India and other countries. In India, nearly 75 percent of utilizable water resources are used in irrigation. Also due to intensive agriculture and changes in climatic patterns, water scarcities are becoming frequent for agriculture, industry, energy and households. Therefore, it is imperative to implement cropping patterns that can lead to reduced agricultural water demand, while preserving farmers' economic returns. To consider this issue, a detailed analysis of crop water demand has been carried out at district level in Chapter 13, and suitable crop plan has been suggested for entire basin at district level.

For writing this book, a vast analysis of data has been done. The data agencies, such as Central Pollution Control Board (CPCB), Central Water Commission (CWC), State Pollution Control Boards (SPCBs), Planning Commission of India, Survey of India, Agricultural Departments, Indian Council of Agricultural research (ICAR), Census Department, etc are duly acknowledge for supplying the good quality of data. Published literatures and documents were also used while writing the manuscript and are duly referred at the last of each chapter. If by chance the reference is missing, we are highly acknowledged to them.

The book will be useful to the engineers, agricultural scientists, environmentalist, planners, managers and administrators who are concerned with water resources.

Authors would like to take this opportunity to express their deep gratitude to our colleagues for their constructive suggestions and encouragement throughout the work. Special thanks are also due to Dr. R. C. Trivedi, Former Director, Central Pollution Control Board; Dr. Ajay Pradhan, Managing Director, DHI (India) Water & Environment Pvt Ltd, New Delhi; Dr. R. Dalwani, Director and Mr. S. K. Srivastava, Deputy Director of National River Conservation Directorate, Ministry of Environment & Forests, Government of India; Mr. Praveen K. Gupta, Former Director, Central Water Commission; Mr. Nitin Joshi, Research Scholar, Indian Institute of Technology Roorkee; and Mr. Prashant Eknath Kadam, Mr. Pankaj Sinha and Ms. Sagarika Rath from DHI (India) Water & Environment Pvt Ltd, New Delhi for their support and kind cooperation.

New Delhi, India — Raveendra Kumar Rai
New Delhi, India — Alka Upadhyay
Roorkee, India — C. Shekhar P. Ojha
College Station, Texas — Vijay P. Singh

Contents

List of Figures

List of Tables

Chapter 1
River Basins of India and Water Resources

Abstract This chapter deals with a brief description of river basins of India and their water resources, including two ways of river basin classification. The basis for classification comprises geological and topographical conditions, and catchment size. Both surface and ground water resource potential are summarized in order to understand the baseline of river basins of India.

A drainage basin is defined by the point on the stream or river, meaning the land area that drains into this point, called outlet. This land area receives water from precipitation (i.e., rainfall or snowfall) and drains downhill into river, lake, reservoir, sea or ocean. The drainage basin acts like a funnel, collecting all the water within the area covered by the basin and channeling it into a waterway. It is the drainage basin where much of the hydrologic action takes place. Indeed it can be called as the natural laboratory of hydrology. India is the land of various water bodies running from north to south and east to west.

1.1 River Systems

India's rivers are classified as Himalayan, peninsular, coastal, and inland-drainage basin rivers. Himalayan rivers are snow fed and maintain a high to medium rate of flow throughout the year. The heavy annual average rainfall amounts in the Himalayan catchments further add to their flow. During the monsoon months of June to September, the catchments are prone to flooding. The peninsular rivers are rainfed, whereas coastal rivers are found in the western part of India and they are short and periodic. Coastal streams, especially in the west, are short and episodic. Rivers of the Inland system, centered in western Rajasthan, are few and frequently disappear in years of scant rainfall. The majority of South Asia's major rivers flow through broad, shallow valleys and drain into the Bay of Bengal. The river systems of India, including the Indus basin, are shown in Fig. 1.1.

1.1.1 Himalayan Rivers

The principal Himalayan Rivers are the Indus, the legendary Ganga and the Brahmaputra. These rivers are both naturally snow fed and rainfed and hence perennial throughout the year. Himalayan rivers discharge approximately 70% of

R.K. Rai et al., *The Yamuna River Basin*, Water Science and Technology Library 66,
DOI 10.1007/978-94-007-2001-5_1, © Springer Science+Business Media B.V. 2012

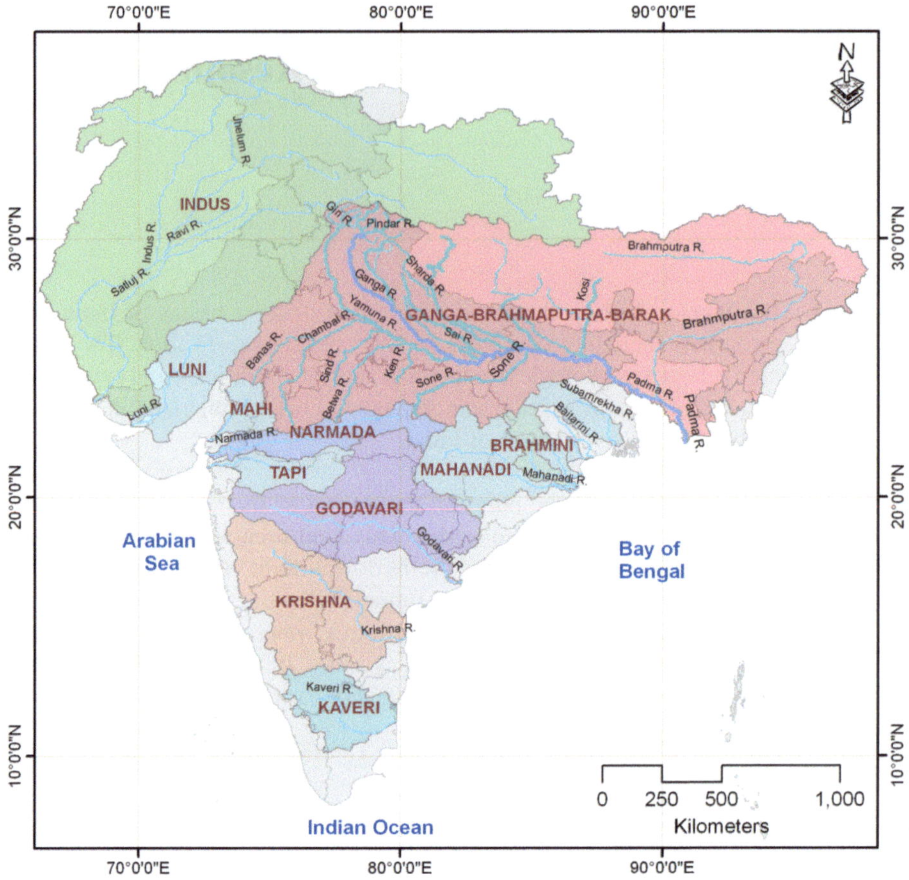

Fig. 1.1 Important river basins of India, including part of Pakistan

their flow into the sea. The other rivers that chalk out their origin and course in the Himalayan ranges include: Sutlej, Chenab or Chandra Bhaga, Beas, Ravi, Jhelum, Yamuna, and Spiti. The principal Himalayan Rivers in India are given in Table 1.1.

The Indus originates near Mansarovar in Tibet, which flows through Kashmir and enters in Pakistan and finally terminates in the Arabian Sea near Karachi. During the course of travel, a number of tributaries, such as the Sutlej, the Beas, the Ravi, the Chenab and the Jhelum, join the Indus. These tributaries flow through India.

The Ganga River basin, India's largest, includes approximately 25% of the nation's area; it is bounded by the Himalayas in the north and the Vindhya Range to the south. The Ganga has its source in the glaciers of the Greater Himalayas, which form the frontier between India and Tibet in northwestern Uttar Pradesh (now called Uttarakhand). Many Indians believe that the source of the legendary Ganga, and several other important Asian rivers, lies in the sacred Mapam Yumco Lake (known to

Table 1.1 Principal Himalayan rivers of India (Jain et al., 2007)

Group No.	Drainage area (km^2)	Name	Himalayan area (km^2)
I	>256,000	Indus	265,728
II	128,000–256,000	Brahmaputra	253,952
III	38,400–128,000	Kosi	61,184
		Karnali	52,736
		Sutlej	47,360
IV	25,600–38,400	Gandak	37,376
		Jhelum	33,280
		Manas	30,720
		Chenab	26,880
		Raidak	26,112
V	12,800–25,600	Ganga	22,784
		Luhit	20,480
		Subansiri	17,920
		Kali	16,128
		Beas	14,336
		Dibang	12,800
VI	<12,800	Tista	12,228
		Yamuna	11,520
		Ravi	7,936
		Rapti	7,680
		Ramganga	6,656
		Baghmati	3,840

the Indians as Manasarowar Lake) of western Tibet located approximately 75 km northeast of the India-China-Nepal tripoint. In the northern part of the Ganga River basin, practically all of the tributaries of the Ganga are perennial streams. However, in the southern part, located in the states of Rajasthan and Madhya Pradesh, many of the tributaries are not perennial.

Of all the rivers in India, the Brahmaputra has the greatest volume of water because of heavy annual rainfall in its basin. At Dibrugarh the annual rainfall averages 2,800 mm, and at Shillong it averages 2,430 mm. Rising in Tibet, the Brahmaputra flows south into Arunachal Pradesh after breaking through the Great Himalayan Range and dropping rapidly in elevation. It continues to fall through gorges impassable by man in Arunachal Pradesh until finally entering the Assam Valley where it meanders westward on its way to joining the Ganga in Bangladesh.

1.1.2 Coastal Rivers

Coastal regions in India are primarily designated by the areas that lie close to or nearby sea and oceans. The coastal regions comprise the states lying by the Bay of Bengal and Indian Ocean; and are Odissa, Andhra Pradesh, Tamil Nadu, Kerala, Karnataka, Maharashtra, Goa, and part of Gujrat. The coastal rivers, which can almost be regarded as streams, especially on the west coast, are short in length,

Table 1.2 Important coastal rivers of India

States	Rivers
West Bengal Coastal Rivers	Subarnarekha, Kharkai, Kamgasabati
Andhra Pradesh Coastal Rivers	Vamsadhara, Nagavati, Sharada
Tamil Nadu Coastal Rivers	Thamirabarani, Palar, Vaigai, Vellar, Vasishta, Sweta
Karnataka Coastal Rivers	Netravati, Sharavathi, Aghanashini rivers
Maharashtra Coastal Rivers	Shastri, Gad, Vashishti, Savitri, Patalganga, Ulhas, Thane Creek, Vasai Creek, Mithi, Oshiware, Dahisar, Tansa, Vaitarna, Surya

episodic, ephemeral and have small catchments. There are as many as six hundred rivers on the west coast. Some of the important coastal rivers are given in Table 1.2.

1.1.3 Deccan Rivers

The southern section of India is predominated by the Deccan plateau, the Western Ghats, the Eastern Ghats, the Vindhyan range and the Satpura range. The Deccan region is also known as Peninsular India, which mainly holds the vast plateau regions of the western and southern India. The Deccan Rivers are primarily rain-fed and as a consequence immensely fluctuate in volume. All of these rivers are non-perennial in nature. Peninsular rivers contribute 30% of the total surface water resources of India. The Deccan or Peninsular rivers behave according to the providential wish, at times overwhelming in their course and at times, running dry most of the year. The major rivers in this region are Godavari, Krishna, Cauveri, Mahanadi, Narmada, Tapi, Indravati, Tungabhadra and Bhima. Among these Narmada and Tapi are the west flowing rivers which terminate into the Arabian Sea. Pennar, Damodar, Sharavati, Netravati, Bharathapuzha, Periyar, and Pamba rivers are other secondary rivers of this region. Most of these rivers are east flowing into the Bay of Bengal except the Pamba. These rivers, generally seasonal and dependent on rainfall, cause trouble during the occurrence of rainfall.

The Mahanadi, rising in the state of Madhya Pradesh, is an important river in the state of Orissa. In the upper drainage basin of the Mahanadi, which is centered on the Chhattisgarh Plain, periodic droughts contrast with the situation in the delta region where floods may damage the crops in what is known as the rice bowl of Orissa. Hirakud Dam, constructed in the middle reaches of the Mahanadi, has helped in alleviating these adverse effects by creating a reservoir.

The source of the Godavari is northeast of Bombay (Mumbai in the local Marathi language) in the state of Maharashtra, and the river follows a southeasterly course for 1,400 km to its mouth on the Andhra Pradesh coast. The Godavari River basin area is second in size only to the Ganga; its delta on the east coast is also one of the country's main rice-growing areas. It is known as the "Ganga of the South," but its discharge, despite the large catchment area, is moderate because of the medium level of annual rainfall, for example, about 700 mm at Nasik and 1,000 mm at Nizamabad.

The Narmada and the Tapti are the only major rivers that flow into the Arabian Sea. The Narmada rises in Madhya Pradesh and crosses the state, passing swiftly through a narrow valley between the Vindhya Range and spurs of the Satpura Range. It flows into the Gulf of Khambhat (or Cambay). The shorter Tapti follows a generally parallel course, between 80 and 160 km to the south of the Narmada, flowing through the states of Maharashtra and Gujarat on its way into the Gulf of Khambhat.

The Krishna rises in the Western Ghats and flows east into the Bay of Bengal. It has a poor flow because of low level of rainfall in its catchment area i.e. 660 mm annually at Pune. Despite its low discharge, the Krishna is the third longest river in India.

The source of the Kaveri is in the state of Karnataka, and the river flows southeastward. The waters of the river have been a source of irrigation since antiquity; in the early 1990s, an estimated 95% of the Kaveri was diverted for agricultural use before emptying into the Bay of Bengal. The delta of the Kaveri is so mature that the main river has almost lost its link with the sea, as the Kollidam, the distributary of the Kaveri, bears most of the flow.

1.1.4 Inland Drainage Basins

The rivers in drainage basins inland consist of rivers in western Rajasthan. These rivers are very few in numbers and they flow for a very short period of time. The rivers in this category are Sambhar which vanish in the desert sands, and the Luni that drains into the Rann of Kutch. Another major river of the western region is the Mahi, which originates from the northern slope of the Vindhyan range in Madhya Pradesh. It enters the southeastern portion of Rajasthan and flows through the Banswara district, and finally terminates into the sea by a wide estuary near Cambay.

1.2 Classification of Indian River Basin Based on Drainage Area

On the basis of drainage area of the rivers, the Indian river can be classified into three groups: (i) major river basins: $A > 20{,}000\ km^2$, (ii) medium river basins: $2{,}000 \leq A \leq 20{,}000\ km^2$, and (iii) minor river basins: $A < 2{,}000\ km^2$. According to this classification, the number of major and medium river basins are 12 and 46, respectively (Jain et al., 2007) and these contribute nearly 92% of the total runoff in the country. The remaining 8% runoff is contributed by the minor rivers. Of the major rivers, the Ganga-Brahmaputra-Meghana (GBM) has the biggest drainage area of 1.10 million km^2, which is more than 43% of the total geographical area of the country. The catchment area of medium rivers is about 0.25 million km^2, and Subernarekha is the largest among them having a drainage area of 19,300 km^2. The drainage area of major and medium river basins, length of rivers and their origin are summarized in Tables 1.3 and 1.4, respectively.

Table 1.3 Major river basins of India (CWC, 1990; MoWR, 2004)

S. No.	Name	Origin	Length (km)	Area (km^2)	States falling
1	Indus	Mansarovar (Tibet)	1,114+	321,289+	J&K, Punjab, Himachal Pradesh, Rajasthan and Chandigarh
2	(a) Ganga	Gangotri (Uttarkashi)	2,525+	861,452+	Uttarakhand, Uttar Pradesh, Himachal Pradesh, Haryana, Rajasthan, Madhya Pradesh, Bihar, West Bengal, and Delhi
	(b) Brahmaputra	Kailash Range (Tibet)	916+	194,413+	Arunachal Pradesh, Assam, Meghalaya, Nagaland, Sikkim, West Bengal, Mizoram, and Tripura
	(c) Barak and Meghna			41,723+	Assam, Meghalaya, Nagaland, Manipur, Mizoram, and Tripura
3	Sabarmati	Aravalli Hills (Rajasthan)	371	24,674	Rajasthan and Gujarat
4	Mahi	Dhar (Madhya Pradesh)	583	34,842	Rajasthan, Madhya Pradesh and Gujarat
5	Narmada	Amarkantak (Madhya Pradesh)	1,312	98,796	Madhya Pradesh, Maharashtra and Gujarat
6	Tapi	Betul (Madhya Pradesh)	724	65,145	Madhya Pradesh, Maharashtra and Gujarat
7	Brahmini	Ranchi (Jharkhand)	799	39,033	Madhya Pradesh, Bihar and Odissa
8	Mahanadi	Nazri Town (Madhya Pradesh)	851	141,589	Madhya Pradesh, Maharashtra, Bihar, Chattisgarh and Odissa
9	Godavari	Nasik (Maharashtra)	1,465	312,812	Maharashtra, Andhra Pradesh, Madhya Pradesh, Odissa and Pondicherry
10	Krishna	Mahabaleshwar (Maharashtra)	1,401	258,948	Maharashtra, Andhra Pradesh and Karnataka
11	Pennar	Kolar (Karnataka)	597	55,213	Andhra Pradesh and Karnataka
12	Cauvery	Coorg (Karnataka)	800	81,155	Tamil Nadu, Karnataka, Kerala and Podicherry
		Total		2,528,084	

Table 1.4 Medium river basins of India (CWC, 1990; MoWR, 2004)

S. No.	Name	Origin	State	Length (km)	Area (km^2)
West Flowing Rivers					
1	Ozat	Kathiawar	Gujarat	128	3,189
2	Shetrunji	Dalkania	Gujarat	182	5,514
3	Bhadar	Rajkot	Gujarat	198	7,094
4	Aji	Rajkot	Gujarat	106	2,139
5	Dhadhar	Panchmahal	Gujarat	135	2,770
6	Purna	Dhosa	Maharashtra	142	2,431
7	Ambika	Dangs	Maharashtra	142	2,715
8	Vaitarna	Nasik	Maharashtra	171	3,637
9	Dammanganga	Nasik	Maharashtra	143	2,357
10	Ulhas	Raigarh	Maharashtra	145	3,864
11	Savitri	Pune	Maharashtra	99	2,899
12	Sastri	Ratnagiri	Maharashtra	64	2,174
13	Washishthi	Ratnagiri	Maharashtra	48	2,239
14	Mandvi	Belgaum	Karnataka	87	2,032
15	Kalinadi	Belgaum	Karnataka	153	5,179
16	Gangavati or Bedti	Dharwar	Karnataka	152	3,902
17	Sharavati	Shimoga	Karnataka	122	2,209
18	Netravati	Dakshina Kannada	Karnataka	103	3,657
19	Chaliar or Baypore	Elamtalvi Hills	Kerala	169	2,788
20	Bharathapuzha	Annamalai Hills	Tamil Nadu	209	6,186
21	Periyar	Sivajini Hills	Kerala	244	5,398
22	Pamba	Devarmalai	Kerala	176	2,235
East Flowing Rivers					
23	Burhabalang	Mayurbahanj	Odissa	164	4,837
24	Baitarni	Keonjhar	Odissa	365	12,789
25	Rushikulya	Phulbani	Odissa	146	7,753
26	Bahuda	Ramgiri village	Odissa	73	1,248
27	Vamsadhara	Kalahandi	Odissa	221	10,830
28	Nagavali	Kalahandi	Odissa	217	9,410
29	Sarda	Vishakhapatnam	Andhra Pradesh	104	2,725
30	Eleru	Vishakhapatnam	Andhra Pradesh	125	3,809
31	Vogarivagu	Guntur	Andhra Pradesh	102	1,348
32	Gundlakamma	Kurnool	Andhra Pradesh	220	8,494
33	Musi	Nellore	Andhra Pradesh	112	2,219
34	Paleru	Nellore	Andhra Pradesh	104	2,483
35	Muneru	Nellore	Andhra Pradesh	122	3,734
36	Swarnamukhi	Koraput	Odissa	130	3,225
37	Kandleru	Vinukonda	Andhra Pradesh	73	3,534
38	Kortalaiyar	Chinglepet	Tamil Nadu	131	3,521
39	Palar	Kolar	Karnataka	348	17,871
40	Varahandi	North Arcot	Tamil Nadu	94	3,044
41	Ponnaiyar	Kolar	Karnataka	396	14,130
42	Vellar	Chithri Hills	Tamil Nadu	193	8,558
43	Vaigai	Madurai	Tamil Nadu	258	7,031
44	Pambar	Madurai	Tamil Nadu	125	3,104
45	Gundar	Madurai	Tamil Nadu	146	5,647
46	Vaippar	Tirunolvolli	Tamil Nadu	130	5,288
47	Tambraparni	Tirunolvolli	Tamil Nadu	130	5,969
48	Subarnarekha	Nagri/Ranchi	Jharkhand	395	19,296
			Total		248,505

1.3 Water Resources of India

The total utilizable water resources of the country are assessed as 1,086 km^3. A brief description of surface and groundwater water resources of India is given below.

1.3.1 Surface Water Resources

In the past, several organizations and individuals have estimated water availability for the nation. Recently, the National Commission for Integrated Water Resources Development estimated the basinwise average annual flow in Indian river systems as 1953 km^3 (NCIWRD, 1999). The basinwise surface water resources are given in Table 1.5.

Table 1.5 Basin wise surface water resources potential (km^3)

S. No.	River basins	Average annual flows (NCIWRD, 1999)	Utilizable flows
1	Indus (Catchment area in Indian territory)	73.31	46.00
2	(a) Ganga	525.02	250.00
	(b) Brahmaputra	629.05	24.00
	(c) Meghna (Barak)	48.36	NA
3	Godavari	110.54	76.30
4	Krishna	69.81	58.00
5	Cauvery	21.36	19.00
6	Pennar	6.32	6.86
7	Mahanadi	66.88	49.99
8	Brahmani and Baitarni	28.48	18.30
9	Subarnarekha	12.37	6.81
10	Sabarmati	3.81	1.93
11	Mahi	11.02	3.10
12	Narmada	45.64	34.50
13	Tapi	14.88	14.50
14	East flowing and river from Mahanadi to Godavari and Krishna to Pennar	22.52	13.11
15	East flowing rivers between Pennar and Kanyakumari	16.6	16.73
16	West flowing rivers of Kutch and Saurashtra including Luni	15.10	14.98
17	West flowing rivers from Tapi to Tadri	87.41	11.94
18	West flowing rivers from Tadri to Kanyakumari	113.53	24.27
19	Area of inland drainage in Rajasthan desert	NA	NA
20	Minor rivers draining to Myanmar (Burma) and Bangladesh	31.00	
	Total	1,952.8	690.00

The utilizable water resource is the quantum of withdrawable water from its place of natural occurrence. Within the limitations of physiographic conditions and socio-political environment, legal and constitutional constraints and the technology of development available at present, utilizable quantity of water from surface flow has been assessed by various authorities differently. The utilizable annual surface water of the country is 690 km^3 (Table 1.5). In a majority of river basins, the percent utilization is quite high and is within the range of 50–95% of utilizable surface resources (IWRS, 1996; Kumar et al., 2005). There is considerable scope for increasing the utilization of water in the Ganga–Brahmaputra basins by construction of storages at suitable locations in neighboring countries.

1.3.2 Ground Water Resources

The annual potential natural groundwater recharge from rainfall in India is about 342.43 km^3, which is 8.56% of the total annual rainfall of the country. The annual potential groundwater recharge augmentation from canal irrigation system is about 89.46 km^3. Thus, the total replenishable groundwater resource of the country is assessed as 431.89%. After allotting 15% of this quantity for drinking, and 6 km^3 for industrial purposes, the remaining can be utilized for irrigation purposes. Thus, the available groundwater resource for irrigation is 361 km^3, of which utilizable quantity (90%) is 325 km^3. The estimates made by the Central Groundwater Board (CGWB) of the total replenishable groundwater resource, provision for domestic, industrial and irrigation uses and utilizable groundwater resources for future use are given in Table 1.6. The basin-wise per capita water availability varies between 13,393 m^3 per annum for the Brahmaputra–Barak basin to about 300 m^3 per annum for the Sabarmati basin. The basin-wise groundwater potential of the country is given in Table 1.7.

Due to increasing population, water demand for various uses such as domestic, food, industrial and power will be increased. For the year 2050, NCIWRD (1999) has estimated that the total water demand in the country will be 973–1,180 km^3, whereas the utilizable surface and ground water resources are 690 km^3 and 396 km^3, respectively. In spite of the spatial variability of utilizable water resources in the country, India will become a highly water stressed country, if better water management plan is not implemented at the watershed level.

Table 1.6 Groundwater resources of India (km^3/annum)

1	Total replenishable groundwater resource	432
2	Provision for domestic, industrial and other uses	71
3	Available groundwater resource for irrigation	361
4	Utilizable groundwater resource for irrigation (90% 361 km^3)	325
5	Total utilizable groundwater resources (sum of 1–4)	396

Table 1.7 Basin wise groundwater resources potential (in km^3/annum)

S. No.	Basin	Total replenishable GW resources	Provision for domestic, industrial and other uses	Available GW for irrigation	Net draft	Balance GW Potential	Level of GW development (%)
1	Brahmani and Baitarni	4.05	0.61	3.44	0.29	3.16	8.45
2	Brahmaputra	26.55	3.98	22.56	0.76	21.80	3.37
3	Chambal Composite	7.19	1.08	6.11	2.45	3.66	40.09
4	Cauvery	12.30	1.84	10.45	5.78	4.67	55.33
5	Ganga	170.99	26.03	144.96	48.59	96.37	33.52
6	Godavari	40.65	9.66	30.99	6.05	24.94	19.53
7	Indus	26.49	3.05	23.43	18.21	5.22	77.71
8	Krishna	26.41	5.58	20.83	6.33	14.50	30.39
9	Kutch and Saurashtra Composite	11.23	1.74	9.49	4.85	4.64	51.14
10	Madras and South Tamil Nadu	18.22	2.73	15.48	8.93	6.55	57.68
11	Mahanadi	16.46	2.47	13.99	0.97	13.02	6.95
12	Meghna	8.52	1.28	7.24	0.29	6.95	3.94
13	Narmada	10.83	1.65	9.17	1.99	7.18	21.74
14	Northeast Composite	18.84	2.83	16.02	2.76	13.26	17.20
15	Pennar	4.93	0.74	4.19	1.53	2.66	36.60
16	Subarnarekha	1.82	0.27	1.55	0.15	1.40	9.57
17	Tapi	8.27	2.34	5.93	1.96	3.97	33.05
18	Western Ghat	17.69	3.19	14.50	3.32	11.18	22.88
	Total	431.43	71.08	360.35	115.21	245.13	31.97

References

CWC (1990) Storages in river basins of India. Report No. 62/90, Central Water Commission, New Delhi, India

IWRS (1996) Theme paper on 'Inter Basin Transfers of Water for National Development – Problems and Prospects'. Indian Water Resources Society, Roorkee

Jain SK, Agarwal PK, Singh VP (2007) Hydrology and Water Resources of India. Springer-Verlag, New York. ISBN-10 1-4020-5179-4

Kumar R, Singh RD, Sharma KD (2005) Water Resources of India. Curr Sci 89(5):794–811

MoWR (2004) Water Resources of India. Ministry of Water Resources, Government of India, New Delhi

NCIWRD (1999) Integrated water resources development – A plan for action. Report of The National Commission for Integrated Water Resources Development, Ministry of Water Resources, New Delhi

Chapter 2
Salient Features of the Yamuna River Basin

Abstract In this chapter, salient features of the Yamuna River basin are presented, including mythological importance, catchment area description within states, and general river flow regime. Major abstractions, including water-related issues at various locations in the main river stem, are also discussed. The drainage area of the Yamuna River basin encompasses six states, accordingly the inter-state agreement has been formulated for water sharing. A summary of the inter-state agreement is also included in the chapter. Consequences of various abstraction points in the main river are duly addressed in order to identify critical stretches.

The Yamuna River is the largest tributary of the Ganga River (Fig. 2.1). It is also considered as a sacred river in India. According to the mythological legends, Yamuna was the daughter of Surya (the Sun God), and sister of Yama (the God of Death). A popular belief is that those who take a dip in its holy waters are not tormented by the fears of death. The Yamuna River is intimately connected to Lord Krishna's life. Lord Krishna sanctified the Yamuna River from the beginning of his *avtar*

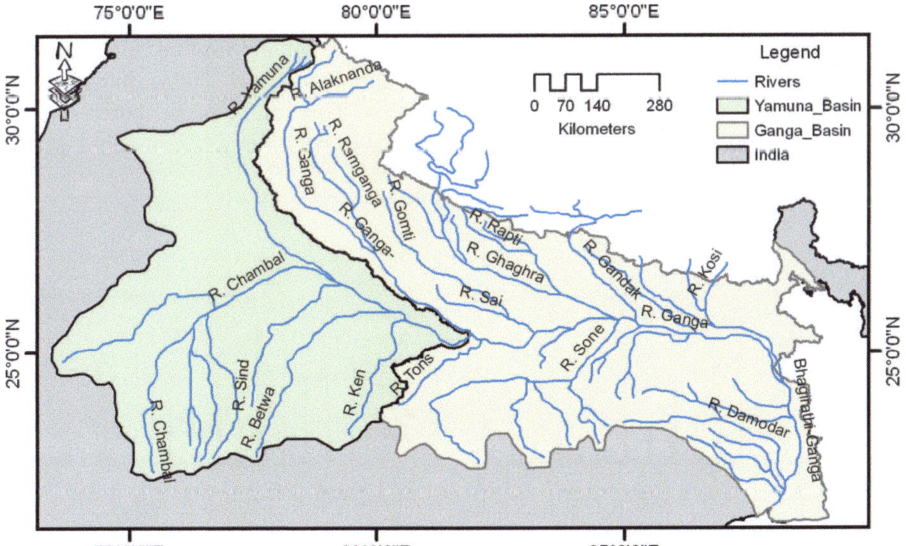

Fig. 2.1 Gangatic River system including the Yamuna River

R.K. Rai et al., *The Yamuna River Basin*, Water Science and Technology Library 66,
DOI 10.1007/978-94-007-2001-5_2, © Springer Science+Business Media B.V. 2012

(incarnation) in the world. While his father Vasudeva was crossing the Yamuna with baby lord Krishna for a safe place at the other bank of the river, the river was in spate. Legends say that the moment the rising water touched the feet of Lord Krishna, the flood receded.

The Yamuna River originates from the Yamunotri Glacier near Banderpoonch peaks (30°59′ N and 78°27′ E) in the Mussourie range of the lower Himalayas at an elevation of about 6,387 m above mean sea level in the district of Uttarkashi (Uttarakhand). Arising from the source, the Yamuna River flows through a series of valleys for about 200 km in lower Himalayas and then emerges into Indo-Gangetic plains. At this reach, the gradient of the river is steep and the entire geomorphology of the valley has been carved by the erosive action of the river water. In the head-water reach of 200 km, the Yamuna draws water from several major streams. The combined stream flows through the Shivalik range of Himachal Pradesh and Uttarakhand and enters into plains at Dak Pathar in Uttarakhand where the river water is regulated by a weir and is diverted into a canal for power generation (http://www.cwc.nic.in). From Dak Pathar, the Yamuna flows through the famous Sikh religious shrine of Poanta Sahib and reaches Hathnikund in the Yamuna Nagar district of Haryana. At Hathnikund, river water is diverted into Western Yamuna Canal (WYC) and Eastern Yamuna Canal (EYC) for irrigation and drinking water supply of Delhi. The catchment of river up to the Hathinikund headworks is nearly 12,950 km^2. At this point the normal high flow is of the order of 7,079 m^3/s, whereas the minimum flow is 70 m^3/s. During lean season, practically no water flows in the river downstream of Hathnikund barrage and the river remains dry in several stretches between Hathnikund and Delhi. Only approximately 160 cusec (4.53 m^3/s) is released downstream of Hathnikund for ecological considerations which vanish a after short distance due to losses (Fig. 2.2). Upstream of Hathnikund barrage, the ratio Q_{max}/Q_{min} (where Q_{max} is the flood peak discharge and Q_{min} is the lowest flow rate in the year) is about 92 (with mean annual maximum and minimum of 5,042.8 and 54.5 m^3/s (Figs. 2.3 and 2.4) based on the data from 1994–2008), in comparison to 33 of Indus, 34 of Ganga and 4 of Amazon. This might be attributed due to the lesser glacier area in the Yamuna catchment (i.e., 125 km^2) as compared to the Indus basin (i.e. 14,043 km^2).

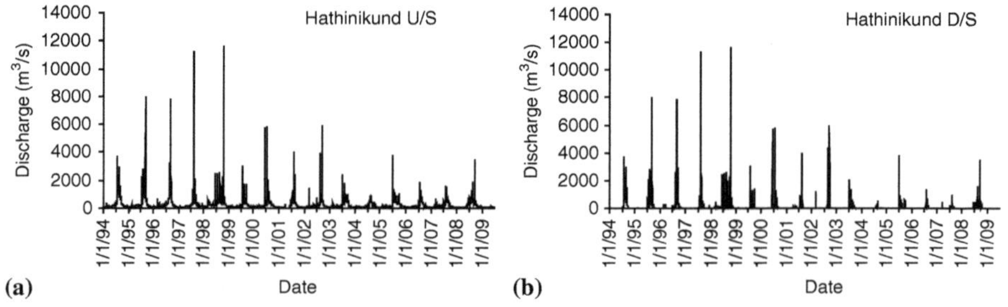

Fig. 2.2 Discharge in the Yamuna River downstream of the Hathnikund barrage

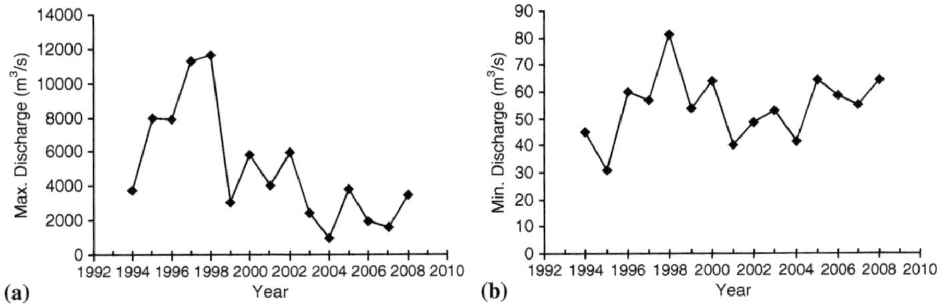

Fig. 2.3 Annual maximum and minimum discharge upstream of the Hathnikund barrage

Fig. 2.4 Ratio of Q_{max} and Q_{min} upstream of the Hathnikund barrage

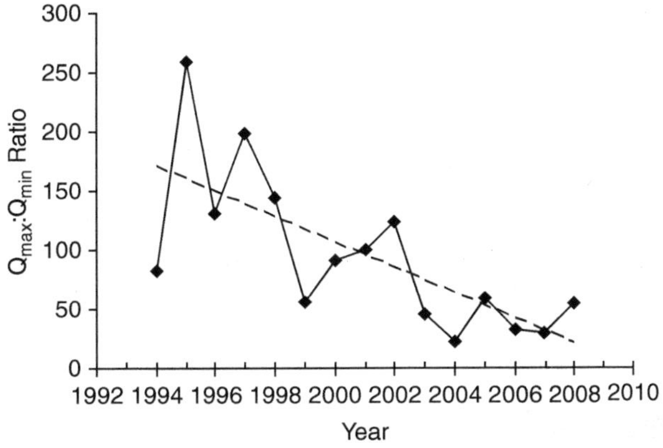

Further, increasing trend in the minimum discharge (Fig. 2.3) and declining trend in the ratio of Q_{max} and Q_{min} (Fig. 2.4) can be seen based on the data for the period of 1994–2008 which might be due to the reduction in precipitation upstream of the Hathinikund barrage. This large ratio is indicative of a wide temporal variation in the flow. During non-monsoon period the entire inflow at Hathnikund is diverted to canal systems leaving the river dry. A canal, known as Sutlej Yamuna Link (SYL) canal, joins the Satluj with Yamuna for transferring water to Haryana and Delhi. This link canal irrigates the command areas of Sirsa branch of WYC and Hissar major distributary. Yamuna River enters Delhi near Palla village after traversing for about 224 km. At Palla, the flow in the river is received through WYC for the Delhi water supply at Wazirabad. During the dry period the flow upstream of the Wazirabad barrage is approximately 10–11 m^3/s and is used as raw water for Wazirabad water works and practically no water is released in the river downstream of the barrage (Fig. 2.5).

The Yamuna River flow downstream of the Wazirabad barrage consists largely of untreated or partially treated domestic and industrial wastewater contributed by numerous drains along with the water transported by Haryana Irrigation Department

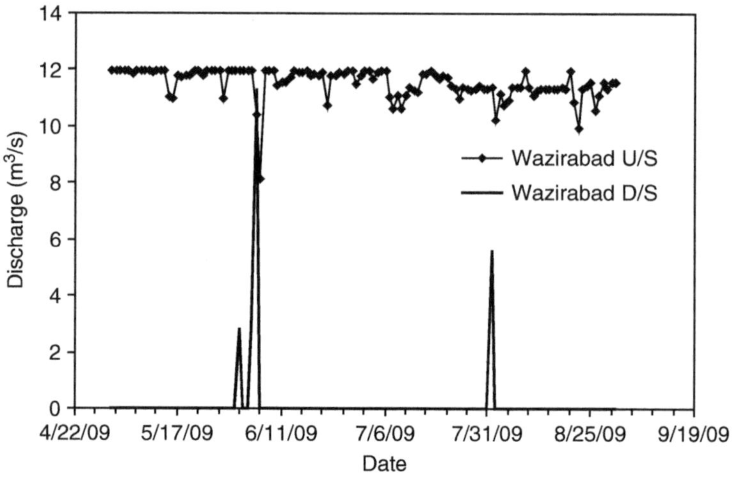

Fig. 2.5 Discharge in the Yamuna River downstream of Wazirabad barrage (1/5/2009–28/8/2009)

from WYC to the Agra Canal via Nazafgarh drain and the Yamuna. About 22 km downstream of the Wazirabad barrage, the Yamuna water is diverted into the Agra canal for irrigation upstream of the Okhla barrage. Generally, water flow through the barrage during the dry season is small with 105 cusec (2.86 m^3/s) being released for ecological considerations. Whatever water flows in the river beyond the Okhla barrage is contributed through domestic and industrial wastewaters generated by East Delhi, NOIDA, Sahibabad and joins the river through the Shahdara drain.

2.1 Tributaries of the Yamuna River

In the upper reaches of Yamuna River, there are several hill streams joining together to form the main stream (http://www.cwc.nic.in). There are four main rivers that join Yamuna in the higher Himalayan ranges, and these are Rishi Ganga, which joins on the right bank of Yamuna, whereas Unta and Hanuman Ganga join on left bank. In the lower Himalayan ranges the Yamuna River receives Kamal, Tons, Giri and Bata on its right bank and receives Aglag and Asan on its left bank. After the Okhla barrage, the important tributary joining the Yamuna River on the left bank is Hindon. However, Hindon River hardly contributes to the Yamuna River except during the monsoon season. All the water available in the Hindon River plus the water released from Upper Ganga Canal is diverted into the Agra canal through Hindon cut via Okhla barrage. The Chambal, Betwa, Sind, Ken and Mandakini are the important tributaries joining the Yamuna on the right bank. The main tributaries of the Yamuna along with location of major cities are depicted in Fig. 2.6.

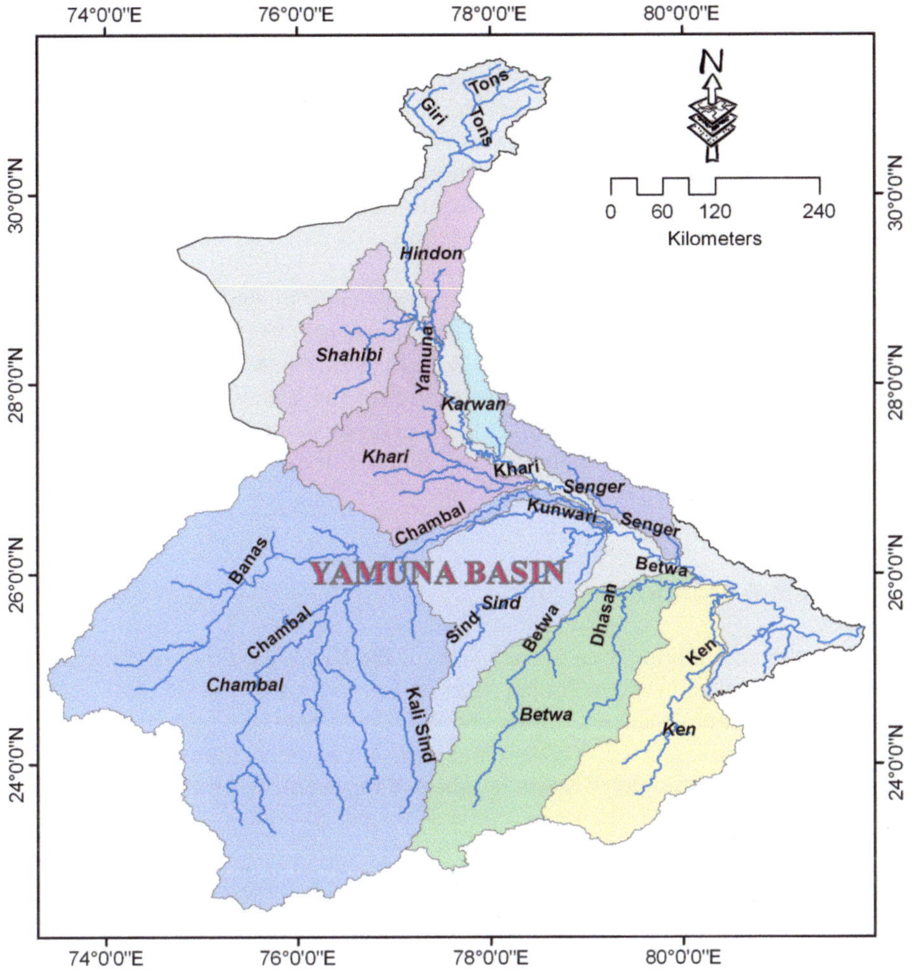

Fig. 2.6 Important tributaries and sub-catchments of Yamuna River basin

Among all these tributaries, Tons in hills and Chambal in plains are the most important tributaries in terms of their discharges. The Tons is the principal source of water in the mountainous range and generally carries more water than mainstream. In plains, during non-monsoon period, River Chambal contributes about 5–10 times more water to the Yamuna than its own flow. However, since the year 2003, there is a significant reduction in the water quantity that Chambal River discharges into Yamuna River.

2.2 Yamuna River Catchment Area

The total catchment of the Yamuna River is 366,223 km^2 (catchment basin area in various states equals 345,848 km^2; the water spread area of Yamuna River is 20,375 km^2), which is 42.5% of the total Ganga basin area and 10.7% of the total geographical landmass of the country. The catchment of the Yamuna River system covers parts of Uttarakhand, Uttar Pradesh, Himachal Pradesh, Haryana, Rajasthan, Madhya Pradesh and National Capital Territory – Delhi. The state-wise details of catchment areas are presented in Table 2.1.

The percentage contribution of the various states to the catchment area of Yamuna River are (http://www.cwc.nic.in): Uttaranchal 1.1%; Uttar Pradesh 20.4%; Himachal Pradesh 1.7%; Haryana 6.1%; Rajasthan 29.7%; Madhya Pradesh 40.6% and NCT–Delhi 0.4%. The sub-basin tributaries of the Yamuna contributes 70.9% of catchment area and remaining 29.1% account for direct drainage into the Yamuna River through other small tributaries. The major sub-catchments of the Yamuna basin are shown in Fig. 2.6.

2.3 Water Allocation

Based on the contributing area of each state to the basin, the flows in the Yamuna River were regulated between Punjab (later on Haryana) and Uttar Pradesh according to May 1954 agreement and subsequent MOU of 1994 (made effective in 1995). The available supplies on the WJC system despite the augmentation of supplies from the Bhakra system through NBK link fall short of requirement. The 1954 Agreement and 1994–1995 MOU on sharing of Yamuna Water is given as follows.

The brief agreement made on March 1954 is:

Agreement made on the 12th day of March 1954; effective the Ist of April 1950 held binding on both parties for a period of 50 years form 1/4/1950

Table 2.1 Catchment area details of Yamuna River

State/Territory	Contributing area in Yamuna basin (km^2)	Area in major sub-basin (km^2)					
		Hindon	Chambal	Sind	Betwa	Ken	Others
Uttarakhand	3,771	–	–	–	–	–	3,771
Uttar Pradesh	70,437	70,83	452	748	14,438	3,336	44,380
Himachal Pradesh	5,799						5,799
Haryana	21,265	–	–	–	–	–	21,265
Rajasthan	102,883	–	79,495	–	–	–	23,388
Madhya Pradesh	140,208	–	59,838	25,131	33,502	21,090	647
NCT-Delhi	1,485	–	–	–	–	–	1,485
Total	345,848 (100%)	7,083 (2.0%)	139,785 (40.5%)	25,879 (7.5%)	47,940 (13.9%)	24,426 (7.1%)	100,735 (29.1%)

In the event of the river discharge being less than the total indent of the two canals, i.e. Western Canal and Eastern Canal; Supplies at Tajewala will be distributed as follows:

Total river supplies at Tajewala (cusecs) say "X"	Supplies to WJC (entitlement) (cusecs)	Supplies to EJC (entitlement) (cusecs)	Remarks
Up to 5,890	$2X/3 - 47$	$X/3 - 47$	WJC first
5,890–8,790	$X - 2010$	2010	WJC first
8,790–9,280	6780	$X - 6780$	WJC first
9,280–10,900	$X - 2500$	2500	

When the supplies are in excess of 10,900 cusecs the Punjab and U.P. to utilise the excess over 10,900 cusecs in the ratio of 2:1.

This arrangement has been in force since the formation of Haryana on 1.11.66.

Whereas, the memorandum of understanding between different states are as follows:

Memorandum of Understanding between Uttar Pradesh, Haryana, Rajasthan, Himachal Pradesh and National Capital Territory of Delhi Regarding Allocation of Surface Flow of Yamuna.

1. WHEREAS the 75% dependable notional virgin flow in the Yamuna River up to Okhla has been assessed as 11.70 Billion Cubic Meters (BCM) and the mean year availability has been assessed as 13.00 BCM.
2. AND WHEREAS the water was being utilized by the Basin States ex-Tajewala and ex-Okhla for meeting the irrigation and drinking water needs without any specific allocation.
3. AND WHEREAS a demand has been made by some Basin States on this account and the need for a specified allocation of the utilizable water resources of Yamuna River has felt for a long time.
4. AND WHEREAS to maximize the utilization of the surface flow of Yamuna River a number of storage projects have been identified.
5. AND WHEREAS the States have agreed that a minimum flow in proportion of completion of upstream storages going up to 10 cumecs shall be maintained downstream of Tajewala and downstream of Okhla Headwork throughout the year from ecological considerations, as upstream storages are built up progressively in a phased manner.
6. AND WHEREAS it has been assessed that a quantum of 0.68 BCM may not be utilizable due to flood spills.
7. NOW THEREFORE, considering their irrigation and consumptive drinking water requirements, the Basin States agree on the following allocation of the utilizable water resources of Yamuna River assessed on mean year availability.

Subject to the following:

(i) Pending construction of the storages in the upper reaches of the river, there shall be an interim seasonal allocation of the annual utilization flow of Yamuna River as follows:

	Seasonal allocation of Yamuna waters (BCM)			
States	July–Oct	Nov–Feb	March–June	Annual
Haryana	4.107	0.686	0.937	5.730
Uttar Pradesh	3.216	0.343	0.473	4.032
Rajasthan	0.963	0.070	0.086	1.119
Himachal Pradesh	0.190	0.108	0.080	0.378
Delhi	0.580	0.068	0.076	0.724
Total	9.056	1.275	1.652	11.983

Provided that the interim seasonal allocations will be distributed on a ten day basis. Provided further that the said interim seasonal allocations shall get progressively modified, as storages are constructed, to the final annual allocations as indicated in para 7 above.

(ii) Separate agreement will be executed in respect of each identified storage within the framework of overall allocation made under this agreement.

(iii) The allocation of available flows amongst the beneficiary States will be regulated by the Upper Yamuna River Board within the overall framework of this agreement.

Provided that in a year when the availability is more than the assessed quantity, the surplus availability will be distributed amongst the States in proportion to their allocations.

Provided also that in a year when the availability is less than the assessed quantity, first the drinking water allocation of Delhi will be met and the balance will be distributed amongst Haryana, U.P., Rajasthan and H.P. in proportion to their allocations.

8. This agreement may be reviewed after the year 2025, if any of the basin States so demands.
9. We place on record and gratefully acknowledge assistance and advice given by the Union Minister of Water Resources in arriving at this expeditious and amicable settlement.

2.4 Uses of Yamuna River Water

Water is one of the essential requirements of life. In the modern age it also plays a significant role in various economic activities. The higher growth rate is reflected during good monsoon period and availability of good amount of water in the river. The various uses of river water can be kept into two major groups. In one group the water is abstracted and transported away from the natural water bodies for beneficial uses and are called abstractive uses or uses involving collection and transportation. The other is just opposite of the first, in which withdrawal and transportation of water are not required but the water is utilized. It is known as non-abstractive or in-situ water uses.

2.4.1 Abstractive Uses

The river water is abstracted at different locations for varied uses. At two places, i.e., Hathnikund/Tajewala and Okhla, the water abstraction is significant. The annual abstraction at various locations is presented in Table 2.2 and the percent use of abstracted water for various purposes is presented in Fig. 2.7, whereas, the sharing of Yamuna River water by various states is depicted in Fig. 2.8. The various abstractive uses of river water are as below (http://www.cpcb.nic.in/).

2.4.1.1 Domestic Water Supplies

The large urban centers located along river banks and where suitable ground water is not available, water is abstracted for drinking water supplies after suitable treatment. The urban agglomerations, like Delhi, Mathura, Agra and Allahabad, use the Yamuna water significantly for domestic water supplies. The water abstracted for domestic water supply at various locations is presented in Table 2.2. At Wazirabad, Delhi the entire river water is diverted for this purpose along with the increase in demand of water for drinking purposes. Along with the population increase, there are plans at various locations to withdraw more and more water from the river.

Table 2.2 Water abstraction from Yamuna River

S. No.	Location	River water abstraction (approx. MLD)	Abstraction use
1	Hathinikund	20,000	Irrigation, drinking water supply and others
2	Wazirabad	1,100	Drinking water supply
3	Wazirabad to Okhla Stretch	5,000	Irrigation and others
4	Okhla to Etawah Stretch	400	Irrigation, drinking water supply and others
5	Etawah to Allahabad Stretch	475	Irrigation, drinking water supply and others

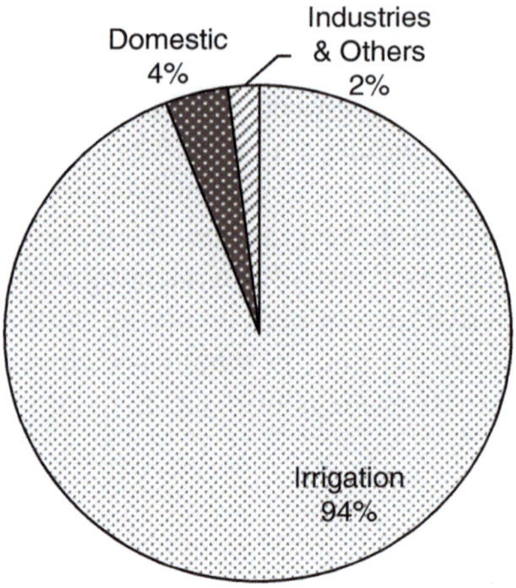

Fig. 2.7 Percentage water abstractions from Yamuna River for various uses

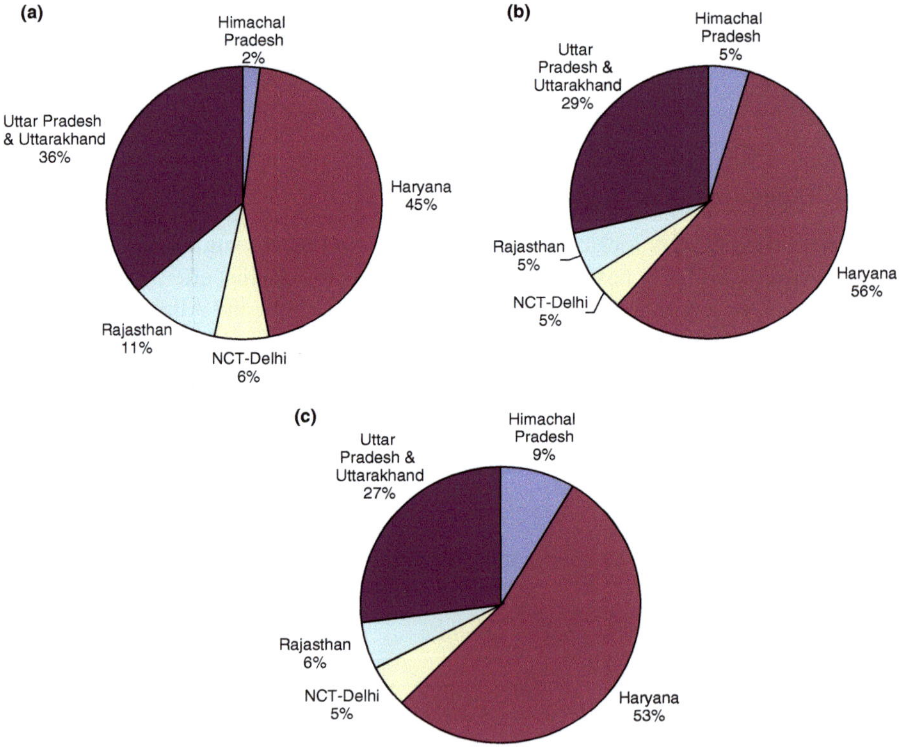

Fig. 2.8 State wise water use of Yamuna River during (**a**) monsoon (July to October); (**b**) summer (March to June); (**c**) winter (November to February)

2.4.1.2 Irrigation

Irrigation is an important use of Yamuna River water. It is estimated that about 92% of Yamuna River water is used for irrigation. In the entire Yamuna basin the irrigated land is about 12.3 million hectares and approximately half of it (about 49%) is irrigated exclusively from surface water. At present there are four irrigation canals transporting the Yamuna River water to the command areas.

2.4.1.3 Western Yamuna Canal (WJC)

This canal originates from the right bank of Yamuna River at the Hathnikund/ Tajewala barrage. The capacity of main canal is 163 m^3/s and irrigates an area of about 486,000 ha annually in Haryana State. This irrigation system is more than 100 years.

2.4.1.4 Eastern Yamuna Canal (EJC)

This canal takes off from the left bank of the Yamuna River at the Hathnikund/ Tajewala barrage. The canal is about 206 km long and having a capacity of 85 m^3/s. It irrigates an area of about 191,000 ha annually in Uttar Pradesh.

2.4.1.5 Agra Canal

Agra canal originates from the right bank of Yamuna River at Okhla barrage. The canal is 163 km long and carries a discharge of about 63.5 m^3/s. It irrigates about 138,000 ha of land mainly in two districts of Mathura and Agra in Uttar Pradesh.

2.4.1.6 Gurgaon Canal

This canal is an interstate project between Rajasthan and Haryana and takes off from Agra canal at a distance of around 8 km from its offtake at the Okhla barrage. The water flow capacity of this canal is 14.15 m^3/s and the land irrigated by this canal is about 40,000 ha.

2.4.2 In-Stream Uses

2.4.2.1 Hydropower

The total potential for hydropower development in the entire Yamuna basin is about 1,300 MW. The present utilization is only one third of the total potential. There are a few new schemes which are at various stages of construction:

2.4.2.2 Fisheries

Fish is not a popular traditional food for the people residing in the Yamuna basin area except in some tribes and castes and collectively called fishing community. This is the reason that the pisciculture is neither practiced on large scale nor undertaken in an organized manner in the area. However, the entire river stretch and tributaries are being utilized for fishing in an unorganized manner. There is a large scope of farming for fish and other aquatic animals in stretches of Yamuna River.

2.4.2.3 Growing Aquatic Plants

The most prevalent aquatic plants in Yamuna River is the water hyacinth. In West Bengal this plant is used as cattle fodder and composted an-aerobically on a large scale. In Kerala too there are various schemes to generate biogas from water hyacinth. Unfortunately, these practices are yet to be undertaken in the Yamuna basin.

2.4.2.4 Navigation

The most of the river channel of Yamuna River and its tributaries are not suitable for Navigation. Low flow of the river further restricts this activity. At few locations the boats are playing on a need based basis, mainly for crossing the river. Earlier, the timber logs and sleepers were floated down from the Upper Himalayan areas but now this practices is also replaced by road transportation. There is a scope to use the Yamuna River stretch between Agra and Allahabad for navigation.

2.4.2.5 River Bathing and Washing

River bathing and washing is one of the most important uses of river water in the country. The Hindu culture and the other cultures of Indian origin are generally considered to be river oriented. Bathing is an essential part of various Hindu rituals. Bathing in flowing water and that too on rivers like the Ganga, the Yamuna, the Narmada, the Godawari, etc. is considered superior to bathing in house with well or tap water. On religious and cultural occasions millions of people take bath especially near religious towns in a congested stretch of the river within the span of a few hours. The river water is also used for washing clothes and utensils by nearby communities, particularly by the poor inhabitants. Some of the prominent bathing centers at Yamuna River are Yamunotri, Kalpi, Paonta Sahib, Delhi, Mathura-Vrindavan, Agra, Bateshwar, Etawah and Allahabad on Yamuna River, Kota on Chambal, Orcha on Betwa and Ujjain on Shipra River.

2.4.2.6 Recreational Uses

In the Yamuna basin the rivers are used very little for recreational value due to unsuitable conditions like rocky river bed and low water depth, and water sports like boating have a vast potential in future specially at urban centers and at various barrage sites in the reservoir formed by the barrages.

2.4.2.7 Cattle Bathing and Washing

The cattle in most of the towns and villages along the rivers are regularly taken toward the river for drinking and bathing. It is estimated that about 70% of the total cattle population in the Yamuna basin uses flowing water of river and canals for bathing and watering purposes directly. These cattle activities impart substantial impact on water quality. This occurs not only through direct discharge of urine, dung and washed off organic inorganic materials but the bottom sediments are also churned up because of cattle wading.

Besides these uses, the river Channel of Yamuna River, particularly in Delhi stretch is also used for the transportation of water for irrigation from one water body to another or from one place to another. The Yamuna water is being transported from the Western Yamuna Canal to the Gurgaon canal via the Najafgarh drain, Yamuna River and the Agra canal. Similarly, the Ganga River water from the upper Ganga canal is transported to the Agra canal for irrigation in districts of Mathura and Agra via Hindon River, the Hindon Cut Canal and Yamuna River. The dilution that the rivers receive through this means of water transportation significantly affects the water quality of drain, canal and river.

Chapter 3
Climate of the Basin

Abstract This chapter describes important hydro-meteorological variables and their spatial analysis, including rainfall, temperature, wind, humidity, evapotranspiration. Apart from these variables, other derived variables relevant to water resources and agricultural planning, such as seasonal rainfall, number of rainydays, onset of effective monsoon, and aridity index are also critically analyzed. From a hydrological design point of view, quantile estimates of annual rainfall and daily maximum rainfall are treated. Based on the aridity index, which is an indicator of the annual soil moisture deficit, the climate classification of the basin is discussed.

India has four distinct seasons, winter: December-January; spring (Basant): February-March; summer: April-June; rainy season (or Monson): July-September; fall (Sharad): September-November. Hydro-meteorologically Yamuna basin is a heterogeneous basin having great spatial variation of rainfall pattern ranging from 200 to 2,350 mm. However, the average annual rainfall varies between 400 and 1,500 mm. The entire basin comes under the influence of the south-west monsoon and a major part of the rainfall is received between June and September. Winter rainfall is scanty. In the basin, rainfall distribution increases from North-West to the South-East direction. Broadly, climate of the basin can be classified as: (a) humid in the upstream Himalayan catchment; (b) semi-arid in north-west to western catchments; (c) sub-humid in south-west catchments and catchments located on the left bank of the river. The mean maximum temperature is 24–42.5°C and the mean minimum temperature is –1.0 to 11.0°C.

3.1 Rainfall

In the Yamuna River basin, rainfall is the only input to the hydrologic cycle. However, in the upper Himalayan part of the basin, the river receives water from the glacier melt runoff. During the non-monsoon period, water availability in the river is either due to glacier melt runoff and base flow. Excluding the upper Himalayan segment, the Yamuna River basin is rainfed and perennial behaviour of the river is due to upstream storages and base flow.

R.K. Rai et al., *The Yamuna River Basin*, Water Science and Technology Library 66,
DOI 10.1007/978-94-007-2001-5_3,

3.1.1 Average Rainfall

Considering the one degree records of daily rainfall for the period of 1951–2003 collected from India Meteorological Department (IMD, the average annual rainfall of the basin varies from 645 to 1,175 mm with annual basin average of approximately 906 mm. The spatial variation of the annual rainfall pattern in the basin is shown in Fig. 3.1. It is apparent that the Yamuna basin comparatively receives heavy rainfall in the extreme Northern catchments of Shiwalik range (i.e. Himalayan catchments, namely Tons, Giri, Upper Yamun and Asan, Pabbar sub-catchments) followed by extreme Southern sub-catchments of Betwa, Ken,

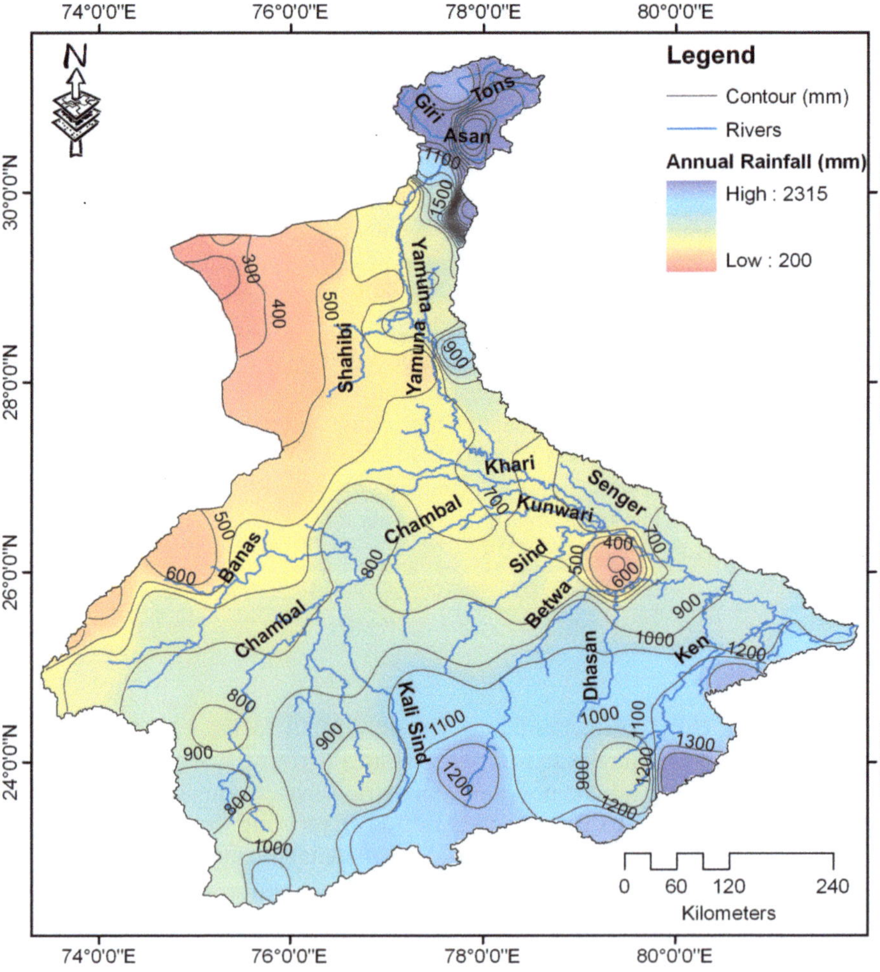

Fig. 3.1 Spatial variation of mean annual rainfall in the basin

Table 3.1 Mean rainfall distribution in the Yamuna basin

Mean values (mm)	Period of analysis	Betwa	Ken	Sind	Chambal	Hindon	Khari	Tons	Giri	Himalayan upper	Yamuna basin
Annual Rainfall	1951–2003	1,064.9	1,125.0	848.3	783.7	887.7	644.2	831.1	1,045.1	1,175.5	906.7
75% Probability of Exceedance	1951–2003	892.0	928.9	705.0	667.1	752.1	548.5	583.1	853.0	994.0	796.3
Annual Rainfall	1994–2003	1,074.1	1,116.3	828.7	749.3	780.9	589.3	1,081.4	1,010.0	1,257.5	894.8
Monsoon Rainfall	1951–2003	972.1	1,025.6	769.5	717.4	734.0	572.1	323.3	579.1	707.3	756.4
Monsoon Rainfall	1994–2003	981.9	1,021.0	756.0	673.7	639.2	517.1	667.1	626.4	853.0	754.9
Winter Rainfall	1951–2003	26.0	29.9	19.4	10.9	53.8	19.5	188.4	175.5	165.8	47.6
Winter rainfall	1994–2003	23.5	26.6	16.9	13.8	62.5	23.3	135.5	145.7	138.9	46.4
Post Monsoon Rainfall	1951–2003	50.8	51.2	44.3	41.5	47.2	29.6	88.9	88.3	89.4	50.4
Post Monsoon Rainfall	1994–2003	48.6	50.5	39.6	47.6	33.8	25.0	50.5	51.4	50.7	41.1

Baghein, and Mandakini rivers. Annual rainfall in the eastern and central parts of the catchments is quite less and within the range of 300 to 600 mm.

The sub-catchment-wise rainfall was also estimated using the Theissen polygon method and are summarized in Table 3.1. The temporal variation of the weighted annual rainfall for each sub-catchment is shown in Fig. 3.2 which also include trends in annual rainfall patterns of catchments. To determine the general trend in the rainfall pattern in the basin, besides the whole record of 1951–2003, the last ten year of record (i.e., 1994–2002) are further analysed and compared with 53 years of record. The catchment-wise summarized results of analysis are given in Table 3.1. Based on this analysis, it is observed that the average rainfall in the basin has been reduced, however, the detailed statistical evidence of presence of trends and periodicities are presented in Chapter 4. Besides the analysis of annual rainfall, seasonal rainfall pattern is also studied and results are summarized in Table 3.1. It is apparent from the Table 3.1 that seasonal rainfall other than the Monsoon season is quite less, except for the upper Himalayan catchments.

The monthly distribution of the rainfall pattern for the catchments of the Yamuna basin as well as the percentage distributions are given in Tables 3.2 and 3.3.

3.1.2 Probability Analysis of Annual Rainfall

For water resources planning and management, the magnitude of rainfall at 75% probability of exceedance is generally used. This analysis is performed based on the well established Weibull's plotting position formula which can be expressed as follows:

$$E = \frac{m}{N+1} \times 100\% \tag{3.1}$$

where E is the probability of non-exceedance, m is the rank of the observation when arranged in ascending order, and N is the number of observation in the sample. The probability of exceedance can be estimated using the following equation:

$$P = 100 - E \tag{3.2}$$

where P is the probability of exceedance. Using these equations [eqs. (3.1) and (3.2)], catchment rainfall at 75% probability of exceedance is estimated for all the catchments of the Yamuna basin, as given in Table 3.1.

3.1.3 Quantile Estimates of Daily Maximum and Annual Rainfall

Quantile estimates of various return periods are important for designing the water resources system. For the Yamuna River basin, the quantiles were estimated for daily maximum rainfall and annual rainfall using the daily grid data for the period of 1951–2003. The quantile were estimated using Weibull's plotting position

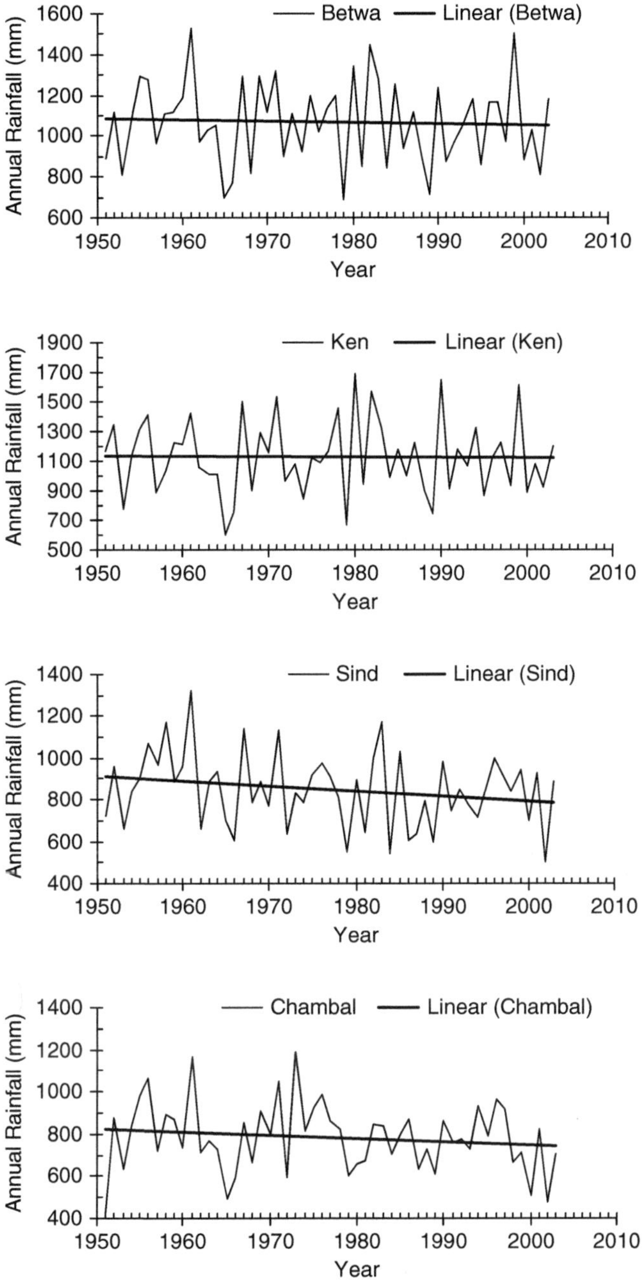

Fig. 3.2 Annual rainfall along with the trend and 75% probability rainfall

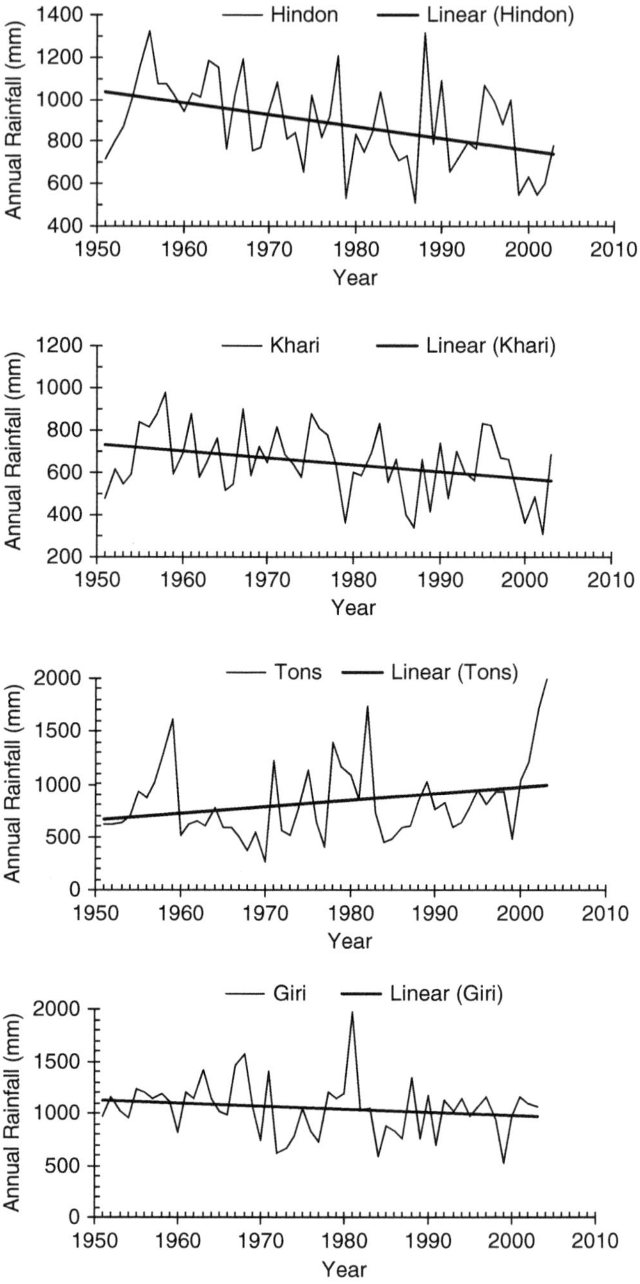

Fig. 3.2 (continued)

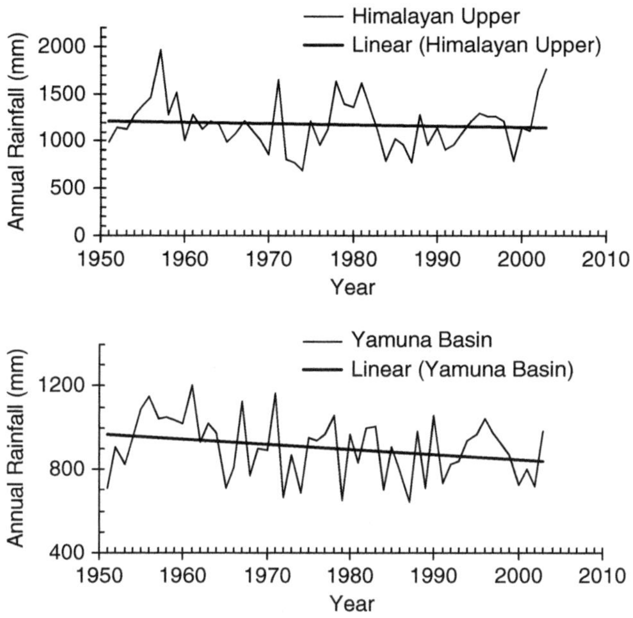

Fig. 3.2 (continued)

Table 3.2 Monthly distribution of mean rainfall (mm) [1951–2003] in the Yamuna basin

Month	Betwa	Ken	Sind	Chambal	Hindon	Khari	Tons	Giri	Himalayan upper	Yamuna basin
Jan	15.3	16.9	10.7	6.2	27.8	9.8	91.0	93.4	81.9	24.6
Feb	10.7	13.0	8.8	4.7	26.1	9.8	97.5	82.1	83.9	23.0
Mar	6.0	7.4	4.7	3.6	21.5	6.0	100.4	94.1	92.2	20.9
Apr	3.1	3.9	2.8	3.0	11.6	4.4	70.2	48.5	56.5	12.3
May	6.8	7.0	7.6	7.4	19.6	12.5	59.7	59.7	64.3	19.0
Jun	100.1	112.4	71.6	78.5	78.7	48.6	52.8	89.6	101.3	81.9
Jul	316.1	319.9	262.6	254.4	248.6	199.7	104.6	210.4	244.5	257.6
Aug	370.0	392.4	283.5	262.1	281.0	225.6	93.5	177.9	231.5	279.2
Sep	185.9	200.8	151.8	122.4	125.7	98.2	72.4	101.2	130.0	137.7
Oct	33.3	34.2	30.7	25.7	28.5	20.5	29.1	28.2	36.3	29.5
Nov	8.8	8.6	6.3	8.7	4.6	3.8	16.5	14.5	13.9	7.4
Dec	8.7	8.5	7.3	7.1	14.1	5.4	43.3	45.5	39.3	13.5
Annual	1,064.9	1,125.0	848.3	783.7	887.7	644.2	831.1	1,045.1	1,175.5	906.7

Monsoon: June–September; *Post-monsoon*: October–December; *Winter*: January–February; *Summer*: March–May

Table 3.3 Monthly percentage distribution of mean rainfall (1951–2003) in the Yamuna basin

Month	Betwa	Ken	Sind	Chambal	Hindon	Khari	Tons	Giri	Himalayan Upper	Yamuna Basin
Jan	1.4	1.5	1.3	0.8	3.1	1.5	10.9	8.9	7.0	2.7
Feb	1.0	1.2	1.0	0.6	2.9	1.5	11.7	7.9	7.1	2.5
Mar	0.6	0.7	0.6	0.5	2.4	0.9	12.1	9.0	7.8	2.3
Apr	0.3	0.3	0.3	0.4	1.3	0.7	8.5	4.6	4.8	1.4
May	0.6	0.6	0.9	0.9	2.2	1.9	7.2	5.7	5.5	2.1
Jun	9.4	10.0	8.4	10.0	8.9	7.5	6.4	8.6	8.6	9.0
Jul	29.7	28.4	31.0	32.5	28.0	31.0	12.6	20.1	20.8	28.4
Aug	34.7	34.9	33.4	33.4	31.7	35.0	11.3	17.0	19.7	30.8
Sep	17.5	17.9	17.9	15.6	14.2	15.3	8.7	9.7	11.1	15.2
Oct	3.1	3.0	3.6	3.3	3.2	3.2	3.5	2.7	3.1	3.3
Nov	0.8	0.8	0.7	1.1	0.5	0.6	2.0	1.4	1.2	0.8
Dec	0.8	0.8	0.9	0.9	1.6	0.8	5.2	4.4	3.3	1.5
Annual	100	100	100	100	100	100	100	100	100	100

formula given by eq. (3.1). The return period T can be estimated using the following relationship,

$$T = \frac{1}{P} \tag{3.3}$$

where T is the return period (year), P is the probability of exceedance estimated from eq. (3.2). The values of daily maximum and annual rainfall for 2-, 5-, 10-, 25-, and 50-years return periods are given in Table 3.4. However, the contours of these variables are shown in Figs. 3.3, 3.4, 3.5, 3.6, 3.7, 3.8, 3.9, 3.10, 3.11, and 3.12. For the distribution of daily maximum rainfall into shorter period (say hourly), area to point rainfall ratios (Appendix C) can be used.

3.1.4 Onset of Effective Monsoon and Number of Rainydays

The onset of effective monsoon (OEM) is an important governing agro-climatic parameter for planning of *kharif* crops. The OEM is estimated using the Ashokraj criteria (1979), which uses daily rainfall and potential evapotranspiration data. The potential evapotranspiration (PET) can be estimated using the Thornthwaite method (1948). Details of the Thornthwaite method for estimating PET are presented later. A seven days rainfall spell satisfying the following conditions is termed as OEM. These conditions are: (i) first day rainfall in the seven days spells should be more than evaporation of that particular day; (ii) a day with more than 3 mm of rainfall is considered to be a rainy day; (iii) total rainfall during the seven day spell is more than (3× PET + 10) mm; and (iv) at least four out of seven days are rainy days. Considering this methodology, the value of OEM is estimated for the entire grid location (65 grids in the basin) for entire years. Using the annual value of OEM, the mean OEM was estimated for grid locations, and interpolated to derive the spatial pattern using the Kriging method. The spatial pattern of OEM for the basin is shown in Fig. 3.13.

Table 3.4 Quantile estimates of daily maximum and annual rainfall for different grid points

Grid station		Quantile estimates of daily maximum rainfall (mm)					Quantile estimates of annual rainfall (mm)				
Lat	Long	$T = 2$ yr	$T = 5$ yr	$T = 10$ yr	$T = 25$ yr	$T = 50$ yr	$T = 2$ yr	$T = 5$ yr	$T = 10$ yr	$T = 25$ yr	$T = 50$ yr
21.5	75.5	68.1	98.9	123.3	196.8	223.6	701.9	938.5	972.3	10,89.8	1,133.0
22.5	74.5	96.8	151.8	160.3	205.1	230.7	791.6	1,123.0	1,212.3	1,348.0	1,530.7
22.5	75.5	85.4	133.2	148.5	173.6	254.9	861.5	1,124.9	1,307.2	1,395.3	1,600.2
22.5	76.5	85.8	121.3	150.2	162.4	168.8	979.3	1,109.9	1,271.6	1,488.9	1,625.8
22.5	77.5	126.6	176.0	225.9	298.0	326.0	1,125.9	1,491.1	1,633.2	1,835.1	2,021.0
22.5	78.5	138.4	193.9	225.0	337.3	364.1	1,655.0	2,053.1	2,372.2	3,020.6	3,273.5
22.5	79.5	81.9	119.4	163.0	201.5	231.6	1,084.9	1,297.8	1,509.6	1,624.5	1,633.3
23.5	73.5	93.0	151.7	212.0	297.1	425.1	759.1	962.5	1,262.4	1,341.6	1,454.3
23.5	74.5	116.4	196.2	274.5	295.3	304.1	1,005.0	1,281.3	1,456.3	1,645.8	1,683.8
23.5	75.5	84.5	143.3	156.1	208.5	261.1	849.4	1,103.9	1,180.9	1,489.3	1,694.9
23.5	76.5	95.5	146.1	188.3	246.6	250.6	922.7	1,114.9	1,359.8	1,732.0	1,802.7
23.5	77.5	96.3	136.3	188.6	212.8	233.6	1,170.7	1,465.5	1,585.2	1,748.8	1,853.8
23.5	78.5	101.3	148.2	191.3	213.5	221.8	1,192.7	1,470.6	1,675.2	1,929.8	1,978.4
23.5	79.5	90.3	139.1	158.9	184.8	246.5	1,229.5	1,485.7	1,650.0	1,996.5	2,139.6
23.5	80.5	85.1	130.6	170.0	195.0	229.8	1,294.7	1,563.0	1,681.0	1,956.0	2,142.2
24.5	72.5	95.8	177.9	278.4	315.2	332.2	883.5	1,176.1	1,675.2	1,983.0	2,004.9
24.5	73.5	68.3	95.7	121.8	158.3	173.8	638.7	854.3	940.4	1,067.9	1,232.2
24.5	74.5	80.0	111.4	130.2	165.1	195.6	794.7	1,008.7	1,044.1	1,117.5	1,677.2
24.5	75.5	93.1	127.2	141.0	165.9	168.9	825.5	1,057.0	1,169.5	1,301.6	1,317.2
24.5	76.5	87.1	139.9	176.4	206.7	225.4	855.0	1,145.2	1,280.4	1,433.7	2,576.4
24.5	77.5	86.6	143.7	176.6	243.9	251.9	1,026.6	1,279.8	1,396.7	1,488.8	1,524.2
24.5	78.5	68.0	102.8	135.6	168.2	203.5	988.2	1,223.2	1,423.5	1,463.5	1,564.0
24.5	79.5	93.0	135.3	151.1	171.1	177.2	1,125.7	1,273.3	1,578.8	1,877.4	1,949.9
24.5	80.5	96.1	118.3	142.9	183.7	204.7	1,130.2	1,367.9	1,480.4	1,610.9	1,709.3
24.5	81.5	102.5	136.2	159.8	197.1	224.0	1,088.5	1,325.9	1,447.0	1,530.0	1,713.7
24.5	82.5	83.8	128.7	150.2	184.5	281.4	1,101.7	1,281.4	1,460.4	1,691.3	1,875.1
25.5	73.5	80.7	113.4	179.0	234.3	260.8	544.9	803.7	869.6	1,059.5	1,385.3

Table 3.4 (continued)

Grid station		Quantile estimates of daily maximum rainfall (mm)					Quantile estimates of annual rainfall (mm)				
Lat	Long	$T = 2$ yr	$T = 5$ yr	$T = 10$ yr	$T = 25$ yr	$T = 50$ yr	$T = 2$ yr	$T = 5$ yr	$T = 10$ yr	$T = 25$ yr	$T = 50$ yr
25.5	74.5	74.7	103.8	137.2	236.5	247.8	632.4	840.7	919.7	980.7	1,130.0
25.5	75.5	78.6	106.2	131.6	144.6	158.9	672.9	850.3	934.6	1,147.9	1,237.7
25.5	76.5	72.5	108.2	129.2	154.6	168.4	776.8	937.4	1,034.3	1,354.2	1,450.1
25.5	77.5	91.7	124.8	152.3	201.2	215.7	866.7	1,061.0	1,230.0	1,262.4	1,450.8
25.5	78.5	109.2	145.8	179.5	200.4	214.8	897.1	1,092.2	1,235.5	1,361.4	1,566.8
25.5	79.5	80.4	123.2	142.1	169.2	177.8	1,030.5	1,162.6	1,360.9	1,654.6	1,709.4
25.5	80.5	96.1	137.1	177.6	205.2	211.4	957.4	1,127.6	1,296.0	1,591.0	1,728.7
25.5	81.5	74.0	112.6	125.4	166.2	172.7	988.7	1,179.6	1,380.1	1,506.8	1,548.5
25.5	82.5	102.2	140.1	174.6	218.2	236.6	1,073.2	1,245.0	1,355.2	1,386.3	1,428.0
26.5	73.5	54.4	76.9	101.4	167.3	284.9	365.3	594.3	663.7	842.2	885.3
26.5	74.5	65.4	101.0	146.8	170.0	376.1	462.8	613.7	721.5	872.3	912.5
26.5	75.5	65.5	90.0	111.0	173.6	212.6	525.5	759.0	823.8	957.9	997.6
26.5	76.5	65.8	92.9	119.9	162.9	241.6	637.3	809.0	965.9	1,060.6	1,087.0
26.5	77.5	66.8	95.8	111.9	164.6	234.0	634.6	913.5	955.1	1,060.6	1,084.8
26.5	78.5	78.9	98.6	129.7	192.3	217.9	751.7	956.2	1,048.9	1,271.7	1,403.9
26.5	79.5	68.8	123.0	138.2	161.9	164.9	808.3	975.0	1,120.2	1,351.0	1,458.1
26.5	80.5	81.6	113.9	135.2	154.7	161.5	883.4	1,176.6	1,384.7	1,493.3	1,846.7
26.5	81.5	74.3	109.0	126.7	159.8	169.1	976.1	1,259.3	1,370.2	1,567.6	1,777.6
27.5	74.5	43.5	75.4	113.1	120.9	142.0	324.3	501.3	593.6	848.2	897.7
27.5	75.5	60.5	87.9	106.9	131.1	157.2	489.1	622.8	724.4	807.8	930.2
27.5	76.5	69.3	87.7	101.6	136.4	176.4	630.6	780.3	903.3	943.3	1,005.0
27.5	77.5	67.1	92.5	102.5	123.0	143.7	608.7	753.8	794.6	828.9	1,002.1
27.5	78.5	57.2	82.1	91.8	129.6	273.2	751.6	951.9	1,011.1	1,059.5	1,098.9
27.5	79.5	69.2	102.0	134.7	176.7	193.6	805.6	1,064.8	1,139.7	1,188.7	1,522.8
27.5	80.5	93.0	131.5	167.1	204.9	221.0	1,007.9	1,241.5	1,454.9	1,703.8	1,755.8
28.5	74.5	55.6	78.2	120.0	198.1	215.6	296.5	491.3	600.5	651.5	732.1
28.5	75.5	50.1	75.4	101.1	151.9	175.6	400.9	532.1	607.2	824.9	924.6

Table 3.4 (continued)

Grid station		Quantile estimates of daily maximum rainfall (mm)					Quantile estimates of annual rainfall (mm)				
Lat	Long	$T = 2$ yr	$T = 5$ yr	$T = 10$ yr	$T = 25$ yr	$T = 50$ yr	$T = 2$ yr	$T = 5$ yr	$T = 10$ yr	$T = 25$ yr	$T = 50$ yr
28.5	76.5	60.3	76.6	87.3	117.2	139.2	478.0	685.1	767.0	937.3	961.5
28.5	77.5	73.8	115.1	135.1	171.3	176.9	735.0	915.1	1,038.7	1,110.4	1,151.2
28.5	78.5	66.2	93.2	108.1	141.9	176.2	785.9	1,071.5	1,146.9	1,236.6	1,273.0
29.5	74.5	46.5	66.4	91.3	128.7	133.9	246.9	344.5	392.1	486.6	581.3
29.5	75.5	58.6	79.9	100.0	121.9	137.0	378.5	514.9	560.9	668.6	773.6
29.5	76.5	51.4	79.0	99.7	108.0	136.8	374.3	555.5	626.7	803.9	820.4
29.5	77.5	77.6	102.7	112.6	154.8	165.0	839.2	1,017.8	1,107.7	1,274.5	1,342.2
29.5	78.5	108.9	165.6	188.0	242.2	325.9	1,140.0	1,425.4	1,622.6	1,800.2	1,906.9
30.5	75.5	69.4	94.7	119.6	149.2	206.1	502.6	703.6	822.1	980.9	1,011.2
30.5	76.5	78.2	115.1	142.1	150.2	264.1	764.9	911.3	1,021.8	1,449.5	1,546.2
30.5	77.5	118.0	160.0	177.2	138.4	219.3	1,490.3	1,975.0	2,282.2	2,982.7	3,269.7
30.5	78.5	90.6	147.6	188.9	253.8	531.9	1,615.3	1,973.3	2,244.1	2,318.8	3,527.1
31.5	76.5	67.9	105.8	124.7	149.0	171.8	1,230.2	1,554.6	1,717.7	1,876.4	2,052.9
31.5	77.5	54.4	78.9	104.4	140.7	154.4	1,063.1	1,187.7	1,375.4	1,552.3	1,914.2
31.5	78.5	49.0	85.2	119.9	225.1	338.8	756.6	1,098.4	1,334.7	1,737.0	1,961.3
31.5	79.5	52.5	77.0	103.8	170.4	195.7	1,040.0	1,206.1	1,326.4	1,435.1	1,548.3

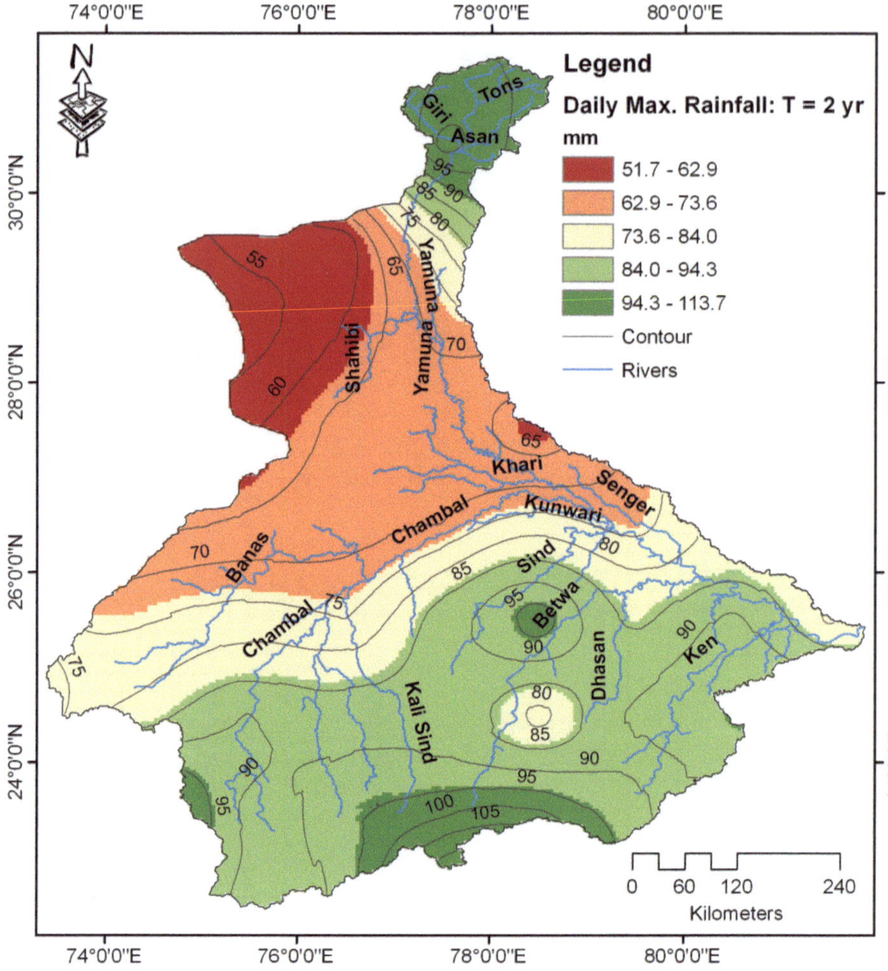

Fig. 3.3 Daily maximum rainfall for the return period of 2 years (data period: 1951–2003)

The number of raindays from daily rainfall data is derived following the criteria of India Meteorological Department (IMD). The criterion is: "a day having rainfall of more than 3 mm is referred as the rainyday". Using this criterion, the mean values of numbers of annual, monsoon and non-monsoon rainydays are derived and depicted in Figs. 3.14, 3.15, and 3.16, respectively.

3.2 Temperature

Temperature is one of the driving climatic variables that govern the hydrological cycle. It directly affects the magnitude of snow melt runoff, evapotranspiration rate, water use, and water quality in the river. The Yamuna basin experiences a wide

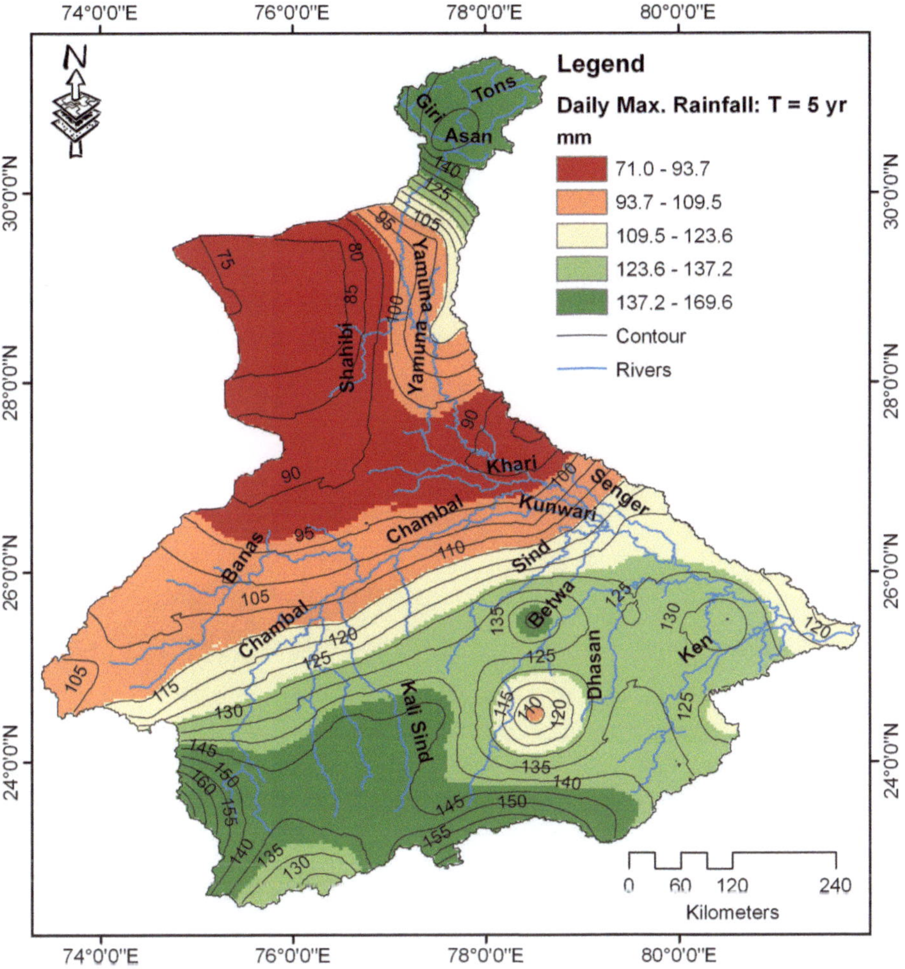

Fig. 3.4 Daily maximum rainfall for the return period of 5 years (data period: 1951–2003)

range of spatio-temporal variability in the temperature. Mean maximum and mean minimum temperature in the basin spatially varies between 24.25 and 42.46°C and –1.05 to 10.81°C, respectively. The spatial pattern of mean maximum and minimum temperature is depicted in Figs. 3.17 and 3.18. The mean maximum and minimum isotherms are also included in Figs. 3.17 and 3.18.

Based on the available records for the period of 1983–2008 collected from National Climatic Data Center (http://www.ncdc.noaa.gov), the highest recorded daily maximum and minimum temperatures observed at different location are also identified and upscaled to the basin level using the Kriging method, and are depicted in Figs. 3.19 and 3.20. The catchment-wise estimated mean maximum and mean minimum temperature are also given in Table 3.5.

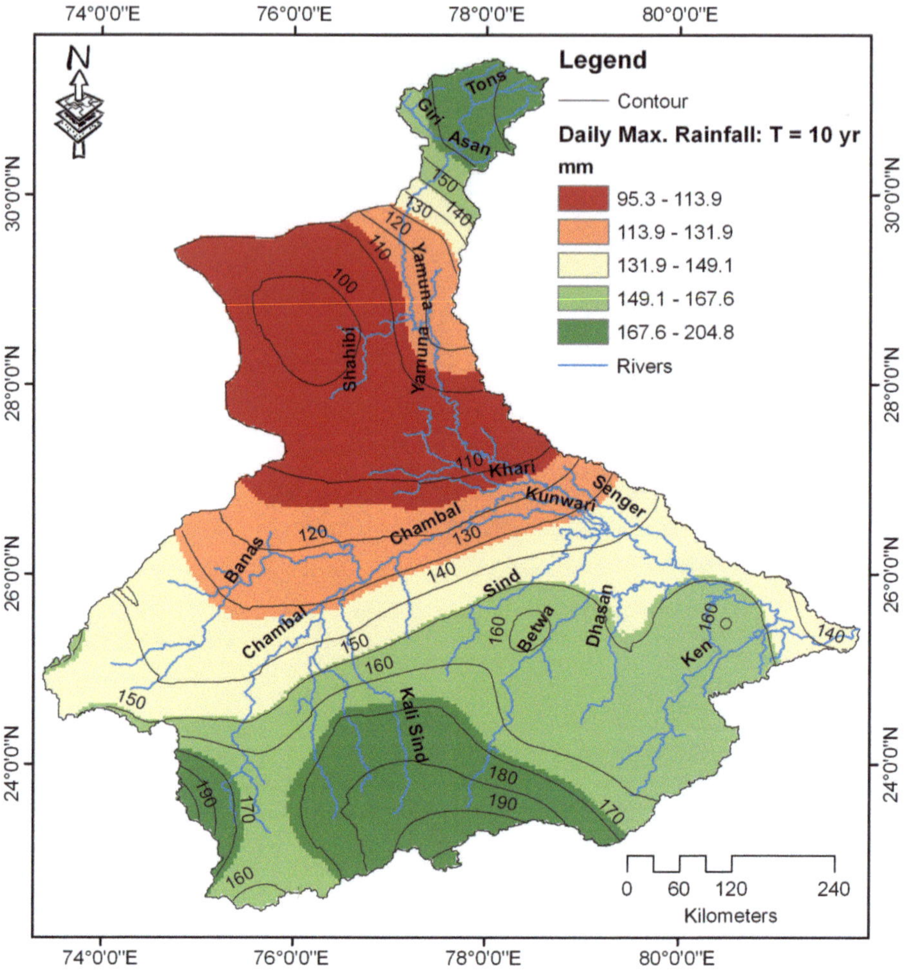

Fig. 3.5 Daily maximum rainfall for the return period of 10 years (data period: 1951–2003)

3.3 Wind

Using daily values of wind speed for the period of 1980–2008 from the National Climatic Data Center (http://www.ncdc.noaa.gov) for a few stations in the Yamuna basin, average and maximum wind speed were derived and interpolated at the basin scale. The generated contours of annual mean wind speed for the Yamuna basin are depicted in Fig. 3.21.

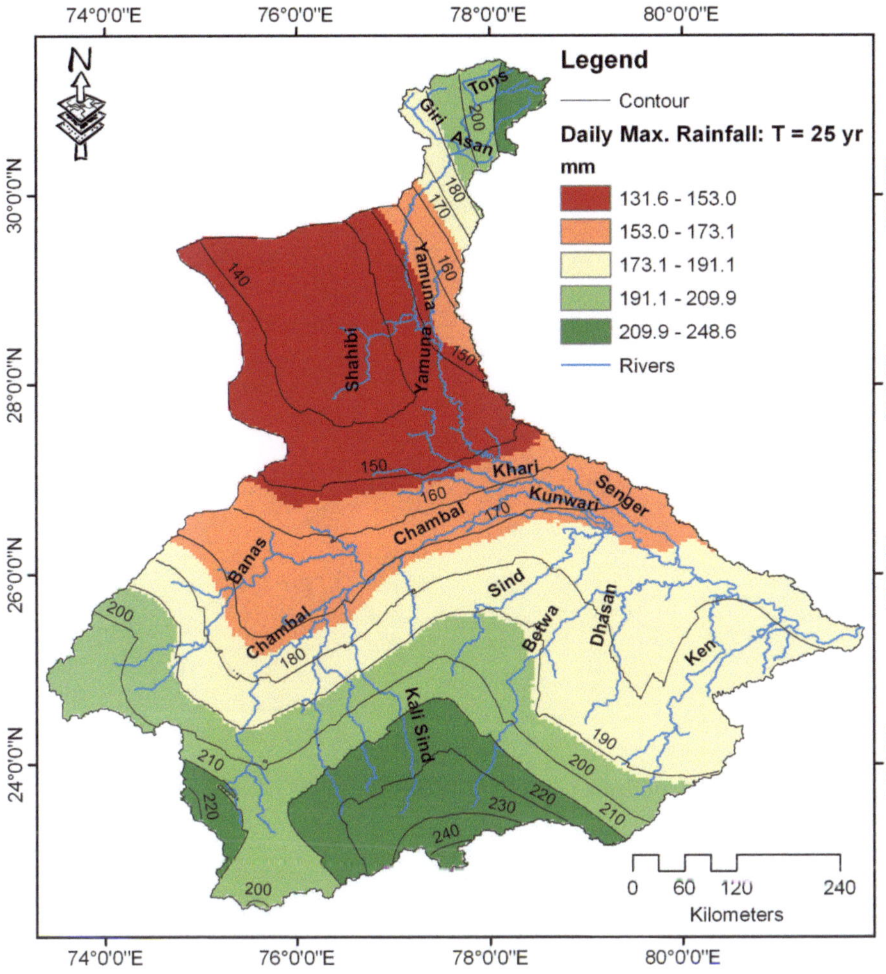

Fig. 3.6 Daily maximum rainfall for the return period of 25 years (data period: 1951–2003)

3.4 Humidity

Direct measurement of humidity is not easy; therefore, it is presented through relative humidity. The relative humidity (RH) is referred to as the ratio of actual vapor pressure to the vapor pressure at saturation. Mathematically, RH is expressed as:

$$\mathrm{RH} = \frac{e_\mathrm{a}}{e_\mathrm{s}} \times 100\,\% \tag{3.4}$$

where e_a is the actual vapor pressure and is estimated at dew point temperature as follows:

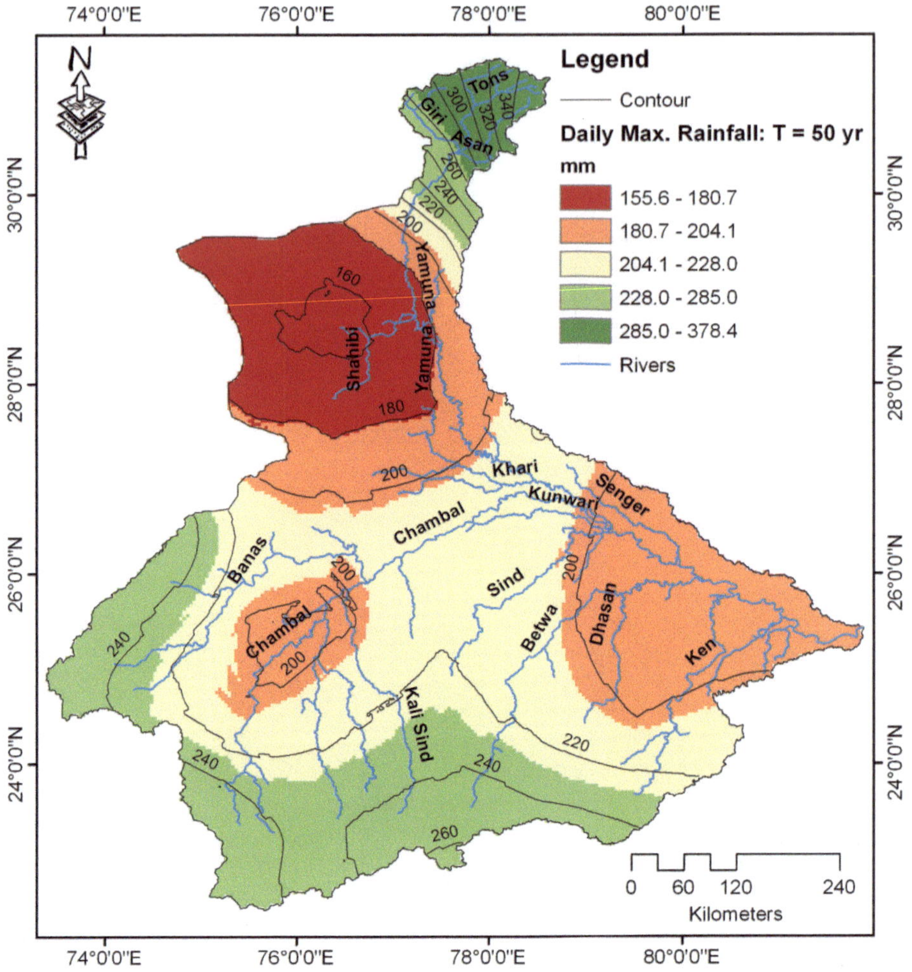

Fig. 3.7 Daily maximum rainfall for the return period of 50 years (data period: 1951–2003)

$$e_a = 0.6108 \times \exp\left\{\frac{17.27\,T_{dew}}{T_{dew} + 237.3}\right\} \tag{3.5}$$

where T_{dew} is the dew point temperature (°C).

The vapor pressure at saturation is estimated as follows:

$$e_s = \left[e_0\left(T_{max}\right) + e_0\left(T_{min}\right)\right]/2 \tag{3.6}$$

$$e_0\left(T_{max}\right) = 0.6108 \times \exp\left\{\frac{17.27\,T_{max}}{T_{max} + 237.3}\right\} \tag{3.7}$$

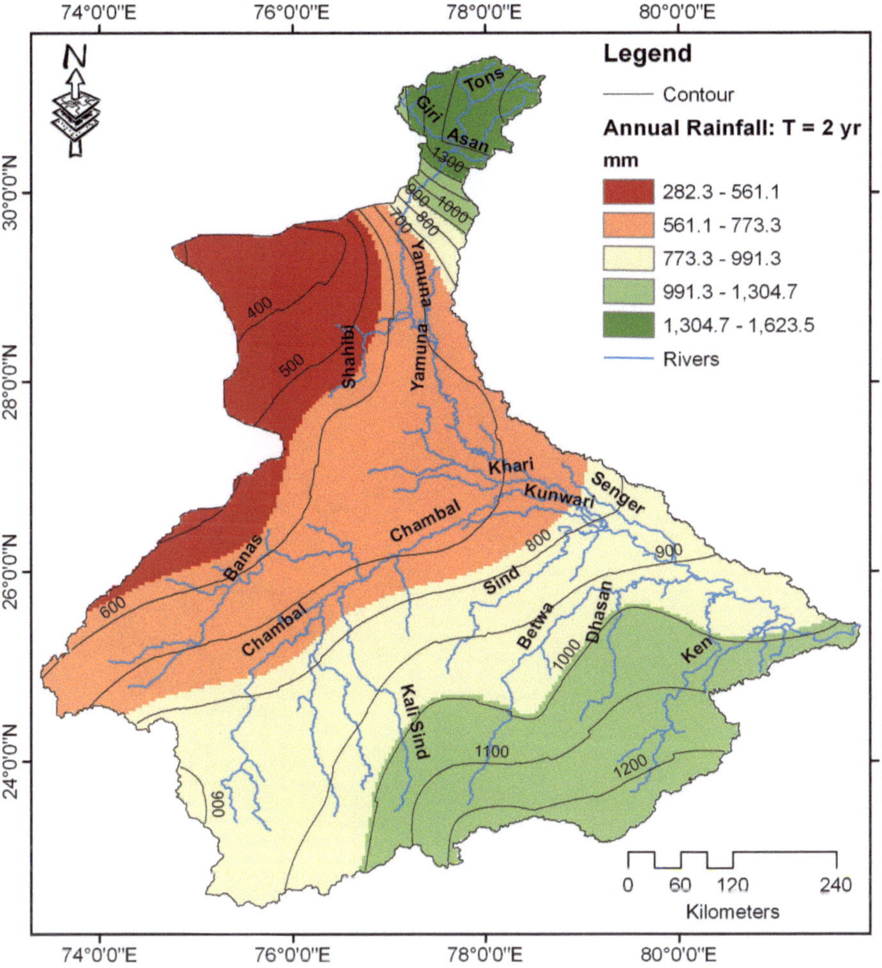

Fig. 3.8 Annual rainfall for the return period of 2 years (data period: 1951–2003)

where $T_{\max}$ and $T_{\min}$ are the daily maximum and minimum temperatures, respectively (°C).

Using daily values of dew point temperature, minimum and maximum temperatures available from National Climatic Data Center (http://www.ncdc.noaa.gov) for a few stations in the Yamuna basin, the relative humidity is computed from eq. (3.4) with auxiliary equations [eqs. (3.5), (3.6), and (3.7)], and interpolated at the basin scale using the Kriging interpolation technique. The derived spatial variation of estimated mean RH is shown in Fig. 3.22.

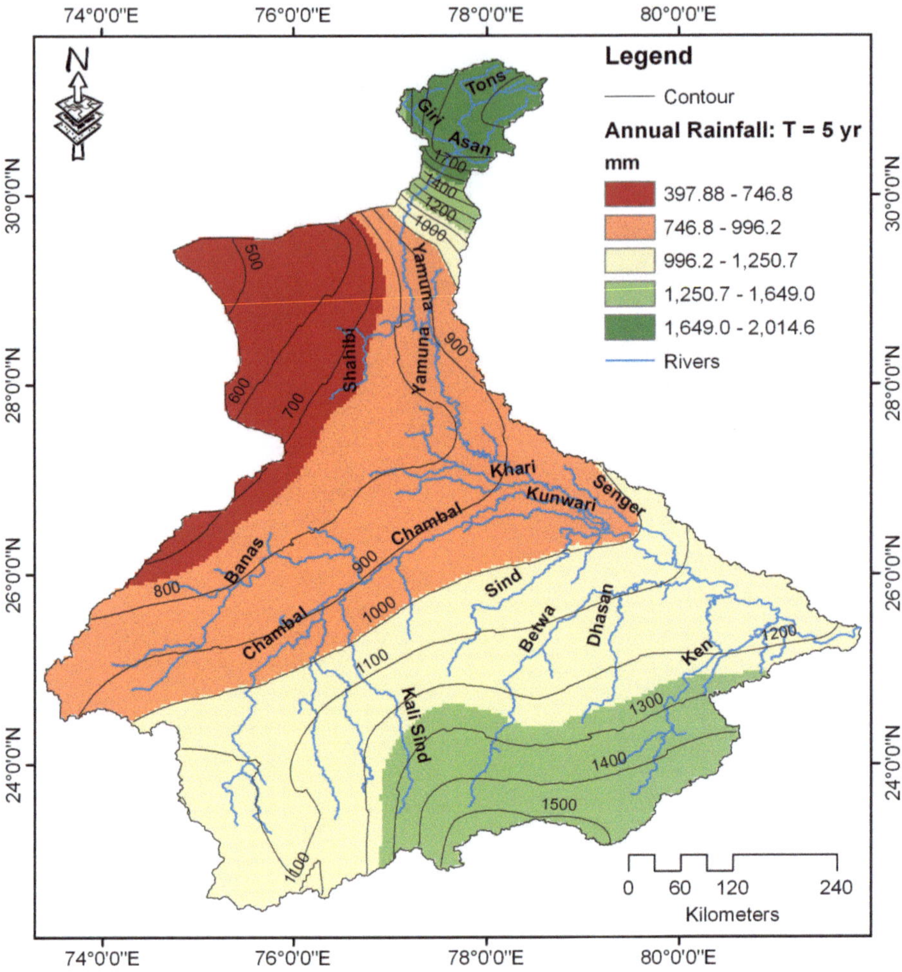

Fig. 3.9 Annual rainfall for the return period of 5 years (data period: 1951–2003)

3.5 Evapotranspiration

Evapotranspiration is a process of mass transfer due to heat energy. It is the combined effect of evaporation as well as transpiration. Evaporation can be defined as the process in which liquid water is converted to water vapor from the evaporating surface, such as lakes, rivers, pavements, soil and vegetation; whereas, transpiration is a process of vaporization of liquid water from plant tissues and the vapor removal from the atmosphere. On the other hand, the evapotranspiration rate from a reference surface, not short of water, is called the reference crop evapotranspiration or reference evapotranspiration and is denoted as ET_0. The reference surface is a hypothetical grass reference crop that closely resembles an extensive surface of green,

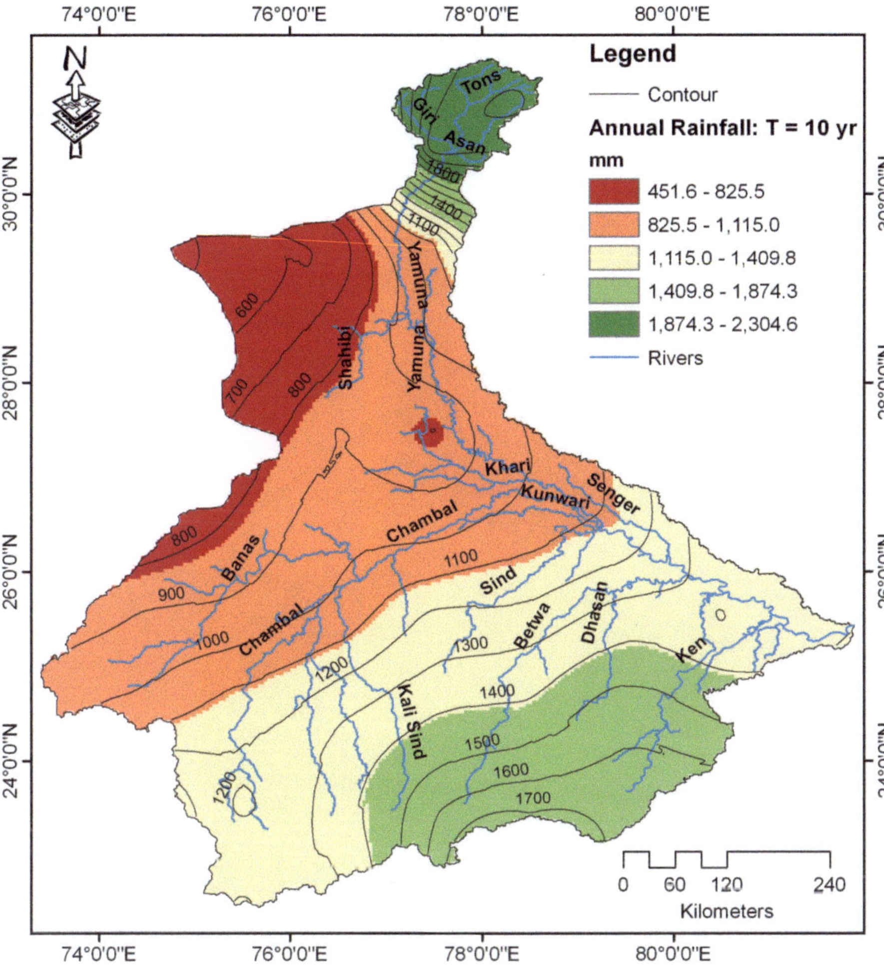

Fig. 3.10 Annual rainfall for the return period of 10 years (data period: 1951–2003)

well-watered grass of uniform height, actively growing and completely shading the ground.

The estimation of ET_0 requires several climatic parameters, such as weather parameters (i.e., radiation, air temperature, wind speed, humidity); crop characteristics (type, variety, development stage); and management and environmental aspects. These data are difficult to manage for such a large basin. Under such circumstances, FAO-56 procedure based on the Penmann-Montieth approach has become limited in field application. Therefore, the Hargreaves method of ET_0 has been used with reliable accuracy and is described as follows. Hargreaves equation has a tendency to under-predict under high wind speed conditions ($u > 3$ m/s) and over-predict under conditions of high relative humidity.

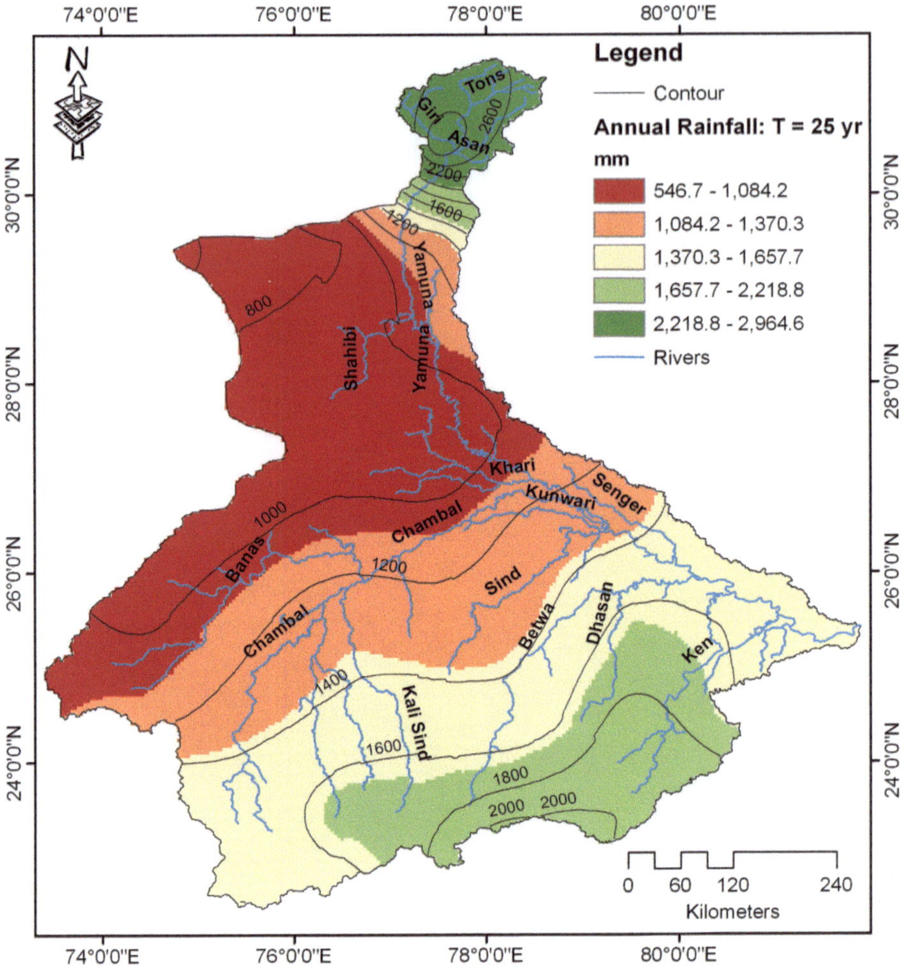

Fig. 3.11 Annual rainfall for the return period of 25 years (data period: 1951–2003)

$$ET_0 = 0.0023\ (T_{\text{mean}} + 17.8)\ (T_{\max} - T_{\min})^{0.5} \times R_a \tag{3.8}$$

where ET_0 is the reference evapotranspiration (mm d^{-1}); T_{mean}, $T_{\max}$, and $T_{\min}$ are the daily mean, maximum and minimum temperatures (°C); and R_a is the extra-terrestrial radiation for each day (mm d^{-1}). R_a for each day of the year and for different latitudes can be estimated as follows:

$$R_a = 0.408 \times \frac{24\,(60)}{\pi} G_{sc} d_r \left[\omega_s \sin(\varphi) \sin(\delta) + \cos(\varphi) \cos(\delta) \sin(\omega_s)\right] \tag{3.9}$$

where: R_a is the extra-terrestrial radiation (mm d^{-1}); G_{sc} is the solar constant (0.0820 MJ m^{-2} min^{-1}); d_r is the inverse relative distance (Earth-Sun); ω_s is the

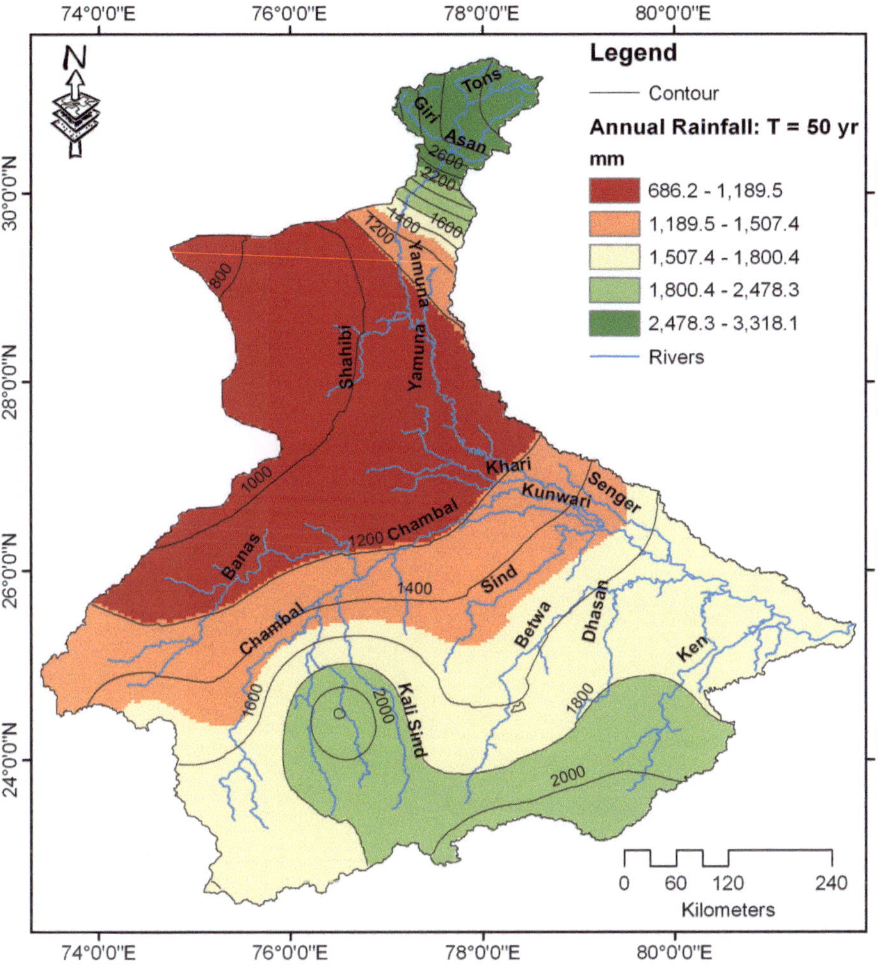

Fig. 3.12 Annual rainfall for the return period of 50 years (data period: 1951–2003)

sunset hour angle (rad); φ is the latitude (rad); and δ is the solar declination (rad) [rad $= \pi \times$ decimal deg/180]. The inverse relative distance Earth-Sun, d_r and the solar declination, δ, is given as:

$$d_r = 1 + 0.033 \cos\left(\frac{2\pi}{365}J\right) \tag{3.10}$$

$$\delta = 0.409 \sin\left(\frac{2\pi}{365}J - 1.39\right) \tag{3.11}$$

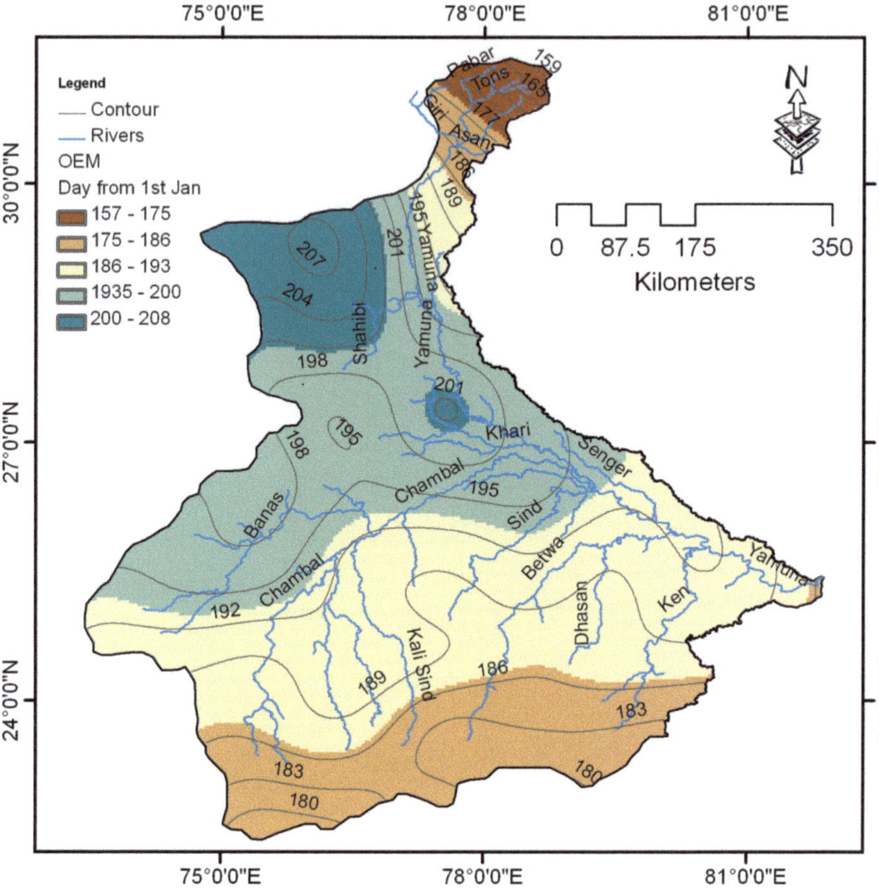

Fig. 3.13 Date of onset of effective monsoon for the Yamuna basin

where J is the number of days in the year between 1 (1st January) and 365 or 366 (31st December). The sunset hour angle ω_s is given by:

$$\omega_s = \arccos\left[-\tan(\varphi)\cdot\tan(\delta)\right] \tag{3.12}$$

Using the temperature data for the period of 1951–2002 and latitude of the district centroid, the daily reference crop evapotranspiration is estimated for the 52 year period (1951–2002). The spatial variation of estimated mean annual ET_0 is depicted in Fig. 3.23 which also consists of the contours of ET_0. Approximately, 75% of the basin area has the annual ET_0 varying between 1,536–1,624 mm.

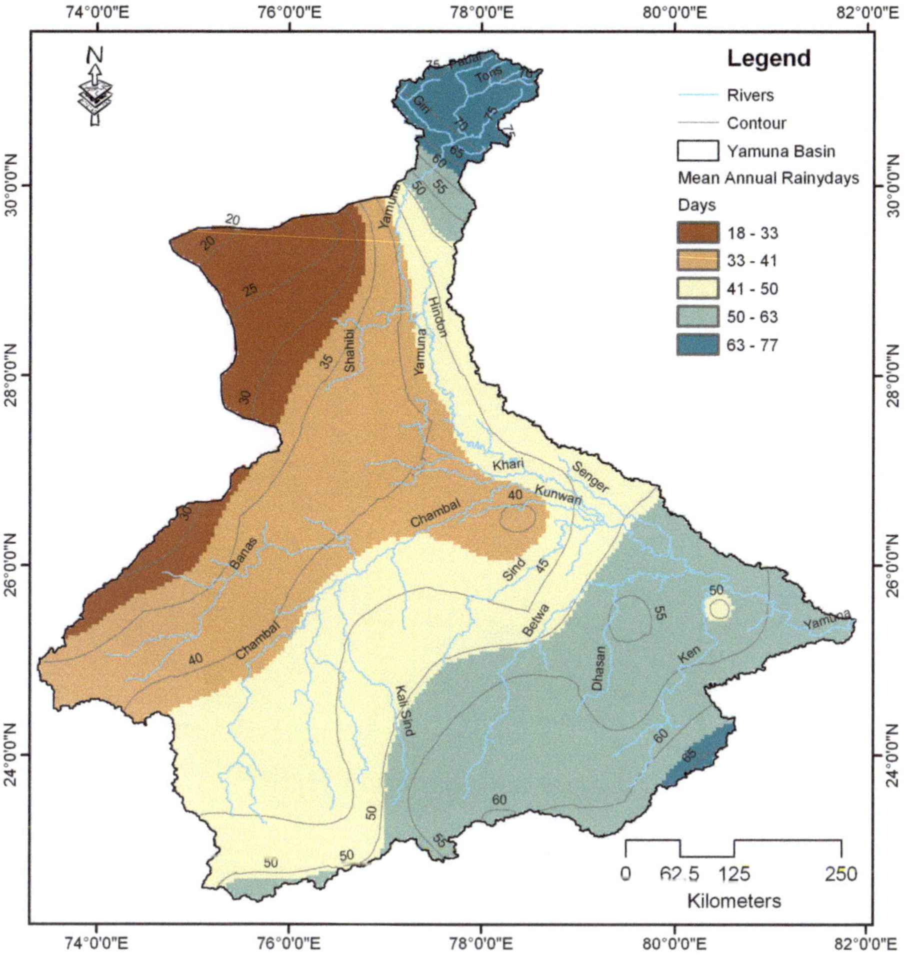

Fig. 3.14 Number of annual rainydays in the Yamuna basin

3.6 Aridity Index and Climate Classification

Aridity Index (AI), an indicator of the annual moisture deficit and degree of dryness, can be defined as the ratio of annual rainfall to annual potential evapotranspiration. Numerically, the aridity index (AI) can be expressed as follows (UNEP, 1993):

$$AI = \frac{\text{Annual Rainfall}}{\text{Annual Potential Evapotranspiration}} \tag{3.13}$$

Both terms appearing in eq. (3.13) are in millimeter. Using the UNEP (1993) criteria, climate can be classified as: hyper arid ($AI < 0.05$), arid ($0.05 < AI < 0.20$), semi arid ($0.20 < AI < 0.50$), dry sub humid ($0.50 < AI < 0.65$), and humid ($AI > 0.65$).

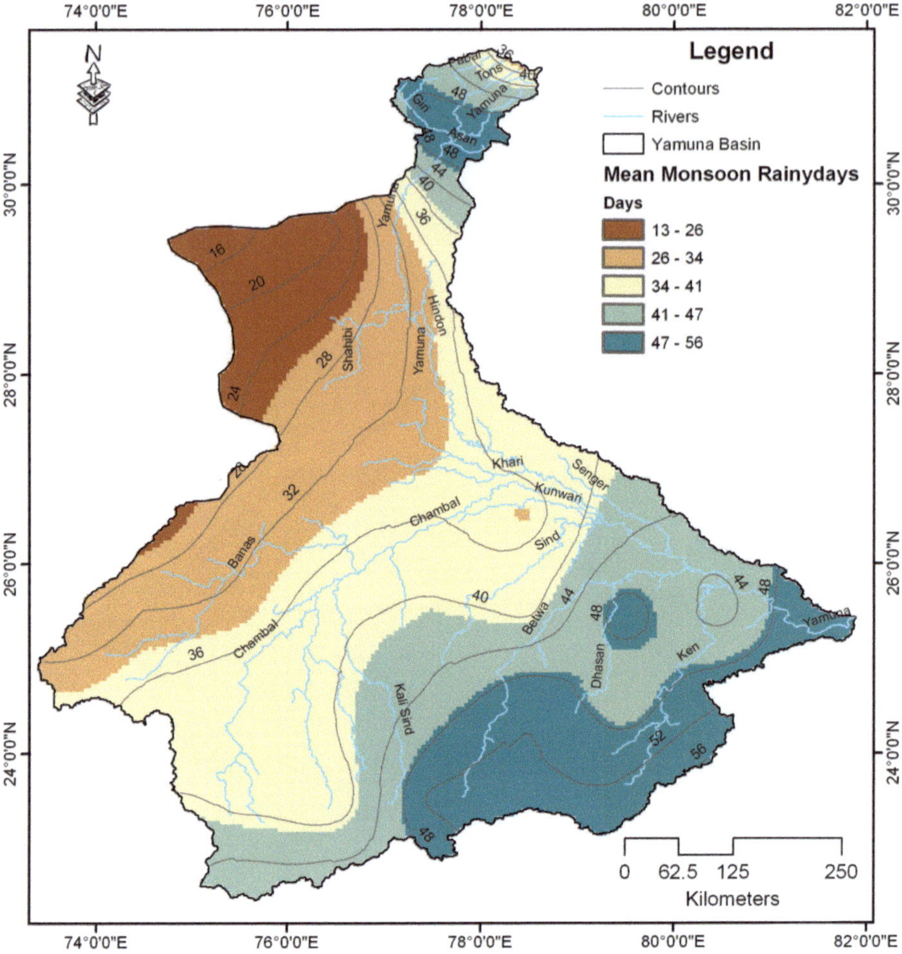

Fig. 3.15 Number of monsoon rainydays in the Yamuna basin

The potential evapotranspiration (PET) is a representation of the environmental demand for evapotranspiration and represents the evapotranspiration rate of a short green crop, completely shading the ground, of uniform height and with adequate water status in the soil profile. It is a reflection of the energy available to evaporate water, and of the wind available to transport the water vapor from the ground up into the lower atmosphere. Evapotranspiration is said to equal potential evapotranspiration when there is ample water. The value of potential evapotranspiration is generally estimated using the Thornthwaite equation (Thornthwaite, 1948). The method is based on the annual temperature efficiency index J, defined as the sum of 12 monthly values of the mean monthly temperature T_{mean} as follows:

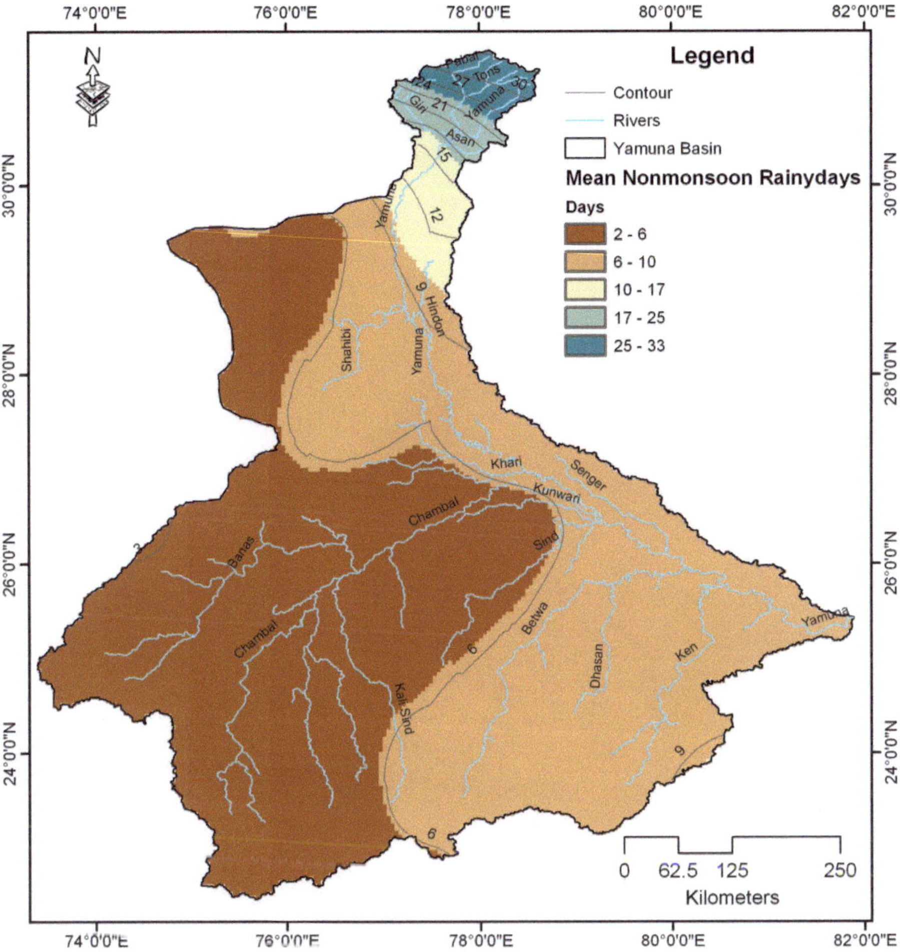

Fig. 3.16 Number of non-monsoon rainydays in the Yamuna basin

$$\mathrm{PET}_i\left(0^\circ\right) = 1.6\left(\frac{10\,T_i}{J}\right)^c \tag{3.14}$$

where:

$$c = 0.000000675\,J^3 - 0.0000771\,J^2 + 0.0179\,J + 0.49239 \tag{3.15}$$

$$J = \sum_{i=1}^{12} I_i \tag{3.16}$$

$$I_i = \left(\frac{T_i}{5}\right)^{1.514} \tag{3.17}$$

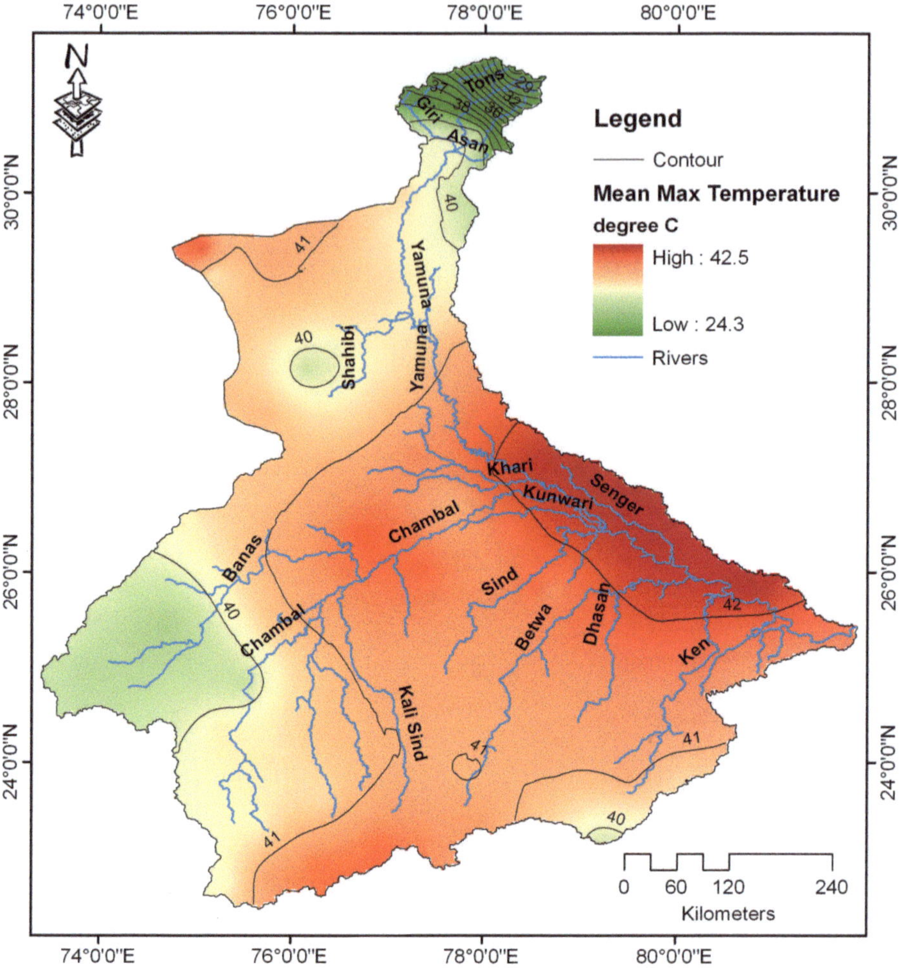

Fig. 3.17 Spatial variation of mean maximum temperature

where T_i is the mean monthly temperature (°C) for the ith month, and $PET_i(0°)$ is the potential evapotranspiration for the ith month at 0° latitude [$PET_i(0°)$]. For other latitudes, PET can be estimated as follows:

$$PET_i = k \times PET_i\left(0^\circ\right) \tag{3.18}$$

Where k is the latitude correction factor which depends on the day light hours. It can be defined as:

$$k = \frac{N}{N'} \tag{3.19}$$

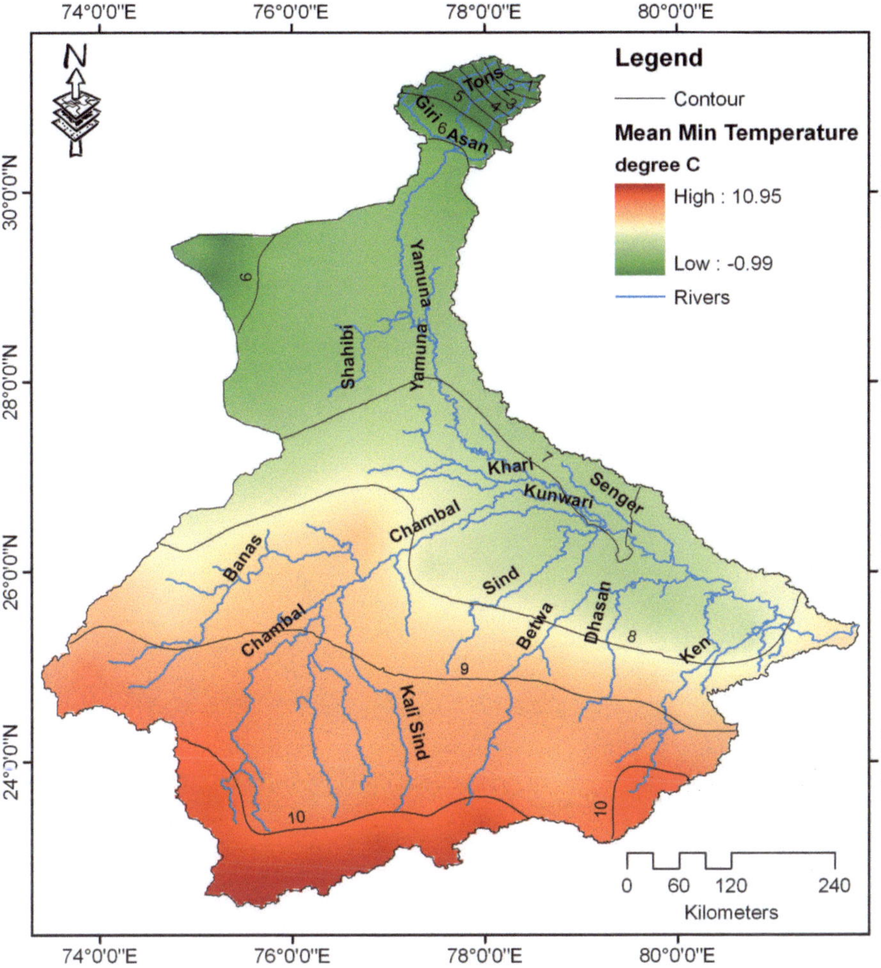

Fig. 3.18 Spatial variation of mean minimum temperature

where, N = day light hours at latitude other than 0°, and N' = day light hours at 0° latitude. The value of N can be obtained as:

$$N = \frac{24}{\pi} \times \omega_s. \tag{3.20}$$

$$\omega_s = \arccos\left[-\tan(\varphi)\tan(\delta)\right] \tag{3.21}$$

$$\delta = 0.409 \sin\left(\frac{2\pi}{365}J' - 1.39\right) \tag{3.22}$$

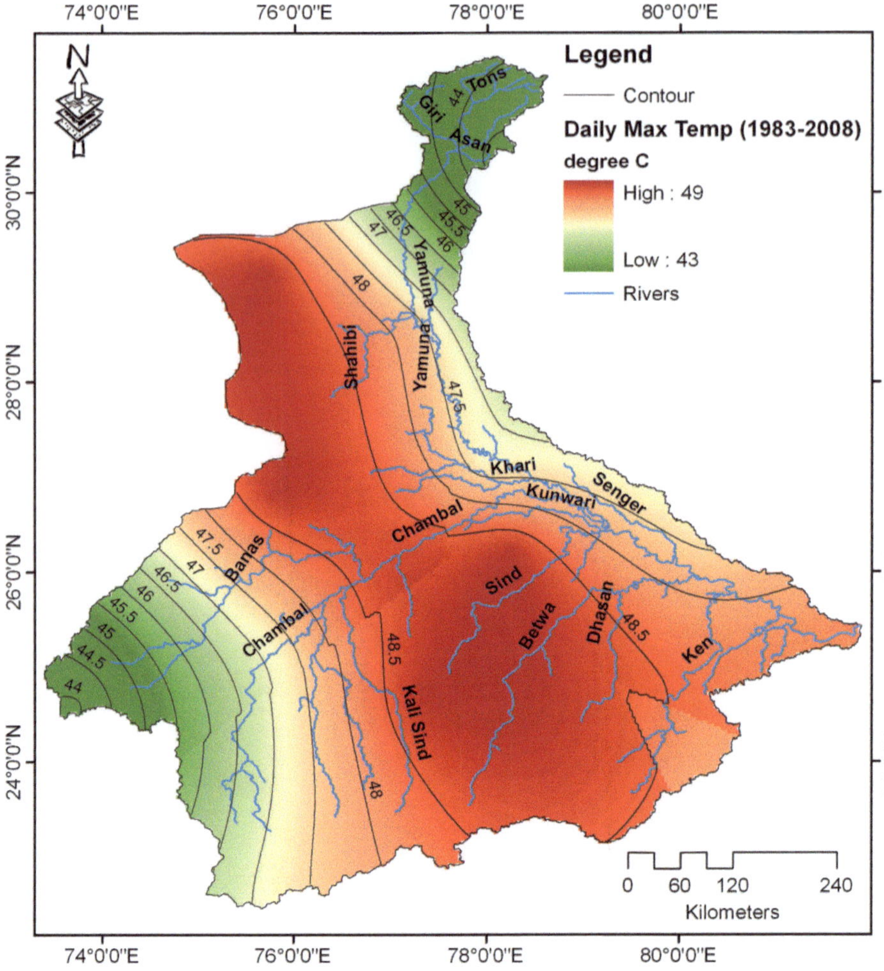

Fig. 3.19 Spatial variation of daily maximum temperature (1983–2008)

where J' is the number of days in the year between 1 (1 January) and 365 or 366 (31 December), and φ is the latitude in radians. In the estimation of monthly PET, J' is taken in the middle of the each month (Table 3.6).

Using eq. (3.13), aridity index is estimated for the entire grid locations of the Yamuna basin, and interpolated using the Kriging method. Using the aridity index, the climate of the Yamuna basin varies from arid to humid (Fig. 3.24). Based on Fig. 3.24, a majority of the catchment area in the basin falls under semi-arid climate followed by dry sub-humid and humid climate.

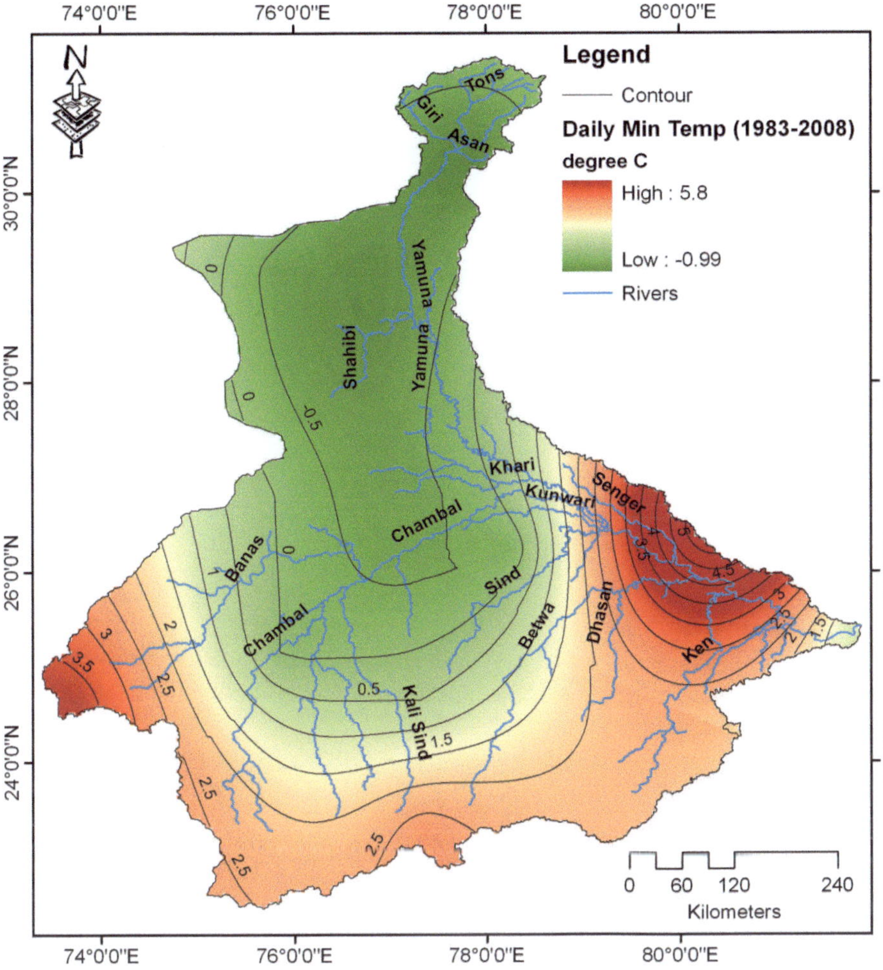

Fig. 3.20 Spatial variation of daily minimum temperature (1983–2008)

Table 3.5 Catchment wise mean temperatures

S. No.	Basin/sub-catchment	Maximum temperature (°C)	Minimum temperature (°C)
1	Upper Himalaya	34.72	3.53
2	Betwa	41.53	8.80
3	Ken	41.28	8.90
4	Sind	42.00	7.86
5	Chambal	40.80	9.08
6	Hindon	40.31	6.55
7	Khari	41.19	7.33
8	Yamuna basin	40.52	7.60

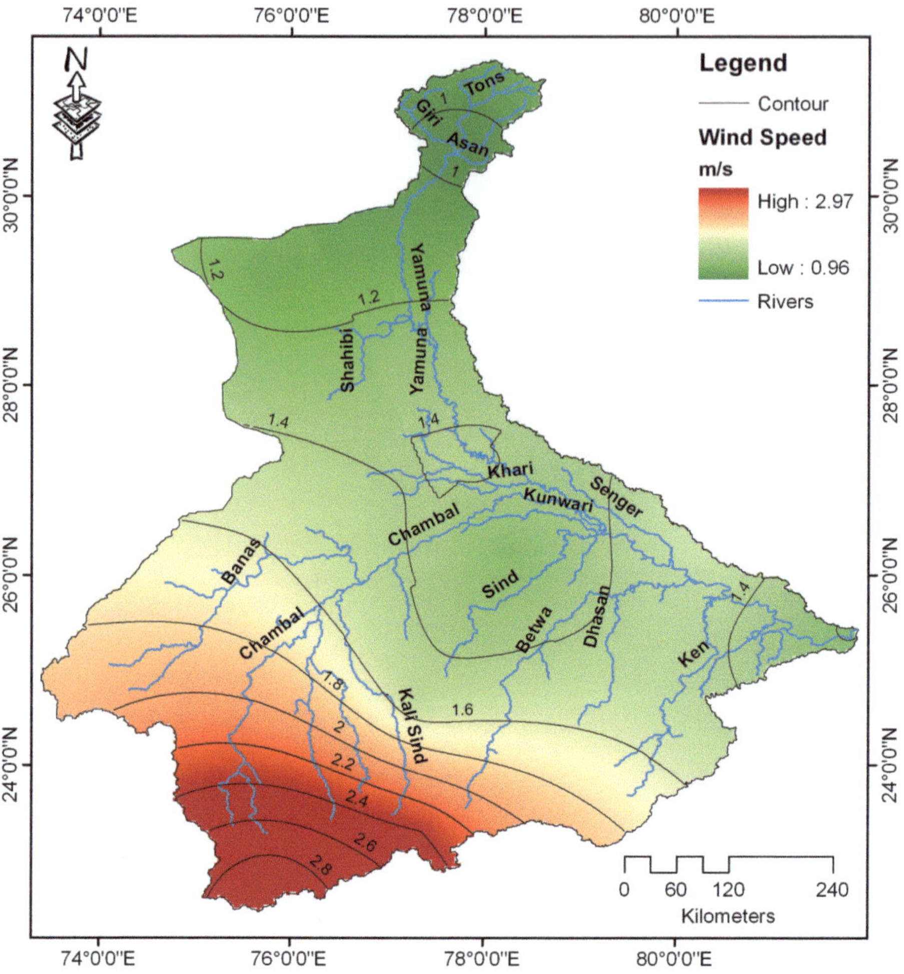

Fig. 3.21 Spatial variation of average wind speed in Yamuna basin (1980–2009)

3.7 Concluding Remarks

Based on the above analysis, Yamuna basin has a wide spatial variability in the climate. Rainfall, temperature and evapotranspiration along with the aridity index are highly variable. Therefore, planning of water or natural resources is quite difficult, when the whole basin is considered as a single hydrologic unit. Under such circumstance, planning should be made at the watershed scale. However, based on topography, soil characteristics and climate, homogeneous watersheds can be identified for regional planning.

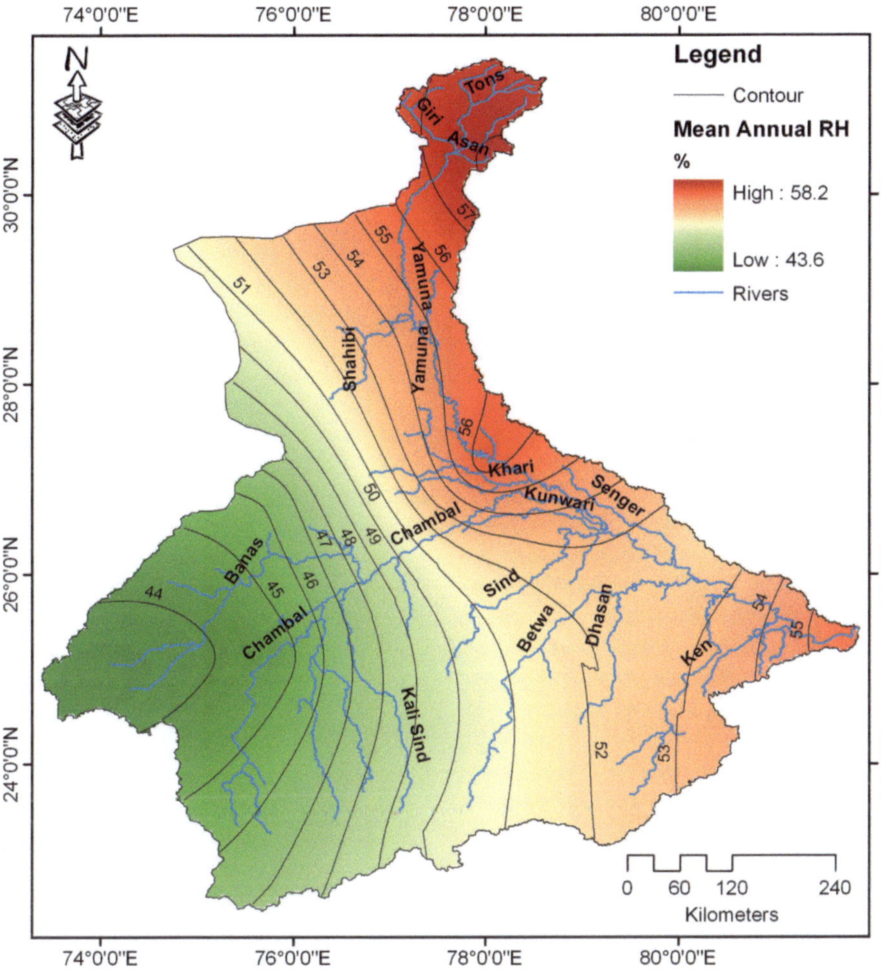

Fig. 3.22 Spatial variation of average relative humidity in Yamuna basin (1980–2009)

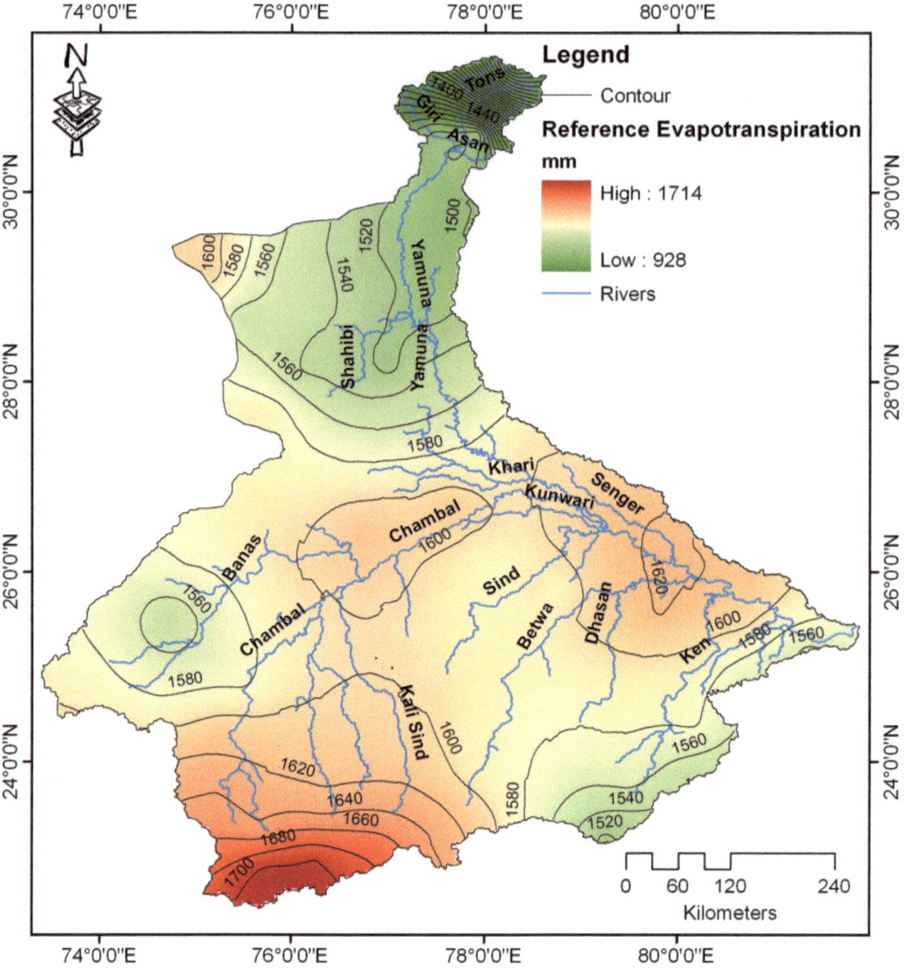

Fig. 3.23 Spatial variation of ET_0

Table 3.6 J' value used for the estimation of the solar declination, δ (rad)

Month	Jan	Feb	Mar	Apr	May	June	July	Aug	Sep	Oct	Nov	Dec
J' (Normal)	15	46	74	105	135	166	196	227	258	288	319	349
J' (Leap)	15	46	75	106	136	167	197	228	259	289	320	350

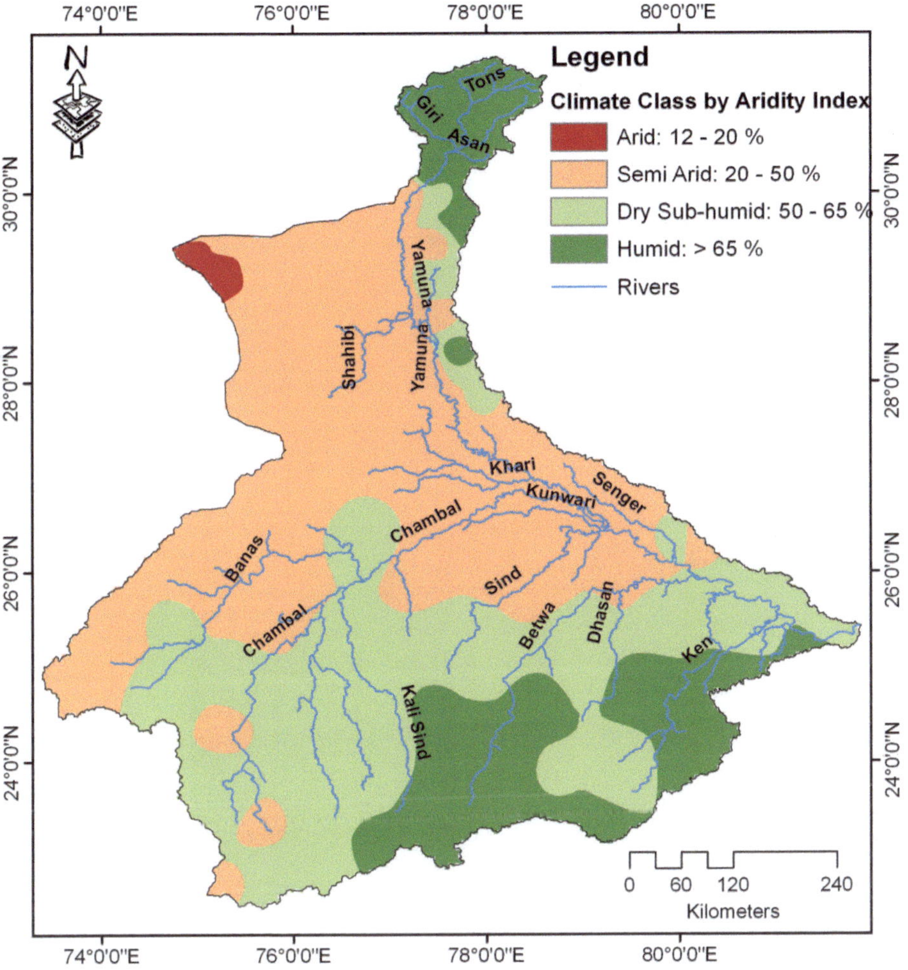

Fig. 3.24 Climate classification of Yamuna basin

References

Ashokraj PC (1979) Onset of effective monsoon and critical dry spell-A computer based forecasting technique. Indian Agricultural Research Institute, ICAR, New Delhi, Bulletin No. 11, India

Thornthwaite CW (1948) An approach toward a rational classification of climate. Geogr Rev 38:55–94

UNEP (1993) World Atlas of Desertification. The United Nations Environment Programme (UNEP), London

Chapter 4
Pattern Changes in Climatic Variables

Abstract This chapter describes a methodology for identifying the nature and attributes of the time series of climatological variables, which is usually the first step for water resources planning and management. A detailed description of the statistical approach used to investigate the presence and extent of persistence, trend and periodicity in climatic time series is presented. Results of persistence and trend analysis are reported spatially for the basin. The hydro-climatic variables considered for analysis are annual rainfall; monsoon and non-monsoon rainfall; annual, monsoon and non-monsoon rainy days; onset of effective monsoon; and aridity index.

Climate is the most important driving parameter that causes year-to-year variability in socio-economic and environmental systems including the availability of water resources. It affects the development and planning of water resources schemes, such as flood prevention and control, drought management, food and fiber production, etc. Further, any change in climate will increase the uncertainty in water resources planning. Apart from this, changes in climatic pattern will have profound effects and consequences for natural and agricultural ecosystems and for society as whole. These changes could even alter the location of the major crop production regions on the earth (Reddy and Hodges, 2000). The shifting from "normal weather", with its associated extreme events will surely change the zones of crop adaptation and cultural practices required for successful crop production. Climate and weather induced instability in food and fibre supplies will alter social and economic stability and regional competitiveness (Reddy and Hodges, 2000). Therefore, analysis of hydro-climatic variables such as rainfall, potential evapotranspiration, etc., becomes a prerequisite to understand the climatic changes.

In recent years, there has been a considerable concern about the possibility of global warming and climatic changes. Alteration in our climate is governed by a complex system of atmospheric, land surface and oceanic processes and their interactions. Atmospheric processes also result in an increase in surface-level ultraviolet radiation and changes in temperature and rainfall pattern. Human activities on the other hand are responsible for changes in ecosystems due to increased emissions rate of CO_2 and other green house gases. The evidence using state-of-art computer models incorporating as much of the theoretical understanding of the earth's weather suggests that global warming is occurring along with shifting patterns of rainfall and incidents of extreme weather events (IPCC, 2007). It has been demonstrated that global surface warming has been taking place at a rate of 0.74 ± 0.18°C

R.K. Rai et al., *The Yamuna River Basin*, Water Science and Technology Library 66,
DOI 10.1007/978-94-007-2001-5_4, © Springer Science+Business Media B.V. 2012

over the period of 1906–2005 (IPCC, 2007) and it is expected to be more in the next century than what has occurred during the past 10,000 years (IPCC WG II, 2007). The increased atmospheric moisture content associated with warming might be expected to increase the global mean precipitation.

The global annual land mean precipitation shows a small, but uncertain, upward trend of approximately 1.1 mm per decade (uncertainty ±1.5 mm) over 1901–2005. During the 20th century, precipitation has generally increased from latitudes 30° to 85°N over land; but notable decreases have occurred between latitudes 10°S and 30°N in the last 30–40 years. In western Africa and southern Asia, linear trends in the rainfall decrease during 1900–2005 are 7.5% per century (significant statistically at <1% level), whereas much of northwest India shows an increase in rainfall with more than 20% per century (IPCC, 2007). At lower latitudes, especially in seasonally dry and tropical regions, the crop productivity is projected to decrease for even small local temperature increases (1–2°C), which would increase food risk (IPCC WG II, 2007). Based on the data for the period of 1901–2005, it is demonstrated that all-India mean annual temperature has been rising at 0.05°C/decade, with maximum temperature at +0.07°C/decade and minimum temperature at +0.02°C/decade (Kothawale and Kumar, 2005). As a result, the diurnal temperature range shows an increase of 0.05°C/decade. However, in northern India, the average temperature is falling at a rate of −0.38°C unlikely to rise in all-India average temperature (i.e., at +0.42°C/Century) (Arora et al., 2005).

Change in the precipitation pattern is also reported for the 21st century. Kripalani et al. (2007) reported an increasing trend in South-Asian mean monsoon precipitation associated with intensification of land–ocean pressure gradient during the establishment phase of the monsoon. At a regional scale, diversified precipitation pattern have been investigated by various scientists/agencies. A declining trend in precipitation has been observed over Greece (Giakoumakis and Baloutsos, 1997), Canadian Prairies (Gan, 1998), Bologna in Italy (Ventura et al., 2002); whereas North Carolina (Boyles and Raman, 2003), Mainland Spain (Mosmann et al., 2004) experienced rising trends. No trend has been identified in precipitation over the Iberian Peninsula (Serrano et al., 1999) and Japan (Xu et al., 2003). Besides these, a comprehensive review of precipitation changes has also been carried out at global and regional levels, and decreasing precipitation trends have been identified in Russia, Kazakhstan, China and Thailand with a global increase in variance (Dore, 2005). Probable causes of change in rainfall may be due to: (a) global climate shift (Baines, 2006) or weakening global monsoon circulation (Chase et al., 2003; Duan and Yao, 2003; Pant, 2003); (b) reduction in forest cover (Meher-Homji, 1991; Hingane, 1996; Kothyari et al., 1997; Chen et al., 2001; Lawton et al., 2001; Nair et al., 2003; Ray et al., 2003; Avissar and Werth, 2005; Gupta et al., 2005; Ray et al., 2006; etc.) and change in land use, including introduction of irrigated agriculture (Pielke, 2001; Douglas et al., 2006; Ramankutty et al., 2006; Pielke et al., 2007; etc.); and (c) increasing aerosol due to anthropogenic activities (Ackerman et al., 2000; Ramanathan et al., 2001; Ramanathan et al., 2005; Sarkar and Kafatos, 2004; etc.).

In India, a diverse climate with large variations in rainfall is experienced. Rainfall variability of the Indian monsoons has been extensively studied by several investigators (Kripalani and Kulkarni, 1997, 2001; Kulkarni et al., 2007). An extensive review of studies on rainfall patterns in India excluding the Himalayan region, is well presented by Basistha et al. (2009). Based on these studies, a consensus regarding evidence of insignificant trend in annual rainfall series can be reported, though divergences do exist depicting an increasing (Parthasarathy and Dhar, 1976a) or decreasing trend (Singh et al., 2005). Regional analyses of data reveal increasing rainfall trends over the Indus, the Ganga, the Brahmaputra, the Krishna and the Cauvery basin (Singh et al., 2005), the west coast (Alvi and Koteswaram, 1985), north Andhra Pradesh and northwest India (Kumar et al., 1992), Rohtak and Kurukshetra of Haryana (Lal et al., 1992), Delhi (Rao et al., 2004), west Madhya Pradesh (Parthasarathy and Dhar, 1976b), coastal Orissa (Senapati and Misra, 1988) and peripheries of the Rajasthan desert (Pant and Hingane, 1988), but decreasing trends in the central Indian basins of Sabarmati, Mahi, Narmada, Tapi, Godavari and Mahanadi (Singh et al., 2005), east Madhya Pradesh and adjoining areas, northeast India (Zveryaev and Aleksandrova, 2004) and parts of Gujarat (Kumar et al., 1992), south Kerala (Singh et al., 1989; Singh and Soman, 1990) and central north Indian divisions (Subbaramayya and Naidu, 1992). Singh et al. (2007) have indicated increasing trends in annual rainfall and relative humidity in north Indian River basins, though the least variation has been observed in monsoon rainfall. Studies by Kothyari and Singh (1996) and Kothyari et al. (1997) report a decreasing rainfall trend over the Ganga basin, beginning around the second half of the 1960s. This is elaborated by Singh and Sontakke (2002) to be increasing over western Indo-Gangetic Plains and decreasing (statistically insignificant, though) over the central part. IINC (2004) depicts a decreasing summer monsoon rainfall trend over the state of Uttarakhand.

Exploring changes in rainfall pattern in the Indian Himalayas during the 20th century, Basistha et al. (2009) have found that the most probable year of change in annual as well as monsoon rainfall in the region is 1964. There is an increasing trend up to 1964 (corroborating with all India and nearby plains), followed by a decreasing trend in 1965–1980 (exclusive to this region). In the entire region, changes are most conspicuous over the Shivaliks and the southern part of the Lesser Himalayas. Chase et al. (2003) indicated a consistent reduction in the intensity of all tropical monsoon systems since 1950. A decrease in monsoon precipitation in the central Himalayas (Tibet, China) has been identified from early 1920s to the present (Duan and Yao, 2003). However, a substantial global climate shift has been reported in the late 1960s affecting South America–Africa (Baines, 2006). Apart from the rainfall pattern, other meteorological parameters show significant variations. Jhajharia et al. (2008) have analyzed the temporal characteristics of pan evaporation (Epan) under the humid conditions for northeast India and reported general decreasing trends in pre- and monsoon seasons. A declining trend in sunshine hours and wind speed, and increasing trend in relative humidity is also reported for most of the stations. Bandyopadhyay et al. (2009) reported similar results for India in case of reference evapotranspiration (ETo) for the period of 1971–2002, which is mainly caused by a significant increase in relative humidity and a consistent significant decrease in the wind speed throughout the country. However, a general increase in rainfall has not been found in recent years.

Considering the aforesaid studies, it can be stated that there is no consensus among the studies. Possibly, it may be due to the analysis of climatic variables at the regional scale, which induces loss of spatial information about the variables. Therefore, for water resources planning, it seems to be logical and compulsory to analyze the hydro-climatic variables at small scale that it can be rationalized at the river basin scale. Though the change in climate is governed by the complex system of atmosphere and oceanic processes and their interaction, but due to limitations on the availability of wide variety of atmospheric data, this study focuses on the analysis of indicative hydro-climatic parameters to demonstrate the climate change or changes in weather patterns.

In this chapter, the Yamuna River basin is investigated to demonstrate the climatic fluctuation/variability which exhibits high spatio-temporal variability in terms of climate (semi-arid to humid subtropical climates) and topography. Parameters that directly influence water resources and agricultural planning (viz., annual, monsoon and non-monsoon rainfall; number of annual, monsoon and non-monsoon rainydays; onset of effective monsoon; and aridity index) are considered.

4.1 Statistical Test for Climatological Data

To define the climatic fluctuations exhibited in the hydro-climatic time series is important. From a statistical point of view, the study of climatic fluctuations is a problem of time series analysis. Statistical evidence of persistence in such time series is equated with evidence of bona fide climatic fluctuations and is said to be dependent. In many instances, a time series is generally not statistically independent but is comprised of persistence, cycles, trends or other non-random components. A steady and regular movement in a time series through which the values are on average either increasing or decreasing is termed a trend. This type of behavior can be local, in which case the nature of the trend is subject to change over short intervals of time, or, on the other hand, it can be visualize a global trend that is long lasting. If a trend in a hydrologic time series appears, then it is, in effect, part of a low frequency oscillatory movement induced by climatic factors or through the change in land use and catchment characteristics. Looking into the importance of the deterministic components of climatic series, statistical tests for persistence, trend and periodicity are discussed, and analyzed for the hydro-climatic variables of the Yamuna River basin.

4.1.1 Persistence

For climatic variability and changes, the definition of persistence given by WMO (1966) is commonly used. According to this definition, persistence is a "tendency for successive values of the series to "remember" their antecedent values, and to be influenced by them." The value of r_1 has been used to detect the possible persistence in the observed year-to-year variations of normalized anomaly series and to examine its nature and magnitude. The approach proposed by WMO (1966) and Matalas (1967) is widely used in many studies related to long-term climatic variations (Rodhe and Virji, 1976; Granger, 1977; Ogallo, 1979; Anyadike, 1993; Drosdowsky, 1993; Nicholson and Palao, 1993; Türkes, 1998; Türkes et al., 2002). To test the persistence in the climatological time series normalized anomaly of the time series is used, which is obtained as:

$$X_t = (x_t - \bar{x}) / \sigma \tag{4.1}$$

where X_t is the normalized anomaly of the series, x_t is the observed time series, and $\bar{x}$ and σ are the long-term mean and standard deviation of annual/seasonal time series. All serial correlation coefficients of the normalized climatic series are computed for lags $L = 0$ to m, where m is the maximum lag (i.e. $m = n/3$); n is the length of the series. The serial correlation coefficient is computed from eq. (4.2):

$$r_L = \frac{\sum_{t=1}^{n-L} (X_t - \bar{X}_t) \cdot (X_{t+L} - \bar{X}_{t+L})}{\left[\sum_{t=1}^{n-L} (X_t - \bar{X}_t)^2 \cdot \sum_{t=1}^{n-L} (X_{t+L} - \bar{X}_{t+L})^2\right]^{1/2}} \tag{4.2}$$

where r_L is the lag-L serial correlation coefficient of the series. To test the significance of serial correlation, eq. (4.3) is used (Yevjevich, 1971):

$$(r_L)_{t_g} = \frac{-1 \pm t_g \, (n - L - 1)^{1/2}}{n - L} \tag{4.3}$$

where, $(r_L)_{t_g}$ is the normally distributed value of r_L, t_g is the normally distributed statistic at g level of significance. The value of t_g are 1.645, 1.965 and 2.326 at a significance level of 0.10, 0.05 and 0.01, respectively.

The hypothesis test is carried out at the 0.05 significance level as this level is sufficient from an engineering point of view. The "null" hypothesis of the randomness of climatic series against the serial correlation is rejected for the large value of r_1. If r_1 of the series is not statistically significant or is significant but has a negative sign, it is assumed that the series does not contain the persistence and the appropriate null continuum is termed as "white noise". On the other hand, the persistence in the time series is characterized by a positive serial correlation. In this case, a "Markov red noise" type "null" continuum is ensured using r_2 and r_3 (WMO, 1966). In addition to this significant negative r_1 is very likely to be indicative of high-frequency oscillations, whereas significant positive r_1 is likely to be indicative of low-frequency fluctuations and persistence in climatic series. Therefore, in the study serial correlation coefficients up to lag-3 are assessed.

4.1.2 Trend

There are several approaches for detecting the trend in the time series. These approaches can be either parametric or non-parametric. Parametric methods assume the data should normally distributed and free from outliers. On the other hand, non-parametric methods are free from such assumptions. The most popularly used non-parametric tests for detecting a trend in the time series is the Mann-Kendall (MK) test (Mann, 1945; Kendall, 1955). It is widely used for different climatic variables (Hirsch et al., 1982; Hirsch and Slack, 1984; Lettenmaier et al., 1994; Gan, 1998; Lins and Slack, 1999; Douglas et al., 2000; Burn and Elnur, 2002; Yue et al., 2002; Yue and Pilon, 2004; Burn et al., 2004; Zhang et al., 2005; Aziz and Burn, 2006; Chen et al., 2007; etc.).

For the original Mann-Kendall test, the time series must be serially independent in nature. However, in many real situations, the observed data are serially dependent (i.e., autocorrelated). The autocorrelation in the observed data results in misinterpretation of trend test results. Cox and Stuart (1955) stated that "positive serial correlation among the observations would increase the chance of significant error, even in the absence of a trend". A closely related problem that has been studied is the case where seasonality exists in the data (Hirsch et al., 1982). By dividing the observations into separate classes according to the season and then performing the Mann-Kendall trend test on the sum of the statistics from each season, the effect of seasonality can be eliminated. This modification is called the seasonal Mann-Kendall test (Hirsch et al., 1982; Hirsch and Slack, 1984). Although the seasonal test eliminates the effect of seasonal dependence, it does not account for the correlation in the series within the season (Hirsch and Slack, 1984). The same problem exists when yearly time series is considered for analysis, as it significantly autocorrelated. Therefore, in this paper, the original Mann-Kendall test is described along with its modified versions that accounts for the serially dependent in the data.

4.1.2.1 Mann-Kendall Test

The Mann-Kendal (MK) test (Mann, 1945; Kendall, 1955) searches for a trend in a time series without specifying whether the trend is linear or nonlinear (Khaliq et al., 2009). The Mann-Kendall test for detecting monotonic trends in hydrologic time series is described by Yue et al. (2002). It is based on the test statistics S, which is defined as:

$$S = \sum_{i=1}^{n-1} \sum_{j=i+1}^{n} \operatorname{sgn}\left(x_j - x_i\right) \tag{4.4}$$

where x_j are the sequential data values, n is the length of the data set, and

$$\text{sgn}\ (t) = \begin{cases} 1, & \text{for}\ t > 0 \\ 0, & \text{for}\ t = 0 \\ -1, & \text{for}\ t < 0 \end{cases} \tag{4.5}$$

The value of S indicates the direction of trend. A negative (positive) value indicates a falling (rising) trend. Mann and Kendall have documented that when $n \geq 8$, the test statistics S is approximately normally distributed with mean and variance as follows:

$$E\,(S) = 0 \tag{4.6}$$

$$\text{Var}\ (S) = \frac{1}{18}\left[n\ (n-1)\ (2n+5) - \sum_{i=1}^{m} t_i\,(t_i - 1)\,(2t_i + 5)\right] \tag{4.7}$$

where m is the number of tied groups and t_i is the size of the ith tie group. The standardized test statistics Z is computed as follows.

$$Z_{\text{MK}} = \begin{cases} \frac{S-1}{\sqrt{\text{Var}\ (S)}}, & \text{for}\ S > 0 \\ 0 \quad , & \text{for}\ S = 0 \\ \frac{S+1}{\sqrt{\text{Var}\ (S)}}, & \text{for}\ S < 0 \end{cases} \tag{4.8}$$

The standardized Mann-Kendall statistics Z follows the standard normal distribution with zero mean and unit variance. If $|Z| \geq Z_{1-(\alpha/2)}$, the null hypothesis about no trend is rejected at the significance level α (10% in this study).

4.1.2.2 Modified Mann-Kendall Test

For Modified Mann-Kendall's test, statistic S tends to normality for large n, with mean and variance given by:

$$E\,(S) = 0 \tag{4.9}$$

$$\text{Var}\ (S) = n\ (n-1)\ (2n+5)\,/18 \tag{4.10}$$

Statistic S is given by eq. (4.8):

$$S = a_{ij} = \sum_{i=1}^{n-1}\sum_{j=i+1}^{n} \text{sgn}\ \left(x_j - x_i\right) \tag{4.11}$$

With the same mean and variance as in eqs. (4.9) and (4.10), a modified version of the Mann-Kendall test which is robust in the presence of autocorrelation is proposed based on the modified variance of S given by eq. (4.12):

$$V^*\,(S) = \text{Var}\ (S) \cdot \frac{n}{n_s^*} = \frac{n\,(n-1)\,(2n+5)}{18} \cdot \frac{n}{n_s^*} \tag{4.12}$$

where n/n_s^* represents a correlation due to the autocorrelation in the data. The n/n_s^* term is evaluated using eq. (4.13):

$$\frac{n}{n_s^*} = 1 + \frac{2}{n(n-1)(n-2)} \times \sum_{i=1}^{n-1} (n-i)(n-i-1)(n-i-2)\,\rho_s(i) \tag{4.13}$$

In eq. (4.13), n is the actual number of the observations, and $\rho_s(i)$ is the autocorrelation function of the ranks of observations. The advantage of using eqs. (4.12) and (4.13) for the evaluation of variance of S is that there is no need of either normalized data or their autocorrelation function. The autocorrelation of ranks of observations $\rho_s(i)$ is related with the parent autocorrelation function and is given as follows (Kendall, 1955):

$$\rho(i) = 2\sin\left(\frac{\pi}{6}\rho_s(i)\right) \tag{4.14}$$

Inverse of eq. (4.14) can therefore be used to evaluate the autocorrelation of rank $\rho_s(i)$ that appears in eq. (4.13) and is given by eq. (4.15):

$$\rho_s(i) = \frac{6}{\pi}\sin^{-1}\left(\frac{\rho(i)}{2}\right) \tag{4.15}$$

The significance of the trends is tested by comparing the standardized test statistics Z.

$$Z = \frac{S}{[V^*(S)]^{0.5}} \tag{4.16}$$

A significant level of $\alpha = 0.1$ for the autocorrelation of ranks $\rho_s(i)$ is used, which produces the best overall empirical significance level. The serial correlation of ranks of the series and their significance test can be computed from eqs. (4.2) and (4.3).

4.1.2.3 Mann-Kendall Test with Pre-whitening

An alternate approach to perform trend analysis of a time series in the presence of serial correlation using the Mann-Kendall test is to first remove the serial correlation from data and then apply the test. Several approaches have been suggested for removing the serial correlation from a data set prior to applying the test. The pre-whitening approach is most common which involves computation of serial correlation and removing the correlation if the calculated serial correlation is significant at the 0.05 significance level (Burn and Elnur, 2002). The pre-whitening is accomplished as follows:

$$X_t' = x_{t+1} - r_1 \times x_t \tag{4.17}$$

where x_t = original time series with autocorrelation for time interval t; X_t' = the pre-whitened time series; and r_1 = the lag-1 autocorrelation coefficient. This pre-whitened series is then subjected to the Mann-Kendall test [i.e., eqs. (4.4) to (4.8)] for detecting the trend.

4.1.3 Periodicity

Periodicity is one of the deterministic components in the time-series. Most of the climatic, atmospheric and hydrological time-series would consist of a combination of stochastic and deterministic components. The power spectrum is a method of analysis that was developed to handle the problem of periodicity in variations of natural events observed in time, such as in climatological and hydrological time series. Power spectrum analysis, also called generalized harmonic analysis, is derived from the principles first developed by Wiener (1930, 1949). It is based on the premise that the time series are not necessarily composed of a finite number of oscillations each with a discrete wavelength, but rather they consist of virtually an infinite number of small oscillations spanning a continuous distribution of wavelengths. The spectrum therefore, gives the distribution of variations in a time series over a continuous domain of all possible wavelengths.

Procedures for computing the power spectra may vary. Here, in this study an approach described in WMO (1966), developed by Tukey (1950) and Blackman and Tukey (1958) is employed. A detailed description of this approach can also be found in various textbooks (Blackman and Tukey, 1958; Jenkins and Watts, 1968; Julian, 1967). It can be summarized through the following steps:

(i) First, all serial correlation coefficients of normalized climatic series (eq. (4.1)) are computed for lags from $L = 0$ to m, where m is the maximum lag ($m = n/3$). The serial correlation coefficient can be computed using eq. (4.2).
(ii) Using the values of r_L, the "raw" spectral estimates, $\hat{s}_k$, are computed using the following set of equations:

$$\hat{s}_0 = \frac{1}{2m}(r_0 + r_m) + \frac{1}{m}\sum_{L=1}^{m-1} r_L \tag{4.18}$$

$$\hat{s}_k = \frac{r_0}{m} + \frac{2}{m}\sum_{L=1}^{m-1} r_L \operatorname{Cos}\left(\frac{\pi kL}{m}\right) + \frac{1}{m} r_m (-1)^k\text{; for } k = 1, 2, \ldots, m-1 \tag{4.19}$$

$$\hat{s}_m = \frac{1}{2m}\left[r_0 + (-1)^m r_m\right] + \frac{1}{m}\sum_{L=1}^{m-1} (-1)^L r_L \tag{4.20}$$

Smallest is the value of k longest will be the wavelength of the spectrum, i.e., the shortest wavelength is achieved at $k = m$.

(iii) The raw spectrum $\hat{s}_k$ is then smoothened with a 3-term weighted average. For smoothing, procedure suggested by Hanning is used (WMO, 1966):

$$s_0 = \left(\hat{s}_0 + \hat{s}_1\right)/2 \quad (4.21)$$

$$s_k = \left(\hat{s}_{k-1} + 2\,\hat{s}_k + \hat{s}_{k+1}\right)/4; \quad \text{for } k = 1,\ 2,\ \ldots\ldots.,\ m-1 \quad (4.22)$$

$$s_m = \left(\hat{s}_{m-1} + \hat{s}_m\right)/4 \quad (4.23)$$

The averaging procedure is performed to derive a constant estimate of the final spectrum in terms of $m + 1$ discrete estimates (WMO, 1966).

4.1.3.1 Statistical Significance Test

The procedure for evaluating the results of power spectrum analysis mentioned in WMO (1966) is described below:

(i) A "null" hypothesis continuum is fitted to the computed spectrum. To start with, the significance of the lag-1 serial correlation coefficient r_1 of the climatic series is tested at the 90% confidence level. The "null" hypothesis of the randomness of climatic series against the serial correlation is rejected for the large value of $(r_1)_t$. If r_1 is not significantly different from zero, then series is regarded to be free from persistence. In this case, the appropriate null continuum is "white noise". In other words, a horizontal straight line, the value of which is everywhere equal to the average of the values of all the $m + 1$ "raw" spectral estimates (i.e., $\bar{s}$) in the computed spectrum (i.e., $S_k = \bar{s}$), is taken as the most suitable theoretical approach.

(ii) If the computed r_1 is positive and statistically significant, serial correlation coefficients for lag-2 and lag-3 are checked to see whether they approximate the exponential relations $r_2 \cong r_1^2$ and $r_3 \cong r_1^3$ (WMO, 1966). If these relations are ensured with the computed serial coefficients, the approximate "null" continuum is assumed as the simple "Markov red noise", whose shape depends on unknown value of the lag-1 serial correlation coefficient for a population ρ. Then the "null" continuum can be created by following an approximate procedure. By assuming that the sample r_1 is an unbiased estimation of ρ, various choices of the Harmonic number of k between $k = 0$ to m are assessed:

$$S_k = \bar{s}\left(\frac{1 - r_1^2}{1 + r_1^2 - 2\,r_1\,\mathrm{Cos}\left(\frac{\pi k}{m}\right)}\right) \quad (4.24)$$

where $\bar{s}$ is the average of all $m + 1$ "raw" spectral estimates $\hat{s}_k$ in the computed spectrum. The resulting values of S_k can be superposed on the sample spectrum, and a smoothed curve passes through these values to reach the required null continuum.

(iii) If r_1 is statistically significant but a few serial correlation coefficients for higher lags do not show the required exponential relations (i.e., $r_2 \cong r_1^2$ and $r_3 \cong r_1^3$) with r_1, then doubt arises as to whether the simple Markov-type persistence is the dominant form of non-randomness in the series of climatic observations. Nevertheless, WMO (1966) suggested that this procedure could be continued with just as before to compute the red noise continuum for r_1.

(iv) At this stage of the power spectrum analysis a first choice of the null continuum is made, and this selected continuum is superposed on the studied spectrum. In this case, it would be possible to make an assessment of the spectrum for its consistency with the chosen continuum. Then, the value of each spectral estimate s_k is compared with the local value of the null continuum.

The statistic associated with the each spectral estimate is the ratio of the magnitude of the spectral estimate to the local magnitude of the continuum (red noise continuum). Tukey (1950) found that the quantity of this ratio is distributed as Chi-square divided by the degree of freedom. The degree of freedom, ν, of each estimate of a computed spectrum is given as follows:

$$\nu = (2n - m/2)/m \tag{4.25}$$

The ratio of any sample spectral estimate s_k to its local value of the red noise continuum is then compared with the critical percentage-point levels of χ^2/ν distribution for the proper ν value. This comparison produces the required statistical significance level. The χ^2 value can be obtained from standard statistical books.

(v) In a sample spectrum, critical percentage-point levels of the χ^2/ν distribution, e.g., the 0.95 confidence level, is the same for all spectral estimates s_k. The confidence limits are finally derived by multiplying the "null" continuum (i.e., S_k) with the χ^2/ν.

(vi) Finally, the cycle associated in the time series is computed as:

$$P = 2m/L \tag{4.26}$$

4.2 Application

4.2.1 Data Processing

The climatic series subjected to the analysis are: (i) annual rainfall; (ii) monsoon and non-monsoon rainfall; (iii) annual rainydays; (iv) monsoon and non-monsoon rainydays; (v) onset of effective monsoon (OEM); and (vi) annual aridity index (AI). For rainfall data, one degree grid data of daily rainfall for the period 1951–2002 maintained by the India Meteorological Department (IMD) is used. Sixty five rain-grids points that fall under the Yamuna basin with 1° buffer are used in the analysis. The monsoon rainfall is derived from daily data for the period July to October and the rest of the period is considered as non-monsoon. The estimation procedure of OEM, rainydays, and the aridity index are presented in Chapter 3.

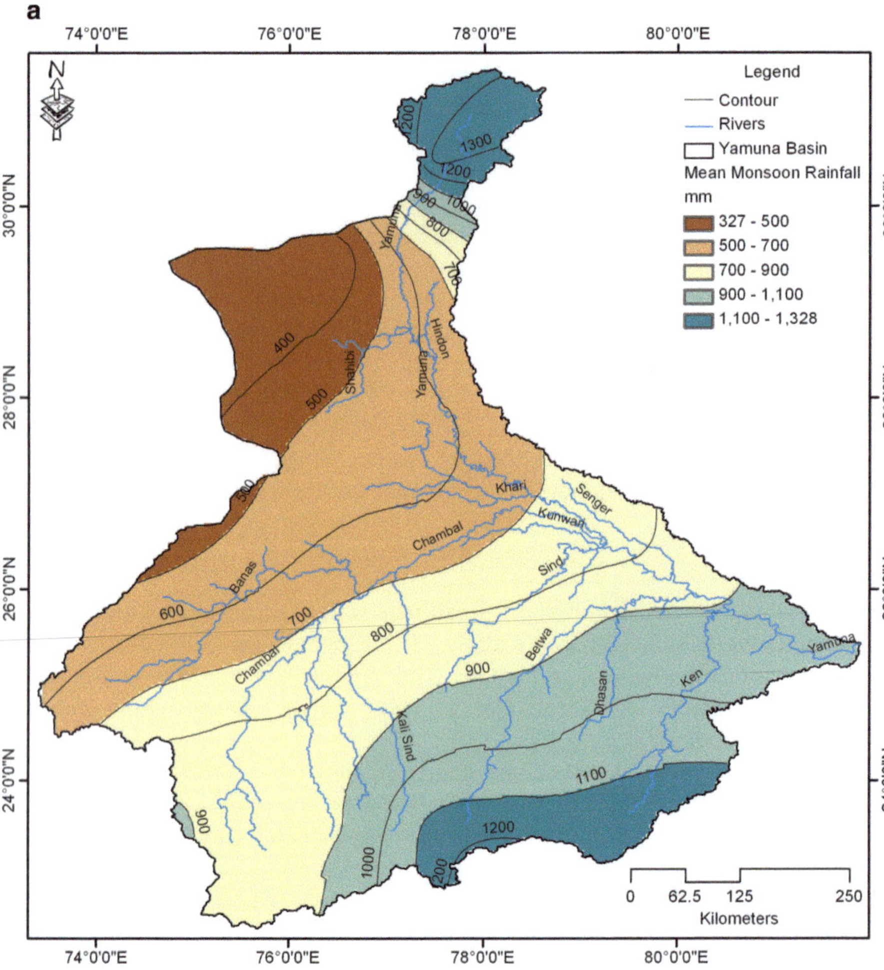

Fig. 4.1 (a) Spatial variability of monsoon rainfall in the Yamuna basin

The spatial distribution pattern of these hydro-climatic parameters is depicted in Chapter 3 except for monsoon mean and non monsoon mean rainfall. The monsoon and non-monsoon rainfall pattern is shown in Figs. 4.1a and 4.1b.

4.3 Results and Discussions

4.3.1 Persistence

Persistence is evident in a long time series of climatic observations characterized by a positive serial correlation. Significant negative r_1 is likely to be indicative of high-frequency oscillations, whereas significant positive r_1 is likely to be indicative of

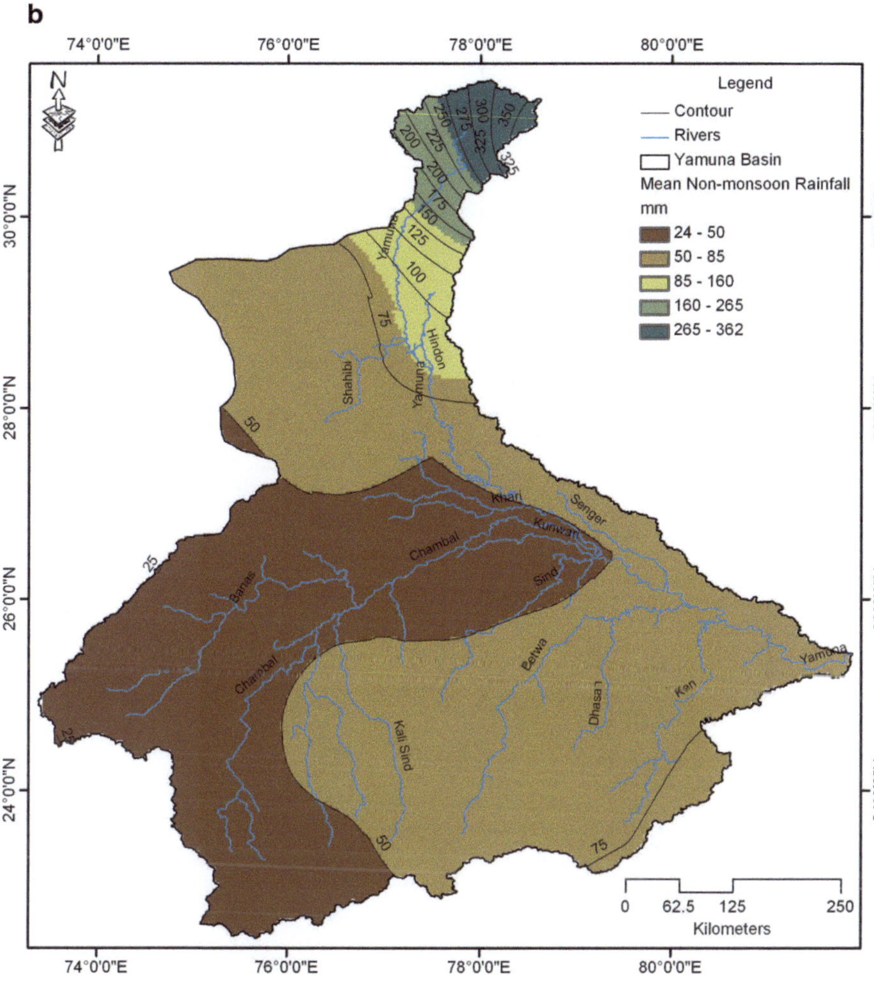

Fig. 4.1 (**b**) Spatial variability of non-monsoon rainfall in the Yamuna basin

low-frequency fluctuations and persistence in climatic series. This feature is further elaborated in the following section (i.e. "periodicity"). In the study, serial correlation for all the lags (i.e. $L = 0$ to m) is computed for all the grid points.

However, serial correlation coefficients up to lag 3 is also assessed. Serial correlation coefficient up to lag-3 is plotted for all the climatic variables except non-monsoon, although all the variables are analyzed. These plots are shown in Figs. 4.2, 4.3, and 4.4, which give the spatial distributions pattern of lag-1 to lag-3 serial correlation coefficients for the considered variables over the Yamuna River basin. A Kriging method is used as an interpolation technique to produce the spatial distribution pattern of the variables.

It is evident from Figs. 4.2, 4.3, and 4.4 that most of the grid point data show statistically insignificant serial correlation coefficient (SC) for all the hydro-climatic

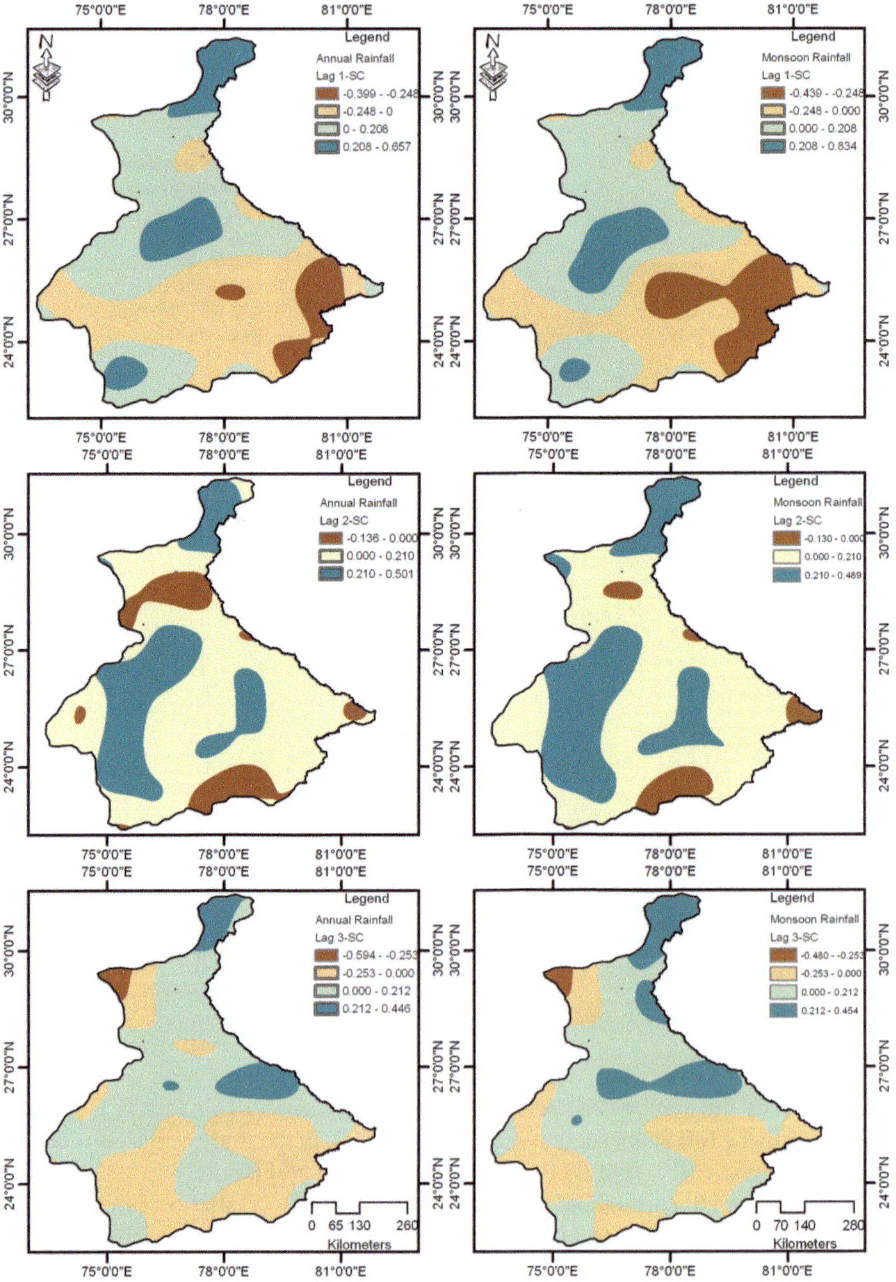

Fig. 4.2 Spatial distribution pattern of serial correlation of annual mean and monsoon mean rainfall up to lag 3

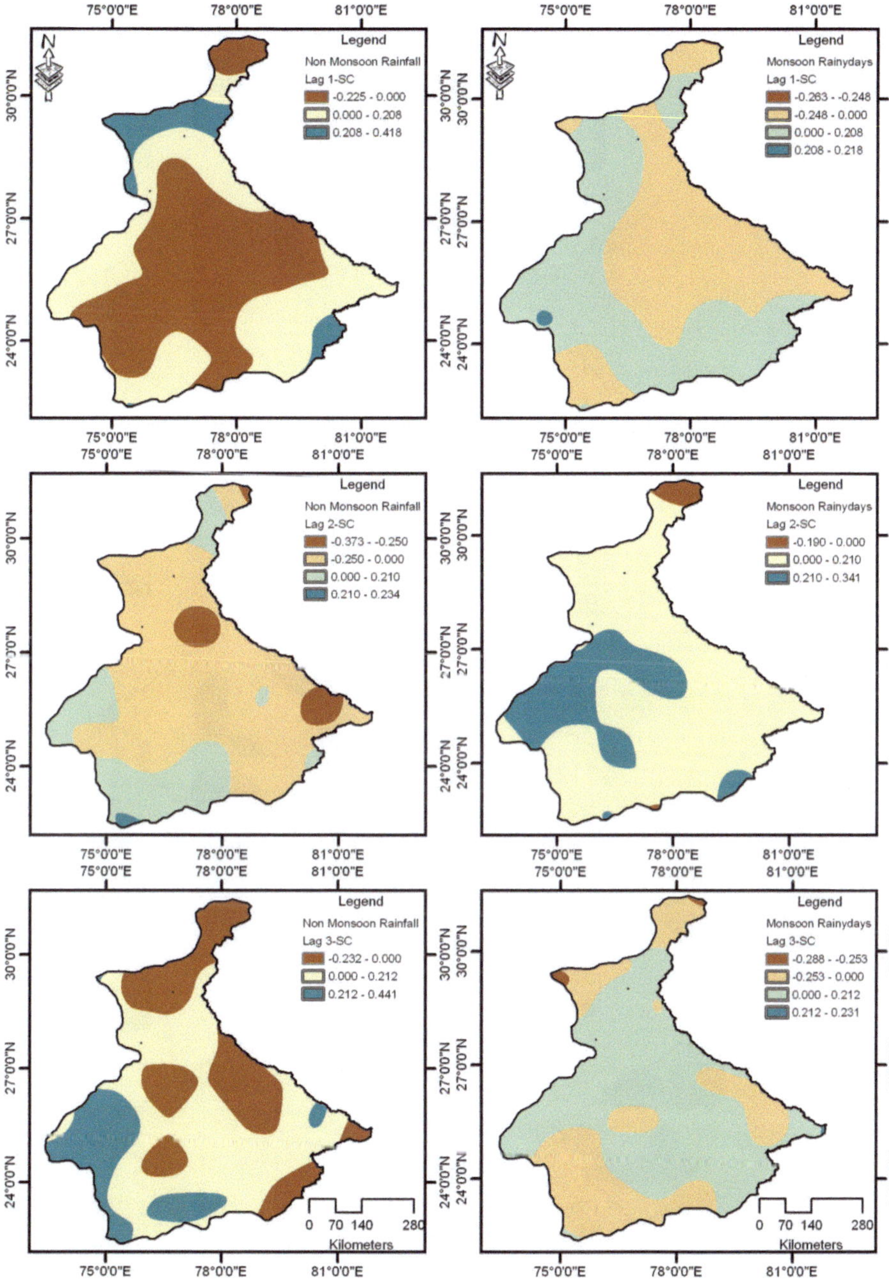

Fig. 4.3 Spatial distribution pattern of serial correlation of non-monsoon rainfall and monsoon rainydays up to lag 3

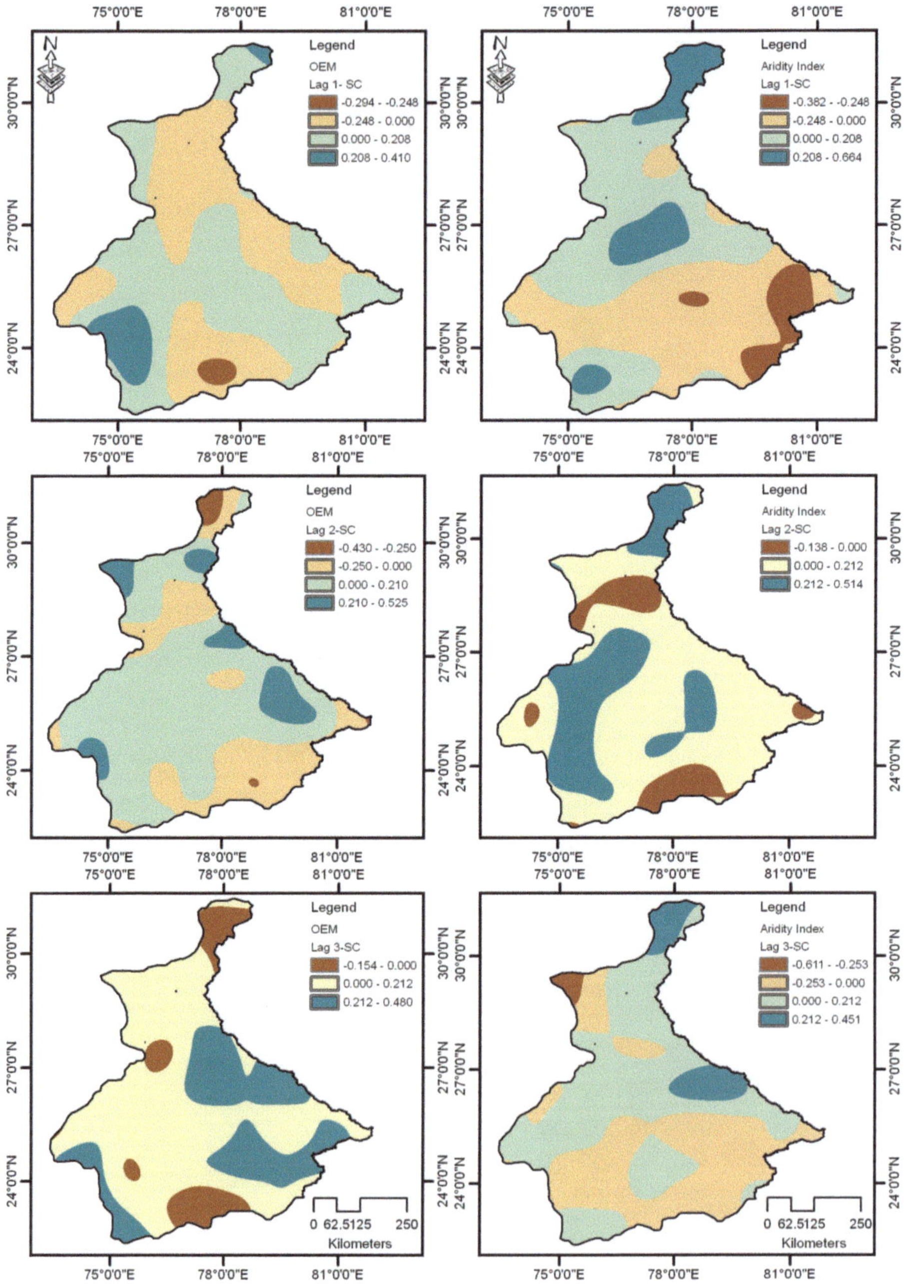

Fig. 4.4 Spatial distribution pattern of serial correlation of OEM and aridity index up to lag 3

variables (i.e. annual, monsoon and non monsoon rainfall; annual, monsoon and non monsoon rainydays; OEM and aridity index). However, a few grid data of annual and monsoon rainfall near the central and northern parts of the basin show the existence of persistence characterized by significant lag-1 SC; whereas, a few grid data points in the south-east part of the basin shows high frequency variability as characterized by significant negative lag 1-SC (Fig. 2.2). The spatial distribution of lag 1-SC of non monsoon (Fig. 2.3) indicates the presence of persistence in the few grid data in the northern part of the basin. Lag-1 SC of annual rainfall and aridity index show similar distribution pattern, which might be due to the more dependence of aridity index on annual rainfall (Figs. 4.2 and 4.4).

Based on Figs. 4.2, 4.3, and 4.4, a little positive spatial coherence characterized by lag-1 SC is identified for annual and monsoon rainfall, annual rainydays and aridity index. Looking into the Figs. 4.2, 4.3, and 4.4 and by comparing the spatial pattern of lag-2 SC with lag-1 SC, it was indicated that coherent area of significant positive lag-2 SC is greater than that of lag-1 SC. The lag-2 SC coefficients are mostly positive but insignificant for all the variables except for the non-monsoon rainfall, non-monsoon rainydays and OEM. On the other hand, by analyzing the lag-3 SC of the hydro-climatic variables, it is evident that the percentage coherent area is significantly reduced for all the variables. However, it is an increase in the case of non-monsoon rainfall, non-monsoon rainydays and OEM. Apart from this, a detailed table of the serial correlation coefficients up to lag-3 is also prepared for all the variables. But serial correlation coefficients up to lag-3 are presented for monsoon rainfall, rainydays and OEM only (Tables 4.1, 4.2, and 4.3), though the summarized table (Table 4.4) is prepared to visualize the overall temporal characteristics of the variables.

4.3.2 Trend Analysis

In the trend analysis, time series of the entire variable from 1951 to 2002 is subjected to three non-parametric statistical tests, viz. Mann-Kendall test, Modified Mann-Kendall test and Mann-Kendall test with pre-whitening of series. Two later tests are basically designed for the series having a significant lag-1 SC. Comparative results were obtained from all the three tests for all the considered variables. However, the sample results of trend analysis for monsoon rainfall and OEM are presented (Tables 4.5 and 4.6). These Tables (Tables 4.5 and 4.6) also include the value of lag-1 SC along with their lower and upper limits at the 95% confidence level. Based on Tables 4.5 and 4.6, it is apparent that the times series with significant negative lag-1 SC do not have the significant trend.

Table 4.1 Results of serial correlation (lag 1 to 3) and power spectrum analysis for monsoon rainfall (r_1 = lag-1 SC, r_2 = lag-2 SC, r_3 = lag-3 SC, L = lag corresponding to significant spectral estimates, Cycles = periodicity (2 *m*/*L*), Cont. = null continuum, WN = white noise, RN = Markov red noise)

					Cycles		
S. No.	r_1	r_2	r_3	L	90%	95%	Cont.
1	0.00	−0.18	−0.18	7	4.9		WN
				8	4.3		
2	0.05	−0.01	0.04	2	17.0		WN
3	0.05	0.02	0.08	2	17.0		WN
4	−0.09	0.11	−0.05				WN
5	−0.12	−0.07	−0.05				WN
6	0.40	0.37	0.42	12	2.8		RN
7	−0.04	0.03	−0.12				
8	−0.04	−0.02	0.18	11	3.1		RN
				12		2.8	WN
9	−0.05	0.13	0.02	2		17.0	
10	0.22	0.24	0.01	1	34.0		
				2		17.0	WN
				3	11.3		
11	0.15	0.23	0.01	2		17.0	WN
				3		11.3	
12	−0.03	−0.03	0.04				WN
13	−0.02	−0.11	−0.02				WN
14	−0.34	0.07	0.01	14		2.4	WN
				15		2.3	
15	−0.18	−0.02	0.12				WN
16	−0.14	0.10	−0.01				
17	0.07	0.05	−0.01	2	17.0		WN
18	−0.20	0.15	−0.03	14	2.4		WN
				15		2.3	
19	−0.08	0.39	−0.11	2		17.0	WN
				3		11.3	
				15	2.3		
				16		2.1	
20	−0.04	0.17	0.13	2		17.0	WN
				3		11.3	
21	−0.22	0.23	0.13	2	17.0		WN
				3	11.3		
				13	2.6		
				14		2.4	
22	−0.21	0.20	0.01	12	2.8		WN
				13	2.6		
				14	2.4		
23	−0.22	0.20	−0.09	14	2.4		WN
				15		2.3	
24	−0.37	0.07	−0.03	13	2.6		WN
				14		2.4	
				15		2.3	
25	−0.21	0.09	−0.13	15		2.3	WN

Table 4.1 (continued)

S. No.	r_1	r_2	r_3	L	Cycles		Cont.
					90%	95%	
				16	2.1		
26	−0.15	0.20	0.13	2	17.0		WN
				12		2.8	
27	0.04	0.08	−0.04	2	17.0		WN
				5	6.8		
28	0.20	0.43	0.21	1		34.0	WN
				2		17.0	
29	0.23	0.20	0.06	2		17.0	WN
				3		11.3	
30	−0.26	0.12	−0.01	14		2.4	WN
				15		2.3	
31	−0.27	0.25	−0.10	13	2.6		WN
				14		2.4	
				15		2.3	
				16	2.1		
32	−0.25	0.10	−0.08	14		2.4	WN
				15		2.3	
33	−0.41	0.07	0.00	13		2.6	WN
				14		2.4	
				15		2.3	
34	−0.02	−0.06	−0.06	7	4.9		WN
35	−0.12	0.11	−0.04	15		2.3	WN
36	0.00	0.02	0.04				WN
37	0.20	0.18	−0.07	1		34.0	WN
				2		17.0	
38	0.12	0.28	0.09	1	34.0		WN
				2	17.0		
39	0.35	0.30	0.27	2	17.0		RN
				12	2.8		
40	0.26	0.14	0.22	12		2.8	RN
				13		2.6	
41	0.11	0.23	0.31	1	34.0		WN
42	0.02	0.12	0.27				WN
43	−0.14	0.06	0.06	13		2.6	
				14		2.4	
44	−0.13	0.15	0.04	13	2.6		WN
				14		2.4	
				15	2.3		
45	0.40	0.30	0.13	13	2.6		RN
46	0.18	0.04	0.19	2		17.0	WN
47	0.11	0.24	0.10	2	17.0		WN
				3	11.3		
48	0.14	0.15	0.06	4	8.5		WN
				5	6.8		
49	−0.10	−0.01	0.13	12	2.8		WN
50	−0.07	0.06	0.09	13		2.6	WN
51	0.06	0.04	−0.16	4	8.5		WN

Table 4.1 (continued)

S. No.	r_1	r_2	r_3	L	Cycles 90%	Cycles 95%	Cont.
				5		6.8	
52	0.15	0.09	−0.17	4	8.5		WN
53	0.07	−0.02	0.07	3	11.3		WN
				4	8.5		
54	−0.03	0.04	0.26	13		2.6	WN
55	0.21	0.34	0.23	1		34.0	WN
56	0.17	0.22	0.11	2		17.0	WN
				3		11.3	
57	0.13	0.22	0.20	13	2.6		WN
58	0.13	0.29	−0.01	14	2.4		WN
59	−0.12	−0.11	0.06	13	2.6		WN
				14	2.4		
60	0.59	0.48	0.41	14	2.4		WN
61	0.44	0.28	0.13	14	2.4		RN
62	0.28	0.14	0.14	14	3.4		RN
63	−0.25	−0.01	0.06	14		2.4	WN
				15	2.3		
64	0.40	0.20	0.01	15		2.3	RN
65	0.48	0.41	0.24	15	2.3		RN

Table 4.2 Results of serial correlation (lag 1 to 3) and power spectrum analysis for monsoon rainydays (r_1 = lag-1 SC, r_2 = lag-2 SC, r_3 = lag-3 SC, L = lag corresponding to significant spectral estimates, Cycles = periodicity (2 *m*/*L*), Cont. = null continuum, WN = white noise, RN = Markov red noise)

S. No.	r_1	r_2	r_3	L	Conf. Level 90	Conf. Level 95	Cont.
1	0.10	0.11	0.00	2	17.0	0.0	WN
2	-0.03	0.14	0.14				WN
3	0.01	0.10	0.00				WN
4	−0.04	0.20	−0.04				WN
5	0.07	−0.12	−0.18	6	5.7	0.0	WN
				7	4.9	0.0	
6	0.08	0.15	0.06	2	17.0	0.0	WN
7	0.32	0.12	−0.09				RN
8	0.10	0.26	0.19	1	34.0	34.0	WN
				2	17.0	0.0	
				13	2.6	0.0	
9	−0.04	0.15	−0.01	12	2.8	0.0	WN
10	−0.07	0.16	−0.13				WN
11	0.06	0.15	−0.09	2	17.0	0.0	WN
12	0.04	0.18	0.01	2	17.0	17.0	WN
13	0.02	0.03	−0.08	5	6.8	0.0	WN
14	0.09	0.27	0.10	1	34.0	0.0	WN
				2	17.0	17.0	
15	0.01	0.15	0.14	2	17.0	17.0	WN
				14	2.4	0.0	

Table 4.2 (continued)

					Conf. Level		
S. No.	r_1	r_2	r_3	L	90	95	Cont.
16	−0.01	0.01	0.08	12	2.8	2.8	WN
17	0.04	0.09	0.02	2	17.0	0.0	WN
18	0.21	0.22	−0.03	1	34.0	0.0	WN
				2	17.0	17.0	
				3	11.3	0.0	
19	0.07	0.07	−0.13	2	17.0	0.0	WN
20	0.03	0.28	0.08	2	17.0	17.0	WN
21	−0.05	0.11	0.12	2	17.0	0.0	WN
				14	2.4	0.0	
22	0.13	0.14	0.12	2	17.0	17.0	WN
23	0.01	0.04	0.06				WN
24	0.14	0.12	0.07	5	6.8	0.0	WN
25	0.01	0.11	0.18	14	2.4	0.0	WN
26	0.09	0.20	0.09	1	34.0	34.0	WN
27	0.16	0.28	0.06	1	34.0	34.0	WN
				2	17.0	17.0	
28	0.07	0.32	0.02	1	34.0	34.0	WN
				2	17.0	17.0	
29	−0.05	0.11	−0.01	14	2.4	0.0	WN
30	−0.06	0.17	0.00	14	2.4	2.4	WN
31	−0.08	0.20	0.10	14	2.4	0.0	WN
32	−0.03	0.11	0.05				WN
33	−0.05	0.05	−0.06				WN
34	−0.01	0.12	0.15	14	2.4	0.0	WN
35	−0.24	0.10	0.06	14	2.4	2.4	WN
				15	2.3	2.3	
36	0.11	0.12	0.00	2	17.0	17.0	WN
37	0.15	0.28	0.15	1	34.0	34.0	WN
				2	17.0	17.0	
38	0.08	0.24	0.05	2	17.0	17.0	WN
39	−0.07	0.21	0.06	2	17.0	17.0	WN
40	−0.09	0.27	0.06	2	17.0	0.0	WN
				14	2.4	0.0	
				15	2.3	0.0	
41	−0.07	0.18	0.00	15	2.3	0.0	WN
42	−0.03	0.11	−0.01	5	6.8	0.0	WN
				14	2.4	0.0	
43	−0.05	0.06	0.03	15	0.0	0.0	
44	0.06	0.05	0.08	5	6.8	0.0	WN
45	0.13	0.12	0.18	1	34.0	34.0	WN
				2	17.0	17.0	
46	0.19	0.17	0.13	1	34.0	34.0	WN
				2	17.0	17.0	
47	0.02	0.21	0.09	2	17.0	17.0	WN
				3	11.3	0.0	
48	−0.08	0.12	0.11				WN
49	−0.13	0.09	0.06	14	2.4	2.4	WN

Table 4.2 (continued)

					Conf. Level		
S. No.	r_1	r_2	r_3	L	90	95	Cont.
50	−0.01	0.14	0.11	2	17.0	0.0	WN
				14	2.4	0.0	
51	0.14	0.07	−0.01	2	17.0	0.0	WN
				5	6.8	0.0	
				14	2.4	0.0	
52	0.11	0.05	−0.06	3	11.3	11.3	WN
				4	8.5	8.5	
53	0.08	0.11	0.15				WN
54	−0.11	0.20	−0.01				WN
55	0.10	0.11	0.16				WN
56	−0.01	0.19	−0.02	2	17.0	0.0	WN
				3	11.3	0.0	
57	0.04	0.16	0.06	2	17.0	0.0	WN
58	0.37	0.28	0.08				RN
59	0.05	0.16	0.09	1	34.0	0.0	WN
60	0.01	0.08	−0.06				WN
61	0.08	0.07	−0.10	5	6.8	0.0	WN
62	0.05	−0.02	−0.01				WN
63	−0.18	0.27	−0.09	15	2.3	0.0	WN
				16	2.1	0.0	
64	0.28	0.10	0.03				RN
65	0.43	0.36	0.27	14	2.4	0.0	RN

Table 4.3 Results of serial correlation (lag 1 to 3) and power spectrum analysis for OEM (r_1 = lag-1 SC, r_2 = lag-2 SC, r_3 = lag-3 SC, L = lag corresponding to significant spectral estimates, Cycles = periodicity (2 *m*/*L*), Cont. = null continuum, WN = white noise, RN = Markov red noise)

					Conf. Level		
S. No.	r_1	r_2	r_3	L	90	95	Cont.
1	0.10	0.06	0.08	4	8.5	0.0	WN
2	0.08	−0.07	0.38	10	3.4	3.4	WN
3	0.05	−0.03	0.42	1	34.0	0.0	WN
				2	17.0	0.0	
				12	2.8	0.0	
				13	2.6	2.6	
4	0.03	0.06	0.12				WN
5	0.00	−0.18	−0.03	7	4.9	4.9	WN
				8	4.3	4.3	
6	0.17	0.07	0.06	1	34.0	0.0	WN
7	0.23	−0.04	−0.03	6	5.7	0.0	WN
				9	3.8	0.0	
8	0.12	−0.09	−0.06				WN
9	0.04	0.13	0.34	1	34.0	34.0	WN
10	0.32	0.13	0.12	7	4.9	0.0	RN

Table 4.3 (continued)

S. No.	r_1	r_2	r_3	L	Conf. Level 90	95	Cont.
11	−0.14	−0.05	0.01				WN
12	−0.29	0.06	−0.12	15	2.3	0.0	WN
				16	2.1	2.1	
				17	2.0	2.0	
13	−0.10	−0.21	−0.02	9	3.8	0.0	WN
				10	3.4	3.4	
14	0.01	−0.19	0.17	8	4.3	0.0	WN
				9	3.8	3.8	
				10	3.4	0.0	
15	0.04	0.03	0.06				WN
16	0.07	−0.02	−0.11				WN
17	−0.10	−0.15	0.21	10	3.4	0.0	WN
				11	3.1	3.1	
				12	2.8	0.0	
18	0.31	0.29	0.30				RN
19	0.32	0.04	−0.01	8	4.3	4.3	
20	−0.04	0.04	0.14	11	3.1	0.0	WN
21	0.02	0.04	0.21				WN
22	0.05	−0.10	0.35	9	3.8	3.8	WN
23	0.14	−0.06	0.24	9	3.8	0.0	WN
24	0.00	−0.13	0.24				WN
25	0.17	−0.15	0.18	8	4.3	4.3	WN
				9	3.8	3.8	
26	−0.16	−0.05	−0.08	10	3.4	0.0	WN
27	−0.11	0.03	0.15	13	2.6	2.6	WN
28	0.12	0.10	0.15				WN
29	0.03	0.12	0.14	14	2.4	2.4	WN
30	0.03	0.09	0.10	14	2.4	0.0	WN
31	0.10	0.13	0.22				WN
32	−0.10	0.26	0.14	13	2.6	2.6	WN
33	0.01	0.20	0.21	12	2.8	0.0	WN
				13	2.6	2.6	
34	0.02	−0.21	0.17	8	4.3	0.0	WN
				9	3.8	0.0	
35	−0.12	−0.25	0.20	9	3.8	3.8	WN
				10	3.4	3.4	
36	0.04	−0.16	0.09				WN
37	0.17	0.08	0.16	1	34.0	0.0	WN
				2	17.0	0.0	
38	0.16	0.11	0.08	2	17.0	0.0	WN
39	−0.04	0.14	0.10	14	2.4	0.0	WN
40	0.14	0.02	0.33	1	34.0	0.0	WN
				9	3.8	0.0	
41	−0.19	−0.02	0.22	13	2.6	0.0	WN
				14	2.4	0.0	
42	0.09	0.29	0.29	1	34.0	0.0	WN
				13	2.6	0.0	

Table 4.3 (continued)

					Conf. Level		
S. No.	r_1	r_2	r_3	L	90	95	Cont.
43	−0.07	−0.13	0.17	9	3.8	3.8	WN
				13	0.0	0.0	
44	−0.19	−0.14	−0.02	14	2.4	2.4	WN
45	−0.01	−0.01	0.19				WN
46	0.03	−0.08	0.10				WN
47	−0.07	0.00	−0.04	8	4.3	4.3	WN
				9	3.8	0.0	
48	0.00	0.19	0.32	1	34.0	0.0	WN
				13	2.6	0.0	
49	−0.03	0.29	0.19	13	2.6	0.0	WN
50	0.03	−0.10	0.22	9	3.8	3.8	WN
				10	3.4	0.0	
51	−0.14	−0.11	0.09	9	3.8	0.0	WN
				13	2.6	0.0	
52	0.05	0.19	0.17	1	34.0	34.0	WN
				2	17.0	0.0	
53	−0.10	−0.02	0.20	11	3.1	0.0	WN
54	−0.15	−0.19	0.13	6	5.7	0.0	WN
				11	3.1	3.1	
				12	2.8	2.8	
				13	2.6	0.0	
55	0.19	0.30	0.07	1	34.0	34.0	WN
56	−0.05	0.06	0.08				WN
57	−0.07	0.33	0.07	2	17.0	0.0	WN
				15	2.3	2.3	
58	0.02	0.03	−0.25	4	8.5	8.5	WN
				5	6.8	6.8	
59	0.01	0.13	0.08	13	2.6	0.0	WN
60	0.05	−0.28	−0.04	8	4.3	4.3	WN
				9	3.8	3.8	
61	0.12	0.12	−0.13	2	17.0	0.0	WN
62	0.16	−0.03	0.02				WN
63	0.12	−0.19	−0.08	5	6.8	0.0	
				6	5.7	5.7	WN
64	−0.06	−0.19	0.08	8	4.3	0.0	WN
				9	3.8	3.8	
65	−0.04	−0.14	0.13	9	3.8	0.0	WN

The spatial pattern of lag-1 SC of the hydro-climatic variables is depicted in Figs. 4.2, 4.3, and 4.4. The summarized statistics of trend analysis is given in Table 4.7. Though the trend results are produced using all the tests but the spatial pattern of Z-statistics was presented using the results obtained from the modified Mann-Kendall test. For pattern analysis, the Kriging method is used for the interpolation of data. The spatial pattern of the trend statistics for the hydro-climatic variables is shown in Fig. 4.5 (annual, monsoon, and non-monsoon

Table 4.4 Summarized result of persistence analysis

S. No.	Variable	No. series sowing significant positive			No. (%) series sowing persistence	No. (%) series sowing significant negative		
		r_1	r_2	r_3		r_1	r_2	r_3
1	Annual Rainfall	13 (20.00)	16 (24.60)	5 (7.69)	13 (20.00)	3 (4.61)	0 (0.00)	0 (0.00)
2	Monsoon Rainfall	12 (18.46)	19 (29.23)	9 (13.84)	12 (18.46)	6 (9.23)	0 (0.00)	0 (0.00)
3	Non-Monsoon Rainfall	13 (20.00)	2 (3.07)	14 (21.50)	13 (20.00)	0 (0.00)	3 (4.61)	1 (1.54)
4	Annual Rainydays	17 (26.15)	12 (18.46)	2 (3.07)	17 (26.15)	0 (0.00)	0 (0.00)	2 (3.07)
5	Monsoon Rainydays	5 (7.69)	14 (21.54)	1 (1.54)	5 (7.69)	0 (0.00)	0 (0.00)	0 (0.00)
6	Non-Monsoon Rainydays	22 (33.84)	2 (3.07)	10 (15.38)	22 (33.84)	0 (0.00)	2 (3.07)	2 (3.07)
7	OEM	4 (6.15)	6 (9.23)	14 (21.5)	4 (6.15)	1 (1.54)	1 (1.54)	0 (0.00)
8	Aridity Index	13 (20.00)	16 (24.60)	4 (6.15)	13 (20.00)	3 (4.61)	0 (0.00)	0 (0.00)

Values in the parenthesis are the percentage of total time series

Table 4.5 Lag-1 serial correlation and Mann-Kendall's Z-statistics of Monsoon rainfall of the Yamuna River basin [$r_1(l) = -0.248$ and $r_1(u) = 0.208$]

Grid Location			Z-MK		
Lat	Long	r_1	Org	Mod	PW
21.5N	75.5E	0.003	0.3	0.3	0.3
22.5N	74.5E	0.052	−0.655	−0.655	−0.655
22.5N	75.5E	0.047	−0.513	−0.513	−0.513
22.5N	76.5E	−0.09	−0.923	−0.923	−0.923
22.5N	77.5E	−0.121	−0.702	−0.702	−0.702
22.5N	78.5E	**0.396**	**−4.34**	0.021	**−3.338**
22.5N	79.5E	−0.039	0.016	0.016	0.016
23.5N	73.5E	−0.036	−0.971	−0.971	−0.971
23.5N	74.5E	−0.049	−0.371	−0.371	−0.371
23.5N	75.5E	**0.218**	−0.608	0.017	−1.16
23.5N	76.5E	0.152	−1.302	−1.302	−1.302
23.5N	77.5E	−0.029	1.065	1.065	1.065
23.5N	78.5E	−0.018	−0.339	−0.339	−0.339
23.5N	79.5E	**−0.344**	−0.813	0.061	−1.239
23.5N	80.5E	−0.177	−0.687	−0.687	−0.687
24.5N	72.5E	−0.144	−0.915	−0.915	−0.915
24.5N	73.5E	0.066	−0.229	−0.229	−0.229
24.5N	74.5E	−0.203	−0.734	0.01	−0.734
24.5N	75.5E	−0.084	−0.592	−0.592	−0.592
24.5N	76.5E	−0.041	0.024	0.024	0.024

Table 4.5 (continued)

Grid Location			*Z*-MK		
Lat	Long	r_1	Org	Mod	PW
24.5N	77.5E	−0.217	−0.418	−0.418	−0.418
24.5N	78.5E	−0.211	−1.176	−1.176	−1.176
24.5N	79.5E	−0.224	0.103	0.103	0.103
24.5N	80.5E	**−0.371**	−0.418	−0.418	−0.718
24.5N	81.5E	−0.21	0.86	0.86	0.86
25.5N	73.5E	−0.149	−0.166	−0.166	−0.166
25.5N	74.5E	0.04	0.229	0.229	0.229
25.5N	75.5E	0.202	***−2.951***	***−2.951***	***−2.951***
25.5N	76.5E	**0.227**	−1.586	−1.586	**−1.996**
25.5N	77.5E	**−0.255**	−0.039	−0.039	−0.481
25.5N	78.5E	**−0.272**	−1.207	0.009	**−2.123**
25.5N	79.5E	−0.247	−1.192	−1.192	−1.192
25.5N	80.5E	**−0.409**	−0.797	−0.034	−1.097
25.5N	81.5E	−0.022	−0.387	−0.387	−0.387
25.5N	82.5E	−0.123	1.065	1.065	1.065
26.5N	73.5E	−0.004	0.292	0.292	0.292
26.5N	74.5E	0.204	−0.181	−0.181	−0.181
26.5N	75.5E	0.119	−1.405	−1.405	−1.405
26.5N	76.5E	0.346	−1.981	0.005	−2.359
26.5N	77.5E	0.257	−2.044	−2.044	−2.438
26.5N	78.5E	0.107	−1.681	−1.681	−1.681
26.5N	79.5E	0.022	−2.486	−2.486	−2.486
26.5N	80.5E	−0.144	−1.57	−1.57	−1.57
26.5N	81.5E	−0.128	−1.413	−1.413	−1.413
27.5N	74.5E	0.401	−1.349	0.015	−1.176
27.5N	75.5E	0.181	−1.199	−1.199	−1.199
27.5N	76.5E	0.105	−1.933	−1.933	−1.933
27.5N	77.5E	0.144	−2.178	−2.178	−2.178
27.5N	78.5E	−0.1	−0.955	−0.955	−0.955
27.5N	79.5E	−0.072	−1.665	−1.665	−1.665
27.5N	80.5E	0.062	−2.075	−2.075	−2.075
28.5N	75.5E	0.145	0.039	0.039	0.039
28.5N	76.5E	0.071	−0.174	−0.174	−0.174
28.5N	77.5E	−0.025	−1.247	−1.247	−1.247
28.5N	78.5E	0.211	−3.717	−3.717	−3.764
29.5N	76.5E	0.169	−1.618	−1.618	−1.618
29.5N	77.5E	0.13	−3.306	−3.306	−3.306
29.5N	78.5E	0.125	−2.044	−2.044	−2.044
30.5N	76.5E	−0.123	0.055	0.055	0.055
30.5N	77.5E	0.593	−1.223	0.013	−1.381
30.5N	78.5E	0.444	−2.517	0.046	−1.996
31.5N	76.5E	0.281	−1.428	0.015	−1.507
31.5N	77.5E	−0.253	−1.081	−1.081	−1.87
31.5N	78.5E	0.395	2.888	0.04	1.728
31.5N	79.5E	0.475	1.476	−0.013	0.245

r_1 = lag-1 SC, *Z-MK* = Mann-Kendall's *Z*-statistics, *Org* = original MK test, *Mod* = modified MK test, *PW* = MK test with pre-whitening of data

Table 4.6 Lag-1 serial correlation and Mann-Kendall's Z-statistics of OEM of the Yamuna River basin [$r_1(l) = -0.248$ and $r_1(u) = 0.208$]

Grid Location			Z-MK		
Lat	Long	r_1	Org	Mod	PW
21.5N	75.5E	0.099	1.320	1.321	1.321
22.5N	74.5E	0.084	1.785	1.785	1.785
22.5N	75.5E	0.050	1.335	1.335	1.335
22.5N	76.5E	0.028	***1.990***	***1.991***	***1.991***
22.5N	77.5E	0.001	0.917	0.918	0.918
22.5N	78.5E	0.168	***2.354***	***2.356***	***2.356***
22.5N	79.5E	**0.234**	**2.110**	**2.111**	1.184
23.5N	73.5E	0.117	***3.776***	0.056	***3.777***
23.5N	74.5E	0.035	***2.678***	***2.679***	***2.679***
23.5N	75.5E	**0.323**	**2.472**	**2.473**	1.097
23.5N	76.5E	−0.136	1.517	1.519	1.519
23.5N	77.5E	**−0.291**	1.146	0.031	0.963
23.5N	78.5E	−0.101	1.406	1.407	1.407
23.5N	79.5E	0.014	1.763	1.766	1.766
23.5N	80.5E	0.037	***2.433***	***2.433***	***2.433***
24.5N	72.5E	0.074	***1.998***	***1.999***	***1.999***
24.5N	73.5E	−0.097	1.240	1.241	1.241
24.5N	74.5E	**0.307**	**3.658**	0.020	**1.965**
24.5N	75.5E	**0.320**	**2.441**	−0.025	1.515
24.5N	76.5E	−0.043	1.651	1.652	1.652
24.5N	77.5E	0.019	***3.792***	***3.793***	***3.793***
24.5N	78.5E	0.051	***3.160***	***3.162***	***3.162***
24.5N	79.5E	0.136	***3.106***	***3.108***	***3.108***
24.5N	80.5E	0.004	***2.782***	***2.786***	***2.786***
24.5N	81.5E	0.167	2.938	2.939	2.939
25.5N	73.5E	−0.157	0.537	0.538	0.538
25.5N	74.5E	−0.106	1.509	1.509	1.509
25.5N	75.5E	0.115	***2.566***	***2.566***	***2.566***
25.5N	76.5E	0.030	***2.812***	***2.813***	***2.813***
25.5N	77.5E	0.030	***2.173***	***2.174***	***2.174***
25.5N	78.5E	0.096	***3.238***	***3.239***	***3.239***
25.5N	79.5E	−0.104	***3.100***	***3.107***	***3.107***
25.5N	80.5E	0.014	***2.964***	***2.966***	***2.966***
25.5N	81.5E	0.020	***2.647***	***2.648***	***2.648***
25.5N	82.5E	−0.118	1.358	1.359	1.359
26.5N	73.5E	0.042	−1.153	−1.154	−1.154
26.5N	74.5E	0.172	0.569	0.569	0.569
26.5N	75.5E	0.155	***2.559***	***2.559***	***2.559***
26.5N	76.5E	−0.036	***1.991***	***1.993***	***1.993***
26.5N	77.5E	0.144	***3.002***	***3.003***	***3.003***
26.5N	78.5E	−0.188	***2.606***	***2.607***	***2.607***
26.5N	79.5E	0.085	***3.502***	***3.507***	***3.507***
26.5N	80.5E	−0.065	***2.141***	***2.141***	***2.141***
26.5N	81.5E	−0.194	0.924	0.924	0.924
27.5N	74.5E	−0.007	0.079	0.079	0.079
27.5N	75.5E	0.032	1.193	1.193	1.193

Table 4.6 (continued)

Grid Location			Z-MK		
Lat	Long	r_1	Org	Mod	PW
27.5N	76.5E	−0.069	0.371	0.372	0.372
27.5N	77.5E	0.003	***3.270***	***3.271***	***3.271***
27.5N	78.5E	−0.025	***3.341***	***3.343***	***3.343***
27.5N	79.5E	0.026	1.808	1.808	1.808
27.5N	80.5E	−0.141	1.011	1.012	1.012
28.5N	75.5E	0.046	1.125	1.131	1.131
28.5N	76.5E	−0.100	0.537	0.537	0.537
28.5N	77.5E	−0.153	0.134	0.134	0.134
28.5N	78.5E	0.194	***4.447***	0.000	***4.448***
29.5N	76.5E	−0.047	1.047	1.096	1.096
29.5N	77.5E	−0.071	1.936	1.937	1.937
29.5N	78.5E	0.024	0.672	0.673	0.673
30.5N	76.5E	0.009	1.698	1.699	1.699
30.5N	77.5E	0.051	***2.156***	***2.157***	***2.157***
30.5N	78.5E	0.118	0.387	0.387	0.387
31.5N	76.5E	0.158	***2.267***	***2.268***	***2.268***
31.5N	77.5E	0.115	1.216	1.216	1.216
31.5N	78.5E	−0.056	−0.348	−0.348	−0.348
31.5N	79.5E	−0.035	0.963	0.963	0.963

r_1 = lag-1 SC, Z-MK = Mann-Kendall's Z-statistics, Org = original MK test, Mod = modified MK test, PW = MK test with pre-whitening of data

rainfall; and annual, monsoon, and non-monsoon rainydays) and Fig. 4.6 (OEM and aridity index). The patterns are classified based on the 90% confidence level (i.e., $Z < -1.645$; $-1.645 < Z < 0.0$; $0.0 < Z < 1.645$; and $Z > 1.645$). These interval ($Z < -1.645$; $-1.645 < Z < 0.0$; $0.0 < Z < 1.645$; and $Z > 1.645$) show the significant negative trend, insignificant negative trend, insignificant positive trend, and significant positive trend at the 90% confidence interval, respectively.

Based on analysis (Fig. 4.5 and Table 4.7), it can be observed that annual rainfall and monsoon rainfall values show an overall declining trend, although the scattered patches of the basin shows insignificant rising trend. The trend statistics for annual rainfall pattern and annual rainydays, and monsoon rainfall and monsoon rainy days are −0.832 and −0.362, and −0.874 and −0.62, respectively. On the other hand, the magnitude of non-monsoon rainfall has been insignificantly increased though few grid data show the significant rising trend in the north and north-west portion of the basin, which can be evident from the average Z-statistic of +0.209 and +0.258 for non-monsoon rainfall and rainy days, respectively. The spatial distribution pattern of rainydays is more or less similar to that of rainfall. Most of the area in the central to South-east and central to south experiences declination in rainfall. Since OEM is dependent on the rainfall pattern and potential evapotranspiration, and both the variables have shown a general falling trend in the basin. Therefore, an increasing trend in OEM is the resulting effect of rainfall and evapotranspiration patterns. The overall mean Z-statistic for OEM is +1.81. Similarly, the aridity index

Table 4.7 Summary of trend analysis based on Modified Mann-Kendall test

		Mean value			No. of series indicating significant lag-1 serial correlation	No. of series indicating significant			
S. No	Variable	Spatial mean	Range	SD		Rising trend	Falling trend	No trend	Overall trend
1	Annual rainfall	893.3	372.2–1653.5	280.3	13		6 (8)	59 (57)	– ve
2	Monsoon rainfall	806.8	327.0–1568.9	259.7	12		8 (13)	57 (52)	– ve
3	Non-monsoon rainfall	86.5	19.5–479.0	100.1	13	3 (2)		62 (63)	+ ve
4	Annual rainydays	88	46.0–160.0	25	17	4 (2)	7 (11)	54 (52)	– ve
5	Monsoon rainydays	70	43–96	15	5		5 (6)	60 (59)	– ve
6	Non-monsoon rainydays	17	1–72	14	22	6 (2)	1 (2)	58 (61)	+ ve
7	OEM	190	173–207	7	4	29 (30)		36 (35)	+ ve
8	Aridity Index (AI)	38.0	14.9–74.6	12.9	13		6 (8)	59 (57)	– ve

Values in the parenthesis are the results from Mann-Kendall with pre-whitening test

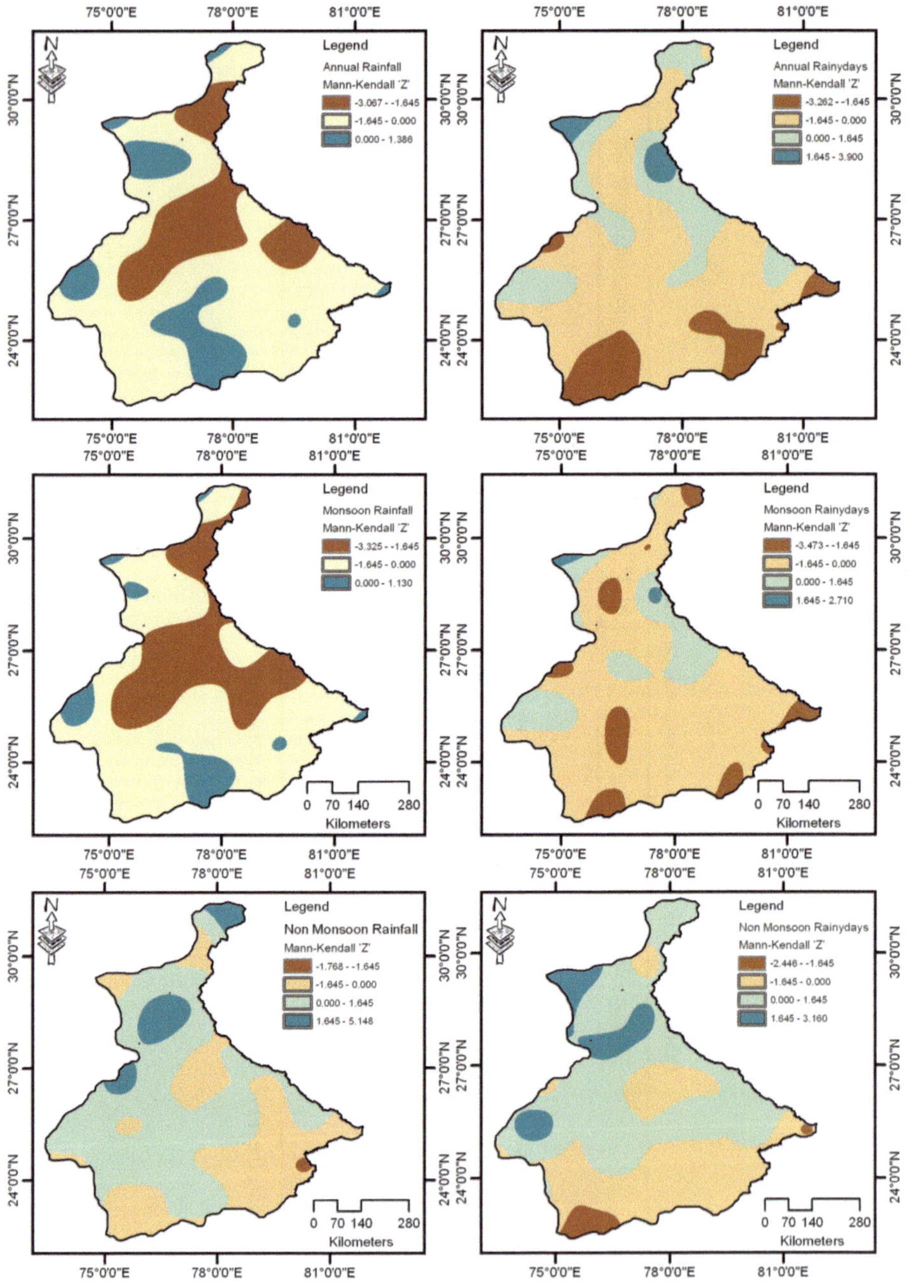

Fig. 4.5 Spatial distribution pattern of Mann-Kendall's *Z*-statistics of rainfall (annual, monsoon and non-monsoon) and rainydays (annual, monsoon and non-monsoon)

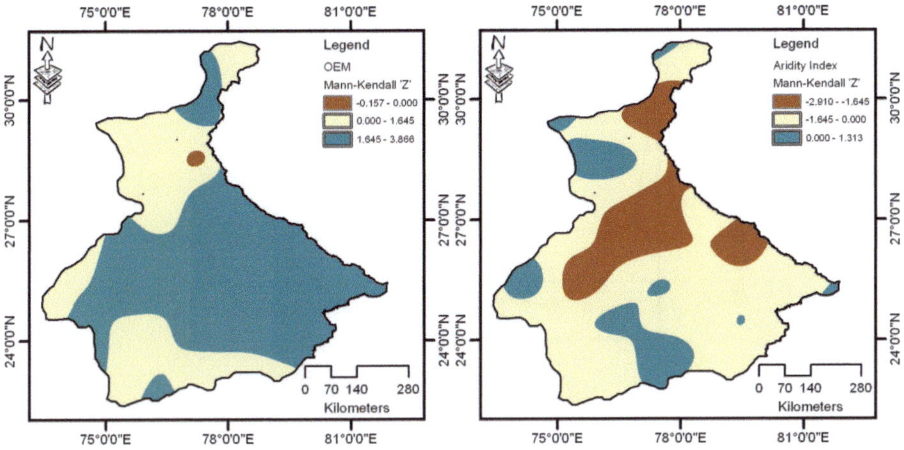

Fig. 4.6 Spatial distribution pattern of Mann-Kendall's *Z*-statistics of OEM and aridity index

is the ratio of annual rainfall and potential evapotranspiration. The falling trend observed in the aridity index is associated with the overall falling trend in the rainfall pattern of the Yamuna River basin (Table 4.7 and Fig. 4.5). The overall mean *Z*-statistic value for the aridity index is −0.78. Based on the trend analysis, it can be stated that the water resources potential of the Yamuna River basin is declining.

4.3.3 Periodicity

The significance of spectral estimates is evaluated at 90 and 95% confidence levels of the appropriate null continuum (red or white noise). The relationship between the correlogram and spectral estimates is shown in Fig. 4.7 for sample time series of annual rainfall of the Yamuna River basin. If a time series has persistence, the spectrum changes over all the wavelengths and the amplitude of the spectrum has a decreasing trend from long to short wavelengths (i.e. corresponding to increasing order of lags); and the spectrum is termed as "red noise" (Fig. 4.7a). For the spectrum of a time series having persistence with necessary exponential relationship between r_1, r_2 and r_3 (i.e., Markov-type persistence), the appropriate null appropriate is assumed to be a Markov red noise continuum (Fig. 4.7b). A series characterized by an insignificant positive lag-1 SC or a series that has a significant positive lag-1 SC but not a simple Markov-type, and any series with a negative lag-1 serial correlation coefficient is evaluated as a white noise continuum (Figs. 4.7c and 4.7d). From Fig. 4.7d, it may be stated that times series with negative lag-1 SC shows the high frequency variability (i.e. $L = 14$, and $P = 2\,m/L = 2.4$ years). For all the variables, power spectrum plots and the tables are prepared. However, for presentation sample power spectrum plots of monsoon rainfall, OEM and AI are given (Figs. 4.8, 4.9, and 4.10). To evaluate the power spectrum results, generally

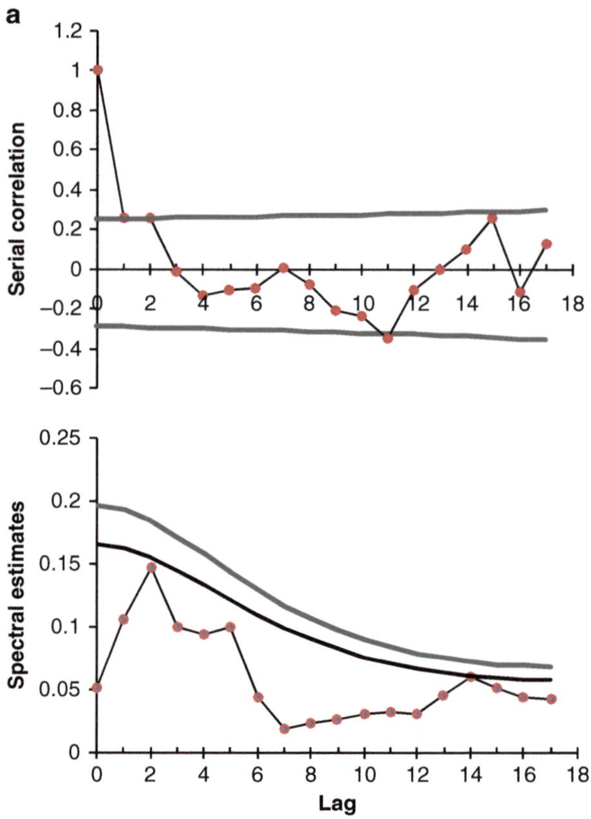

Fig. 4.7 (**a**) Relation between correlogram and spectral estimates: Series with persistence (red noise)

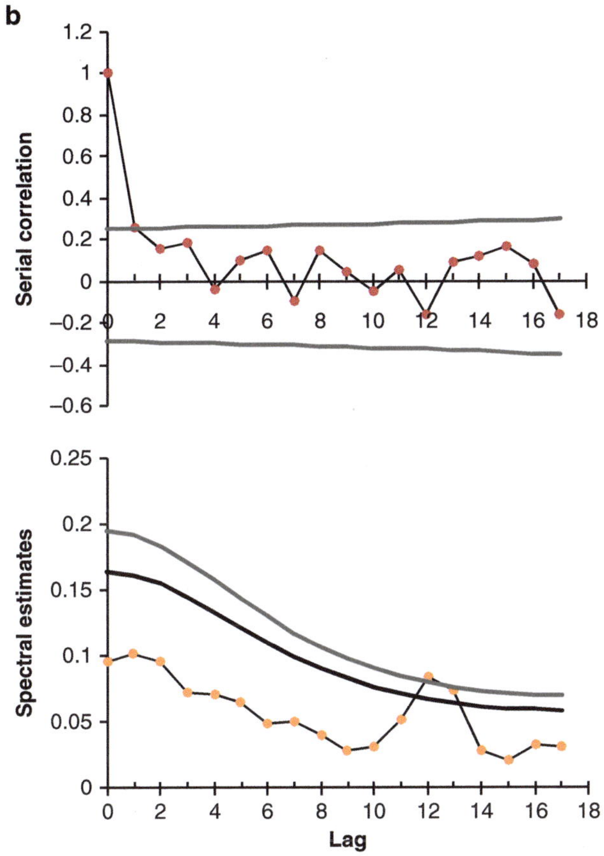

Fig. 4.7 (**b**) Relation between correlogram and spectral estimates: Series with persistence and sowing necessary exponential relationship (Markov red noise)

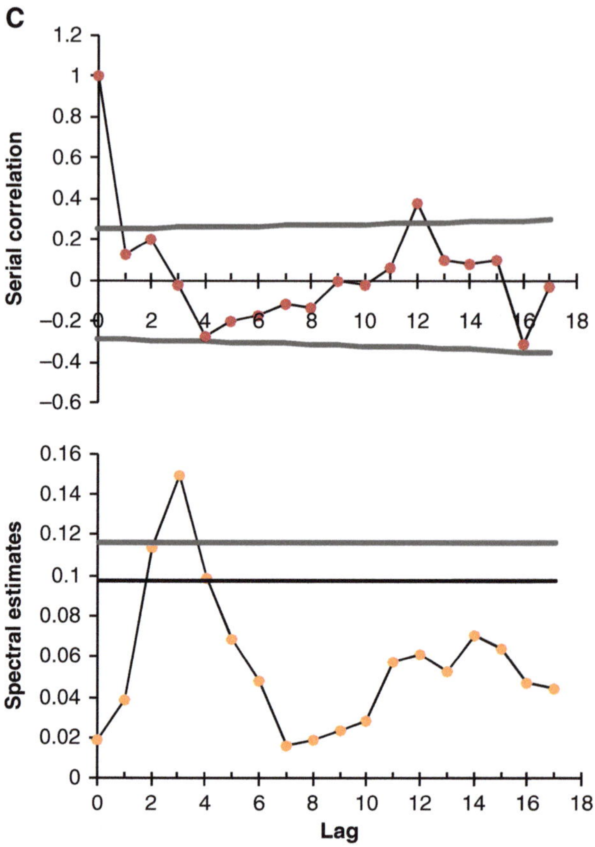

Fig. 4.7 (**c**) Relation between correlogram and spectral estimates: Series with insignificant positive lag-1 SC (white noise)

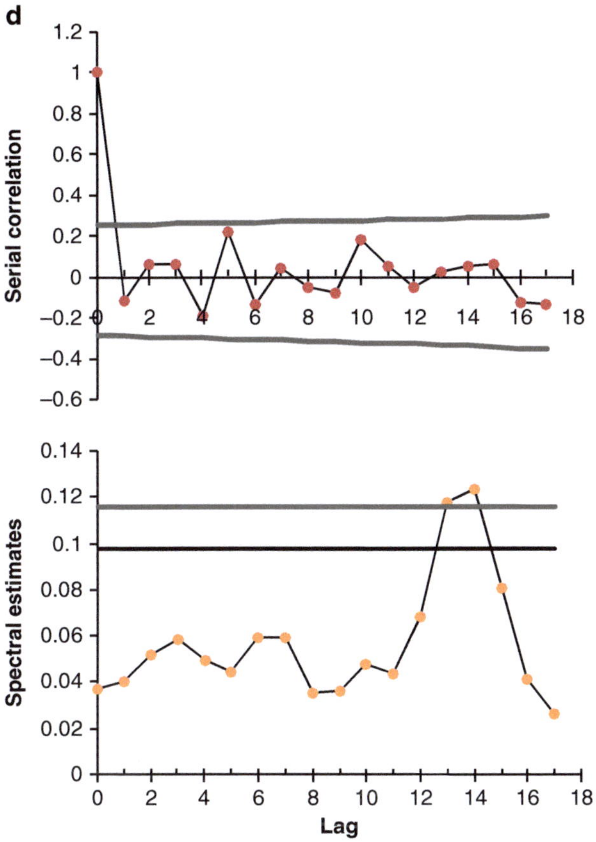

Fig. 4.7 (**d**) Relation between correlogram and spectral estimates: Series with negative lag-1 SC (white noise)

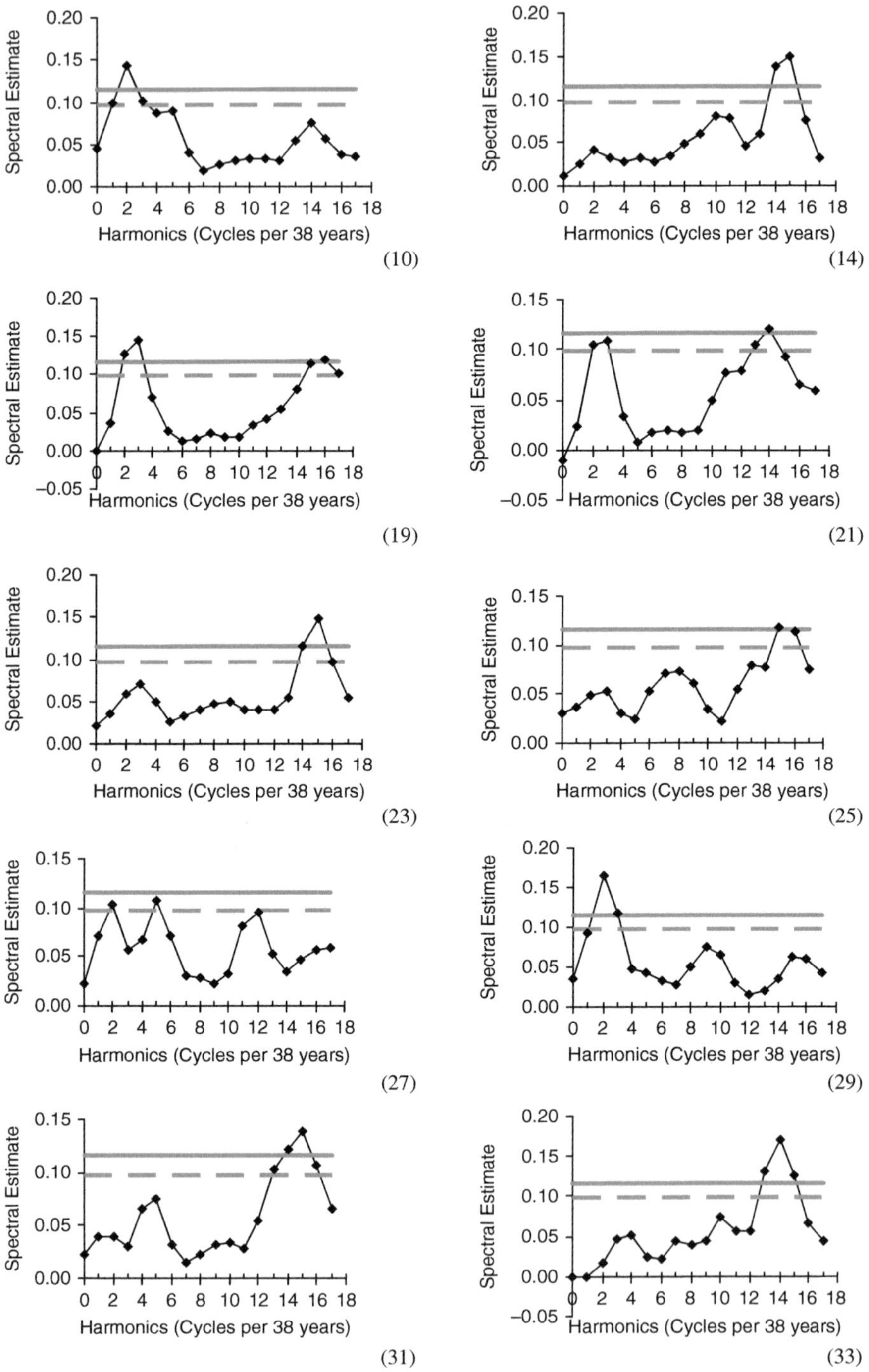

Fig. 4.8 Power spectrum plot for monsoon rainfall (——: 95% confidence limit; - - - - - -: 90% confidence limit)

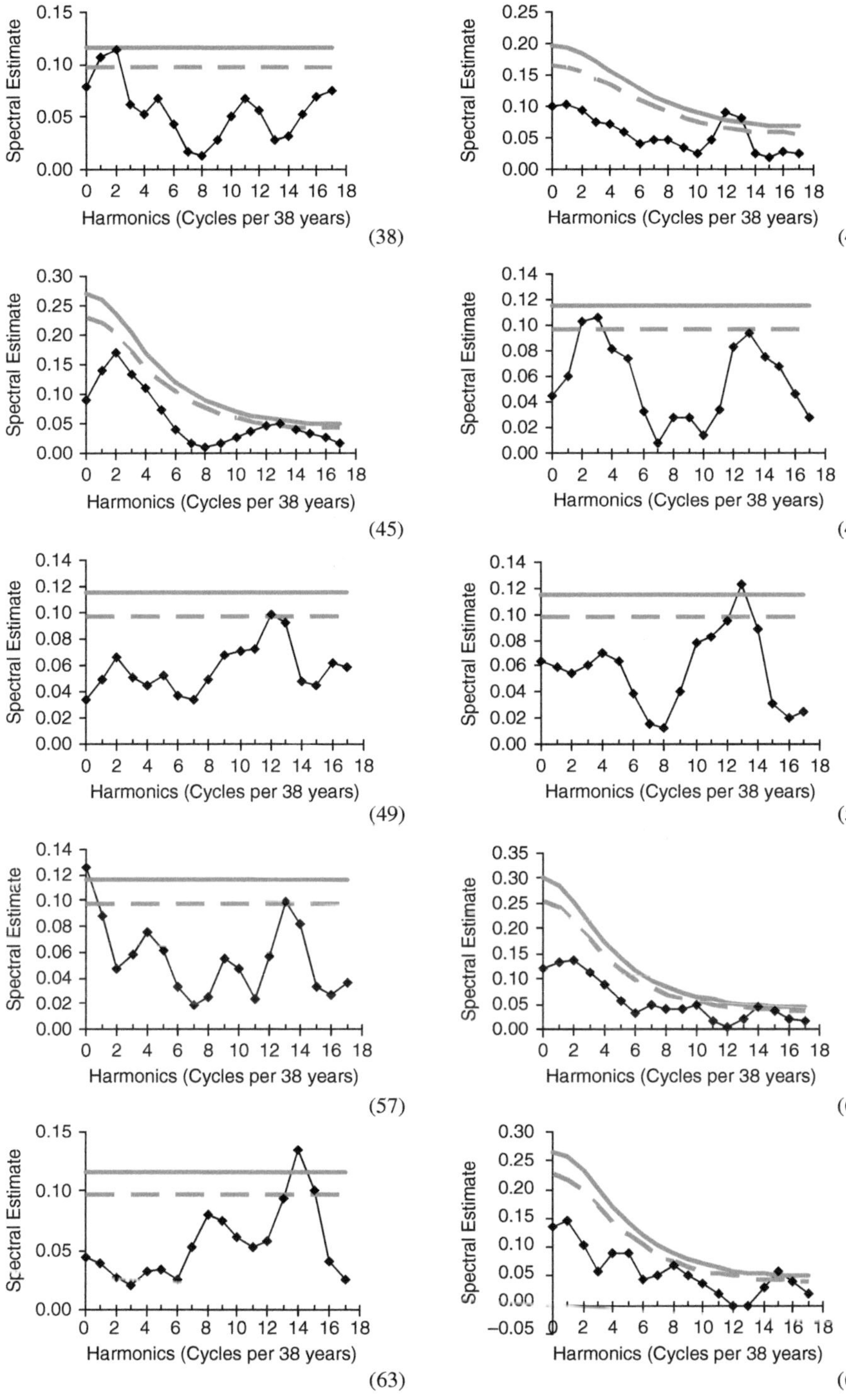

Fig. 4.8 (continued)

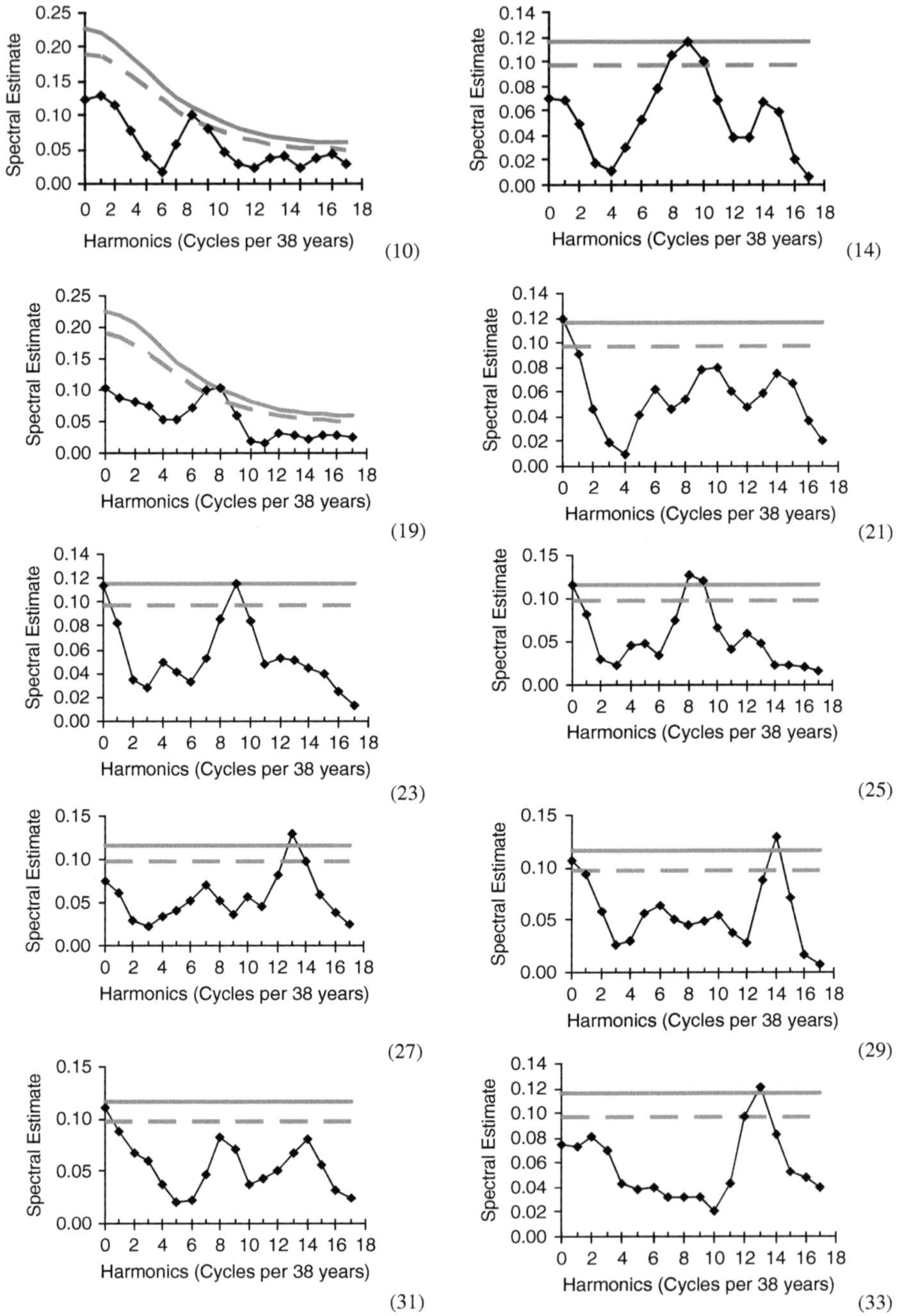

Fig. 4.9 Power spectrum plot for OEM

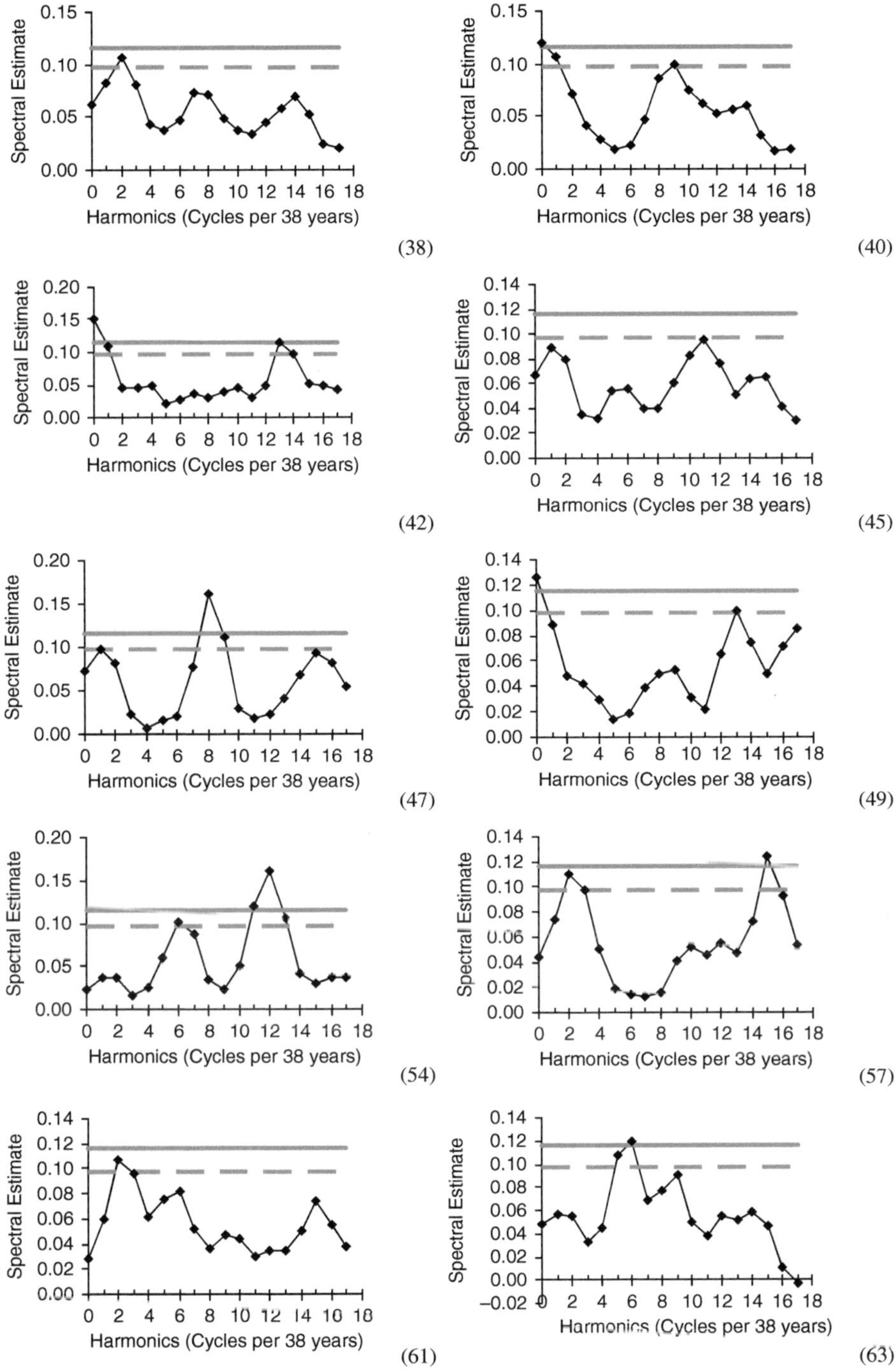

Fig. 4.9 (continued)

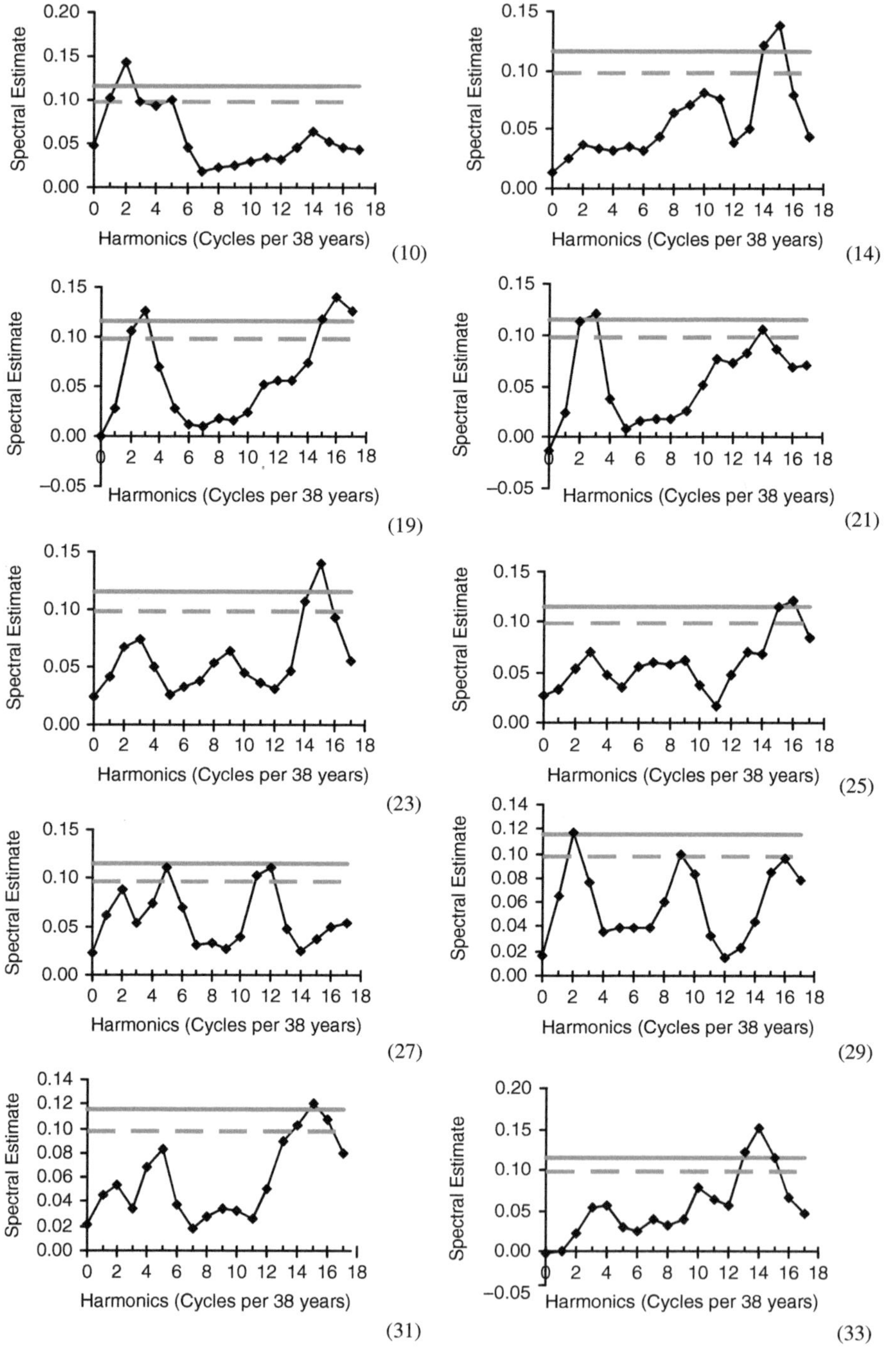

Fig. 4.10 Power spectrum plot for aridity index (AI)

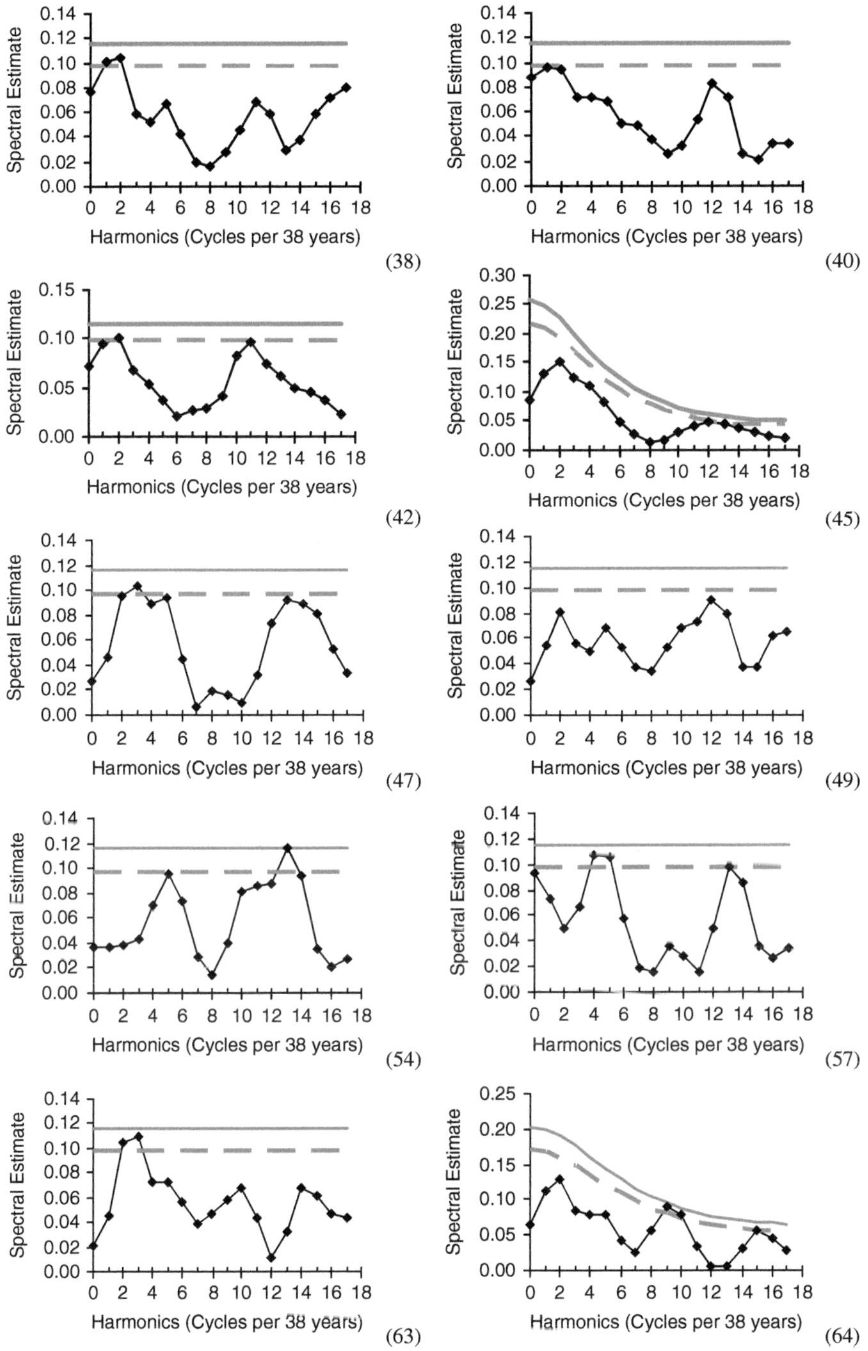

Fig. 4.10 (continued)

period values are computed using eq. (4.26) (i.e., $P = 2\ m/L$) for all the time series. The computed values of the period along with their significance level for monsoon rainfall, rainydays and OEM are given in Tables 4.1, 4.2, and 4.3 which also comprise serial correlation coefficients up to lag-3. Analysis revealsthat the short-term period fluctuation of 2.0–4.9 years is dominant in the annual and monsoon rainfall, whereas medium and long-term period of 5.0–34.0 years is dominant in case of non-monsoon rainfall. This short term periodic behavior (high frequency) of rainfall pattern in the Yamuna River basin should be kept in mind while preparing the water resources plan. In the case of number of rainydays, the long-period is dominated over the short- periodicity. Results of the power spectrum analysis of OEM and AI indicate that OEM and AI have dominating short-period fluctuations of 2.0–3.0 years, which lead to a high probability of meteorological and agricultural droughts and will frequently disturb the planning of *Kharif* crops in the Yamuna River basin.

4.4 Concluding Remarks

Based on the analysis, following conclusion can be drawn:

(i) There is a considerable difference in the monsoon and non-monsoon rainfall patterns in terms of persistence and periodicity.
(ii) Approximately 20% rainfall time series shows the presence of persistence characterized by lag-1 serial correlation.
(iii) The presence of serial correlation in the time series significantly affects Mann-Kendall's trend analysis.
(iv) The original Mann-Kendall test overestimates the presence of significant trend in the series than the modified Mann-Kendall and Mann-Kendall with pre-whitening tests. Based on the overall trend results, the original Mann-Kendall test results in approximately 37% more presence of significant trend than the modified Mann-Kendall test.
(v) Overall falling trend is observed in the annual rainfall, monsoon rainfall, annual rainydays, monsoon rainydays, and AI. In sixty five grid locations, the spatial mean of Mann-Kendall's *Z*-statistics for annual rainfall and monsoon rainfall were −0.832 and −0.874, respectively. Regardless of a few series having significant trends, most of the series have higher values of *Z*-statistic.
(vi) An increasing trend with overall mean *Z*-statistic of +1.81 is observed in OEM which shows the gradual delay in the onset of effective monsoon in the Yamuna River basin.
(vii) It may be remarked that the declining monsoon rainfall and number of monsoon rainydays along with the delay in the onset of effective monsoon (i.e. rising trend in OEM) should be of the great concern while preparing the river basin management plan.
(viii) It is observed that high frequency fluctuations associated with short-period cycle is dominating over the basin for annual and monsoon rainfall, OEM

and AI. However, for non-monsoon rainfall low frequency fluctuations are identified.

(ix) There is a good consistency between the lag-1 serial correlation and results of the power spectrum analysis. A positive lag-1 serial correlation coefficient gives low frequency fluctuations, whereas a negative one is an indicator for high frequency (Fig. 4.7).

References

Ackerman AS, Toon OB, Stevens DE, Stevens DE, Heymsfield AJ, Ramanathan V, Welton EJ (2000) Reduction of tropical cloudiness by soot. Science 288:1042–1047

Alvi SMA, Koteswaram, P (1985) Time series analyses of annual rainfall over India. Mausam 36(4):479–490

Anyadike RNC (1993) Seasonal and annual rainfall variations over Nigeria. Int J Climatol 13:567–580

Arora M, Goel NK, Singh P (2005) Evaluation of temperature trends over India. Hydrol Sci J 50(1):81–93

Avissar R, Werth D (2005) Global Hydroclimatological Teleconnections resulting from tropical deforestation. J Hydrometeorol 6:134–145

Aziz OIA, Burn DH (2006) Trends and variability in the hydrological regime of the Mackenzie River Basin. J Hydrol 319(1–4):282–294

Baines PG (2006) The late 1960s Global Climate Shift and its influence on the Southern Hemisphere. Proceed 8 ICSHMO, INPE, Brazil, pp 1477–1482

Bandyopadhyay A, Bhadra A, Raghuwanshi NS, Singh R (2009) Temporal trends in estimates of reference evapotranspiration over India. J Hydrol Eng 14(5):508–515

Basistha A, Arya DS, Goel NK (2009) Analysis of historical changes in rainfall in the Indian Himalayas. Int J Climatol 29:555–572

Blackman RB, Tukey JW (1958) The measurement of power spectra. Dover, New York, NY

Boyles RP, Raman S (2003) Analysis of climate trends in North Carolina (1949–1998). Environ Int 29:263–275

Burn DH, Elnur MA (2002) Detection of hydrologic trends and variability. J Hydrol 255:107–122

Burn DH, Cunderlink JM, Pietroniro A (2004) Hydrological trends and variability in the Liard river basin. Hydrol Sci J 49(1):53–67

Chase TN, Knaff JA, Pielke RA, Kalnay E (2003) Changes in global Monsoon circulations since 1950. Nat Haz 29:229–254

Chen TC, Hoon JH, Croix KJS, Takle ES (2001) Suppressing impacts of the Amazonian deforestation by the Global Circulation Change. Bul Am Meteorol Soc 82(10):2209–2216

Chen H, Guo S, Xu CY, Singh VP (2007) Historical temporal trends of hydro-climatic variables and runoff response to climate variability and their relevance in water resource management in the Hanjiang basin. J Hydrol 344:171–184

Cox DR, Stuart A (1955) Some quick tests for trend in location and dispersion. Biometrika (42):80–95

Dore MHI (2005) Climate change and changes in global precipitation patterns: what do we know. Environ Int 31:1167–1181

Douglas EM, Vogel RM, Kroll CN (2000) Trends in floods and low flows in the United states: Impact of spatial correlation. J Hydrol 240 (1–2):90–105

Douglas EM, Niyogi D, Frolking S, Yeluripati JB, Pielke RA, Niyogi N, Vorosmarty CJ, Mohanty UC (2006) Changes in moisture and energy fluxes due to agricultural land use and irrigation in the Indian monsoon belt. Geophys Res Let 33:L14403. doi:10.1029/2006GL026550)

Drosdowsky, W (1993) An analysis of Australian seasonal rainfall anomalies: 1950–1987. II: Temporal variability and teleconnection patterns. Int J Climatol 13:111–149

Duan K, Yao T (2003) Monsoon variability in the Himalayas under the condition of Global Warming. J Meteorol Soc Jpn 81(2):251–257
Gan TY (1998) Hydroclimatic trends and possible climatic warming in the Canadian Prairies. Water Resour Res 34(11):3009–3015
Giakoumakis SG, Baloutsos G (1997) Investigation of trend in hydrological time series of the Evinos River Basin. Hydrol Sci J 42(1):81–88
Granger OE (1977) Secular fluctuations of seasonal precipitation in lowland California. Mon Weather Rev 105:386–397
Gupta A, Thaplial PK, Pal PK, Joshi PC (2005) Impact of deforestation on Indian Monsoon – a GCM sensitivity study. J Indian Geophys Union 9(2):97–104
Hingane LS (1996) Is a signature of Socio-economic impact written on the climate. Climatic Change 32:91–102
Hirsch RM, Slack JR (1984) A nonparametric trend test for seasonal data with serial dependence. Water Resour Res 20(6):727–732
Hirsch RM, Slack JR, Smith RA (1982) Techniques of trend analysis for monthly water quality data. Water Resour Res 18(1):107–121
IINC (2004) India's Initial National Communication to the United Nations Framework Convention on Climate Change. Ministry of Environment and Forests, Government of India
IPCC (2007) Climate change 2007: the physical science basis. In: Solomon S, Qin D, Manning M, Chen Z, Marquis M, Averyt KB, Tignor M, Miller HL (eds) Contribution of working group I to the fourth assessment report of the intergovernmental panel on climate change. Cambridge University Press, Cambridge
IPCC WG II (2007) Climate change 2007: impacts, adaptation and vulnerability. Working group II contribution to the intergovernmental panel on climate change, Fourth assessment report, Summary for policymakers
Jenkins GM, Watts DG (1968) Spectral analysis and its applications. Holden-Day, San Francisco, CA
Jhajharia D, Shrivastava SK, Sarkar D, Sarkar S (2008) Temporal characteristics of pan evaporation trends under the humid conditions of northeast India. Agri Forest Meteorol doi:10.1016/j/agrformet.2008.10.024
Julian PR (1967) Variance spectrum analysis. Water Resour Res 3:831–845
Kendall MG (1955) Rank correlation methods. Charles Griffin, London
Khaliq MN, Ouarda TBMJ, Gachon P, Sushma L, St-Hilaire A (2009) Identification of hydrological trends in the presence of serial and cross correlations: a review of selected methods and their application to annual flow regimes of Canadian rivers. J Hydrol 368:117–130
Kothawale DR, Kumar R (2005) On the recent changes in surface temperature trends over India. Geophys Res Lett 32:L18714. doi:10.1029/2005GL023528
Kothyari UC, Singh VP (1996) Rainfall and temperature trends in India. Hydrol Process 10(3):357–372
Kothyari UC, Singh VP, Aravamuthan V (1997) An investigation of changes in rainfall and temperature regimes of the Ganga Basin in India. Water Resour Manage 11:17–34
Kripalani RH, Kulkarni A (1997) Rainfall variability over southeast Asia-connections with Indian monsoon and ENSO extremes: new perspectives. Int J Climatol 17(11):1155–1168
Kripalani RH, Kulkarni A (2001) Monsoon rainfall variations and teleconnections over South and East Asia. Int J Climatol 21(5):603–616
Kripalani RH, Oh JH, Kulkarni A, Sabade SS, Chaudhari HS (2007) South Asian summer monsoon precipitation variability: coupled climate model simulations and projections under IPCC AR4. Theor Appl Climatol 90(3–4):133–159
Kulkarni A, Sabade SS, Kripalani RH (2007) Association between extreme monsoon and the dipole mode over the Indian sub-continent. Met Atmos Phys 95:255–268
Kumar RK, Pant GB, Parthasarathy B, Sontakke NA (1992) Spatial and subseasonal patterns of the long-term trends of Indian summer monsoon rainfall. Int J Climatol 12(3):257–268

Lal B, Duggal YM, Ram P (1992) Trends and periodicities of monsoon and annual rainfall of districts of Haryana State and Delhi. Mausam 43(2):137–142

Lawton RO, Nair US, Pielke RA, Welch RM (2001) Climatic impact of tropical lowland deforestation on nearby Montane cloud forests. Science 294:584–587

Lettenmaier DP, Wood EF, Wallis JR (1994) Hydro-climatological trends in the continental United States 1948–1988. J Clim 7(4):586–607

Lins HF, Slack JR (1999) Streamflow trends in the United States. Geophys Res Lett 26(2):227–230

Mann HB (1945) Non-parametric tests against trend. Econometrica 13:245–259

Matalas NC (1967) Time series analysis. Water Resour Res 3:817–829

Meher-Homji VM (1991) Probable impact of deforestation on hydrological processes. Climatic Change 19:163–173

Mosmann V, Castro A, Fraile R, Dessens J, Sanchez JL (2004) Detection of statistically significant trends in the Summer precipitation of Mainland Spain. Atmos Res 70:43–53

Nair US, Lawton RO, Welch RM, Pielke RA (2003) Impact of land use on Costa Rican tropical montane cloud forests: sensitivity of cumulus cloud field characteristics to lowland deforestation. J Geophys Res 108(D7):4206–4219

Nicholson SE, Palao IM (1993) A re-evaluation of rainfall vari- ability in the Sahel, Part I. Characteristics of rainfall fluctuations. Int J Climatol 13:371–389

Ogallo L (1979) Rainfall variability in Africa. Mon Weather Rev 107:1133–1139

Pant GB (2003) Long-term climate variability and change over monsoon Asia. J Indian Geophys Union 7(3):125–134

Pant GB, Hingane LS (1988) Climatic changes in and around the Rajasthan desert during the 20th century. J Climatol 8(4):391–401

Parthasarathy B, Dhar ON (1976a) A study of trends and periodicities in the seasonal and annual rainfall of India. Indian J Meteorol Hydrol Geophys 27(1):23–28

Parthasarathy B, Dhar ON (1976b) Studies of trends and periodicities of rainfall over Madhya Pradesh. Proc Indian Nat Sci Acad 42(1A):73–80

Pielke RA (2001) Influence of the spatial distribution of vegetation and soils on the prediction of cumulus convective rainfall. Rev Geophys 39(2):151–177

Pielke RA, Adegoke J, Beltrán-Przekurat A, Hiemstra CA, Lin J, Nair US, Niyogi D, Nobis TE (2007) An overview of regional landuse and land-cover impacts on rainfall. Tellus B 59(3):587–601

Ramanathan V, Crutzen PJ, Keihl JT, Rosenfeld D (2001) Aerosols, climate and the hydrological cycle. Science 294:2119–2124

Ramanathan V, Chung C, Kim D, Bettge T, Buja L, Keihl JT, Washington WM, Fu Q, Sikka DR (2005) Atmospheric brown clouds: impacts on South Asian climate and hydrological cycle. Proc Natl Acad Sci USA 102(15):5326–5333

Ramankutty N, Delire C, Snyder P (2006) Feedbacks between agriculture and climate: an illustration of the potential unintended consequences of human land use activities. Glob Planet Change 54:79–93

Rao GSP, Jaswal AK, Kumar MS (2004) Effects of urbanization on meteorological parameters. Mausam 55(3):429–440

Ray DK, Nair US, Welch RM, Han Q, Zeng J, Su W, Kikuchi T, Lyons TJ (2003) Effects of land use in Suthwest Australia: 1. Observations of cumulus cloudiness and energy fluxes. J Geophys Res 108(D14):4414

Ray DK, Nair US, Lawton RO, Welch RM, Pielke RA (2006) Impact of land use on Costa Rican tropical montane cloud forests: sensitivity of orographic cloud formation to deforestation in the plains. J Geophys Res 111:D02108. doi:10.1029/2005JD006096

Reddy KR, Hodges HF (2000) Climate change and global crop productivity. CAB International Publishing, Wallingford

Rodhe H, Virji H (1976) Trends and periodicities in East African rainfall data. Mon Weather Rev 104:307–315

Sarkar S, Kafatos M (2004) Interannual variability of vegetation over the Indian sub-continent and its relation to the different meteorological parameters. Remote Sens Environ 90:268–280
Senapati PC, Misra D (1988) Distribution of rainfall in coastal region of Orissa. Ind J Pow Riv Val Dev 38(6):193–198
Serrano A, Mateos VL, Garcia JA (1999) Trend analysis of monthly precipitation over the Iberian Peninsula for the Period 1921–1995. Phys Chem Earth 24(1–2):85–90
Singh N, Soman MK (1990) Some aspects of hydroclimatic fluctuations in Kerala, India. Ind J Pow Riv Val Dev 40(5–6):75–84
Singh N, Sontakke NA (2002) On climatic fluctuations and environmental changes of the Indo-Gangetic Plains, India. Climatic Change 52(3):287–313
Singh N, Krishna Kumar K, Soman MK (1989) Some features of the periods contributing specified percentages of rainfall to annual total in Kerala, India. Theor Appl Climatol 39(3):160–170
Singh N, Sontakke NA, Singh HN, Pandey AK (2005) Recent trend in spatiotemporal variation of rainfall over india – an investigation into basin-scale rainfall fluctuations. IAHS-AISH Publ 296:273–282
Singh P, Kumar V, Thomas T, Arora M (2007) Changes in rainfall and relative humidity in river basins in northwest and central India. Hydrol Process. doi:10.1002/hyp.6871
Subbaramayya I, Naidu CV (1992) Spatial variations and trends in the Indian monsoon rainfall. Int J Climatol 12(6):597–609
Tukey JW (1950) The sampling theory of power spectrum estimates. In: Symposium on applications of autocorrelation analysis to physical problems. US Office of Naval Research, NAVEXOS-P-735, Washington, DC, pp 47–67
Türkes M (1998) Influence of geopotential heights, cyclone frequency and Southern Oscillation on rainfall variations in Turkey. Int J Climatol 18:649–680
Türkes M (1999) Vulnerability of Turkey to desertification with respect to precipitation and aridity conditions. Turk J Eng Environ Sci 23:363–380
Türkes M, Sümer UM, Kiliç G (2002) Persistence and periodicity in the precipitation series of Turkey and associations with 500 hPa geopotential heights. Climate Res 21:59–81
Ventura F, Pisa PR, Ardizzoni E (2002) Temperature and precipitation trends in Bologna (Italy) from 1952 to 1999. Atmos Res 61:203–214
Wiener N (1930) Generalized harmonic analysis. Acta Math 55:117–258
Wiener N (1949) Extrapolation, interpolation, and smoothing of stationary time series. MIT Press, Cambridge, MA
WMO (1966) Climatic change. WMO Technical Note 79, World Meteorological Organization, Geneva
Xu ZX, Takeuchi K, Ishidaira H (2003) Monotonic trend and step changes in Japanese precipitation. J Hydrol 279:144–150
Yevjevich V (1971) Stochastic processes in hydrology. Water Resources Publications, Fort Collins, CO
Yue S, Pilon P (2004) A comparison of the power of the *t*-test, Mann-Kendall and bootstrap tests for trend detection. Hydrol Sci J 49(1):21–37
Yue S, Pilon P, Cavadias G (2002) Power of the Mann-Kendall and Spearman's rho test for detecting monotonic trends in hydrological series. J Hydrol 259:254–271
Zhang Q, Jiang T, Gemmer M, Becker S (2005) Precipitation, temperature and runoff analysis from 195–2002 in the Yangtze basin, China. Hydrol Sci J 50(1):65–80
Zveryaev II, Aleksandrova MP (2004) Differences in rainfall variability in the South and Southeast Asian summer monsoons. Int J Climatol 24(9):1091–1107

Chapter 5
Geomorphology and Geology

Abstract A detailed analysis of topography, geomorphology, and geology are presented in this chapter. The analysis comprises the classification of the Yamuna basin based on topography. The tributaries contributing to the Yamuna River are duly addressed. The chapter also provides a discussion of sub-catchment wise geomorphological characteristics, followed by spatial analysis of geological and hydrogeological formations.

Besides climate, topography, geomorphology and geology are the other important characteristics of the basin which are not only important for addressing water resources related issues but also the type of habitation. These parameters are highly important for watershed management, flood and drought estimation/management, water logging, erosion estimation/mitigation, identification of recharge zone, installation of hydropower system, construction of rainwater harvesting structure, other structural development, etc. In this chapter, a detailed description of these parameters is presented. This chapter also includes the description of major tributaries and river morphology of the Yamuna.

5.1 Topographic Characteristics

Topography of the Yamuna basin varies between steep topography in the Himalayan segment to almost flat in the middle and lower segments (Fig. 5.1). The elevation in the Yamuna River basin varies from about 6,320 m above mean sea level (msl) near Yamunotri Glacier to around 100 m (above msl) near the confluence of Yamuna River with River Ganga at Allahabad. The topography of the Yamuna basin can be classified into three groups, i.e., hilly region (more than 600 m above msl); foot hills and Plateau region (300–600 m); and plains and valleys (100–300 m above msl). On the basis of this topographic classification 11,700 km^2 basin area (about 3.19%) can be classified as hilly, while remaining equally divided between plains and plateau regions with 161,231 km^2 and 172,917 km^2, respectively. In the Himalayan segment (upper Yamuna catchment), the drainage system and the characteristics of landforms are closely interdependent and inter-related. The upper Yamuna catchment falls into three defined physiographic belts: the Lesser Himalaya, the Siwalik, and the Doon Valley. On the other hand, catchments of the Chambal, the Sind, the Betwa and the Ken experience undulating topography with

R.K. Rai et al., *The Yamuna River Basin*, Water Science and Technology Library 66,
DOI 10.1007/978-94-007-2001-5_5, © Springer Science+Business Media B.V. 2012

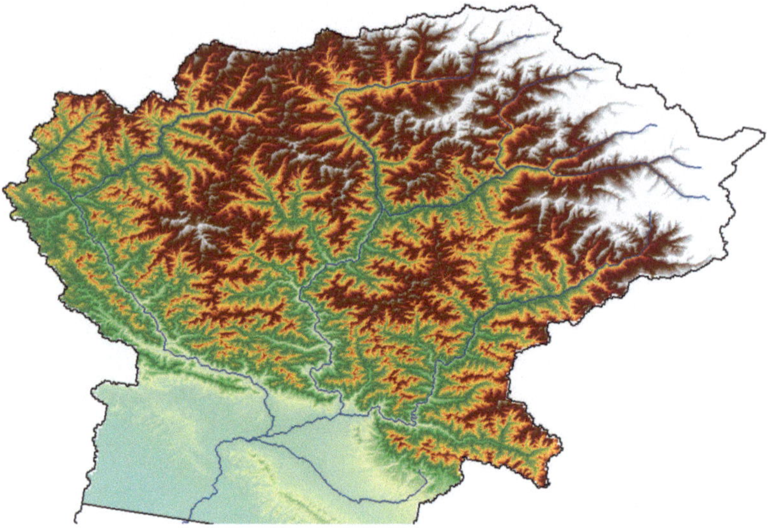

Fig. 5.1 Topography of Himalayan catchment of Yamuna River basin

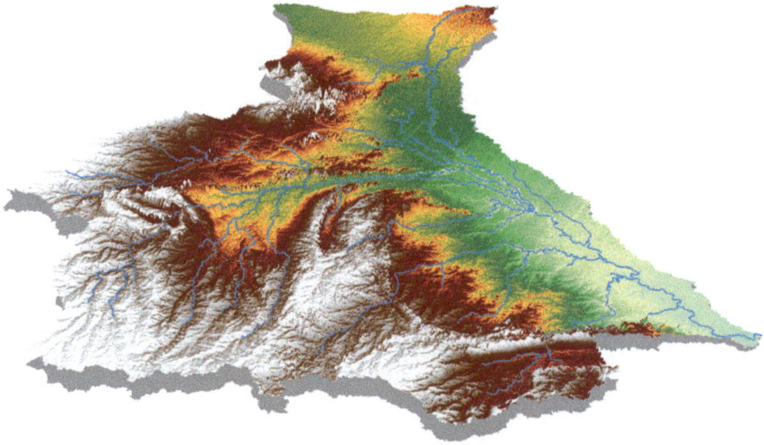

Fig. 5.2 Topography of the Yamuna River basin excluding the Himalayan catchment

ravines (Fig. 5.2). The digital elevation model showing the elevations is shown in Fig. 5.3.

Lesser Himalaya has a mild topography with gentle slopes and deeply dissected valleys which suggest that rivers and streams are still furiously at work. Elevation in this region ranges from 4,000 to 1,000 m. The upper part of this area has high mountains, most of which have seasonal snow capped peaks and glaciated ranges. Prominent glaciers are Bandarpunch, Jamadar Bamak and Deokhera Bamak. The retreating movement of these glaciers and their tributary glaciers may still be observed in the form of "V" shaped valleys with moraines and smooth and

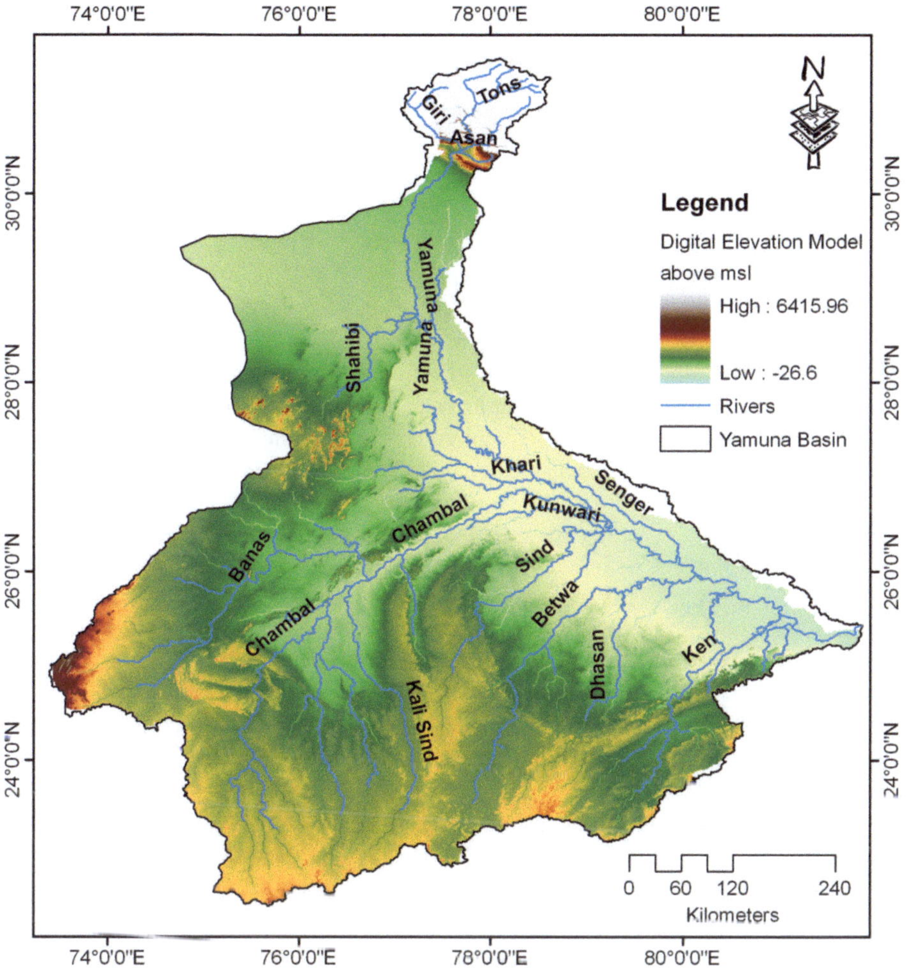

Fig. 5.3 Digital Elevation Model and major tributaries of Yamuna River basin

aggradational slope (Jain et al., 2007). Low to moderately high mountains occur between the altitudes of 1,000 and 3,000 m. Actually, the Lesser Himalayas region is a massive mountainous tract with a series of ridges and spurs divided by river valleys. The slope varies from 25 to 50% but may rise even up to 80%. The rivers and their tributaries have carved out entrenched valleys with steep slopes in higher reaches, while flatter valleys are found in lower reaches. At a number of places, rivers have formed depositional terraces. A large part of this region is made up of sedimentary rock formations encroached by granite and basic volcanic rocks. The northern part has been subjected to low-grade metamorphism.

The Siwalik Hills are formed as a result of intense dissection by fine textured pattern of drainage lines. They are long prominent ridge trending NW-SE with altitudes

Table 5.1 Average slopes in Yamuna River stretches

S. No.	Stretch	Stretch length (km)	Rate of fall (m/km)	Slope (%)
1	Upper Himalaya	25	59.0	5.90
2	Himalaya Stretch	152	19.1	1.91
3	Total Plain Stretch	1,224	0.20	0.02
4	Lower Plain Stretch	768	0.08	0.008

ranging from 750 to 1,500 m. The main ridge is composed of a gentle northern slope and a steeper southern slope. A water divide is located more or less half way through this ridge. It is drained with numerous parallel to sub-parallel streams flowing towards north or south in consequent entrenched channels. The Shiwalik or the outer Himalayan range is a youthful range separated from the lesser Himalayas by the main boundary fault.

The Doon Valley is a long tectonic synclinal structure of the outer Himalaya. It lies within the ranges of lesser Himalaya to the north and the Shiwalik range of the Outer Himalaya towards north. It is a low-lying region between the two ranges with altitudes not more than 500–750 m. The Doon gravels, composed of boulder and gravel beds with this clay bands, constitute the piedmont slopes. Stream frequency and drainage density is low here because of the poor development of drainage, possibly due to porous and permeable characteristics of bed rocks.

In the plains stretch the river flattens gradually with an average slope of 0.02%. The stretch wise average slope of the river is given in Table 5.1 which reflects a remarkable difference in rate of fall in valley profile of the Himalaya and the plain stretch. The maximum rate of fall of 59 m/km is experienced in the first 25 km of river from its origin, while it is minimum (i.e., 0.08 m/km) in the 768 km long tail end starting from Agra.

5.2 Major Tributaries

The important tributaries of the Yamuna River are the Tons, the Chambal, the Hindon, the Sind, the Betwa, the Ken, the Baghein, and the Mandakini. Other small tributaries of the Yamuna River include Rishiganga, the Uma, the Hanuman Ganga, the Giri, the Karan, the Sagar and the Rind. The main Yamuna and the Tons are fed by glaciers, viz., the Bandar Punch Glacier and its branches and originate from the Great Himalayan range. Many smaller streams in the Yamuna basin, viz., the Chautang, the Sahibi, the Dohan, the Kantili, the Bapah and the Banganga end up in the sandy tracts. Major tributaries of the Yamuna River system are shown in Fig. 5.4.

Tons River: The Tons River is the longest tributary of the Yamuna in the upper Himalayan segment. It flows through Garhwal, the western part of the Himalayan

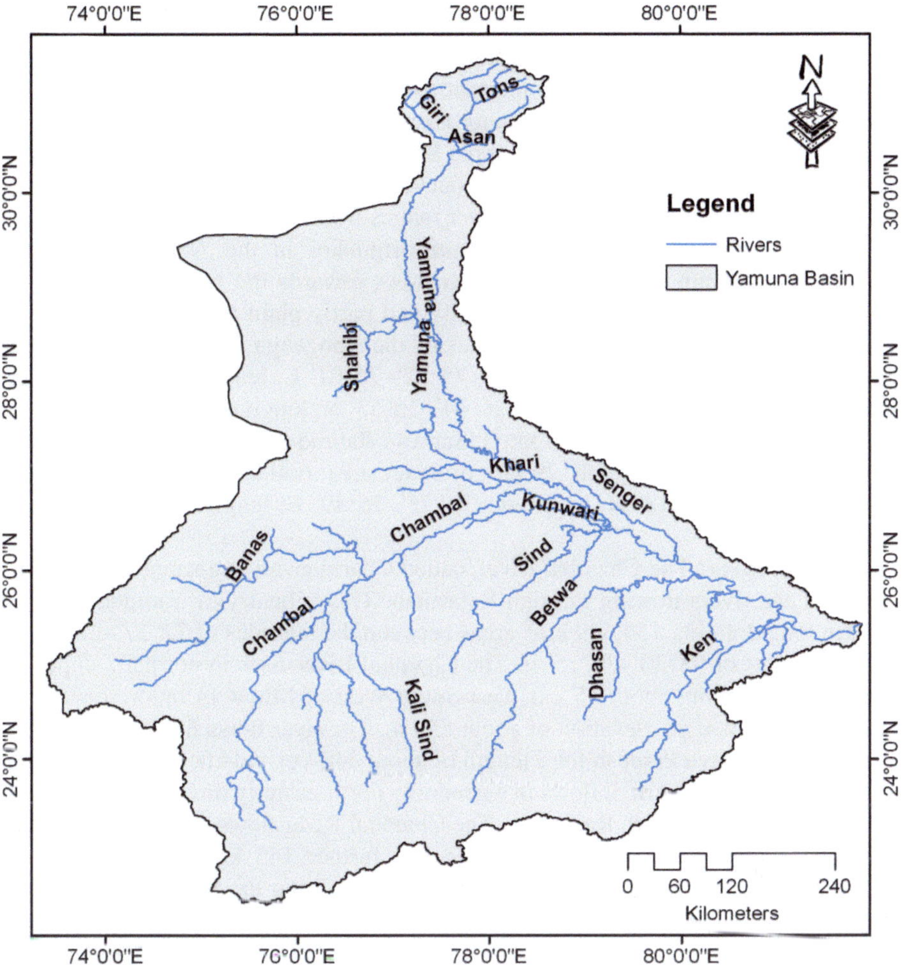

Fig. 5.4 Major tributaries of Yamuna River

state of Uttarakhand. The river originates at an elevation of 3,900 m and joins the Yamuna below Kalsi. In fact, Tons carries more water than Yamuna itself.

Kali River: The Kali River originates from the Doon Valley in the western part of Uttarakhand. The catchment covers a drainage area of approximately 750 km^2 within the latitude of 29° 13′30″ N and longitude of 77°32′45″ E. From its origin up to the confluence with Hindon River, the river travels a distance of approximately 150 km through Saharanpur and Muzaffarnagar, Meerut and Ghaziabad districts. Despite a significant drainage area, mostly lying in plains, the river does not carry any significant flow. The city of Muzaffarnagar is situated on the left bank of the Kali River.

Kali River is a highly polluted river which receives pollutants including heavy metals and industrial sludge from a dozen of towns along the river from Saharanpur to Kannauj. The presence of salt of cyanide has resulted in over 100 species of getting extinct from river over the past 40 years.

Hindon River: Hindon is an important tributary of Yamuna River. In fact, this river is sand-witched between two major rivers: the Ganga on the left and Yamuna on the right. Hindon originates from upper Shiwalik (Lower Himalayas). It is purely a rainfed river with catchment area of about 7,083 km^2. This river has a total run of about 400 km. The width of Hindon River ranges from 20 to 160 m.

Banganga River: The Banganga River originates in the Aravali hills, near Arnasar and Bairath in Jaipur District. It flows towards the south up to the village of Ghat, then east through partly hilly and partly plain terrain. The length of the river is 240 km. The main tributaries of the Banganga River are Gumti Nala (latitude: 25°0.5′–25°53′ N; longitude: 75°52′–76°01′ E; length: 24 km; drainage area: 102 km^2), Suri River (latitude: 26°44′–26°53′ N; longitude: 76°25′–76°30′ E; length: 18 km; drainage area: 76 km^2), Sanwan (latitude: 26°59′–27°22′ N; longitude: 76°16′–76°46′ E; length: 29 km; drainage area: 660 km^2) and Palasan River (latitude: 27°02′–27°18′ N; longitude: 76°25′–76°49′ E; length: 24 km; drainage area: 539 km^2).

Chambal River: The Chambal River, called Charmanvati in ancient times, is the largest of the rivers flowing through Rajasthan. This tributary of Yamuna River is 960 km long draining 134,689 km^2 areas between the latitudes of 22°27′–27°20′ N and longitudes of 73°20′–79°15′ E. The Chambal River rises in northern slopes of Vindhyan Mountain, about 15 km West-South-West of Mhow in Indore district in Madhya Pradesh at an elevation of about 853 m. The river flows first in a northerly direction in Madhya Pradesh for a length of about 346 km and after passing the historic fort of Chaurasigarh, it flows in a generally north-easterly direction for a length of around 225 km through Rajasthan. The Chambal River flows for another 217 km between Madhya Pradesh and Rajasthan and further 145 km between Madhya Pradesh and Uttar Pradesh. The Chambal River enters Uttar Pradesh near the Charak Nagar village and flows for about 32 km before joining the Yamuna River, south east of the Sahon village in Etawah district at an elevation of about 122 m. Three dams, namely, Gandhi Sagar, Rana Pratap Sagar and Jawahar Sagar have been constructed in its basin.

From the source to its confluence with the Yamuna River, the Chambal has a fall of about 732 m. Out of this approximately 305 m is within the first 16 km reach. It falls for another 195 m in the next 338 km, where it enters the gorge past the Chaurasigarh Fort. In the next 97 km of its run from Chaurasigarh Fort to Kota city, the bed falls by another 91 m. In the rest of its 523 km, the river passes through the flat terrain of the Malwa Plateau and later in the Gangetic Plain with an average gradient of 0.21 m/km. River Chambal flows through the deep gorge and has, therefore, a large scope to build the storage capacity. Up to Kota barrage, three large storages have been built namely, Gandhi Sagar, Rana Pratap Sagar and Jawahar Sagar. The Kota barrage is located near the Kota town, where the river emerges from the gorge section into the plateau. The total area draining at the Kota barrage is 27,319 km^2. Important tributaries of the Chambal River are summarized in Table 5.2.

Table 5.2 Description of important tributaries of the Chambal River

S. No.	Name	Tributary/River	Description
1	Alnia	Chambal	Origin: North-western slopes of Mukindwara hills Length: 58 km Confluence: with Chambal near Notana village in Kota Catchment area: 792 km^2
2	Kalisindh	Chambal	Origin: Northern slopes of Vindhyan hills Length: 145 km Confluence: joins Chambal near Nonera village in Kota Catchment area: 7,944 km^2 Extend: Jhalwar and Kota
3	Parwan	Kalisindh	Origin: Malwa Plateau Length: It enter Rajasthan near Kharibor village of Jhalawar district after 186 km run in Madhya Pradesh Confluence: joins Kalisindh near ramgarh village in Kota. Chambal near Notana village in Kota Catchment area: 2,892 km^2
4	Mej	Chambal	Origin: Mandalgarh Tehsil in Bhilwara District Confluence: joins Chambal near Bhaius Khana village in Kota district Catchment area: 5,860 km^2 Extend: Bhilwara, Bundi, Tonk
5	Chakan	Chambal	Origin: formed by the confluence of many small rivulets. Flow direction: South-easterly Confluence: Chambal near Karanpura village in Sawai Madhopur district. Catchment area: 789 km^2 Extend: Sawai Madhopur, Tonk, Bundi, and Kota
6	Parwati	Chambal	Origin: northern slopes of Vindhyan hills in Madhya Pradesh Flow direction: South-easterly Length: 159 km Confluence: Chambal at Pali village in Kota district. Catchment area: 5,001 km^2 Tributaries: Lhasi, Berni, Bethli, Andheri, Retri, Dubraj, Bilas and Kunu
7	Kunu	Parwati	Origin: north of Guna town in Madhya Pradesh Confluence: Chambal at Gordhanpura village in Kota district. Catchment area: 726 km^2 Tributaries: Karal and Rempi
8	Banas	Chambal	Origin: Khamnor hills of the Aravali range (about 5 km from Kumbhalgarh) Length: 512 km Confluence: Chambal at Rameshwar village in Khandar block of Sawai Madhopur district. Catchment area: 45,833 km^2 Tributaries: Berach (R), Menali (R), Kothari (L), Khari (L), Dai (L), Dheel (L), Sohadara (L), Morel (L) and Kalisil (L)

Table 5.2 (continued)

S. No.	Name	Tributary/River	Description
9	Berach	Banas	Origin: in the hills northeast of Udaipur city. Length: 157 km (flows in a hilly region up to Badgaon reservoir and then through plains). Direction: northeast Confluence: Banas near Bigod village in Mandalgarh village of Bhilwara district. Catchment area: 7,502 km^2 Tributaries: Ayar, Wagli Wagon, Gambhiri and Orai
10	Mashi	Banas	Origin: Kishangarh in Ajmer district. Length: 96 km (flows east and then south in partly hilly and partly plain areas of Ajmer and Tonk district). Direction: east Confluence: Banas near Tonk. Catchment area: 6,335 km^2
11	Khari	Banas	Origin: hills near Deogarh in Rajsamand district. Length: 192 km (through Udaipur, Bhilwara and Ajmer district). Direction: northeast Confluence: Banas near Chosala village in Ajmer district. Catchment area: 6,268 km^2
12	Dai	Banas	Origin: southeast slopes of Aravali range near Nasirabad Tehsil of Ajmer district. Length: 96 km Confluence: Banas river near Bisalpur village of Tonk district. Catchment area: 3,015 km^2
13	Morel	Banas	Origin: in the hills near Dharla and Chaibnpura villages in Bassi Tehsil of Jaipur district. Length: 147 km Confluence: Banas river near Hadoli village in Karauli Tehsil of Sawai Madhopur district. Tributaries: Dhund, Kankrauli and Kalisil
14	Kothari	Banas	Origin: in the eastern slopes of the Aravali range near Horera village in Bhilwara district. Length: 151 km Confluence: Banas river near Nandrai village in Bhilwara district Catchment area: 2,341 km^2

Betwa River: The Betwa River is a tributary of Yamuna River. Its basin extends from longitude 77°00′ to 81°00′ E and latitude 23°08′ to 26°00′ N. The Betwa River originates at an elevation of 706 m in the Bhopal district in Madhya Pradesh. After traversing a distance of 590 km, the river joins the Yamuna River near Hamirpur of Uttar Pradesh at an elevation of 106.68 m. The total catchment area of the Betwa

River catchment is 46,580 km^2 of which 31,971 km^2 (68.64%) lies in Madhya Pradesh and 14,609 km^2 (31.36%) lies in Uttar Pradesh. The basin is saucer (bowl) shaped with sandstone hills around the perimeter. The river has 14 principal tributaries out of which 11 are completely in Madhya Pradesh and 3 lie in Madhya Pradesh and partly in Uttar Pradesh. The Halali and Dhasan Rivers are the important tributaries of the Betwa River. The Halali is the largest tributary having a length of 180.32 km.

Ken River: Ken is an inter-state river, flowing through the states of Madya Pradesh and Uttar Pradesh. Its basin lies between north latitudes of 23°20′and 25°20′ and east longitudes of 78°30′ and 80′32′. The river originates near the village Ahirgawan in north-west slopes of Kaimur hills of Jabalpur district of Madhya Pradesh at an altitude of 550 m above mean sea level and joins the Yamuna near Chilla village in Uttar Pradesh at an elevation of about 95 m. The river has a total length of 427 km, out of which 292 km lies in Madhya Pradesh, 84 km in Uttar Pradesh and 51 km forms the common boundary between Madhya Pradesh and Uttar Pradesh. The drainage area of the Ken River catchment is 28,058 km^2, out of which 24,472 km^2 lies in Madhya Pradesh and the balance 3,586 km^2 in Uttar Pradesh. The important tributaries of Ken River are Sonar, Bearma, Kopra, Bewas, Urmil, Mirhasan, Kutni, Kail, Gurne, Patan, Siameri, Chandrawal, Banne, etc. In terms of catchment area, Sonar is the largest tributary which traverses a distance of 227 km and drains approximately 6,550 km^2 areas. The Sonar River catchment fully lies in Madhya Pradesh, extends between 23°20′ and 23°50′ N latitudes and 78°30′–79°15′ E longitudes. It is a leaf shaped elongated catchment, with an average width of about of 40 km. The Sonar catchment is bounded by Bearma catchment in the east side and Dhasan catchment on the west side and the Vindhyan ranges on the south. Sagar and Damoh are the major district falling in the catchment, and parts of Panna, Chhatarpur and Raisen districts also falls in the catchment.

5.3 Geomorphologic Characteristics

Geomorphology of a river basin describes the status of topographic features of the surfaces and streams, and its relationship with geomorphometry provides the geomorphological control on basin hydrology (Rai et al., 2009). Geomorphology reflects the topographic and geometric properties of the watershed and its drainage channel network. It controls the hydrologic processes from rainfall to runoff, and the subsequent flow routing through the drainage network. The role of basin geomorphology in controlling the hydrological response of a river basin is known for a long time. Moreover, for any infrastructural development, it is a very useful tool for the first hand overview of the basin. It is advantageous in the case of laying out the urban drainage and irrigation canal systems, aqueducts; study the physiographic impacts on environment, and selection of silt disposal site, hydropower site, recharge zone, percolation tank, retention tank, dam site, etc. This drainage network analysis of the river basin can provide a significant contribution towards flood management and water logging program (Rai et al., 2009). Earlier works have provided an understanding of basin geomorphology-hydrology relationship through empirical relationships (Snyder, 1938; Horton, 1945; Taylor and Schwartz, 1952; Singh, 1988b; Singh, 1989; Jain and Sinha, 2003; etc.). Snyder (1938) proposed a synthetic unit hydrograph approach (SUH) for ungauged basin as a function of catchment area, basin shape, topography, channel slope, stream density and channel storage; and derived the basin coefficient by averaging out other parameters.

5.3.1 Geomorphologic Parameters

Certain characteristics of the drainage basins reflect hydrologic behavior and are, therefore, useful, when quantified, in evaluating the hydrologic response of the basins. These characteristics relate to either the physical drainage basin of channels. Physical characteristics of the drainage basin include drainage area, basin shape, ground slope, and centroid (i.e. centre of gravity of the basin). Channel characteristics include channel order, channel length, channel slope, channel profile, and drainage density.

5.3.1.1 Basin Order and Channel Order

Drainage areas may be characterized in terms of the hierarchy of stream ordering. The order of the basin is the order of its highest-order channel. An inspection of a drainage basin channel network reveals that as one traces the flow from one of the uppermost channels in the basin towards the outlet. This upper most channel joins another channel, which in turn joins another channel and so on. The first order channel is defined as those channels that receive water entirely from surface overland flow and does not have tributaries. The junction of two first order streams forms the second order stream. A second order channel receives flow from the two first order channels and from overland flow from surface and it might receive the flow from another first order stream. Thus, a second order channel must carry much more flow of water than a first order channel. When the two second order streams join, together they form the third order stream, and so on. This scheme of stream ordering is referred to as the Horton-Strahler ordering scheme (Horton, 1945; Strahler, 1957). A watershed is described as first-, second-, third-, or higher order, depending upon the stream order at the outlet (i.e., highest-order stream within the watershed area).

5.3.1.2 Basin Area

Basin area is defined as the area contained within the vertical projection of the drainage divide on a horizontal plane. Some areas in the drainage basin do not contribute to the runoff and are termed as closed drainage. These areas may be lakes, swamps, etc.

5.3.1.3 Basin Shape

The watershed shape may influence the hydrograph shape, especially for small watersheds. For example, if the watershed is long and narrow, then it will take longer time for water to travel from the most extreme point to the outlet and the resulting hydrograph shape is flatter. For more compacted watershed, the runoff hydrograph is expected to be sharper with a greater peak and shorter duration. Numerous symmetrical and irregular forms of drainage areas are encountered in practice. To define the basin shape, a multitude of dimensionless parameters are used to quantify and these are: form factor, shape factor, elongation ratio, circulatory ratio, and compactness coefficient (Table 5.3). These factors involve watershed length, area and perimeter. The watershed length can be defined as the length of the main stream from its source (projected to the perimeter) to the outlet. Clearly, the form factor is less than unity and its reciprocal, the shape factor, is grater than unity. A square drainage basin has the shape factor equal to unity, whereas the long narrow drainage basin would have a shape factor greater than unity. The elongation ratio, circulatory ratio, and compactness coefficient approach to unity as the watershed shape approaches a circle.

5.3.1.4 Basin Slope

Basin slope has a pronounced effect on the velocity of overland flow, watershed erosion potential, and local wind systems. The average basin slope is defined as

$$S = h/L \tag{5.1}$$

where S is the average basin slope (m/m), h is the fall (m), and L is the horizontal distance (m) over which the fall occurs.

Table 5.3 Watershed shape parameters (Singh, 1988b)

Parameter	Definition	Formula	Value
Form factor	$\frac{\text{Watershed area}}{(\text{Watershed length})^2}$	$= A/L^2$	<1
Shape factor, B_s	$\frac{(\text{Watershed length})^2}{\text{Watershed area}}$	L^2/A	>1
Elongation ratio	$\frac{\text{Diameter of circle of watershed area}}{\text{Watershed length}}$	$\frac{1.128A^{0.5}}{L}$	≤1
Circulatory ratio	$\frac{\text{Watershed area}}{\text{Area of circle of watershed area}}$	$\frac{12.57A}{P_r^2}$	≤1
Compactness coefficient	$\frac{\text{Watershed perimeter}}{\text{Perimeter of circle of watershed area}}$	$\frac{0.2821P_r}{A^{0.5}}$	≥1

A watershed area, L watershed length, P_r watershed perimeter

5.3.1.5 Drainage Basin Similarity

Strahler (1957) hypothesized the drainage basin similarity as follows:

$$A/L_b^2 = C \tag{5.2}$$

where A is the area of the drainage basin $[L^2]$, L_b is the length of the basin $[L]$ and C is the constant for basin similarity.

5.3.1.6 Drainage Density

Drainage density is defined as the length of drainage per unit area. This term was first introduced by Horton (1932) and is expressed as

$$D_d = L/A \tag{5.3}$$

or

$$D_d = \frac{\sum_{i=1}^{\Omega} \sum_{j=1}^{N_i} L_{i,j}}{A} \tag{5.4}$$

The above relationships can be simplified as follows

$$D_d = \frac{\bar{L}_1 R_b^{\Omega-1}}{A} \cdot \frac{r^{\Omega} - 1}{r - 1} \tag{5.5}$$

where

$$r = R_l/R_b \tag{5.6}$$

5.3.2 Determination of Horton's Ratio

Three Horton's ratios, namely bifurcation ratio (R_b), stream-length ratio (R_l) and stream-area ratio (R_a) are unique representative parameters for a given watershed and are fixed values for a given watershed system. These parameters play an important role in deriving the geomorphologic instantaneous unit hydrograph and thus the unit hydrograph and flood hydrographs for catchments.

5.3.2.1 Bifurcation Ratio R_b

The number of channels of a given order in a drainage basin is a function of the nature of the surface of that drainage basin. In general, the greater the infiltration of the soil material covering the basin, the fewer will be the number of channels required to carry the remaining runoff water. Moreover, the larger the number of channels of a given order, the smaller is the area drained by each channel order. A dimensionless parameter based on the number of channels with respect to their order is termed as bifurcation ratio and is useful in defining the watershed response. The bifurcation ratio is given as follows:

$$R_b = N_i / N_{i+1} \tag{5.7}$$

where R_b is the bifurcation ratio, N_i and N_{i+1} are the number of streams in order i and i+1 respectively, $i = 1, 2, \ldots, \Omega$ and Ω is the highest stream order of the watershed. The value of R_b for watersheds varies between 3 and 5. This law is an expression of topological phenomenon, and is a measure of drainage efficiency.

5.3.2.2 Stream-Length Ratio R_l

This refers to length of channels of each order. The average length of channels of each higher order increase as a geometric sequence, which can be further explained as follows: the first order channels are the shortest of all the channels and the length increases geometrically as the order increases. This relation is called Horton's law of channel length and can be formulated as follows:

$$\bar{L}_i = \bar{L}_1 \, R_l^{i-1} \tag{5.8}$$

where $\bar{L}_i$ is the average length of channel of order i, R_l is the stream-length ratio. The R_l is mathematically expressed as follows:

$$R_l = \bar{L}_{i+1} / \bar{L}_i \tag{5.9}$$

where

$$\bar{L}_i = \frac{1}{N_i} \sum_{j=1}^{N_i} L_{j,\,i} \tag{5.10}$$

The lengths of channels of a given order are determined largely by the type of soil covering the drainage basin. Generally, more pervious the soil, longer will be the channel length of a given order. Also, higher is the R_l more will be the imperviousness. Generally it varies between 1.5 and 3.5.

5.3.2.3 Stream-Area Ratio R_a

The Channel area of order i, A_i is the area of the watershed that contributes to the channel segment of order i and all lower order channels. It can be quantified as:

$$\bar{A}_i = \bar{A}_1 \, R_a^{i-1} \tag{5.11}$$

where $\bar{A}_i$ is the average area of order i and R_a is the stream area ratio. These are given as follows:

$$R_a = \bar{A}_{i+1} / \bar{A}_i \tag{5.12}$$

$$\bar{A}_i = \frac{1}{N_i} \sum_{j=1}^{N_i} A_{j,\,i} \tag{5.13}$$

where $A_{j,\,i}$ is the total area that drains into the jth stream area of order i.

Table 5.4 Geomorphologic characteristics of the Yamuna basin

Area (km^2)	366,394.0	Minimum elevation (m)	62.0
Perimeter (km)	5,414.7	Relief (m)	6,226.0
Length of basin (km)	1,376.0	Mean slope (m/km)	4.525
Maximum elevation (m)	6,288.0		
Stream characteristics			
Stream order	Number	Length (km)	Area (km^2)
1	6053	36,805.61	23,5948.5
2	2775	17,727.18	29,6485.8
3	890	9,068.76	329,752.3
4	72	5,464.85	34,9845.5
5	21	2,859.14	360,242.5
6	5	7,94.07	363,315.2
7	2	886.99	366,394.3
Ratios			
Bifurcation ratio	4.63	Shape factor	5.17
Length ratio	2.64	Elongation ratio	0.496
Stream-area ratio	4.94	Circulatory ratio	0.157
Horton ratio	0.568	Compactness coefficient	2.523
Drainage density	0.20	Basin slope (m/m)	0.0045
Drainage frequency	0.026	Length of overland flow (km)	2.49
Form factor	0.193		

Table 5.5 Geomorphologic characteristics of the upper Yamuna catchment

Area (km^2)	11,359.8	Minimum elevation (m)	297
Perimeter (km)	718.56	Relief (m)	5,991
Length of basin (km)	210	Mean slope (m/km)	28.5286
Maximum elevation (m)	6,288		
Stream characteristics			
Stream order	Number	Length (km)	Area (km^2)
1	177	964.0	6,988.995
2	39	485.8	9,278.07
3	7	373.4	10,893.92
4	1	120.9	11,359.8
Ratios			
Bifurcation ratio	5.703	Shape factor	3.882
Length ratio	2.945	Elongation ratio	0.572
Stream-area ratio	6.622	Circulatory ratio	0.276
Horton ratio	0.516	Compactness coefficient	1.902
Drainage density	0.171	Basin slope (m/m)	0.02853
Drainage frequency	0.0197	Length of overland flow (km)	2.921
Form factor	0.258		

Table 5.6 Geomorphologic characteristics of the Chambal basin

Area (km^2)	134,689.4	Minimum elevation (m)	87
Perimeter (km)	3,202.1	Relief (m)	1,219
Length of basin (km)	960	Mean slope (m/km)	1.704
Maximum elevation (m)	1,306		
Stream characteristics			
Stream order	Number	Length (km)	Area (km^2)
1	2,260	13,541.5	88,190.4
2	1,412	63,88.9	110,464.7
3	102	2,843.1	120,959.8
4	27	2,064.7	128,594.7
5	8	1,061.9	132,482.9
6	2	193.2	133,165.2
7	1	389.1	134,689.4
Ratios			
Bifurcation ratio	4.76	Shape factor	3.79
Length ratio	1.361	Elongation ratio	0.578
Stream-area ratio	5.12	Circulatory ratio	0.165
Horton ratio	0.286	Compactness coefficient	2.461
Drainage density	0.197	Basin slope (m/m)	0.0017
Drainage frequency	0.028	Length of overland flow (km)	2.543
Form factor	0.263		

Table 5.7 Geomorphologic characteristics of the Sind sub-basin

Area (km^2)	27,940.0	Minimum elevation (m)	93.0
Perimeter (km)	1,308.71	Relief (m)	469.0
Length of basin (km)	440.0	Mean slope (m/km)	1.066
Maximum elevation (m)	562.0		
Stream characteristics			
Stream order	Number	Length (km)	Area (km^2)
1	442	2,562.91	17,395.7
2	80	1,147.96	21,560.2
3	20	738.09	24,352.9
4	6	603.76	26,782.6
5	2	293.2	27,936.3
6	1	2.92	27,946.6
Ratios			
Bifurcation ratio	3.55	Shape factor	6.93
Length ratio	1.86	Elongation ratio	0.428
Stream-area ratio	4.01	Circulatory ratio	0.205
Horton ratio	0.524	Compactness coefficient	2.21
Drainage density	0.191	Basin slope (m/m)	0.00106
Drainage frequency	0.02	Length of overland flow (km)	2.61
Form factor	0.144		

Table 5.8 Geomorphologic characteristics of the Betwa sub-basin

Area (km^2)	43,651.6	Minimum elevation (m)	79.0
Perimeter (km)	1,689.4	Relief (m)	627.0
Length of basin (km)	590.0	Mean slope (m/km)	1.063
Maximum elevation (m)	706.0		
Stream characteristics			
Stream order	Number	Length (km)	Area (km^2)
1	725	4,186.97	28,538.37
2	317	1,922.3	35,381.33
3	32	1,039.55	39,691.11
4	8	513.43	41,701.67
5	2	438.67	43,237.05
6	1	116.8	43,651.64
Ratios			
Bifurcation ratio	3.91	Shape factor	7.97
Length ratio	2.15	Elongation ratio	0.40
Stream-area ratio	4.33	Circulatory ratio	0.192
Horton ratio	0.55	Compactness coefficient	2.28
Drainage density	0.188	Basin slope (m/m)	0.00106
Drainage frequency	0.021	Length of overland flow (km)	2.656
Form factor	0.125		

Table 5.9 Geomorphologic characteristics of the Ken sub-basin

Area (km^2)	28,572.85	Minimum elevation (m)	76
Perimeter (km)	1,411.62	Relief (m)	663
Length of basin (km)	427	Mean slope (m/km)	1.5527
Maximum elevation (m)	739		
Stream characteristics			
Stream order	Number	Length (km)	Area (km^2)
1	486	2,594.4	18,143.7
2	102	1,219.5	22,709.3
3	24	758.9	25,763.4
4	7	255.9	26,759.7
5	2	185.8	27,443.1
6	1	273.4	28,572.4
Ratios			
Bifurcation ratio	3.58	Shape factor	6.38
Length ratio	2.304	Elongation ratio	0.446
Stream-area ratio	4.00	Circulatory ratio	0.180
Horton ratio	0.642	Compactness coefficient	2.356
Drainage density	0.185	Basin slope (m/m)	0.00155
Drainage frequency	0.022	Length of overland flow (km)	2.702
Form factor	0.157		

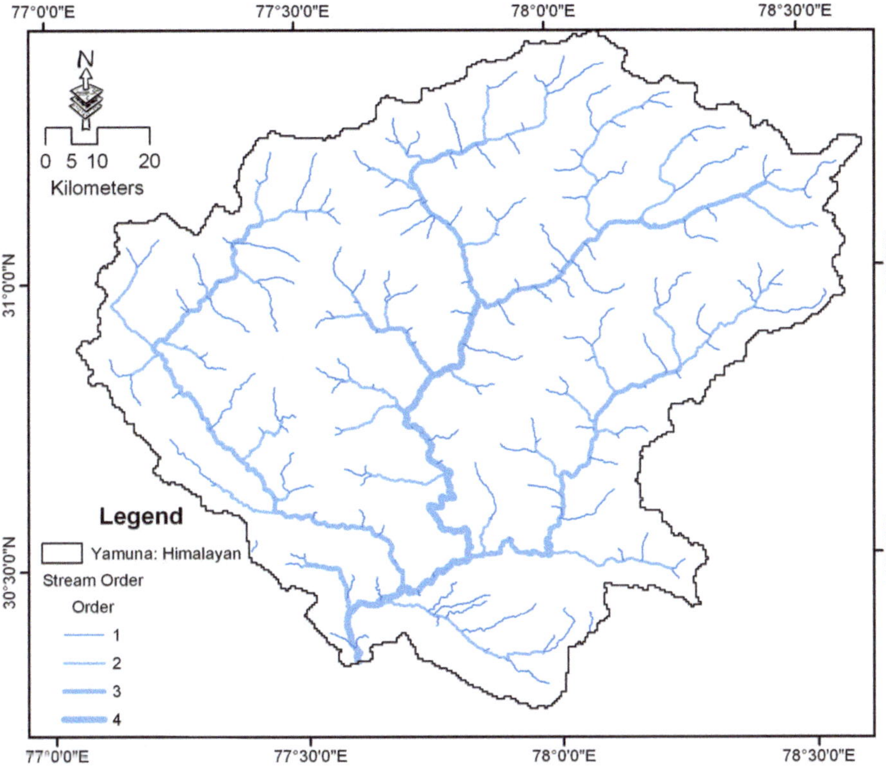

Fig. 5.5 Stream order and drainage network of Upper Yamuna Catchment (Himalayan)

Using the digital elevation model (DEM) of 30 m resolution of the basin, geomorphologic characteristics of the Yamuna basin has been extracted using the ArcGIS and are summarized in Table 5.4. Similarly, the geomorphologic characteristics of other sub-basins, viz., Upper Himalayan Catchment (up to Hathnikund barrage), Chambal, Sind, Betwa and Ken are summarized in Tables 5.5, 5.6, 5.7, 5.8, and 5.9. The stream network of the sub-basins is shown in Figs. 5.5, 5.6, 5.7, 5.8, and 5.9.

5.4 Geology of the Basin

The geological formation of the Yamuna basin has five major classes, viz., Proterozoic, Paleozoic, Mesozoic, Cenozoic and Deccan Volcanics/Sediments (Fig. 5.10). A brief description of these classes is given as follows.

Cenozoic: The Cenozoic "means the new life" is the most recent of the three classic geologic eras and covers the period from 65.5 million years (myrs) to the present. It is marked by the Cretaceous-Tertiary extinction event at the end of the

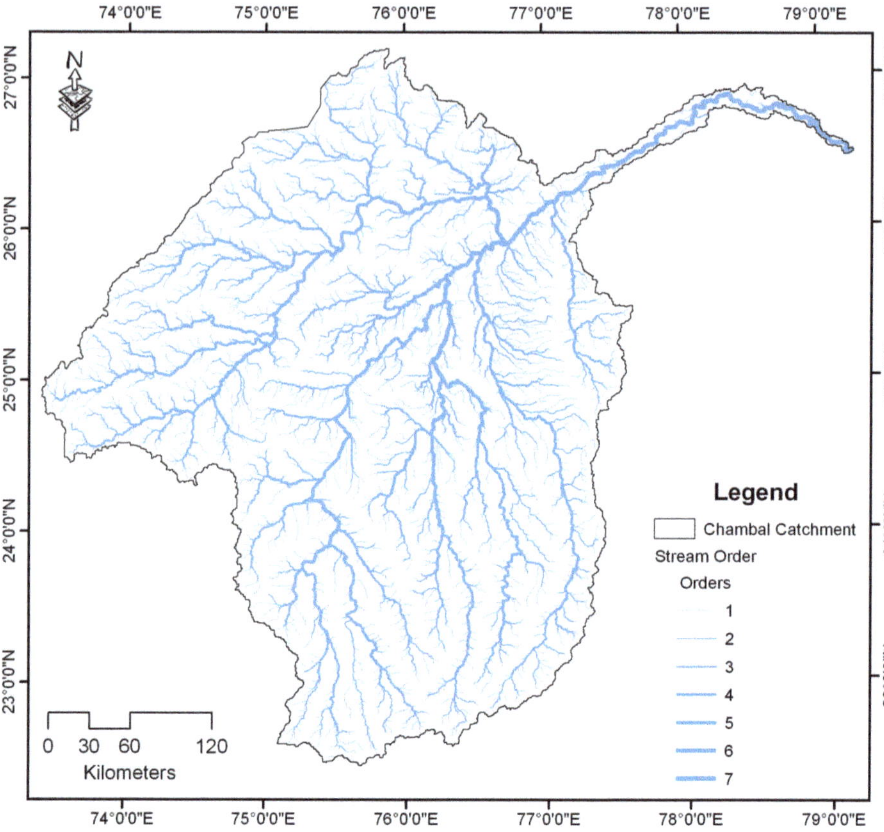

Fig. 5.6 Stream order and drainage network of Chambal sub-basin

Cretaceous that saw the demise of the last non-avian dinosaurs and the end of the Mesozoic era. The Cenozoic era is ongoing. Major Gangetic plain (i.e., duab area of the Ganga-Yamuna system) and trans-Yamuna catchments come under this category. A few parts of the other catchment before the confluence with higher order streams also have this formation. This plain has a deeply incised river with 15–30 m of cliff sections along the river banks. Repeated degradation and aggradations events separated by the stratigraphic discontinuities are recognized in the alluvial sequence, which represents relatively slow rates of accumulation (Sinha et al., 2005).

Mesozoic: The Mesozoic era is an interval of geological time from about 250 myrs ago to about 65 myrs ago. It is called the age of Dinosaurs.

Deccan Volcanics/Sediments: Close to the Mesozoic era is marked by the outcropping of enormous lava flows which spread over vast areas of the Western, Central and Southern India. They issue through long narrow fissures or cracks in the earth's crust, from a large magma basin and are, therefore, called fissure eruptions. The lava spread out far and wide as nearly horizontal sheets, the earliest flows filling up the irregularities of the pre-existing topography. They appear to have erupted sub-aerially. The Deccan traps are thus the most extensive geological formation of

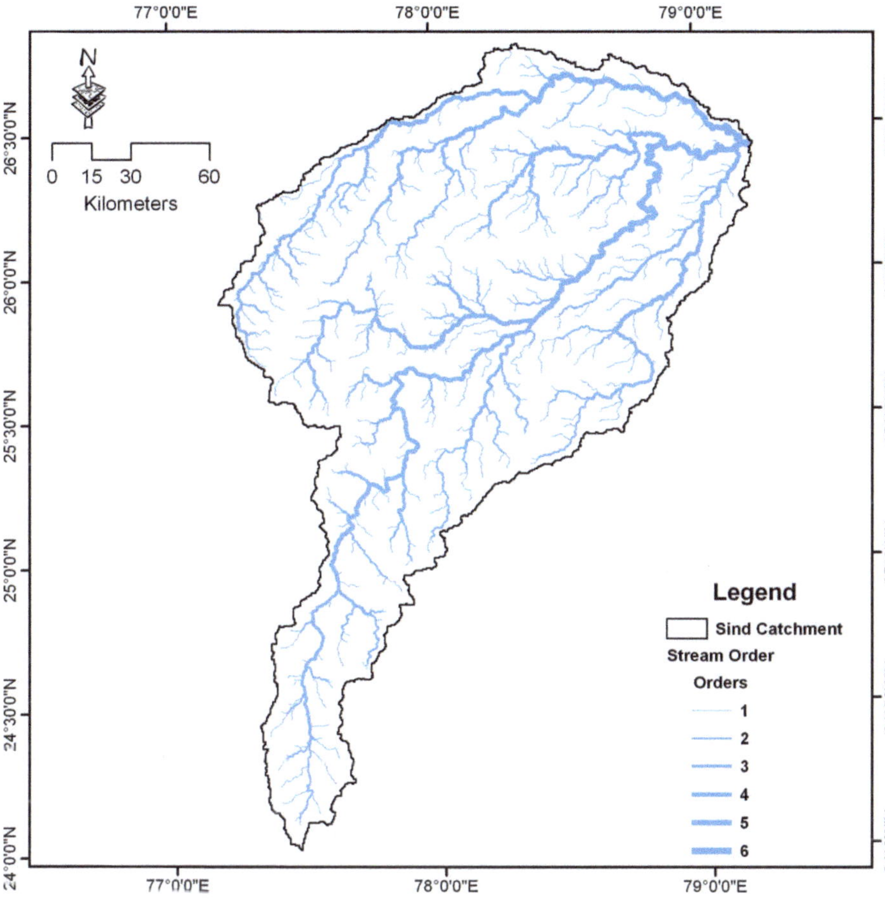

Fig. 5.7 Stream order and drainage network of Sind sub-basin

Peninsular India at present, with the exception of the metamorphic and igneous complex of Achaean age. The existence of Deccan volcanic rocks or sediments is evident in the upstream part of the Chambal, Betwa and Ken catchment, as shown in Fig. 5.10.

Paleozoic: The Paleozoic or Paleozoic era is the earliest of three geological eras of the Phanerozoic Eon. The Paleozoic spanned from roughly 542–251 myrs ago, and is subdivided into six geologic periods; from oldest to youngest these are: the Cambrian, Ordovician, Silurian, Devonian, Carboniferous, and Permian Fish population exploded in the Devonian. During the late Paleozoic, great forests of primitive plants thrived on land forming the great coal beds of Europe and Eastern North America. From the beginning of Paleozoic, shallow seas began to encroach on the continents. As the era proceeded, the marginal seas periodically washed over the stable interior, leaving sedimentary deposits to mark their incursions. In this era, evidence of glaciers is also experienced.

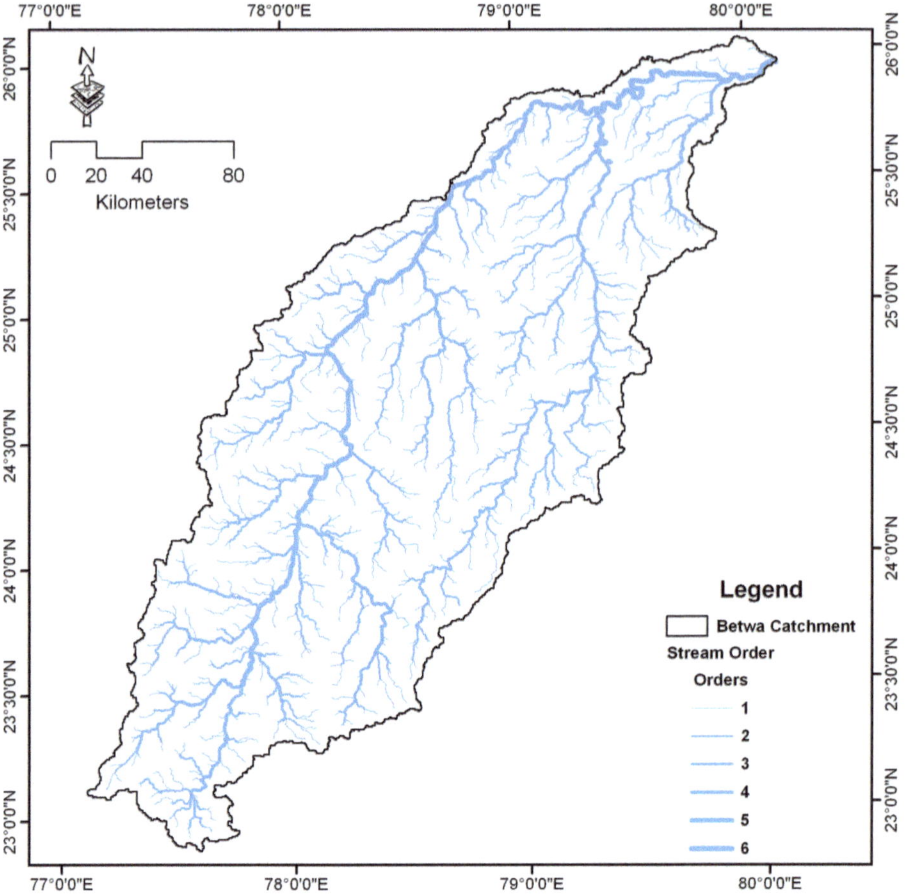

Fig. 5.8 Stream order and drainage network of Betwa sub-basin

Proterozoic: The Vindhyan range is a classical example of Proterozoic intercontinental basin that developed in the central part of the Indian shield along with several other basins such as Cuddapah, Chhattisgarh, etc. The strata are exposed in three major areas: Son Valley, Bundelkhand and Rajasthan. Substantially thick Vindhyan rocks have also been recognized under the Gangetic alluvium. The constituent stratigraphic units of the Vindhyan Supergroup are laterally correlated and vertically stacked in a similar fashion in individual areas, but their stratigraphic and sedimentologic attributes vary from one area to the other. The dominant lithologies include conglomerate, sandstone, shale, limestone as well as volcaniclastics. The strata lie uncomfortably on the Achaean-early Proterozoic gneisses, high grade rocks, granite-greenstones or metasedimentaries of Proterozoic mobile belts. Most parts of the Vindhyas are composed of horizontally bedded sedimentary rocks of ancient age, the contemporaries of the Torridon sandstone of Scotland.

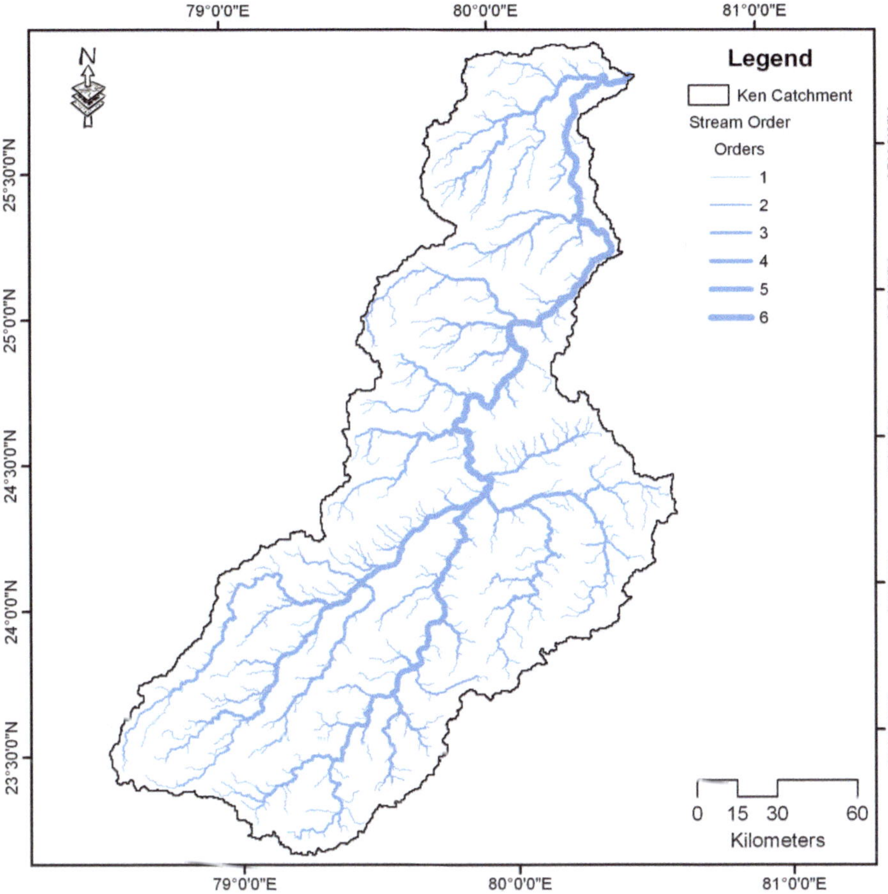

Fig. 5.9 Stream order and drainage network of Ken sub-basin

The Vindhyan system is marked by a major ENE-WSW trending lineament termed Narmada-Son lineament south of which occurs the Satpura orogen. The western margin of the system in Rajasthan is also marked by an NE-SW trending major lineament known as Great Boundary Fault separating the Aravalli-Delhi orogen from the Vindhyan. Nearly half of the basin area belongs to this geological formation category. As sown in Fig. 3.1, significant portions of the Chambal, Sind, Betwa, and Ken catchments are constituted under the Proterozoic era.

In general granitoid gneisses are common in the Bundelkhand, and are generally referred to as the Bundelkhand gneisses. It looks like a typical pink granite in hand specimens, the foliation being very rude, if at all developed. Granulites and crystalline limestone (i.e. marbles) are another important composition of this category which is exposed in the Alwar district of Rajasthan.

Aravalii chains of the Cuddapah system is mainly composed of much indurated and compacted shales, slates, quartzites and limestones. The shales have acquired a

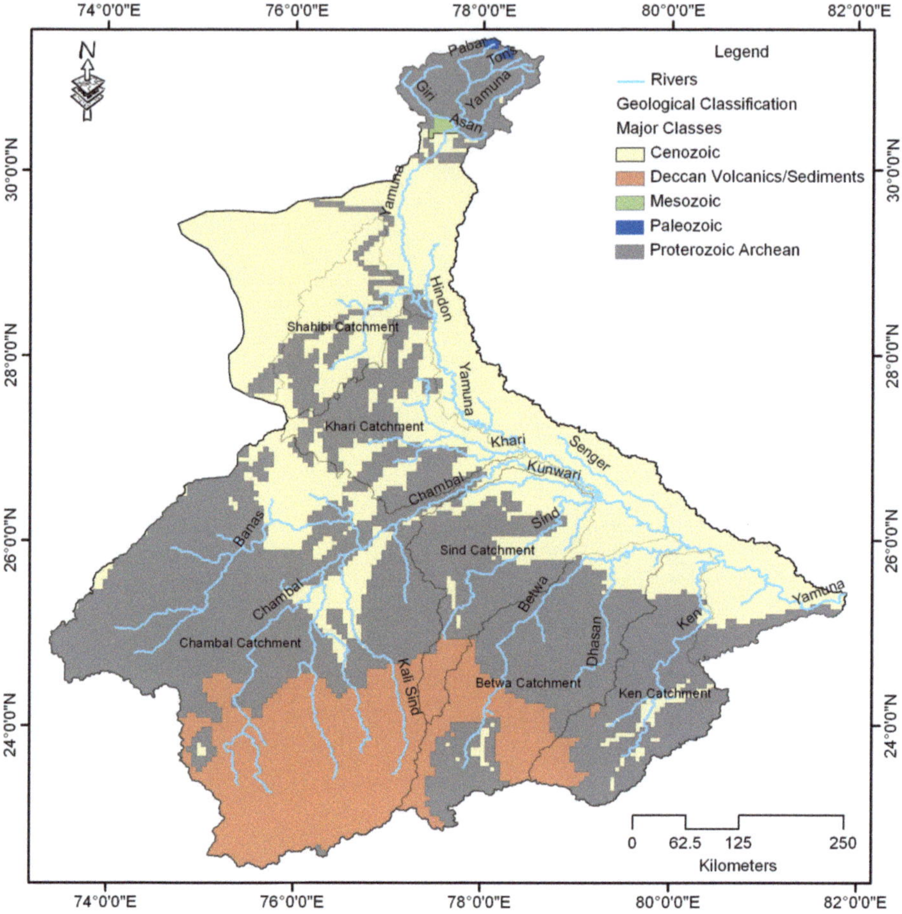

Fig. 5.10 Major classification of geological formation

slaty cleavage, but beyond that there is no further metamorphism (i.e., action of heat and pressure) into phyllites or schists; such secondary minerals as mica, chlorite, andalusite, staurolite, gamets, etc, have not been developed in them; nor are the limestone recrystallized into marbles, as in the Dharwar rocks.

The catchment-wise summary of geological formation is given in Table 5.10. Further to the geology of the basin, a detailed catchment-wise description of lithology and hydrogeology is presented in the following section.

5.5 Lithology and Hydrogeology of the Basin

The Yamuna River system is bounded by the Himalayas on the north and the Vindhyas on the south. In the east, it is separated by the ridge from the main Ganga catchment and on the west by the ridge separating it from the Luni and the

Table 5.10 Catchment wise distribution of geological formation

Sub basin	Predominant formation type	Geological formation	Areal extent (km^2)	Remarks
Chambal/Betwa/Ken/Kali Sind (plateau region)	Hard rock	Crystalline: granite/gneiss, phyllite/schist Deccan trap-basalt sedimentary-lime stone/sand stone	196,232	Alluvium deposits are found in patches near and along the major rivers.
Yamuna/Hindon/Khari/Sahibi, i.e., Yamuna-Ganga and Yamuna-Ghaghar Doab (Indo-Gangatic plain)	Alluvium	Older to recent alluvium sand, silt, gravel, clay, boulder	158,629	

Ghaggar basins. Most parts of catchment in Haryana and Uttar Pradesh lie in the Gangetic alluvial plains. The catchment/region wise lithological and hydrogeological features of the Yamuna basin are described as follows. A brief summary of the hydrogeological characteristics of the basin is given in Table 5.11.

5.5.1 Himalayan Region

A significant part of the Yamuna basin is contained in the Lesser Himalaya. The Yamuna originates from the Yamunotri Glacier at the base of the Bandapunch peak in the Higher Himalaya (Negi, 1991; Dalai et al., 2002); and many tributaries, such as the Tons, the Giri, the Aglar, the Asan and the Bata join in the Himalayan segment. Near its source in the Higher Himalaya, the Yamuna drains mainly the crystallines of Ramgarh and Almora groups (Gansser, 1964). Occurrences of calc-schists and marble with sulphide mineralization have been reported in the areas upstream of Hanuman Chatti (Jaireth et al., 1982). From the Higher Himalayas, the Yamuna flows in the southwest direction and enters the Lesser Himalayas where it drains a variety of lithologies. It flows through a large stretch of quartzite of Berinag Formation. Downstream, it passes through the massive dolomitic limestone and marble of Mandhali and Deoban Formations. The Yamuna then enters a large stretch of the sedimentaries of the Chakrata Formation, Chandpur Formation and Nagthat Formation. Barites occur in the silisiclastic sediments of Nagthat Formations in the Tons river section (Sachan and Sharma, 1993) and in the lower horizons of Krol limestones at Maldeota and Shahashradhara, where they occur as veins (Anantharaman and Bahukhandi, 1984). Southwest of Kalsi, the Yamuna enters the Siwaliks comprising the channel and floodplain deposits by the Himalayan Rivers in the past. In the Lesser Himalaya, occurrences of shales are reported into the Infra Krol, the Lower Tal, the Deoban and the Mandhali Formations (Gansser, 1964; Valdiya, 1980). These are exposed at a number of locations in the Yamuna and

Table 5.11 Hydro-geological features of different formations in terms of groundwater development, use and prospect to recharge

Predominant geological formations zones	Water bearing features	Well yield range (lps)	Feasible well types	Depth range (m)	Areal extent (km^2)
Hilly and Tarai Region	Tarai, Alluvium, Fracture/weather Hard Rock	0.1–15	Hand pumps, Spring, kunds, shallow large diameter dug well to deep tube wells in Tarai, Bhabar zone	Varying	11,528
Younger Alluvium (Unconsolidated formation of sand, silt, clay, deposited closer to major rivers-mainly low land area; Khader)	Porous voids	25–55	Tube well	20–300	63,083
Older Alluvium (Unconsolidated to Semi consolidated layers of sand, gravel, silts, kanker with occasional gravel bed–Upland area Bhanger	Porous voids	10–15	Tube well/Dug well	65–150/20–40	95,546
Consolidated Sedimentary (Sand stone/lime stone, shale, slate)	Weathered/Fractured/Jointed	2–3	Tube well/Dug well	100–150/30–50	52,177
Crystalline rock (Igneous and Metamorphic)-Granite/Gneiss, Phyllite/Schist, Quartzite	Fractured/Weathered	0.5–1.0	Dug well	30–40	55,284
Deccan Trap-Basaltic	Vesicles/Fractures/Joints	0.35–0.9	Dug well	15–20	88,771

the Tons catchments, the largest being at Maldeota and Durmala, around Dehradun, where phosphorite is mined economically. Gypsum occurs in the Krol Formation in the form of pockets and bands (Anantharaman and Bahukhandi, 1984; Valdiya, 1980).

The soils in these regions have been grouped as brownhill soils (Devi, 1992), which form from the weathering of granite, gneiss and schist. Soils on the slope (15–40%) are shallow due to erosion and mass wasting processes and usually have very thin surface horizons.

5.5.2 National Capital Region (NCR) Region

The physiography of the NCR region is characterized by the presence of the Ganga skirting it as its eastern boundary, the Yamuna traversing north-south forming the boundary between Uttar Pradesh and Haryana, and the sand dunes and barren low hills of the Aravalli chain and its outcrops in the west, flat topped prominent and precipitous hills of the Aravalli range enclosing fertile valleys and high table lands in the south-west, and the rolling plains dominated by rainfed torrents in the south. The rest of the region is plain with a gentle slope of north-east to south and south-west.

The rock type exposed in the area belongs to Delhi Super-group of Lower Proterozoic age and consists of Quartzite of the Alwar Group, Phyllite and Slate of the Ajabgarh Group. The Quartzites are massive, thickly bedded, hard, and compact, and highly jointed and are intercalated with thin beds of Phyllite and Slates.

The strike of the beds is NNE-SSW and dip westerly at moderate angles. These rocks are mostly covered by quaternary sediments and are exposed in isolated residual and structural hills and pediments.

These hills are exposed in south and south-west of Delhi at Delhi, Gurgaon, Rewari and Alwar. The rocks near Delhi consist of narrow strike-ridges and are moderately folded and they are over-folded in the southeast as a series of isoclines.

The quartzite and its associated rocks of the Delhi series are traversed by joints and are folded, faulted and fractured. The joints persist deeper down often to about 100 m depth, which is the maximum depth to which most of the tube wells to date exist. The joints and fractures are open, often filled with debris and mutually interconnected; cavities created by the erosion of mica-schist add to the secondary porosity developed by these joints and fractures.

Ground water occurs in the consolidated rocks both in confined and semi-confined conditions. Mostly the top water table zone and the deeper semi-confined aquifer have the same static water level surface because of their interconnections. The depth of water table below ground level varies in general between 5 and 20 m. A number of ground water structures by way of open wells and shallow tube wells (60–70 m deep) exist in this formation in different parts of NCR with varying degrees of discharge rates. On the whole, discharges around 2–3 m^3/h and at times even less (down to one cubic meter) or more (up to 4–5 m^3), with an average of about 15–20 m of the aquifer-zone, are available.

The thickness of the alluvium and the proportion of clean granular zones in aggregate within the same are the two main criteria, which determine the availability of ground water in the alluvium. A major portion of NCR is covered by alluvium, which is fairly thick. It is more than 450 m thick in some parts of Haryana in the upper Yamuna basin (Sonepat, Panipat, etc.). In general, the thickness increases as we proceed towards the northern and eastern areas of NCR away from the hard-rock outcrops, with modifications resulting from tectonics or the bedrock topography sub-surface. Depending upon the thickness of the alluvium, one or more aquifer zones have been identified in the alluvium. The occurrence of ground water sub-area wise, under water tabl, and confined to semi-confined conditions has been described as follows.

Phreatic water surface is generally shallow, about 3–5 m below ground level (mbgl). It could be even as low as one meter or so, in the newer alluvium along the present day flood plains or the low level terraces. Such shallow water levels may be encountered even in the older alluvium. Deeper confined aquifers or medium deep semi-confined aquifers have their piezometric surfaces within about 20–25 mbgl in general.

Open wells, shallow tube wells, gravity wells and deep tube wells are abundant in the areas covered by the alluvium. Their discharges vary anywhere from 18 to 25 m^3/h for about 2–3 m drawdowns in the open wells to about 162 per cubic metre per hour for about 8–12 m drawdowns in the deep tube wells tapping granular zones about 70–100 m in aggregate thickness. As far as ground water quality is concerned, there are a few fresh water pockets in the north-east and the south-east corners of NCR area, otherwise in these areas the Total Dissolved Solids (TDS) are more than the desirable limits and the other quality parameters are within the desirable range. The TDS, nitrate and fluorides are more than the desirable limits in the NCT-Delhi area and most parts of the north-west and south-west portions of NCR. However, in the central part of the north-west zone of NCR, fluorides are within the desirable limits.

5.5.3 Ken and Betwa Catchment

The Ken basin covers the areas of Jabalpur, Sagar, Damoh, Panna, Satna, Chhatarpur and Raisen districts of Madhya Pradesh and Hamirpur and Banda districts of Uttar Pradesh. It is bounded by Vindhyan range in the south, Betwa basin on west, free catchment of Yamuna below Ken on east, the Yamuna River on north. The upper reaches of Ken River are flanked by undulating plateau with sandstone, shale and limestone. Down below recent alluvium is experienced in the river up to Daudhan dam site. The stratigraphy of rock formations found in the region is mostly alluvial soil, Deccan traps, Lameta beds and Vindhyan system (Valdiya et al., 1982).

The Betwa basin covers the areas of Bundelkhand uplands, the Malwa plateau and the Vindhyan scrap lands in the districts of Tikamgarh, Sagar, Vidisha, Raisen, Bhopal, Guna, Shivpuri and Chhatarpur of Madhya Pradesh and Hamirpur, Jalaun, Jhansi and Banda districts of Uttar Pradesh.

The following types of hydrogeological formations are found in the Ken and Betwa basins:

The older metamorphies occur in Panna, Chhatarpur and Tikamgarh districts of Madhya Pradesh. Ground water occurs in them only in the weathered mantle and the fractured zone underlying them. The wells are recorded to be generally up to 25–30 m in depth with water levels in the lean part of the year exceeding 10–15 mbgl. The specific capacity of wells in these formations ranges from 20 to 100 lpm/m of drawdown, where the thickness of the aquifer is commendable. The hydraulic conductivity is generally less than 1 m/d and the specific yield is generally less than 5%.

The purana formations of both Vindhyan and Cuddapah age comprise orthoquartzites, limestones and shale sequence are found in parts of Panna, Raisen and Bhopal districts. The wells located in these areas are easily capable of yielding 100–500 m^3/d for a drawdown of 3 m. The specific capacity is in the range from 100 to 300 lpm/m of drawdown, and the hydraulic conductivity varies from 5 to 15 m/d. Similarly, the specific yield is generally in the range of 5 to 15%.

The Deccan traps cover the Guna, Vidisha, Damoh, Sagar, Bhopal and Raisen districts. They are generally simple and a type of flow, where each flow is separable into vesicular and massive units. These flows are generally 10–20 m in thickness, of which 25–40% is generally vesicular. The characteristics of red bole beds generally form the masker horizons between successive flows. The wells of these areas are capable of yielding 250–750 m^3/d for a drawdown of 3 to 6 m. The specific capacity ranges from 50 to 150 lpm/m of drawdown. The hydraulic conductivity ranges from 5 to 15 m/day. The specific yield in the area is generally in the range of 5 to 10%.

5.5.4 Chambal Catchment

Chambal is the principal tributary of the Yamuna and rises in the Vindhyan range just south of Mhow, at an elevation of 354 m. The Vindhyan range traverses nearly the whole width of Peninsular India with a length of about 1,050 km and an average elevation of some 300 m. The Deccan Trap lava flows cover the major areas in the upstream of the Chambal River. Large inliers of old rocks, belonging to the Archaeans, Vindhyan and Gondwanas, outcrop within the Deccan Trap. The Vindhyan Mountain forms the prominent plateau-like range of sandstones to the north of the Narmada valley, particularly Bundelkhand and Malwa (Krishnan, 1982).

A large part of the soils in the alluvial plains of the Chambal River system (where the annual rainfall is <100 cm) is impregnated with alkaline and saline soils (Sarin et al., 1989). The Chambal catchment exhibits a wide range of land-use types, with large areas of intensely farmed agricultural land, urbanized and industrial regions and open uplands (CBPCWP, 1982). The river water is of variable quality, containing effluents from industries and urban conurbations and sediments, nutrients and agrochemicals from the land surfaces (Rangarajan and Sarin, 2004).

References

Anantharaman MS, Bahukhandi PC (1984) A study of the mineral resources of the Mussoorrie Syncline, Garhwal Lesser Himalaya, Current trends in geology. Sediment Geol Himalaya 5:191–196

Dalai TK, Krishnaswami S, Sarin MM (2002) Major ion chemistry in the headwaters of the Yamuna river system: chemical weathering, its temperature dependence and CO_2 consumption in the Himalaya. Geochim Cosmochim Acta 66:3397–3416

Devi L (1992) Climatic characteristics and water balance (A Study of Uttar Pradesh). Concept Publishing Company, New Delhi

Gansser A (1964) Geology of the Himalaya. Interscience, London

Horton RE (1932) Drainage basin characteristics. Am Geophy Un Trans 13:350–361

Horton RE (1945) Erosional development of streams and their drainage basins: Hydrophysical approach to quantitative morphology. Bull Geol Soc Am 56:275–370

Jain V, Sinha R (2003) Derivation of unit hydrograph from GIUH analysis for a Himalayan river. Water Resour Manage 17:355–375

Jain SK, Agarwal PK, Singh VP (2007) Hydrology and water resources of India. Water Science and Technology Library, Springer

Jaireth SK, Gupta SK, Dave VKS (1982) Mineralogy, geochemistry and geothermometry of sulphide mineralization of the Central Himalayan region of Uttarkashi. U.P. J Himalayan Geol 12:280–294

Krishnan MS (1982) Geology of India and Burma, 6th edn. CBS Publishers & Distributors, New Delhi

Negi SS (1991) Himalayan rivers, lakes and glaciers. Indus Publishing Co., New Delhi

Rai RK, Upadhyay Alka, Sarkar S, Upadhyay A, Singh VP (2009) GIUH based transfer function for Gomti river basin of India. J Spat Hydrol 9(2):29–50

Rangarajan R, Sarin MM (2004) Distribution of rare earth elements in the Yamuna and the Chambal rivers, India. Geochem J 38:551–569

Sachan HK, Sharma KK (1993) Genesis of barite mineralization in the Nigthat Formation, Lesser Himalaya, near Dehradun. J Himalayan Geol 4:165–170

Sarin MM, Krishnaswami S, Dilli K, Somayajulu BLK, Moore WS (1989) Major ion chemistry of the Ganga-Brahmaputra river system: Weathering processes and fluxes to the Bay of Bengal. Geochim Cosmochim Acta 53:997–1009

Singh VP (1988a) Hydrologic systems: vol I, rainfall-runoff modeling. Prentice-Hall, Englewood Cliffs, NJ

Singh VP (1988b) Elementary hydrology. Prentice-Hall, Englewood Cliffs, NJ

Singh VP (1989) Hydrologic systems: vol II, watershed modeling. Prentice-Hall, Englewood Cliffs, NJ

Sinha R, Tonton SK, Gibling MR, Bhattacharjee PS, Dasgupta AS (2005) Late quarternary geology and alluvial stratigraphy of the Ganga basin. Himalayan Geol 26(1):223–240

Snyder FF (1938) Synthetic unitgraphs. Trans Am Geophys Union, 19th Annual Meeting, Part 2, p 447

Strahler AN (1957) Quantitative analysis of watershed geomorphology. Trans Am Geophys Union 38:913–920

Taylor AB, Schwartz HE (1952) Unit-hydrograph lag and peak flow related to basin characteristics. Trans Am Geophys Union 33:235–246

Valdiya KS (1980) Geology of the Kumaun Lesser Himalayas. Wadia Institute of Himalayan Geology, Dehradun

Valdiya KS, Bhatia SB, Gaur VK (eds) (1982) Geology of Vindhyanchal. Hindustan Publishing Co, Delhi

Chapter 6
Soil, Landuse and Agriculture

Abstract This chapter deals with soil characteristics, land use and agricultural activities involved in the Yamuna basin. Spatial analysis of the USDA soil group for characterizing soils in the basin is provided. Landuse classification and biodiversity (i.e., flora and fauna) analysis, followed by agricultural practices, are also included. Following the agro-climatic zoning in the basin, a detailed analysis is presented for crop seasons, crop calendar, cropping patterns, cropping intensity, and fertilizer and pesticide application, including the integrated pest management.

This chapter deals with soil characteristics, landuses and agricultural activities involved in the Yamuna basin. A detailed spatial variation of the above features is also included in the chapter. To start with, a detailed analysis of soil and its spatial variability in the basin has been presented in the following section.

6.1 Soil Classification

Based on geography and geology, the Yamuna basin has three distinct regions: (i) the Hilly and Tarai which is part of Uttarakhand, Himachal Pradesh, and Uttar Pradesh lies in the extreme north side of the basin, (ii) the Indo-Gangetic alluvium region i.e., Yamuna-Ganga and Yamuna–Ghaghar doab (Interfluves) which is the part of Uttar Pradesh, Delhi, Hayana and Rajasthan and lies mainly in the central and eastern side of the basin, and (iii) the plateau region consisting of Deccan trap, Vindhyan range, crystalline rocks which are part of mainly Madhya Pradesh, Rajasthan and Uttar Pradesh, and lies in the south western part of the basin. A detailed description of these features is given in Chapter 5. The geological formations play an important role in the soil formation at the upper surface.

On the basis of soil types there are eight types of soils present in the entire Yamuna River basin (Table 6.1). The majority of soil type is alluvial and covers about 42% of the basin area, whereas calcareous sierozemic soils contribute a minimum with about 0.5% of the basin area. Based on the United States Department of Agriculture (USDA, 1999), the soils of the Yamuna basin can be classified into 35 categories. The spatial pattern of the soil category in the basin is shown in Fig. 6.1, whereas a description of each soil categories is presented in Tables 6.2 and 6.3.

R.K. Rai et al., *The Yamuna River Basin*, Water Science and Technology Library 66,
DOI 10.1007/978-94-007-2001-5_6,

Table 6.1 Soil types in the Yamuna River basin

Soil type	% of total basin area covered	Locations
Red Sandy	2.5	Along Uttar Pradesh-Madhya Pradesh border in districts of Jhansi, Hamirpur, Chhattarpur, etc.
Red and Yellow	5.0	Parts of Jaipur, Alwar, Sawai Madhopur, Banda, Panna districts and along western boundary of basin in Rajasthan
Calcareous sierozemic	0.5	Parts of Mohindergarh and Bhiwani districts in Haryana
Deep Black	5.5	On the southern boundary of the basin in Sehore, Bhopal, Raisen, Vidisha, Sagar and Damoh districts
Medium black	25.5	Most of the basin in MP and strips north of river Chambal in Rajasthan
Mixed red and black	15.0	Chittaurgarh, Bhilwara, Banda, Mandasaur, Shivpuri, Lalitpur, Tikamgarh, Panna and Chattarpur districts
Brown hill	4.0	Hills and foothills in the north
Alluvial	42.0	Plains and valley

6.2 Landuse

Major landuse in the basin can be classified as agriculture, forest, non-agriculture, barren land, permanent pasture, cultivable wasteland, cultivable fallow, etc. (Fig. 6.2). Out of total reported area, more than 50% area is under agriculture followed by the forest area. Other categories of landuses have almost equal distribution in the basin. The existing landuse pattern of the districts that fall under the Yamuna basin have been further analysed to arrive at the catchment-wise landuses. The catchment wise statistics of the landuse category is given in Table 6.4. It can be stated that the Chambal basin has a large forest area in the Yamuna basin followed by the Betwa, Ken and Sind catchments. The percentage of cultivable waste and fallow land are also high in these catchments which follow the similar sequence.

6.2.1 Forests and Protected Areas

Forests provide multiple benefits to environment, people, and animals. The variety of cultural values and symbolic functions ascribed to the forests are as numerous and diverse as the communities and cultures of the region. Forests provide important sources of income. India has 76.84 million ha of recorded forest area in March 1999. This accounts for 23.38% of the total geographic area. The per capita availability of forests in India is 0.07 ha, which is much lower than the world average of 0.8 ha (Table 6.5).

Asia has the highest deforestation rate of any major tropical region in the world. In many parts of the world forests are shrinking at an unprecedented rate, due to agricultural expansion, urban development and accelerating demands for fuel and fibre.

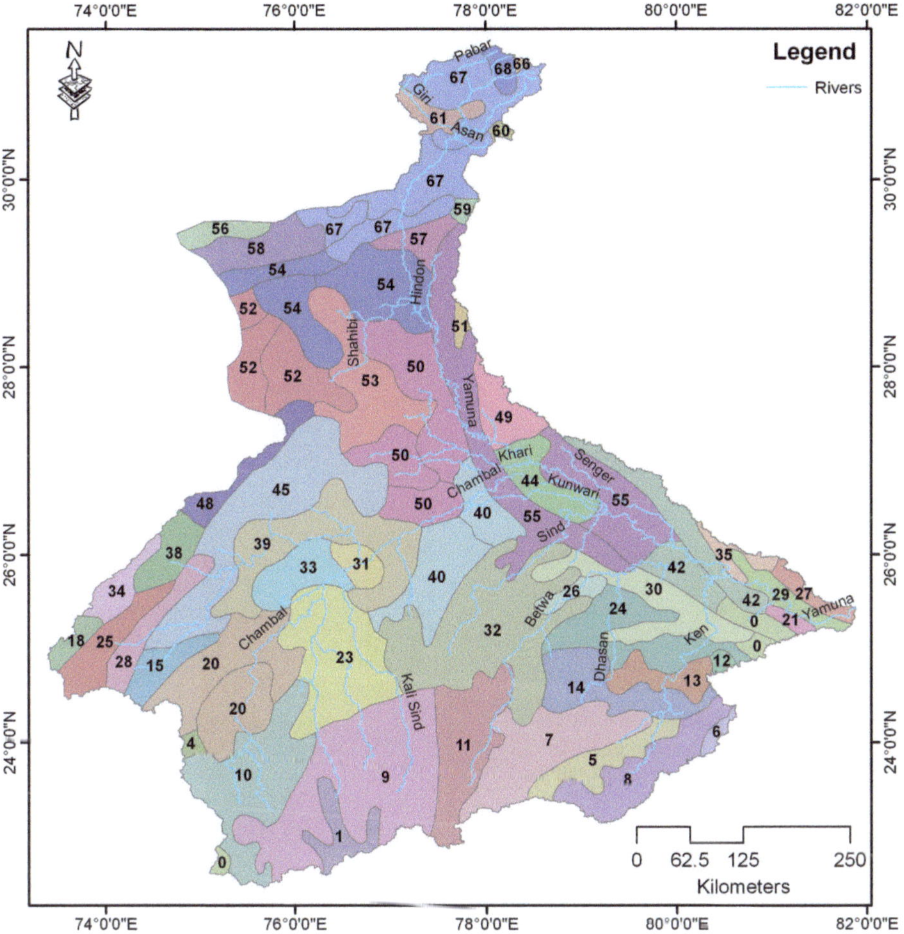

Fig. 6.1 USDA soil classification of Yamuna River basin

Many key forest ecosystems are endangered, with critical consequences for the survival of traditional communities, wildlife, land degradation, soil erosion, flooding, and climatic change.

Forest conservation is one of the most important environmental and social challenges facing the community. Strict forest reserves must be expanded, but the problem can be solved only by implementing sustainable forest management in all forest areas. The Yamuna River basin area also has a rapid increase of population and urbanization. Therefore, care of natural resource and forest is important. State wise we find that conservation steps have been taken by the government for protecting forest lands. Table 6.6 provides state wise forest in Yamuna basin. Tables 6.7 and 6.8 show threatened species and total species in India.

Table 6.2 Description of USDA soils groups

Soil order	Sub-order	Description	Soil order	Sub-order	Description
2	7, 8, 9	Aqualfs + Aquepts + Udalfs	58	1, 2	Orthents + Ochrepts
8	5, 1	Ustalfs + Orthents	59	1, 2, 5	Orthents + Ochrepts + Ustalfs
9	5, 2	Ustalfs + Orchrepts	60	1, 4	Orthents + Rockout Crops
10	5, 6	Ustalfs + Fluvents	64	8, 2	Aquepts + Ochrepts
18	5, 2, 1	Ustalfs + Orchrepts + Orthents	70	2	Ochrepts
20	9	Udalfs	71	2, 14	Ochrepts + Gullied Land
24	9, 2	Udalfs + Ochrepts	72	2, 1	Ochrepts + Orthents
27	10	Orthids	73	2, 12	Ochrepts + Psamments
30	10, 1	Orthids + Orthents	75	2, 1, 4	Ochrepts + Orthents + Rockout Crops
32	10, 6	Orthids + Fluvents	77	2, 8, 5	Ochrepts + Aquepts + Ustalfs
33	11, 6	Aquents + Fluvents	78	2, 3, 5	Ochrepts + Usterts + Ustalfs
35	12	Psamments	79	2, 6, 3	Ochrepts + Fluvents + Ustalfs
36	12, 6	Psamments + Fluvents	86	2, 6, 1	Ochrepts + Fluvents + Orthents
37	12, 6, 10	Psamments + Fluvents + Orthids	92	3	Usterts
38	12, 6, 11	Psamments + Fluvents + Aquents	98	3, 2, 1	Usterts + Ochrepts + Orthents
40	12, 10	Psamments + Orthids	100	3, 2	Usterts + Ochrepts
42	12, 6, 2	Psamments + Fluvents + Ochrepts	103	15	Glaciers and Snowcaps
53	1, 13	Orthents + Argids			

Source: NBSS and LUP

6.2.2 National Parks and Sanctuaries

The National Forest Policy (MoEF, 1988), emphasizes creating massive people's movement through the involvement of village communities living close to the forest in protection and development of forests. Pursuant to this policy, the Government of India issued a notification in June 1990 requesting the State Governments to involve local communities in the management of forests. This led to the development of Joint Forest Management (JFM) programmes. For in situ conservation of bio-diversity of the country, 87 National Parks and 485 Wildlife Sanctuaries have been created so far with a total area of 4.06 million ha and 11.54 million ha, respectively. These together constitute 15.60 million ha and form 4.75% of the geographic area of the country and are referred to as Protected Areas (PA). National Parks and Wildlife sanctuaries in the states falling under the Yamuna basin are shown in Table 6.9.

Table 6.3 USDA soil classification of Yamuna basin

Sub-order No.	Sub-order	Description
1	Orthents	Soil texture is loamy and clayey, surface is erosional, better drained than Aquents, regular decrease in content of organic matter with depth, without podegenic horizons that have retained original structure of parent materials.
2	Ochrepts	Soil organic matter content is low, weakly developed horizons showing alteration of parent materials.
3	Usterts	Soil moisture is limited but present at the time needed by crops, high clay content of cracking nature and dark gray colours, cracks are open for at least 90 cumulative days per year.
4	Rockout Crops	
5	Ustalfs	Soil moisture is limited but present at the time needed by crops, moisture moves through soil to deeper layers only in occasional years, minerals have been strongly weathered, medium to high base status and subsurface horizons of clay accumulation.
6	Fluvents	Soil texture is loamy and clayey (finer in texture than loamy fine sand), soils formed on alluvium, without podegenic horizons that have retained original structure of parent materials.
7	Aqualfs	Wet soils that are seasonally or permanently saturated with moisture which affects plant growth (this may be due to higher water table or low hydraulic conductivity), medium to high base status and subsurface horizons of clay accumulation.
8	Aquepts	Wet soils that are seasonally or permanently saturated with moisture which affects plant growth, usually have cambic horizons, weakly developed horizons showing alteration of parent materials.
9	Udalfs	Moist soils that are not dry for a long time that affect plant growth, the undisturbed soil has a thin horizon darkened by humus, medium to high base status and subsurface horizons of clay accumulation.
10	Orthids	Generally surface is erosional, dry soils with pedogenic horizons, very low organic matter.
11	Aquents	Wet soils that are seasonally or permanently saturated with moisture which affects plant growth, without podegenic horizons that have retained original structure of parent materials.
12	Psamments	Soil texture is loamy fine sand or coarser, subject to the movement by wind if dry, without podegenic horizons that have retained original structure of parent materials.
13	Argids	Usually hot and dry, very low moisture and organic matter.
14	Gullied Land	Gully erosion is prominent, ravine.
15	Glaciers and Snowcaps	Covered with snow.

Source: NBSS and LUP

The periodic assessment of forest cover provides a quantitative measure of the extent of land area under forest/tree cover along with the density. It helps in monitoring the changes in the cover. The forest cover information serves useful purpose for national and state policy planning. The National Forest Policy (MoEF, 1988)

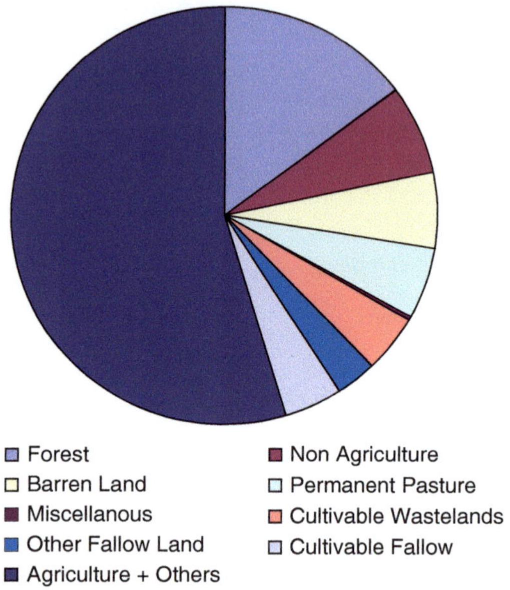

Fig. 6.2 Distribution of landuse categories in the Yamuna basin

envisages bringing one-third of the geographic area of the country under forest/tree cover for maintaining ecological balance and environmental stability.

Following a different classification, the forest cover can be divided into three categories: dense, open and mangroves. The forests are called as dense when the canopy density is 40% and higher; whereas if the canopy density is between 10 and 40%, forests are termed as the Open forests. Besides these two categories, the Mangroves are salt-tolerant forest ecosystems found mainly in the tropical and sub-tropical inter-tidal regions in estuaries and coasts. They are a reservoir of a large number of plants and animal species. Mangroves exhibit a remarkable capacity for salt tolerance, stabilize the shoreline, and act as a barrier against encroachments by the sea. Based on this classification, the Yamuna River basin comprises two classes, namely the dense and opens (Table 6.10).

The forest area can further be classified into reserved and protected forests. The percentage of these two categories in the forest area for the states of Yamuna basin is given in Table 6.11.

The Yamuna basin state has various national parks and sanctuaries some of these are illustrated as follows.

(i) Kalesar National Park and Sanctuary, Yamuna Nagar (100.28 km^2)
(ii) Sultanpur National Park, Gurgaon (1.49 km^2)
(iii) Bhindawas Bird Sanctuary
(iv) Hastinapur Wild life Sanctuary (2,073.00 km^2),
(v) National Chambal Wildlife Sanctuary (635 km^2)
(vi) Okhla Wildlife Sanctuary (4 km^2), Ghaziabad

Table 6.4 Landuse in the Yamuna Basin (in km^2)

Basin/ sub-basin	Shape area	Rural area	Urban area	Reported area	Forest	Non-agric.	Barren	Permanent pasture	Miscellaneous	Cl. wasteland	Other fallow	Cl. fallow
Sahibi	17,561.3	16,962.6	598.63	15,943.3	866.1	1,227.8	715.6	614.7	26.4	208.0	462.0	908.4
Hindon	6,975.74	6,396.5	579.3	7,153.0	491.1	1,067.5	156.7	13.2	50.4	99.0	106.8	116.1
Khri	25,738.0	24,656.0	1,082.0	21,510.4	1,566.1	2,052.5	1,380.2	718.2	34.3	405.2	640.7	1,318.0
Chambal	134,650.4	131,292.6	3,357.9	124,873.6	17,085.8	8,140.4	10,686.7	8,317.4	97.3	7,812.1	4,447.3	5,937.4
Sind	27,940.0	27,188.5	751.5	26,316.2	4,091.8	1,769.4	1,845.4	817.3	74.0	1,281.9	333.5	1,184.5
Betwa	43,637.8	43,429.2	965.3	47,563.9	9,509.6	3,009.0	1,660.8	2,030.3	75.4	2,291.6	883.0	1,251.0
Ken	28,527.1	28,336.6	769.6	30,072.0	7,498.8	1,862.8	797.1	1,285.3	62.4	1,262.5	639.3	829.4
Karwan	4,058.4	4,296.6	187.0	4,587.8	126.2	474.4	95.3	12.7	30.6	112.0	115.5	117.7
Sengar	4,058.4	4,296.6	187.0	4,587.8	126.2	474.4	95.3	12.7	30.6	112.0	115.5	117.7
Yamuna	355,870	356,790	9,309	342,541	50,346	24,180	20,057	18,580	1,039	15,371	9,976	15,082

Table 6.5 Forest cover and per capita availability in different regions/countries

Region/country	Percentage of forest cover to land area (1995)	Per capita forest (ha)
World	26.6	0.64
Asia	16.4	0.1
Africa	17.7	0.7
Europe	41.3	1.3
China	14.3	0.1
Pakistan	2.3	0.01
Bangladesh	7.8	0.02
Japan	66.8	0.2
USA	23.2	0.8
India	15.7	0.06

Source: FAO (1999)

Table 6.6 State wise forest cover in Yamuna basin (Area in km^2)

State/UT	1999 Assessment	1997 Assessment	1995 Assessment	Changes in 1999	Changes in 1997
Delhi	88	26	26	62	0
Haryana	964	604	603	360	1
Himachal Pradesh	13,082	12,521	12,501	561	20
Madhya Pradesh	131,830	131,195	135,164	635	−3,969
Rajasthan	13,871	13,353	13,280	518	73
Uttar Pradesh	34,016	33,994	33,986	22	8

Source: MoEF (1999a)

Table 6.7 Globally threatened animal occurring in India by status category

1994 IUCN Red list threat category					
	Endangered	Vulnerable	Rare	Indeterminate	Insufficiently known
Mammals	13	20	2	5	13
Birds	6	20	25	13	5
Reptiles	6	6	4	5	2
Amphibians	0	0	0	3	0
Fishes	0	0	2	0	0
Invertebrates	1	3	12	2	4
Total	26	49	45	28	24

Source: Groombridge (1993)

(vii) Indira Privadarshini (Asola) Wildlife Sanctuary (13.2 km^2)
(viii) Ranthambore National Park, Sawai Madhopur
(ix) Kumbhalgarh Wildlife Sanctuary, Udaipur
(x) Keoladeo Ghana National Park, Bharatpur
(xi) Sariska Wildlife Sanctuary, Alwar
(xii) Darrah Wildlife Sanctuary, Kota

Table 6.8 Comparison between the number of species in India and the World

	No. of species in India (SI)	No. of species in worlds (SW)	SI/ SW
Mammals	350	4,629	7.6
Birds	1,224	9,702	12.6
Reptile	408	6,550	6.2
Amphibians	197	4,522	4.4
Fishes	2,546	21,730	11.7
Flowering Plants	15,000	250,000	6.0

Table 6.9 National parks and wildlife sanctuaries

	National park		Wildlife sanctuaries		
States	Number	Area (km^2)	Number	Area (km^2)	Total area (km^2)
Delhi	0	0	1	27.60	27.60
Haryana	1	1.43	9	278.32	279.75
Himachal Pradesh	2	1,429.40	32	5,736.85	7,166.25
Madhya Pradesh	11	6,474.69	35	10,704.05	17,178.74
Rajasthan	4	3,856.53	24	5,712.83	9,569.36
Uttar Pradesh	7	5,410.82	29	7,594.54	13,005.36
India	**87**	**40,631.64**	**485**	**115,374.42**	**156,006.06**

Table 6.10 Extent of forest covers in states of Yamuna basin

State/UT	Dense forest, %	Open forest, %	Forest cover, %
Delhi	39.77	60.23	5.93
Haryana	46.58	53.42	2.18
Himachal Pradesh	69.71	30.29	23.50
Madhya Pradesh	61.91	38.09	29.73
Rajasthan	31.06	68.94	4.05
Uttar Pradesh	67.33	32.67	11.55
Total			**19.39**

Source: MoEF (1997)

Table 6.11 Forest area and protected area in states falling under Yamuna basin

States	Reserved forests (%)	Protected forests (%)	Forest to geographic area (%)
Delhi	90.70	8.14	5.73
Haryana	15.88	73.85	3.52
Himachal Pradesh	5.12	89.23	66.52
Madhya Pradesh	61.68	37.37	30.89
Rajasthan	36.78	54.98	9.49
Uttar Pradesh	30.08	60.8	6.98
Uttarakhand	68.74	30.79	54.81
India	51.62	30.78	23.57

Source: MoEF (2003)

Table 6.12 Area under joint forest management

State	No. of JFM committees	Area under JFM ('000 ha)
Haryana	350	60.73
Himachal Pradesh	203	62.00
Madhya Pradesh	12,038	5,800.00
Rajasthan	2,705	235.63
Uttar Pradesh	197	34.59
India	36,130	64

Source: MoEF (1999b, 2002)

The Joint Forest Management Scheme (JFMS) encourages the involvement of village institutions for the forest development. The local institutions engaged in the task are known by different names in different states like Forest Protection Committee (FPC), Village Forest Committee (VFC), *Van Sanrakshan Samiti* (VSS), Village Forest Protection Management Committee (VFPMC) etc. In JFM, the user (local communities) and the owner (Government) manage the resource and share the cost equally. After the National JFM guidelines were issued in 1990, 22 States are now implementing the programme. The JFM programme has evolved to different levels across these States. The area under joint forest management is shown in Table 6.12. Table 6.12 provides information of per capita availability of forests in different categories.

6.2.3 Biodiversity

India is rich in biodiversity as evident that the country is included in the list of 12 mega-diversity nations. Country has a 7% of the world's biodiversity and supports 16 major vegetation types. A wide variety of flora and fauna exists in the Yamuna basin, and their spatial variability largely depends on the geo-physical environment. The state-wise status of the flora and fauna in the Yamuna River basin is given as follows:

6.2.3.1 Haryana

Flora: Haryana forest cover is mostly dry deciduous which comprises pine and thorny shrubs. Chief trees are Kikar *(Acacia nilotica)*, Khair *(Acacia catechu)*, Neem *(Azadirachta indica)*, Shisham or Indian Rosewood *(Dalbergia sissoo)*, Pipal *(Ficus religiosa)*, Mango *(Mangifera indica)*, Jamun *(Syzygium cumini)*, Tamarind *(Tamarindus indica)*, Banyan (*Ficus indicus)*, Sagwan or Teak *(Tectona grandis)*, Ber or Indian Jujube *(Zizyphus mauritiana)*, Mitha Jal or Pillu *(Salvadora indica)*, Khara Jal or Pillu *(Salvador persica)*, Semul, Khejri *(Prosopis cineraria)*, Lahswa (*Cordia dichotoma*), Amla, Dhak (*Butea frondosa)*, Shahtoot or mulberry (*Morus alba atropurpurea)*, eucalyptus, Guava *(Psidium guajava)*, pine and poplar.

The Kalesar forest is full of Sal trees. The Ritha tree can be found in the Morni Hills as well as the Kalesar forest. Another relatively bigger forest is the Saraswati Plantation near Pehowa. An important shrub species of the state is the leafless Kair or Teat *(Capparis decidua)*. Another important plant that has religious and medicinal value, is the *Tulsi*. A beautiful flowering tree is the Kachnar, whose flowers are used for many purposes. Other trees that are found in Haryana are Tun (Red Cedar), Baheda and Bel.

Among small tree species, *Careya arboreal, Holarrhena antidysenterica, Mallotus philippinensis, Murraya exotica, Randia dumetorum, Wrightia tomentosa, Zizyphus mauritiana*, etc. are common.

Shrubs: *Adhatoda sp., Callicarpa macrophylla, Carissa opaca, Clerodendron viscosum, Colebrookia oppositifolia, Euphorbia royleana, Ixora sp., Murraya sp., Woodfordia sr. Zizyphus sp. etc.*

Climbers and Grasses: *Acacia pinnata, Arundo donex, Bauhinia vahlii, Caesalpinia sepiaria, Cenchrus setigerus, Chrysopogon sp., Clematis gouriana, Cymbopogon martini, Oendrocalamus strictus, Oesmostachya bipifJnata, Oioscorea belophylla, Erianthus munja, Heteropogon contortus, Eulolopsis binanta, Ichnocarpus sp., Milletia ovaldolia, Mimosa himalayan, Pueraria tuberosa, Saccharum spontaneum, Smilex sp., Vallaris solanacea, Vetiveria zizanioides* etc.

Ferns Species: *Adiantum lunulatum., Adiantum caudatum, Adiantum cappilisveneris, Athyrium spo, Oryopteris sp., Ophioglossum, Salvia, Azolla, Nephrolepsis, Oyptris, Amphilopteres.*

***Fauna* – Mammals:** Black Buck *(Antelope cervicapra)*, blue bull *(Boselaphus tragocamelus Pallas)*, leopard *(Panthera pardus)*, striped Hyena *(Hyena hyena)*, wild boar *(Sus scrofa)*, mongoose *(Herpestes edwardsi)*, Jackals *(Canis aureus)* and wild dogs *(Cuon alipnus)*, Monkey (*Rhesus macaque*), langur *(Presbytis entellus)*, jungle cat *(Felis chaus)*, the small Indian civet *(Viverricula indica)* and Indian fox are common wild animals. The state has deer like the barking deer *(Muntiacus muntjak)*, Sambar (*Cervus unicolor*), Chital (*Axis axis*) and antelopes like Neelgai (Blue Bull) and black buck. Other common mammal species are gray musk shrew or chachunder *(Suncus murinus)*, common yellow bat, Tickell's bat, five stripped palm squirrel *(Funambulus palmarum)*, Sehi *(Hystrix indica)*, Indian gerbil, common house rat *(Rattus rattus)*, house mouse, Indian Hare (*Lepus nigricollis).*

Birds: Various types of ducks and geese such as *Graylag Goose, Bar-headed Goose, Brahminy Duck, Northern Pintail, Common Shelduck, Mallard, Gadwall, Eurasian Wigeon, Blue winged Teal, Shoveller, Common Pochard, Ferruginous Duck* and *Tufted Duck* can be seen at river and tanks during winter. Some other ducks such as *Comb Duck, Common Teal, Spot-billed Duck, Tree duck* are found throughout the year at suitable habitat. *Dabchick* is also a residential bird. Other Game Birds like *Grey Francolin (Teetar), Black Francolin (Kala Teetar)* and *quails* are also common. *Sand grouse* has been noted as a resident (*Indian Sand grouse*) as well as visitor birds. Common *Green Pigeon (Hariyal)* is found in vicinity of villages chiefly on *Ficus* trees and *Blue Rock Pigeon* occurs in almost all the villages.

Oriental Turtle Dove is a winter visitor, whereas the Laughing Dove is found near cultivated fields.

Among scavenger and predator birds, *Black Kite, Brahmin Kite, White backed Vulture, King Vulture, Tawny Eagle, Greater Spotted Eagle, White Eyed Buzzard Eagle, House Crow* and *Jungle Crow, Vulture, Black-winged Kite, Shikara, Lagger Falcon, Shahin Falcon, Merli, Kestrel, Booted Eagle, Eastern Steppe Eagle, Pale Harrier, Spotted Owlet, Eurasian Eagle* and *Owl.*

The common colorful birds, such as *Indian Roller (Neelkanth), Northern Green Barbet, Coppersmith Barbet, Northern Golden backed Woodpecker, Eurasian Golden Oriole, Alexandrine Parakeet, Rose-ringed Parakeet, Pied Crested Cuckoo, Asian Koel, Kingfishers* such as *Common Kingfisher, White breasted Kingfisher, Indian Pied Kingfisher, Red vented Bulbul, Lal munia (Red Avadat), Purple sunbird, Indian Spotted Munia, White Storks, storks, cranes, White Ibis* and *Crested Bunting* etc., along with National Bird of India, the *Blue Peafowl,* add beauty to the varied wildlife of the state.

Snakes: The poisonous snakes, like common Indian Krait, Russell's Viper, Phoosra and other snakes, like *Blind Snake, Indian Python, John's sand boa, wolf snake, rat snake* and *sand snake* are found in Haryana.

Fish: Feather back fish *Parri, Katla, Mrigal, Chunni, Bata, Siriha, Ghally, Mallee*, and the snake-head fish, *Dolla* and *Curd.*

6.2.3.2 Rajasthan

***Flora*:** Forests are mostly confined to the east of the Aravali range. Vegetation in the desert region is limited to very slow growing stunted trees, thorny shrubs and some grasses. The other natural vegetation type in Rajasthan is ephemeral, occurring only during the monsoon season.

Kejri (*Prosopis cineraria*), ker, akaro (*Calotropis precera*), thor (*Euphorbia caduca*), bordi (*Zizyphus nummularia*), babul (*Acacia nilotica*), anwal (*Cassia aureculata*), Dhak (*Butea monosperma*), peepal (*Ficus religiosa*), banyan (*Ficus benghalensis*), ber (*Zizyphus mauritiana*) and khajur (*Phoenix sylvestris*). salar (*Bowellia seriata*), bamboo (*Dendor calamus strictus*), dhok (*Anogeisrus pendula*), mango (*Mangifera indiaca*)and jamun (*Syzygium cumini*), guggal (*Commiphora*), brahmi (*Bacopa monnieri*), shatawari (*Asparagus*) and adusa (*Adathoda vassica*) are common species.

Among perennial grasses sewan (*Lasiurus sindicus*), dhaman (*Cenhrus ciliaris*), boor (*Cenchrus jwarancusa*) and bharut (*Cenchrus catharficus*) are common.

***Fauna*:** Mammal species in wild life include Black buck (kala hiran), Indian gazelle (chinkara), blue bull (Nilgai), four horned antelope (*Chau singha*), sambar, spotted deer (chital), monkeys (rhesus macaque), langur, Indian tiger (bagh), tigers, panthar (baghera), jungle cat, caracal (*Syiagosh*), jackal, wolf, desert fox, wild boar, sloth bear, common mangoose and smaller Indian mangoose . Other fauna species include Indian python, Indian chameleon and the garden lizard, crocodile and ghariyal. Bharatpur has almost 375 bird species and of great interest is the world's tallest black necked stork standing up to 1.8 m tall and its black and white wings span up to 2.5 m. The rare Indian bustard and the grey partridge are the birds of open scrub forests of Rajasthan.

6.2.3.3 Uttar Pradesh and Uttarakhand

***Flora*:** The plains of Uttar Pradesh have been very rich in natural vegetation which has, however, diminished due to wide-ranging needs of the people. Only a few patches of natural forest are now found scattered here and there in the plains, while such forests are extensively found on a very large scale in sub-mountain and mountain regions in Uttrakhand.

Tropical Moist Deciduous Forests are found in the moist region of Terai. They grow in regions which record 100–150 cm of rainfall annually, have an average temperature between 26 and 27°C and have considerable degree of humidity. A special feature of forests is that deciduous trees of uneven size grow on higher altitude regions. Lower regions have several species interspersed with Bamboo, Climbers, Cand and ever green shrubs. Main trees are Sal, Ber, Gular, Jhingal, Palas, Mahua Semal, Dhak, Amla, Jamun, etc.

Tropical Dry Deciduous Forests are found in all parts of the plains, and usually in central eastern and western regions. The trees are mostly deciduous. Since sunlight reaches the ground in abundance, shrubs and grasses also grow here. Large tracts of these forests have been cleared for cultivation. Important trees are Sal, Palas, Amaltas, Bel, Anjeer etc. Neem, Peepal, Sheesham. Mango, Jamun, Babool, Tamarind, etc. grows along river banks and in other moist regions.

Tropical Thorny Forests are mostly found in south-western parts of the State. Such forests are confined to the areas which have low annual rainfall (50–70 cm), mean annual temperature between 25 and 27°C and low humidity (less than 47%). Widely scattered thorny trees, mainly, Babool, Thorny, legumes and Euphorbias are extensively found here. During rains, short grasses also grow here. The trees are generally small here forming open dry forests. Important trees of the region are Phulai, Khair, Kokke, Dhaman, Danjha, Neem, etc. Various types of resin and gum are obtained from these trees.

***Fauna*:** The population of the wild animals includes various species of animals including *Antelope, Sambhar, Cheetal, Blue bull, Leopard, Hyena, Wild cat*, Shrew, Porcupine, Squirrel, Hare, Mongoose, Cow, Buffalo and Mouse. Other common species found here are Tiger, Panther, Snow Leopard, Sambhar, Cheetal, Kastura, Chinkara, Black Deer, Nilgai, Back-brown Bear, Mountain Goat, Hyena, Hill Dog, Elephant etc.

Among the birds, Fowl, Pheasant, Partridge, Florican, Duck, Goose and Wader are common.

Fish species commonly found are Mahaser, Hilsa, Saul, Tengan, Parthan, Rasela, Vittal, Rohu, Mirgal, Kata, Labi, Mangur, Cuchia, Eel, Einghi, Mirror Carp, and Trout.

Reptiles: Bamania, Pit-viper, Lizard, Goh, Cobra, Tortoise, Krait, Dhaman and Crocodile. *Birds*: Cheel, Vulture, Peacock, Nightingale, Pigeon, Parrot, Owl, Nilkanth and Sparrow.

6.3 Agriculture in the Basin

Agriculture is the dominant sector in India and it constitutes a significant part of the national economy. It contributes 23% of Gross Domestic Product (GDP) and employs about 64% of the workforce. Of course, in recent times, the share of agriculture in the economy is declining and will continue in the near future. Even then agriculture will have important bearings on India's economy, as it helps feed a growing population, employs a large labour force, and provides raw material to agro-based industries. In recent years, it has been confirmed that India has largely been unaffected by the world-wide recession because of its agriculture based economy. India has the potential to produce crops in almost all climatic classes namely, dry, arid, semiarid, sub-humid and humid.

To start with, The Yamuna basin has been analysed with the agro-climatic zones followed by crop seasons, cropping pattern and production.

6.3.1 Agro-climatic Zones

India has a large number of complex agro-climatic conditions. The Planning Commission has demarcated the geographical area of India into 15 agro-climatic regions for resource development. The basis of the zonalization is soil type, climate including temperature and rainfall and its variability, and water resources availability. These are further divided into 72 more homogeneous sub-zones. The 15 agro-climatic zones are given as follows:

1. Western Himalayan Region: J&K, HP, UP, Uttarakhand
2. Eastern Himalayan Region: Assam Sikkim, W. Bengal and all North-Eastern states
3. Lower Gangetic Plains Region: W. Bengal
4. Middle Gangetic Plains Region: UP, Bihar
5. Upper Gangetic Plains Region: UP
6. Trans-Gangetic Plains Region: Punjab, Haryana, Delhi and Rajasthan
7. Eastern Plateau and Hills Region: Maharashtra, UP, Orissa and W. Bengal
8. Central Plateau and Hills Region: MP, Rajasthan, UP
9. Western Plateau and Hills Region: Maharashtra, MP and Rajasthan
10. Southern Plateau and Hills Region: AP, Karnataka, Tamil Nadu
11. East Coast Plains and Hills Region: Orissa, AP, TN, and Pondicherry
12. West Coast Plains and Ghat Region: TN, Kerala, Goa, Karnataka, Maharashtra
13. Gujarat Plains and Hills Region: Gujarat
14. Western Dry Region: Rajasthan
15. The Islands Region: Andaman and Nicobar, Lakshadweep

The Yamuna river basin comprises mainly five agro climatic zones (Fig. 6.3): (i) Western Himalayan Region: Himachal Pradesh, Uttar Pradesh and Uttarakhand; (ii) Upper Gangetic Plains Region: Uttar Pradesh; (iii) Trans-Gangetic Plains Region: Haryana, Delhi and Rajasthan; (iv) Central Plateau and Hills Region:

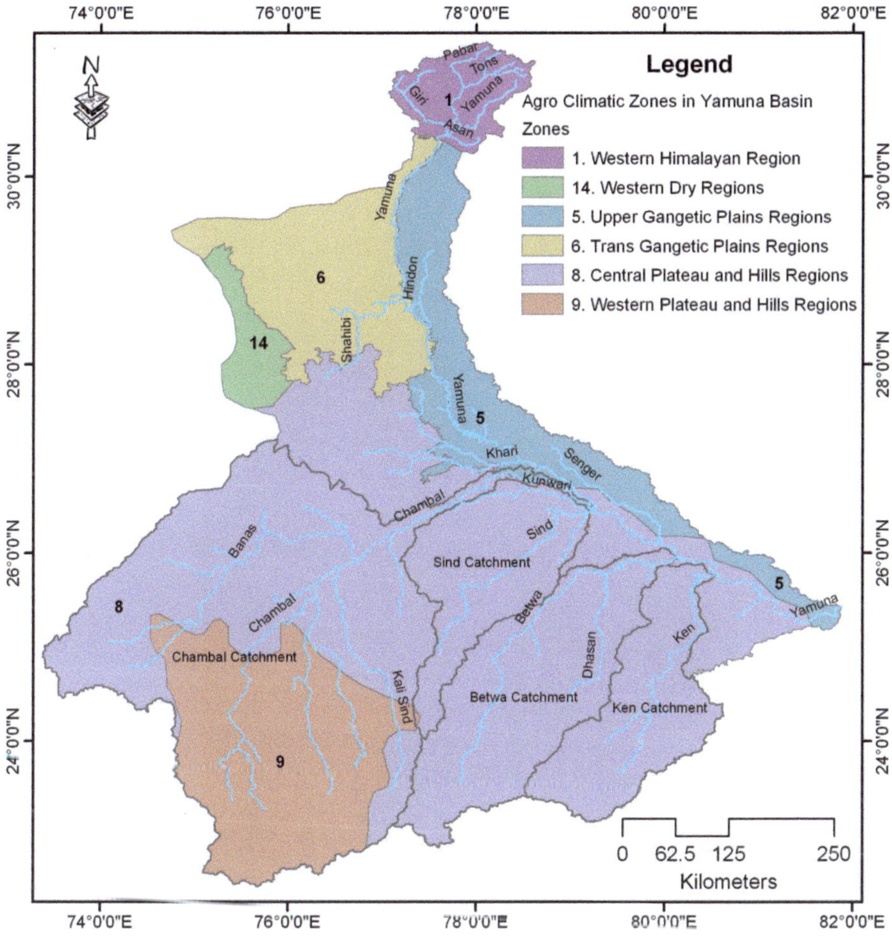

Fig. 6.3 Agro-climatic zone of Yamuna basin

Madhya Pradesh, Rajasthan, and Uttar Pradesh; and (v) Western Plateau and Hills Region: Madhya Pradesh and Rajasthan. However, a major part of the basin area falls under zone 8 (i.e. Central Plateau and Hills Region: MP, Rajasthan, UP). State wise agriculture status is discussed in the following section.

Considering agricultural planning in India, Chowdhury et al. (1993) demarcated another type of 15 classes, which is summarized in Table 6.13. Based on this category the Yamuna River basin comprises three categories: The Satluj-Yamuna Alluvial Plains, the Central Semi-Arid Zone, and Lower Ganga basin.

6.3.2 Crop Seasons

In the basin, three distinct crop seasons can be identified, viz., *Kharif* (July-October), *Rabi* (October-March) and *Zaid* (April-June) and majority of agricultural land are sown more than once in a single calendar year, and therefore

Table 6.13 Agro-climatic zones of India (Chowdhury et al., 1993)

S. No.	Name	Area coverage	Rainfall	Major soil type	Major crops
1	Northern most zone	Jammu and Kashmir	50–75 cm	Podsolic type	Maize and Paddy
2	Eastern Himalayan zone	Areas of eastern Himalayas	About 250 cm	Alluvial and deep black soils	Paddy
3	Western Himalayan zone	Hilly region of western Himalayas (outside Ladakh) and its foothills	Much diversity from Eastern Himalayan zone	Brown hilly and alluvial soils	Paddy
4	Lower Assam Hills	Tripura, Mizoram, Manipur	About 250 cm	–	Paddy
5	Sutlej, Yamuna Alluvial plains	Haryana, Punjab, adjoining areas of Jammu and Kashmir and Uttar Pradesh	Low rainfall zone	Alluvial and calcareous soil	Paddy, cotton
6	The arid west zone	Areas of Thar Desert, Western Rajasthan and adjoining areas	Low rainfall (20–30 cm annually)	Sandy soil	Jowar, Bajra
7	The Central semi-arid zone	Eastern Rajasthan, West Madhya Pradesh and adjoining areas of Uttar Pradesh and Gujarat	Over 75 cm	Medium black to red black soil	Sorghum, Jowar, Maize
8	Lower Ganga basin	Ganga basin in Uttar Pradesh, hills and plains of east Madhya Pradesh	100–150 cm	Medium black to red black soil	Paddy
9	Eastern Ganga and Mahanadi basin	Gangetic plains in Bihar and West Bengal, The Bihar Plateau and Mahanadi basin	Moderate rainfall	Rich alluvial intercepted by red, yellow and lateritic soils	Paddy and Jute
10	Gujarat and neighbourhood	Parts of Gujarat and Maharashtra	Low rainfall (50–70 cm)	Sandy loam to loamy soil	Groundnut, Jowar, Bajra and cotton
11	The northern and central Western Ghats	Coastal areas of Maharashtra and Karnataka	Heavy rainfall	Acidic soil (pH $\leq$3)	Paddy
12	Central highlands	Leeward side of Western Ghats	Drought prone area	Medium black in north and red sandy in south	Jowar, cotton, ground nut
13	Godavari basin	Andhra Pradesh	High rainfall	Sandy coastal alluvial to red soil	Paddy ground nut, Jowar
14	Lower Western Ghats	Kerala, adjoining hills of Tamil Nadu and Karnataka	High rainfall	–	Topioca and Paddy
15	Semi arid Tamil Nadu tract	Areas of Tamil Nadu	High rainfall	Red sandy to deltaic alluvial soils	Jowar, ground nut

the cropping intensity of the basin is more than 100%. A detailed description of cropping intensity is given in the subsequent section. Based on crop seasons, the cropping pattern in the Yamuna basin is very diverse because of large variation in the hydro-climatic, soil and topographic conditions.

6.3.2.1 Cropping Pattern

A diverse cropping system exists in the Yamuna basin. Cropping pattern depends on climate, soil, overall agro-ecological setting and economic conditions. At farmers' level, potential productivity and monetary benefits act as guiding principles, while opting for a particular crop/cropping system. These decisions with respect to choice of crops and cropping systems are further narrowed down under the influence of several other forces related to infrastructure facilities, socio-economic factors and technological developments, all operating interactively at micro-level. Table 6.14 shows details of crops according to cropping season in the basin. However, in most parts of the basin *Kharif* and *Rabi* are the two cropping season. Crops can also be categorized on the basis of life cycle, such as annual biennials and perennial (Table 6.14).

Due to different climatic and soil characteristics, and water availability in the basin, a wide variation in the cropping pattern exists in Yamuna basin. The sub-basin wise cropping pattern is presented in Table 6.15.

6.3.2.2 State Wise Agro-ecosystem

State wise agro ecosystem and cropping pattern (1998–1999) with percentage of cultivated area is shown in Table 6.16 based on which it may be remarked that crop diversity is low in the states of Yamuna basin (MoAg, 2000). Lower the crop

Table 6.14 Crop type according to season and life cycle

S. No.	Season	Crop
1	Kharif	Rice, Bajra, Maize, Cotton, Ground nut, Urd, Moong, Lobia, Sorghum, Sweet potato, Til, Guar, Jute, Sanai, Arhar, Dhaincha, Sugarcane, Soyabean, Lady's finger
2	Rabi	Wheat, Mustard, Lentil, Gram, Joe, Potato, Peas, Barseem, Rijka, Tobacco, Lahi, Jai
3	Zaid	Pumpkin, water melon, melon, bottle gourd, Chili, Tomato, Sunflower, Kheera, Kakri, Mung, Sunflower
Life cycle		
1	Annual	Rice, Wheat, Gram, Dhaincha, Bajra, Moong, Cotton, Ground nut, Mustard, Potato, Sweet Potato, Bottle gourd, Pumpkin, Soyabean
2	Biennials	Beet root, Onion
3	Perennial	Napier grass, Rijka, fruit crops

Source: http://www.krishisewa.com/articles/crops.html

Table 6.15 Cropping pattern in the Yamuna Basin (ha)

Basin/ Sub-Basin	Paddy	Jowar	Maize	Bajra	Arhar	Wheat	Barley	Gram	Potato	Sugarcane	Groundnut	Mustard	Soybean	Cotton	Fruit	Veg
Sahibi	128,363	12,203	1,678	78,880	7,264	406,702	29,434	33,690	442	14,001	25,246	317,486	7	55,681	2,600	17,355
Hindon	80,039	3,455	28,546	8,048	7,395	247,228	6,243	305	6,833	190,080	1,646	7,970	4	258	21,134	17,545
Khri	42,968	6,046	4,738	47,588	8,693	508,938	48,227	11,297	32,476	7,540	27,834	544,117	361	15,285	4,969	57,197
Chambal	67,128	498	35,090	27,770	60,133	1,225,290	66,759	304,471	40,487	10,556	124,862	1,034,715	1,853,005	114,305	15,021	90,032
Sind	58,492	78	5,165	258	59,156	440,795	10,468	86,689	6,776	6,598	42,638	205,271	165,451	1	958	16,130
Betwa	23,337	49	2,117	306	145,474	692,390	19,000	271,202	4,825	9,320	67,692	77,099	538,850	70	1,116	26,197
Ken	64,360	6	22	0	53,377	357,176	5,163	190,438	5,403	17,775	7,596	19,992	167,414	1	2,352	25,503
Karwan	37,127	259	33,652	6,554	7,084	216,751	16,493	1,921	14,267	27,060	120	37,950	0	681	8,343	21,741
Sengar	63,421	246	24,634	3,065	11,554	255,487	18,839	3,284	23,153	6,755	222	51,596	25	600	5,787	31,063
Yamuna	1,352,681	101,460	190,169	246,619	351,229	5,601,043	396,779	869,132	156,664	575,661	239,483	2,593,843	2,579,040	611,744	125,104	362,503

Table 6.16 State-wise agro-ecosystems cropping pattern for 1998–1999 (% of the total cultivated area of the country)

Climatic Zones	State	Crops												
		Rice	Wheat	Jowar	Bajra	Maize	Other coarse cereals	Gram	Tur	Other Pulses	Groundnut	Rapeseed and Mustard	Other Oilseeds	Sugarcane
Arid and Semi Arid	Haryana	2.4	8	Neg	6.6	Neg	2.7	4.3	1.2	1.8	Neg	7.4	2.1	3.2
Humid	HP	Neg	1.4	Neg	Neg	5.1	1.2	Neg	Neg	Neg	Neg	Neg	Neg	Neg
Semi Arid	MP	11.9	17	8.1	1.5	14	9.4	31.7	11.5	21.2	3.4	10.5	23.1	1.2
Arid and Semi Arid	Rajasthan	Neg	10.1	5.4	44.9	15.6	19.9	33.5	Neg	19.5	4.3	45.5	16.1	0.5
Semi Arid and Sub Humid	UP	13.3	33.7	3.3	8.5	14.8	8.9	9.8	12.1	11.4	1.6	5.2	6	48.3
	All India-acreage	44.6	27.4	9.98	9.28	29.54	29.54	8.41	3.47	23.82	7.57	6.6	26.71	4.08

Neg. = negligible; acreage = million hectare
Source: Agricultural Statistics at a Glance (2000)

diversity, lesser will be the crop and economic sustainability of the farmers. For example, Himachal Pradesh and Haryana have low crop diversity followed by Uttar Pradesh, Rajasthan and Madhya Pradesh. Haryana has a high dependency on wheat, mustard and sugarcane. Rajasthan is more dependent on bajra, gram and mustard. Uttar Pradesh has a high dependency on sugarcane, wheat, maize and rice.

The crop time is dependent on temperature, light, and growing environment. The crop calendar is intended as a general reference for growing plants common to any area. It does not include all crops that can be grown, nor does it represent the only way those crops listed can be grown. Table 6.17 provides sowing time of different cereal crops. Sowing time for oilseeds, vegetable crops and pulses has been shown in Tables 6.18, 6.19 and 6.20, respectively (ICAR, 2007).

Besides the above descriptions, a few crops can be sown in more than one season. For example, Soybean and Sunflower crops which can be sown during autumn (Zaid Kharif) as well as spring (Zaid Rabi), though the main crop is spring season.

Table 6.17 Sowing time of cereal crops

S. No	Crop	Sowing time	Reaping time (Transplanting)
1	Wheat	10 Nov–10 Dec	–
2	Barley (Joe)	Irrigated Area: 15 Nov–15 Dec Un-irrigated Area: 20 Oct–7 Nov	–
3	Maize		–
4	Pearl Millet (Bajra)	May–June–July: Northern Plains Apr–May: North West Hilly areas March: North East Hilly areas	July
5	Paddy	June–July	July–August
6	Sorghum (Jowar)	15 June–15 July	–

Source: ICAR (2007)

Table 6.18 Sowing time for oil seeds

S. No.	Crop	Sowing time
1	Indian mustard/ Brown mustard/ Laha	30 Sept–15 Oct
2	Toria/ Lahi	1 Sept–15 Sept
3	Yellow sarson, Brown sarson	25 Sept–15 Oct
4	Taramira/ Rocket salad	Oct
5	Sesamum/ Sesame	Jun–July (Kharif)
6	Ground nut	Mid Jun–July
7	Linseed	Oct–Nov
8	Soybean	Mid Jun–Mid July
9	Castor	Mid June– Mid July
10	Sunflower	Spring: 20 Feb–10 Mar Kharif: July–Aug Rabi: Nov

Source: ICAR (2007)

Table 6.19 Sowing time of vegetable crops

S. No.	Crop	Sowing time	Transplanting time	Harvest period
1	Amaranthus	Feb–July	–	April–Oct
2	Beat Root	Oct–Nov	–	Dec–Feb
3	Bitter Gourd	Feb–March	–	May–July
		June–July		Aug–Oct
4	Bottle Gourd	Feb–March	–	April–June
		June–July		Oct–Dec
5	Brinjal	Jan–Feb	Feb–Mar	April–June
		May–June	June–July	Sept–Nov
		Oct–Nov	Jan	March–May
6	Cabbage	Sept–Oct	Oct–Nov.	Dec–March
7	Capsicum	Nov–Jan	Jan–Feb	April–May
		June–July	July–Aug	Sept–Oct
8	Carrot	Aug–Oct	–	Dec–March
9	Cauliflower – Early	Early June	July	Nov.
10	Cauliflower – Mid season	July–Sept	Aug–Oct	Nov–Jan
11	Cauliflower – Late	Sept–Oct	Oct–Nov	Jan–March
12	Chillies	Nov–Jan	Jan–March	April–June
		May–June	June–July	Sept–Nov
13	Cluster Bean	Feb–March	–	April–June
		June–July		Aug–Oct
14	Cowpea	June–July	–	Aug–Oct
		Feb–March		April–June
15	Cucumber	Feb–March	–	May–July
		June–July		Aug–Oct
16	Dolichos Bean	June–July	–	Oct–Dec
17	Fenu Greek	Sept–Nov		Nov–Feb
18	French Bean	Feb–March	–	April–May
19	Knol-Knol	Sept–Oct.	Oct–Nov.	Dec–Feb
20	Lettuce	Sept–Oct	Oct–Nov	Dec–Feb
21	Muskmelon	Jan–Feb	–	April–June
22	Okra	Feb–March	–	March–June
		June–July		Aug–Nov
23	Onion	Oct–Nov	Dec–Jan	April–June
		May–June	June–July	Oct–Nov
24	Peas-Early	Sept–Oct	–	Nov–Jan
	Peas	Oct–Nov	–	Jan–March
25	Potato	Sept–Nov	–	Jan–March
		Dec–Feb	– /td>	March–April
26	Radish	April–Aug.	–	May–Sept
		Sept–Oct		Nov–Jan
		Nov–Jan		Dec–March
27	Palak	Sept–Nov	–	Nov–Feb
		Feb		March–April
28	Sponge Gourd	Feb–March	–	April–June
29	Ridge Gourd	June–July	–	Aug–Oct
30	Round melon	Feb–March	–	May–June
		June–July		Sept–Oct
31	Tomato	June–Aug	Aug–Sept	Oct–Dec
		Nov–Dec	Dec–Feb	April–June
32	Turnip	Oct–Nov	–	Dec–March
33	Watermelon	Jan–March	–	May–June

Source: ICAR (2007)

Table 6.20 Sowing time of pulses

S. No.	Crop	Sowing time
1	Gram/Chickpea	Unirrigated-15–20 Oct Irrigated–15 Nov
2	Pea	Mid Oct–Mid Nov
3	Lentil	Mid Oct–Mid Nov
4	Moong bean	Summer: Mid April Kharif: July
5	Urd bean	Summer: Mar- April Kharif: July
6	Pigeon pea/Red gram/Tur	Spring: Feb Kharif: Jun-July
7	Cowpea	Spring: Feb last Summer: Mid April Kharif: July

Source: ICAR (2007)

Similarly, maize is the Kharif season crop but can be sown during the spring season, if soil moisture condition is good. Fruit crops, such as Banana and Guava, can be grown as multiple season crops (Kharif and Rabi).

6.3.2.3 Cropping Intensity

Cropping intensity can be defined as number of times a land is cultivated within the single crop calendar year. For example, if a piece of land is sown two times in a single crop calendar year, the cropping intensity will be 200%. To reflect this feature the term cropping intensity is used as an indicator. The spatial variation in the cropping intensity of the basin is derived from the district wise average value of cropping intensity and is shown in Fig. 6.4. It is evident from Fig. 6.4 that the number of crops taken from a particular land per calendar year is more either along the river or in the irrigation command area. It might be due to the availability of surface water through the Yamuna and Ganga canal system. The cropping intensity in the extreme North-West area is very high (i.e. of the order of 200%) and it is due to the support of Bhakhra canal system (i.e. SYL and Narwana Canals). The average cropping intensity of the Yamuna basin is of the order of 141%. The catchment-wise cropping intensity is summarized in Table 6.21.

6.4 Irrigation

Irrigation plays a major role in the integrated water resources management as a majority of water resources is used for the agricultural irrigation. Irrigation is the replacement or supplementation of rainfall with water from other sources in order

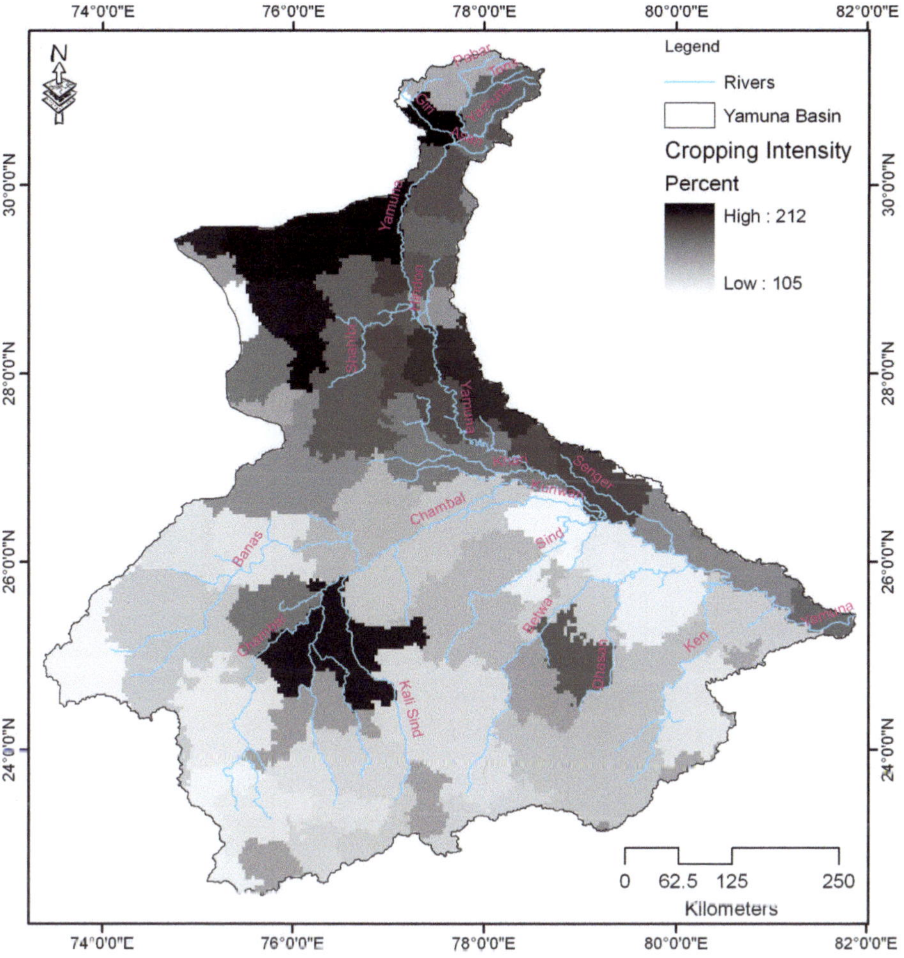

Fig. 6.4 Spatial variation of cropping intensity in the Yamuna basin

to grow crops. In contrast, agriculture that relies only on direct rainfall is sometimes referred to as dry land farming. Irrigation demand is either fulfilled by surface water or groundwater. Irrigation is available from various sources and canals are the major water suppliers from large and small rivers. Dams, barrages, and anicuts are constructed on the rivers to regulate the flow of rivers and large storage reservoirs are developed to use water at times of deficiency. The storage in different basins varies according to its flow. The storage capacity in the Ganga basin is the largest due to its vast size. Major irrigation projects use surface water through canals. The percentage of the net irrigated area to the cultivable area is maximum in the alluvial plais of Ganga and Trans Yamuna plains. Uttar Pradesh ranks highest in area irrigated by canal followed by Andhra Pradesh, Punjab and Haryana. Minor irrigation schemes

Table 6.21 Agriculture and irrigation sources in the Yamuna basin (All the units in ha)

Catchment	Net sown area	Area sown more than once	Grass area	Net irrigated area (NIA)	NIA by canal	NIA by tube well	NIA by other sources	Cropping intensity (%)
Sahibi	1,091,428	625,734	1,717,161	768,541	163,584	604,543	414	164
Hindon	505,223	268,794	774,016	451,932	83,396	360,982	7,554	153
Khri	1,339,520	631,598	1,971,118	1,027,059	102,467	918,136	6,456	148
Chambal	6,234,917	1,423,958	7,658,875	2,775,368	422,908	2,113,435	239,026	125
Sind	1,491,835	328,336	1,820,171	733,527	248,823	421,145	63,560	123
Betwa	2,685,310	630,452	3,315,761	1,212,924	274,703	734,686	203,535	126
Ken	1,583,438	404,388	1,987,825	664,013	100,532	444,868	118,614	127
Karwan	350,343	2,00,049	550,392	316,077	58,049	253,185	4,843	157
Sengar	413,643	239,900	653,543	360,722	93,560	266,397	765	158
Yamuna	18,790,861	6,609,467	25,400,328	10,517,828	2,856,924	7,040,012	620,893	141

Cropping Intensity = 100×[(Net Sown Area + Area Sown More than Once)/Net Sown Area] include groundwater development through the construction of dugwells, deep tube wells, and private shallow tubewells and lift irrigation projects.

The Yamuna River basin can be further classified based on the canal command area into four categories: arable land within command area, arable land outside the command area, water bodies and non-arable land (Fig. 6.5). For crop irrigation different sources are being practiced in the basin. These sources can be categorized as canal, tube well, and others (i.e. tanks, ponds, etc.). Approximately 67% area is irrigated through groundwater resources. Only 27% agriculture water demand is met by canal systems and the rest 6% area is irrigated through tanks, ponds, etc. Details of areas irrigated by different sources are given in Table 6.21. However, the state wise irrigation sources in the Yamuna basin is given in Table 6.22.

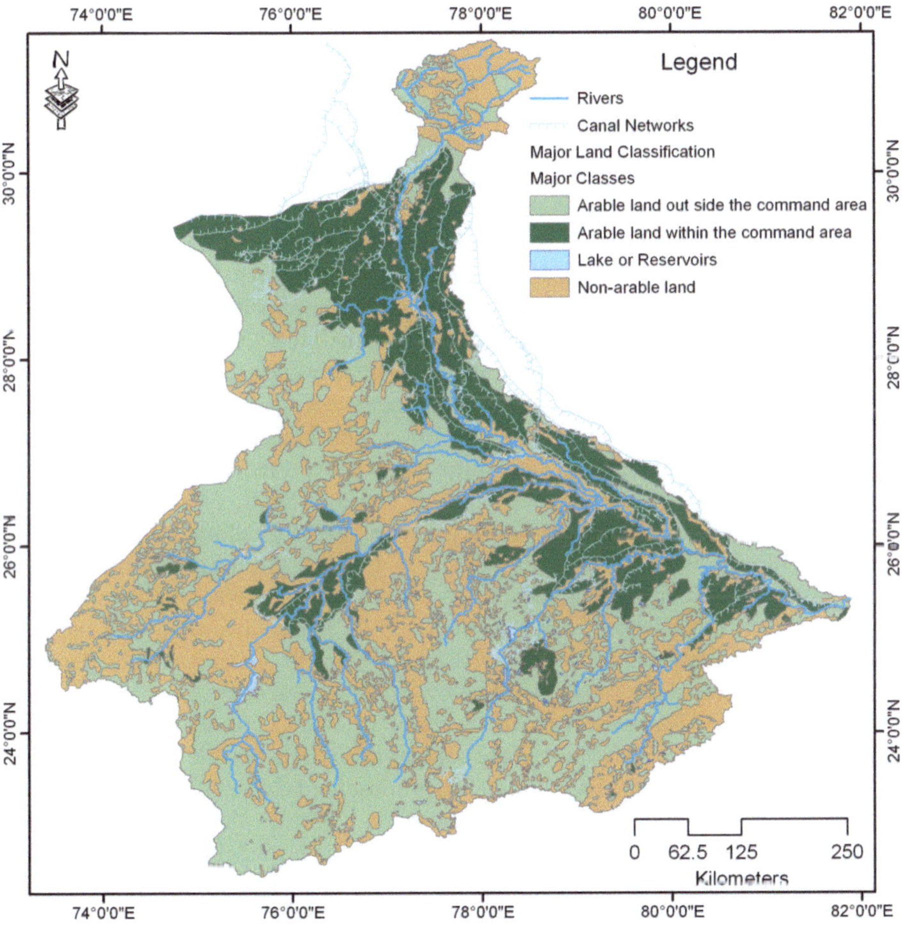

Fig. 6.5 Land classification based on canal command area in Yamuna basin

Table 6.22 State-wise irrigation system in the Yamuna basin

State	Area (1,000 ha)	NSA (1,000 ha)	Total IA (1,000 ha)	% IA	Irrigated area by source (%) Canal	Tank	TW	Well	Other sources
Haryana	4,421	3,696	2,350	63.7	53.5		45.7	0.5	0.3
Himachal Pradesh	5,567	571	93	16.3	2.1	1.2	6.4		90.3
Madhya Pradesh	44,345	19,027	2,656	13.9	43.7	7.0	2.0	40.3	7.0
Rajasthan	34,224	15,660	3,218	20.5	31.8	4.4	5.2	56.9	1.7
Uttar Pradesh[a]		17,226	9,884	57.4	33.7	1.7	54.4	7.0	3.2
Delhi	148	65	52	80	7.7	1.9	84.7	1.9	3.8

[a] Uttar Pradesh including Uttarakhand

6.5 State-Wise Agriculture

6.5.1 Uttar Pradesh

The economy of Uttar Pradesh is predominantly agrarian and the performance of agriculture and allied activities, such as horticulture, animal husbandry, dairying and fisheries, are critical in determining the growth rate of the State. The primary sector (inclusive of mining) contributes 36.8% to the State's income in 2003–2004 and provides employment to 66% of total workers. However, the share of this sector in State income has been progressively declining.

Out of an estimated number of 22.15 million rural households in Uttar Pradesh, 77.4% are farmer households. According to data released by National Sample Survey Organization (NSSO) based on 59th round of NSS, 24% of UP farmers (27% at all India level) de not like farming and feel that agriculture is not profitable. In all 41% farmers in UP (40% at all India level) feel that given a choice, they would take up some other career. This indicates a serious problem, wherein the main protagonist are suffering from low self-esteem and do not believe that what they are doing is worthwhile economically or even socially. In the social hierarchy, farming as a profession now figures considerably low in the social order.

***Indebtedness of Farmers*:** The data released by NSSO provides useful insights regarding indebtedness of farmers in the country as a whole as well as Uttar Pradesh. In Uttar Pradesh, out of 17.16 million farmer households, 6.92 million (40.3%) are reported to be indebted while for the country as a whole, 48.6% (43.42 million) of 89.35 million farmer households were reported to be indebted. The estimated prevalence of indebtedness among farmer households is highest in Andhra Pradesh (82%) followed by Tamil Nadu (74.5%) and Punjab (65.4%). In UP, households with one hectare or less land account for 74% of all farmer households and about 39% of them are indebted. Tables 6.23 and 6.24 show the agricultural production and productivity in Uttar Pradesh as whole. Agricultural production has increased from

Table 6.23 Agricultural production in Uttar Pradesh (000 tonnes)

S.No.	Crop	Year									
		1995–1996	1996–1997	1997–1998	1998–1999	1999–2000	2000–2001	2001–2002	2002–2003	2003–2004	2004–2005
1	Paddy	9,788	11,197	11,678	10,826	12,633	11,679	12,856	9,596	13,022	9,559
2	Total Kharif foodgrains	12,967	14,374	15,082	13,212	15,681	14,998	15,877	12,003	15,996	12,498
3	Wheat	21,077	23,287	22,147	22,781	25,551	25,168	25,498	23,748	25,567	22,514
4	Total Rabi foodgrains	23,705	26,321	24,939	25,612	28,580	27,777	28,310	26,370	28,442	253,05
5	Total Food grains	36,672	40,695	40,021	38,824	44,261	42,775	44,187	38,373	44,438	37,803
6	Oilseeds	1,389	1,520	984	1,070	1,268	1,145	1,110	851	928	946
7	Pulses	2,163	2,591	2,282	2,308	2,551	2,160	2,377	2,182	2,380	2,366

Table 6.24 Crop productivity in Uttar Pradesh (Quintal/ha)

S. No.	Crop	Year									
		1995–1996	1996–1997	1997–1998	1998–1999	1999–2000	2000–2001	2001–2002	2002–2003	2003–2004	2004–2005
1	Rice	18.54	22.21	21.46	19.42	21.85	19.77	21.17	18.4	21.87	17.9
2	Total Kharif foodgrains	16.06	17.74	18.29	16.04	18.73	17.53	18.25	15.7	18.27	16.01
3	Wheat	24.69	27	25.25	25.51	28.03	27.24	27.55	25.9	27.94	25.02
4	Total Rabi foodgrains	21.02	23.39	21.99	22.15	24.55	23.59	24.14	22.8	24.75	22.05
5	Total foodgrains	18.95	21.02	20.43	19.61	22.12	21.04	21.63	20	21.95	19.6
6	Oilseeds	8.67	8.98	6.08	7.01	8.74	8.25	8.69	7.72	8.19	8.4
7	Pulses	7.74	9.32	8.27	8.29	9.57	8.03	8.86	8.26	8.9	8.5

1995 to 2005 for crops except paddy and oilseeds. Productivity has also declined from 2003–2004 to 2004–2005, except for pulses and oilseeds as it is a drought year.

***Irrigation in Uttar Pradesh*:** In Uttar Pradesh, the proportion of net irrigated area to net cropped area has increased from 36.12% in 1966–1967 to 54.89% in 1980–1981 and 76.30% in 2001–2002. However, the net irrigated area by private sources such as private tube well/ pump sets constitute nearly 87.09 lakh hectares, which is 68% of the net irrigated area. There are 35 lakh private tube wells/pump sets in Uttar Pradesh of which about 28 lakh are diesel operated and 7 lakh electricity operated. Canal irrigation provides irrigation facilities to only 27.19 lakh hectares, which is about 21% of net irrigated area. 28,123 State Tube wells account for just 8% of the net irrigated area.

The area irrigated by canals has been continually declining over the years. Further, the water use efficiency in most irrigation systems is low in the range of 30–40% as against an ideal value of 60%. There has also been steady decline in the water table due to over-exploitation of ground water and insufficient recharge from rainwater. Declining trend of groundwater has been observed in 559 of 819 blocks in the State. Between 1987 and 2000, the percentage of groundwater used for irrigation has gone up from 30 to 57.31% for the State as a whole. It is also noteworthy that against created irrigation potential of 31,063.73 thousand hectares, the actual utilization of potential was only 22,404.48 thousand hectares in 2002–2003. This lag of nearly 90 lakh hectares is a cause of serious concern. It could be due to several reasons, such as non-construction of on-farm development works below the outlet, change in cropping patterns to more water intensive crops, loss in live storage due to sedimentation, low water use efficiency due to disrepair of the system etc. Heavy load of silt in river waters and thereby excessive siltation of the canals has reduced the carrying capacity of the canal system and hence there is a decline of water use at tail end areas of canals. Further, the irrigation potential created can be described as "protective irrigation" rather than "assured irrigation". The difference in operating costs of canal irrigation and tube well based irrigation is in the ratio of 1:7 and rising cost of diesel is placing a heavy burden on small and marginal farmers. Consequently, farmers provide only that much irrigation as required to save the crop rather than meet the full requirement of water for the crop.

6.5.2 Madhya Pradesh

Madhya Pradesh has about 73% rural population depending largely on agriculture. Approximately 70% of the geographical area is under cultivation. Of the total gross cropped area cereals covers 42%, pulses 21% and oilseed 21%, commercial (cotton and sugarcane etc.) 3% and the remaining is covered by other crops e.g. vegetables, fruit, fodder and medicinal crops. Of the total gross cropped area 59% is sown in Kharif season, 41% in Rabi season and multiple cropped area is about 24%. The major agriculture and horticulture crops grown in Madhya Pradesh are summarized in Table 6.25.

Table 6.25 Major agriculture and horticulture crops of Madhya Pradesh

S. No	Crop type	Crop
1	Cereals	Wheat, sorghum (jowar), maize and paddy
2	Oilseeds	Soya bean, mustard, linseed
3	Pulses	Gram, pigeon pea (tur), lentil (masoor)
4	Vegetables	Green peas, cauliflower, okra, tomato, onion, potato
5	Fruits	Mango, guava, orange, papaya, banana
6	Spices	Chilies, garlic, coriander, ginger, turmeric

The state share (2004–2005) in the total national production of oilseeds is 18.8% (soybean 50%), pulses 25% (gram 44%) and the state is ranked Number 01 in the country in production of oilseeds and pulses. Madhya Pradesh also leads in spices production with the largest production of garlic, accounting for 37% of the total national production and is second largest producer of coriander in the country.

Among vegetables, Malwa potato has gained reputation for processing into potato chips. The state's share in the total national production of pea is 15%.

***Irrigation in Madhya Pradesh*:** There are several small and large dams constructed in the Chambal, Sind, Ken and Betwa sub-basins of Madhya Pradesh that have increased the irrigation potential of the state. The Rajghat irrigation project is a major interstate development with water from the Rajghat Dam on the Betwa River being shared equally between Uttar Pradesh and Madhya Pradesh. The water allocated to Madhya Pradesh is planned to irrigate 121,450 ha in the districts of Guna, Shivpuri, Tikamgarh, Bhind, Gwalior and Datia.

6.5.3 Rajasthan

Seventy six percent (76%) of the population of Rajasthan reside in the rural area, and their economy is agriculture based. In Rajasthan, cereal crops, such as bajra, juar, wheat and barley cover the largest cultivated area. Salient features of Rajasthan as whole for agriculture is given in Table 6.26. The main crops of the Rajasthan are as follows:

(i) Bajra (*Pennisetum typhoideum*) is consumed by the rural poor, particularly the nomads. Rajasthan is the largest producer of bajra in India.
(ii) Juar (*Sorghum vulgare*) is an important crop during the monsoon.
(iii) Gram (*Cicer arietinum*) is another major pulse crop grown in rabi.
(iv) Wheat (*Triticum sp.*) is cultivated on the irrigated land.
(v) Barley (*Hordeum vulgare*) is the second largest crop in Rajasthan.
(vi) Maize (*Zea mays*) is a stable crop for the Bhil tribes in the Aravallis.
(vii) In northern Rajasthan, maize is a delicacy eaten with butter and the green leaf of the mustard plant.
(viii) Groundnut is a major kharif oilseed crop.

Table 6.26 Salient features of Rajasthan agriculture

S. No.	Item	Year	Unit	
1	Average size of Operational holding	2000–2001	ha	3.65
2	Gross irrigated area to gross cropped area	2007–2008	%	36.42
3	Net irrigated area to net sown area	2007–2008	%	37.69
4	Net area sown to total reporting area	2007–2008	%	49.89
5	Consumption of fertilizers per hectare of cropped area	2006–2007	kg/ha	49.45(F)
6	Average yield per hectare (F)			
	(i) Food grains	2007–2008	kg	1,177
	(ii) Oil seeds	2007–2008	kg	1,045
	(iii) Sugarcane	2007–2008	kg	57,089
	(iv) Cotton (Lint)	2007–2008	kg	397

(ix) Sesame Rajasthan provides the second highest quantum of sesame (*Sesamum indicum*) in India. One of its common uses is as an ingredient in sweetmeats.
(x) Cotton (*Gossypium sp.*) used to make the famous Rajasthan textiles, is a major cash crop.

Irrigation in Rajasthan**:** The Indira Gandhi or Rajasthan canal provides irrigation to the arid western districts of Bikaner and Jaisalmer. Ganganagar district is irrigated with water from the Ganga canal in the Punjab. Irrigation projects have also developed on the Chambal and Luni rivers. The western region grows predominantly kharif (monsoon) crops, while the eastern belt, which has better rains and soil, grows both kharif and rabi (winter) crops. The latter is grown under rain-fed farming conditions or in irrigated areas.

In the central and eastern region of Rajasthan, the Persian wheel method is popular where the ground water table is comparatively high. One Persian wheel can irrigate up to one hectare of land. For groundwater abstraction through pumps, nearly 80% villages have electricity supply. The canal density in the Rajasthan is quite low. Canal command area in the Chambal basin is due to the canal networks of Kota barrage and Bisalpur dam. However, a few parts of the Shahibi and Khari river catchment are partly fed by the Agra canal.

6.5.4 Delhi

Important crops grown in the union territory are wheat, gram, Bajra and Jowar. The important sources of irrigation are tube wells, wells and canals. On the floodplain of Yamuna, agriculture is based on excessive use of groundwater, agrochemicals and

Table 6.27 Irrigated area in Delhi (ha)

Source of Irrigation	1994–1995	1995–1996	1996–1997	1997–1998	1998–1999
Canals	2,304	2,425	2,102	3,179	2911
Tanks	–	–	–	–	–
Wells	36,600	42,306	34,698	37,988	36,159
(a)Tubewells	35,887	41,055	33,481	34,214	34,972
(b)Other Sources	713	1,251	1,217	3,774	1,187
Net Area Irrigated	38,904	44,731	36,800	41,165	39,070
Area Irrigated under more than one crop	11,292	2,724	428	2,836	8,968
Gross Area Irrigated	50,196	47,455	37,228	44,001	48,038

Source: Directorate of Economics and Statistics (2000) Delhi

Table 6.28 Salient features of agriculture in Delhi

Year	Net sown area (ha)	Gross cropped area (ha)	Cropping Intensity
1994–1995	47,409	68,613	145
1999–2000	41,386	60,886	147
2005–2006	23,809	36,041	161

organic manures, which is not sustainable. The revival of earlier water bodies along with creation of new water bodies to "harvest" floodwaters during the monsoon, and the accompanying change in land use from agriculture to semi intensive aquaculture and pasture greater economic returns per unit of land area can be achieved. Tables 6.27 and 6.28 show trends in agriculture in Delhi. It was observed that the net sown area in Delhi has reduced though the cropping intensity has been increased. This trend shows the completely unsustainable use of natural resources.

6.5.5 Haryana

Due to adequate irrigation facilities in state, intensive agriculture is being practiced. Considering agricultural statistics of Haryana, it is evident that the cropping intensity is close to 200% (Table 6.29). The percentage net irrigated area is 82.3%. Haryana contributes 69.11 lakh tone of wheat to central pool. This is mainly due to high infrastructure developed for irrigation.

Haryana state has a good canal density, and most of the cultivated areas are well connected with canal system fed by either the Western Yamuna Canal or Bhakra canal. Apart from this source, groundwater exploitation in the state is very high causing most of the area under dark with respect to the stage ground water development percentage.

Table 6.29 Agricultural statistics of Haryana

Item	
Cultivable area ("000" ha)	3,809
Percent Cultivable area	86.2
Net area sown ("000" ha)	3,566
Percent Net area sown	93.6
Area sown more than once ("000" ha)	2,938
Gross cropped area ("000" ha)	6,504
Cropping Intensity	182.39
Net area irrigated (Total) ("000" ha)	2,936
• By canal	1,331 (45.3%)
• Tube well	1,591 (54.2%)
• Others	14 (0.5%)
Gross irrigated area ("000" ha)	5,446
Intensity of Irrigation	185.49
Percent net irrigated sown area	82.3
Fertilizer consumption (kg/ha.) (2007–2008)	209 (NP ratio 3.2:1)
Foodgrains Prod. (Lakh Tonnes) (2008–2009) Kharif- 44.88 and Rabi- 116.78)	161.66
Contribution of Haryana to central pool of Wheat. (lakh tonnes) 2008–2009	69.11

6.6 Fertilizer and Pesticides Application

The key role of fertilizers and their judicious use in cropping is well understood, when one is familiar with the general facts about plant nutrition. In one study in New Delhi 28 kg N, 4.4 kg P, 41 kg K, 4.9 kg S, 5.9 kg Ca, 4.2 kg Mg, 400 g Fe, 100 g each of Mn and Zn and 30 g Cu are removed from the soil for each tonne of wheat grain produced. Similarly, from the International Rice Research Institute, Philippines it is reported that 10–31 kg N, 1–5 kg P and 8–35 kg K were removed from the soil per tonne of rice grain produced. When crop yields of 5–8 t grain ha^{-1} are taken as with the high yielding varieties of crops, it is not possible for most soils to supply the needed amounts of plant nutrients and that is why fertilizers are needed.

Such heavy removal of plant nutrients from soil leads to the depletion of soil fertility, which shows up in the crop yield decline and lowered factor productivity. For example, in a long-term study at four research centers rice yields have declined by 32% in plots receiving nitrogen and phosphate fertilization; the decline in unfertilized plots has been 57%.

From the dry land rain fed areas where yields are low and very little fertilizer applied, a negative balance between crop removal and plant nutrient application of 6.37 million tonnes yr^{-1} ($N + P_2O_5 + K_2O$) in India has been reported. Thus, contrary to lowering the rates of fertilizer application, Indian farmers may have to think of fertilizing for crop production as well as for soil fertility resilience. Thus, fertilizer has been and will continue to be the main input in India's achieving self sufficiency in food grain production.

Among the seven states of the Yamuna basin, Uttar Pradesh and Haryana are the big users of fertilizer.

***Pesticide*:** The use of synthetic pesticides in agriculture is world-wide still the most widespread method for pest control. Environmental and human health problems related to the use of synthetic pesticides have created an increasing pressure against their use. Therefore, non chemical alternatives for pest control have been developed and modern pesticides have become safer and more specific. Technical developments of the application equipment have kept pace, ensuring an adequate application of these modern products. However, those technical developments regarding modern application equipment and their professional use have not been transferred satisfactorily to field practice. In order to improve these situations, the Agricultural Engineering Service of FAO has initiated a programme to create awareness for the issue on a government level, introduce adequate practical training programmes for farmers and equipment operators and improve equipment quality through the introduction of standards and regular tests.

***Potential Adverse Effects of Pesticides*:** Despite their many advantages, there are some potential hazards or risks when using farm chemicals. These risks may be associated with all chemicals whether they be industrial chemicals, pesticides, household products or even natural chemicals found in the environment. Undesirable side effects of farm chemical use usually stem from a lack of understanding of the impact of chemicals on the environment, compounded by indiscriminate and overuse of the product. These side effects do not always occur when farm chemicals are used and damage does not necessarily result. Some of these effects may be:

- Reduction of beneficial species. Non-target organisms, including predators and parasites of pests, can also be affected by chemical application. The reduction of these beneficial organisms can result in changes in the natural biological balances. Losses of honeybees and other pollinating insects can also be a problem.
- Drift of sprays and vapour during application can cause severe damage and residue problems in crops, livestock, waterways and the general environment. Care in the methods of application and the weather conditions under which it is carried out can reduce drift. Environmental pollution from careless application and runoff can result in wildlife and fish losses. This should be a concern for all of us.
- Residues in food for humans and feed for livestock can be a consequence of direct application of a chemical to the food source, by the presence of pollutants in the environment or by transfer and biomagnifications of the chemical along a food chain. Not all residues are undesirable although good agricultural practice must be observed to prevent unnecessary and excessive levels of residues.
- Ground water contamination by leached chemicals can occur in high use areas if persistent products are used.

- Resistance to the pesticide used can develop in target pests due to the overuse and incorrect use of the chemical.
- Poisoning hazards and other health effects to operators can occur through excessive exposure if safe handling procedures are not followed and protective clothing not worn. Poisoning risks depend on dose, toxicity, duration of exposure and sensitivity.
- Other possible health effects due to indiscriminate use of farm chemicals also concern many people in the community.

The Indian pesticide industry with 85,000 MT of production during year 2007 is ranked second in Asia (behind China) and twelfth globally. The per hectare consumption of pesticide is low in India at 381 grams when compared to the world average of 500 g. Low consumption can be attributed to fragmented land holdings, low level of irrigation, dependence on monsoons, low awareness among farmers about the benefits of usage of pesticides, etc. India, being a tropical country, the consumption pattern is also more skewed towards insecticides which accounted for 64% of the total pesticide consumption in year 2007. Table 6.30 shows state-wise distribution points in the Yamuna basin for pesticides. Consumption in all the concerned states has reduced from year 2002–2003 to 2006–2007 except Uttar Pradesh and Uttarakhand (Table 6.31).

Table 6.30 Distribution points for pesticides (Year 2007)

State	State department	Cooperative	Other institutions	Private	Total
Haryana	–	870	–	6,080	6,950
Himachal Pradesh	570	59	–	942	1,571
Madhya Pradesh	–	503	–	1,557	2,060
Rajasthan	–	951	–	7,337	8,288
Uttar Pradesh	868	443	520	11,857	13,688
Uttarakhand	267	177	–	963	1,407
Delhi	14	08	–	175	197

Source: PMU (2007–2008)

Table 6.31 Consumption of pesticides

States	2002–2003	2003–2004	2004–2005	2005–2006	2006–2007
Haryana	5,012	4,730	4,520	4,560	4,600
Himachal Pradesh	380	360	310	300	290
Madhya Pradesh	1,026	662	749	787	879
Rajasthan	3,200	2,303	1,628	1,008	1,523.20
Uttar Pradesh	6,775	6,710	6,855	6,672	7,022
Uttarakhand	129	147	132	141	139
Delhi	60	56	53	39	57

6.7 Practice of IPM and IPNM

Site-specific nutrient management considers indigenous nutrient supply of the soil and productivity targets capable of sustained high yields on one hand, and assured restoration of soil fertility on the other hand. With this approach, the present food grain production could be achieved from half of the presently irrigated area. Practice of Integrated Pest Management (IPM) and Integrated Pest & Nutrient Management (IPNM) is very little in the basin, still traditional practices for fertilizer and pesticides consumption are on in the entire area.

***Integrated Pest and Nutrient Management (IPNM)*:** The basic purpose of inducting this component has been to popularize and practice the balanced use of fertilizers based on soil testing. In general farmers have been using more and more nitrogenous and phosphatic fertilizers with less emphasis on the use of potash fertilizers and other micronutrients.

***Integrated Pest Management (IPM)*:** Integrated Pest Management, or IPM, "is a sustainable approach to managing pests by combining biological, cultural, physical and chemical tools in a way that minimizes economic, health, and environmental risks" (National Coalition on Integrated Pest Management). Integrated Pest Management strategies have been used for many years in agriculture.

In technical terms, Integrated Pest Management (IPM) is the coordinated use of pest and environmental information with available pest control methods to prevent unacceptable levels of pest damage by the most economical means and with the least possible hazard to people, property, and the environment.

***Integrated Nutrient Management (INM)*:** The main objective of Integrated Nutrient Management (INM) is to ensure adequate availability of quality fertilizers to farmers through periodic demand assessment and timely supply, promoting integrated nutrient management, which is soil test based, judicious and balanced use of chemical fertilizers in conjunction with organic manures and bio-fertilizers, promotion of organic farming and ensuring quality control of fertilizers through implementation of Fertilizer (Control) Order, 1985.

India is the third largest producer and consumer of fertilisers in the world after China and the USA. Against 18.4 million tonnes of fertiliser nutrients (Nitrogen, Phosphorous and Potassium, NPK) consumed during 2004–2005, the nutrient consumption is estimated at 20.2 million tonnes during 2005–2006. The consumption of major fertilisers, namely, Urea, DAP, MOP, SSP and Complexes are estimated to be 22.2, 6.8, 2.6, 2.8 and 6.7 million tonnes, respectively, during 2005–2006. India is, by and large, self sufficient in respect of Urea and about 90% in case of DAP.

The all-India average fertiliser consumption is 96.4 kg/ha of NPK nutrients through there is a wide variation from state to state varying from 197 kg/ha in Punjab, 164 kg/ha in Haryana to less than 10 kg/ha in States like Arunachal Pradesh, Nagaland, Sikkim, etc. Considering the skewed pattern of fertiliser use, Government of India is promoting balanced and integrated use of fertiliser nutrients through various initiatives. As a result, NPK consumption ratio has now improved to 5.7: 2.2: 1 during 2004–2005 from 7: 2.7: 1 during 2000–2001.

***Promotion of Bio-fertiliser*:** With a view to promote the use of bio-fertilisers as an environment-friendly and cheaper source of plant nutrients, Government of India had earlier taken up a "National Project on Development and Use of Bio fertilisers" during the Sixth Plan and was continued till Ninth Plan.

***Organic Farming*:** Organic farming is becoming increasingly popular worldwide and global demand for organic products is increasing rapidly. Organic farming is not only expected to increase the income of farmers but also helps in moving towards sustainable agriculture and environmentally friendly and healthy food. Many farmers are practicing organic cultivation but not in an organized manner properly. Hence, they need proper scientific knowledge, management and certification system.

For promotion of organic farming, it is advocated that an area and cluster based approach be adopted by the state governments. This will help in concentrating activities of target groups and also help in creation of awareness and educating the farmers easily, making a quick assessment of the demand and arranging the supplies etc. There is a need to identify and appoint adequate committed service providers who will act as a technology transfer vehicle to farmers and will serve as a link between certification agencies and farmers.

References

Agriculture Informatics Division, National Informatics Centre, Ministry Of Communication & IT, Government of India, New Delhi

Chowdhury A, Das HP, Singh SS (1993) Agro-climatic classification in India. Mausam 44(1):53–60

Directorate of Economics and Statistics (2000) Department of Agriculture and Cooperation, Ministry of Agriculture, Government of India

FAO (1999) State of World's Forests. FAO Documentation Group Cataloguing in Publication Data, FAO, Rome (Italy)

Groombridge B (1993) The 1994 IUCN Red List of threatened animals. IUCN, Gland, Switzerland

ICAR (2007) Handbook of agriculture. Indian Council of Agricultural Research, New Delhi, India

MoAg (2000) Agricultural Statistics at Glance, Directorate of Economics and Statistics, Department of Agriculture & Cooperation, Ministry of Agriculture, Government of India, New Delhi

MoEF (1988) National forest policy: 1988. Ministry of Environment and Forests, Government of India

MoEF (1997) State of forest report: 1997. Forest Survey of India, Ministry of Environment and Forests, Dehradun, India

MoEF (1999a) State of forest report: 1999. Forest Survey of India, Ministry of Environment and Forests, Dehradun, India

MoEF (1999b) Joint forest management: a decade of partnership. JFM Cell. Ministry of Environment and Forests, Government of India

MoEF (2002) Joint forest management: a decade of partnership. JFM Cell. Ministry of Environment and Forests, Government of India

MoEF (2003) State of forest report: 2003. Forest Survey of India, Ministry of Environment and Forests, Dehradun, India

PMU (2007–2008) Zonal conference of inputs Kharif, Pesticide Monitoring Unit, Department of Chemical and Petro-Chemicals, Ministry of Chemicals and Fertilizers, Government of India

USDA (1999) Soil taxonomy: a basic system of soil classification for making and interpreting soil surveys. Agriculture Handbook No. 436, United States Department of Agriculture

Chapter 7
Socio-economic Status

Abstract This chapter addresses major issues of socio-economic status of the Yamuna River basin, including demography, population growth, urbanization, migration, sex ratio, literacy, classification of workers, agricultural labor and farmers, and population below poverty line. Various health indicators are discussed to assess the health status in the basin. For overall spatial socio-economic status in the basin, various indicators, such as human development index and technical efficiency, are described. Other important parameters, such as livelihood and employment, livestock population, water supply, and water auditing, are also covered.

The World Summit on Sustainable Development (WSSD, 2002) recommended that integrated water resources management and water efficiency planning should be an essential element in all national or regional development strategies. The success of Integrated Water Resources Management (IWRM) and planning depends on institutional and human resources capacity, awareness and leadership.

Monitoring at all levels is important to ascertain progress in meeting the IWRM targets and to alert agencies and organizations capable of providing support to those areas lagging behind and in need of assistance.

The disparity between the areas that are making progress and those that are left behind in the process of implementing IWRM is increasing. Most monitoring efforts to date have tended to rely upon largely subjective assessments and have focused more on planning than implementation aspects. IWRM planning requires a variety of socio-economic data of the basin. Some indicative parameters in the Yamuna basin are elaborated in this chapter.

The Yamuna River basin covers seven states partly or completely, viz. Uttarakhand, Himachal Pradesh, Haryana, Uttar Pradesh, Delhi, Rajasthan and Madhya Pradesh. The basin area comprises 89 districts from these states. The administrative boundary in the basin is shown in Fig. 7.1.

7.1 Demography

Demography results can be used in many different ways, ranging from very general to very specific technical applications. From a strictly statistical viewpoint, the census data represents one of the most important components of the information system and can serve as a basis for many other statistical activities related to food,

R.K. Rai et al., *The Yamuna River Basin*, Water Science and Technology Library 66, DOI 10.1007/978-94-007-2001-5_7, © Springer Science+Business Media B.V. 2012

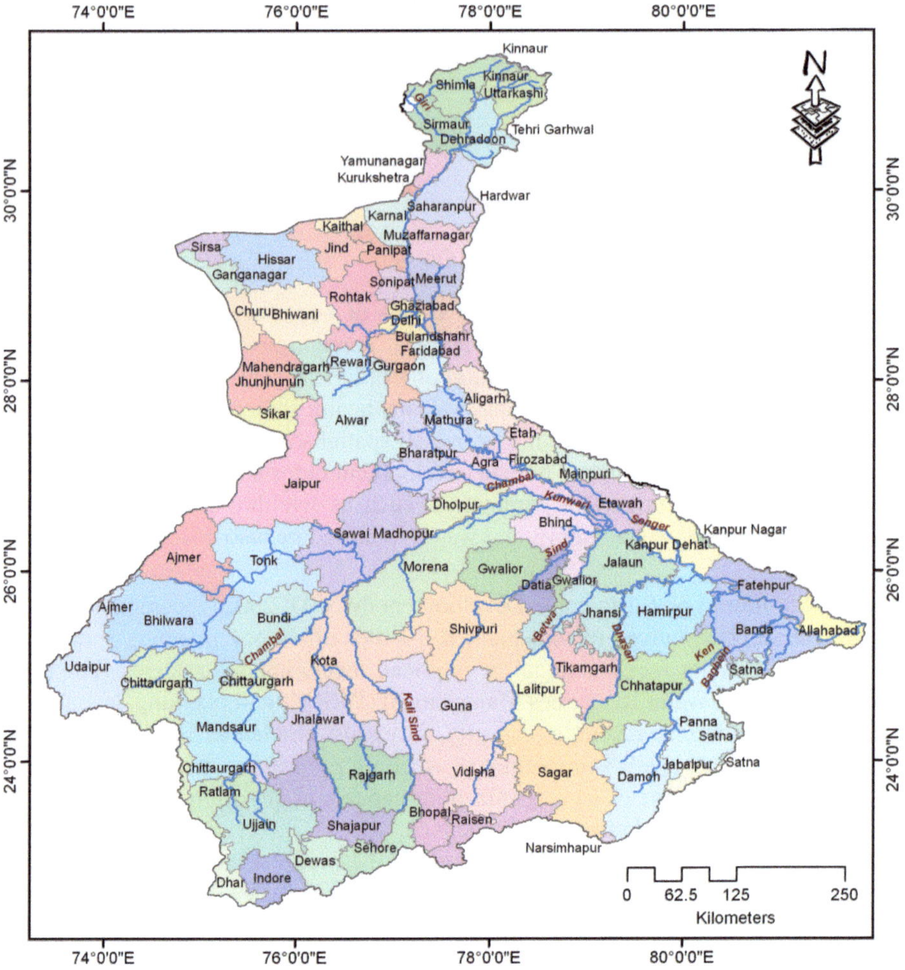

Fig. 7.1 Administrative setup (districts) of the Yamuna River basin

agriculture, education and infrastructure. While the reason for census data collection is no doubt the use of census data for agriculture and infrastructural development planning and formulation of policies, a much wider application of census results is possible.

As stated above, seven states form the drainage area (approximately 3,65,868 km^2) of the Yamuna basin. The state-wise population share estimated from the population data of Census-2001 in the basin is shown in Fig. 7.2. The total population of the basin is estimated to be 127,524,051 (i.e., 127.5 million). Among these states, Uttar Pradesh alone contributes 32.95% of the basin population followed by Madhya Pradesh, Rajasthan, Haryana, Delhi, Uttarakhand and Himachal Pradesh; though the geographical area contribution of Uttar Pradesh (19.19%) in the basin is less as compared to the Madhya Pradesh (38.25%) (Fig. 7.3). By comparing

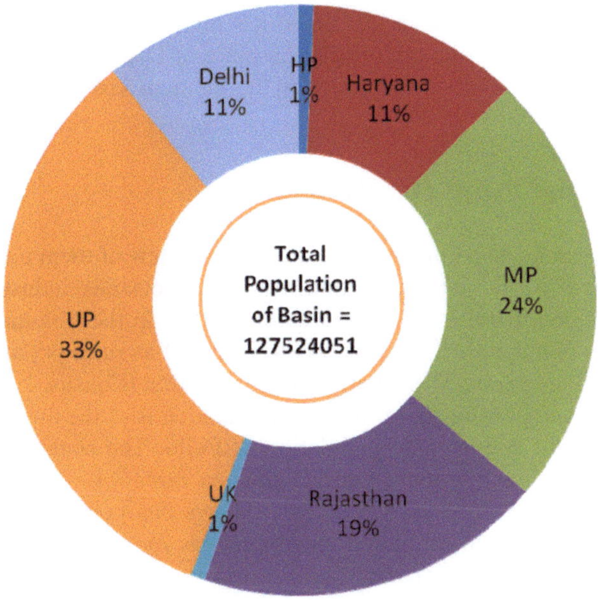

Fig. 7.2 State wise share of population in the Yamuna basin

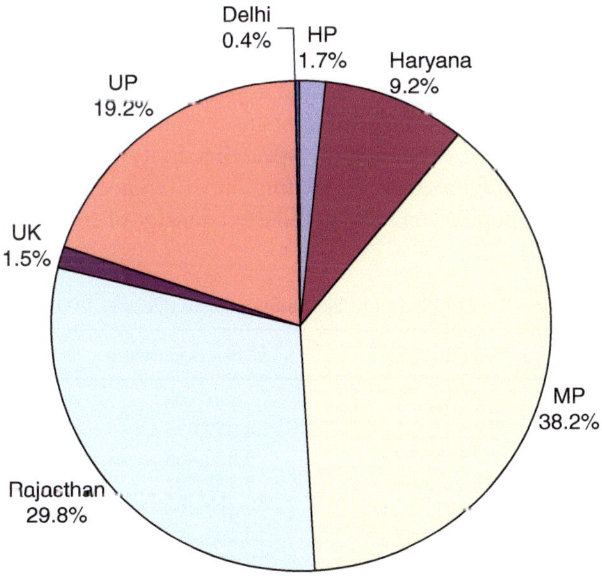

Fig. 7.3 State wise share of area in the Yamuna basin (%)

Figs. 7.2 and 7.3, it can be seen that Delhi has highest population density, which shares only 0.40% area while the population share of Delhi is 11% close to that of Haryana which results in severe pollution in the Yamuna River.

7.1.1 Urban and Rural Population

The Yamuna River basin comprises 92 districts, and a few of them partly contribute to the area. Out of 92 districts, approximately 32 districts have industrial influence. Rural areas are dominantly agricultural communities. In the Yamuna basin, more than half of the population (i.e. 62.3%) is categorized as rural population, and the rest of the population (i.e., 37.6%) are urban population (Census, 2001). However, the national average percentage of the rural population is of the order of 70%. This difference is caused due to large population of Delhi. The state-wise categorized population residing in the Yamuna basin is given in Table 7.1.

Analysis of Census (Census, 1991, 2001) shows that population in rural areas is shrinking mainly due to migration to urban areas for employment opportunity, and will further continue because of better living standard and other reasons. The overall distribution of the total urban population in the basin is the highest in Uttar Pradesh followed by Delhi, Madhya Pradesh, Rajasthan, Haryana, Uttarakhand and Himachal Pradesh as 27.51, 26.85, 19.83, 15.5, 8.91, 1.01 and 0.40%, respectively (Fig. 7.4). The rural population is also highest in Uttar Pradesh (i.e., 36.49%) followed by Madhya Pradesh and Rajasthan (Fig. 7.5).

7.1.2 Population Density

Population density is commonly represented as people per square kilometer. Several of the most densely populated city-states share a relatively small area and a high urbanization level, with an economically specialized city population drawing also on rural resources outside the area, illustrating the difference between high population density and over-population. The population density of India in 2001 was 324

Table 7.1 Population in the Yamuna basin (Census, 2001)

State	Population	Urban population	Rural population
Himachal Pradesh	1,107,815	191,589	916,226
Haryana	14,733,152	4,272,975	10,460,177
Madhya Pradesh	30,363,417	9,512,468	20,850,949
Rajasthan	24,213,470	7,432,741	16,780,729
Uttarakhand	1,033,572	482,496	551,076
Uttar Pradesh	42,227,667	13,193,488	29,034,179
Delhi	13,844,958	12,875,811	969,147
Total	127,524,051	47,961,568	79,562,483

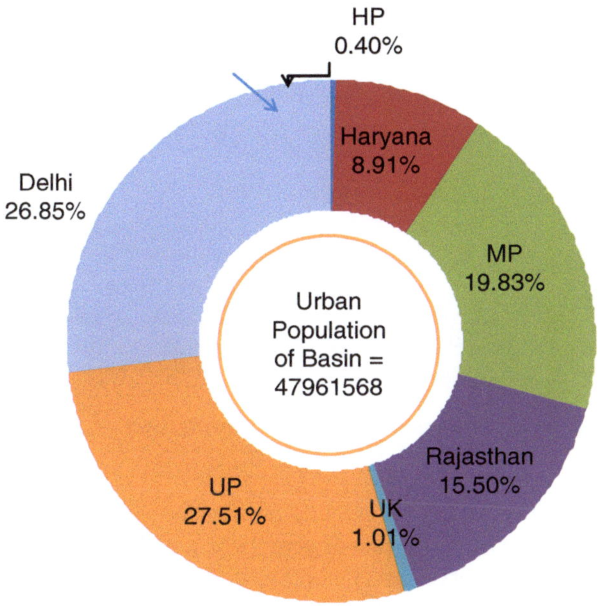

Fig. 7.4 State wise share of urban population in the Yamuna basin (%)

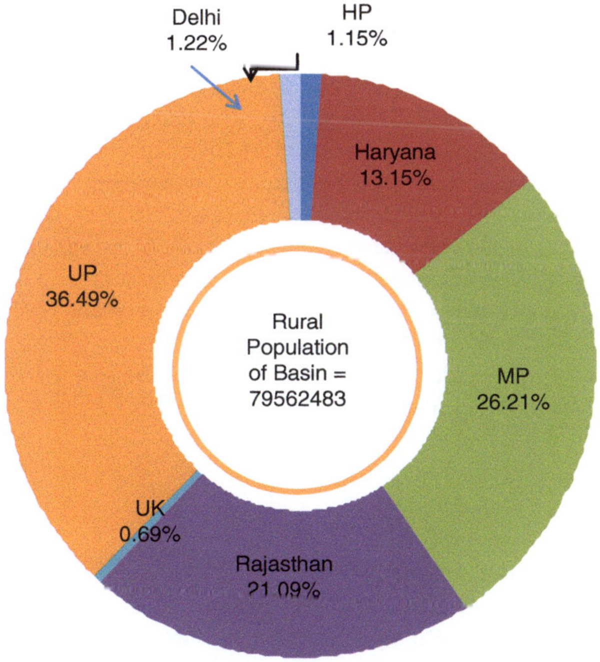

Fig. 7.5 State wise share of rural population in the Yamuna basin (%)

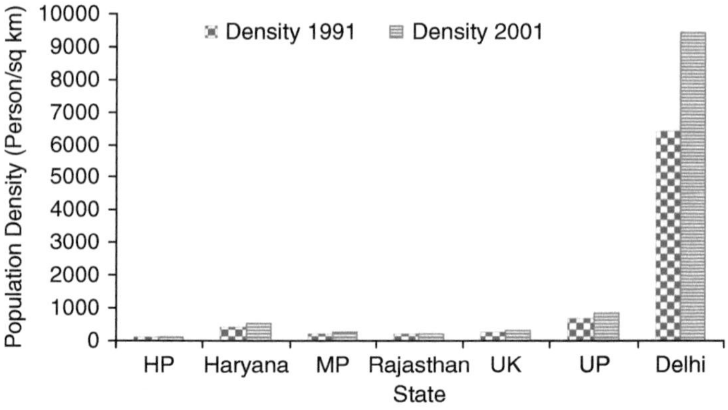

Fig. 7.6 State-wise average population density in the Yamuna basin

people per sq. km. The density of population has been increased in all the States and Union Territories during the census year 1991 and 2001. Among major states, West Bengal is still the most densely populated state with a population density of 903 (Census, 2001). Bihar is now the second most densely populated state, pushing Kerala to the third place.

In the Yamuna basin, Delhi has the highest population density (Fig. 7.6) followed by Uttar Pradesh and Haryana. In the Yamuna River basin, most of the cities are located along the river due to better availability of water resources. Even most of the industrial and urban developments are taking place near the river banks. Similar to states, the population density of the Yamuna basin has been significantly increased during the census year 1991–2001 (Fig. 7.6). The density of Delhi has increased from 6,402 per sq. km (Census, 1991) to 9,412 per sq. km (Census, 2001). It is indeed a great challenge to protect the natural environment of the river and its catchment. Population in Delhi has been so increased in last two decades that all the natural resources have been over-exploited, even the intra-basin natural resource transfer gets shortened for adequate livelihood and eco-system functioning. The spatial pattern of the population density of the basin for the years 1991 and 2001 are shown in Fig. 7.7. It is clear from Fig. 7.7 that population growth is significant in 1991–2001 along the main river course as well as the major tributaries of the Yamuna basin such as Hindon and Chambal.

7.1.3 Population Growth Rate

The decadal population growth is estimated using the population census data. This helps in planning for the infrastructure development. It can also be used for population projections for long term integrated water resources management plan. Again Delhi has the highest decadal growth rate from year 1981 to 1991 and from 1991 to 2001 (Fig. 7.8). However, it has slightly decreased for 1991–2001 than 1981–1991.

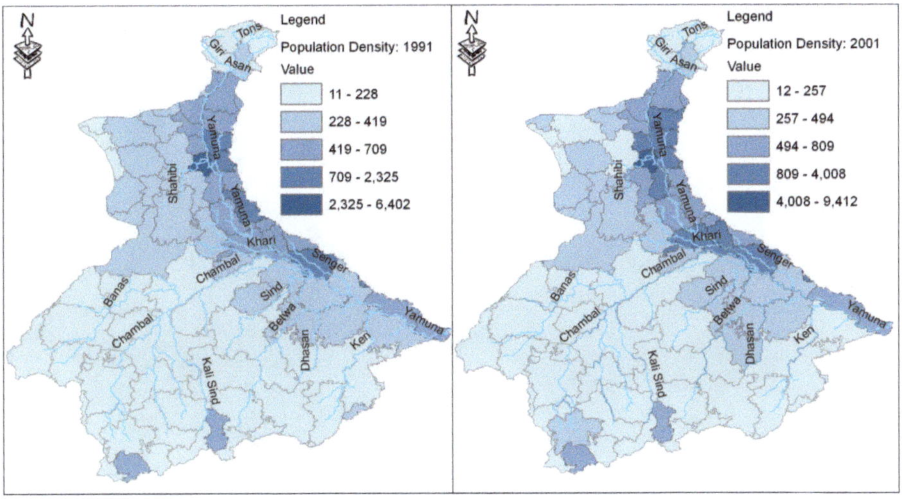

Fig. 7.7 Comparison of population density in the Yamuna River basin

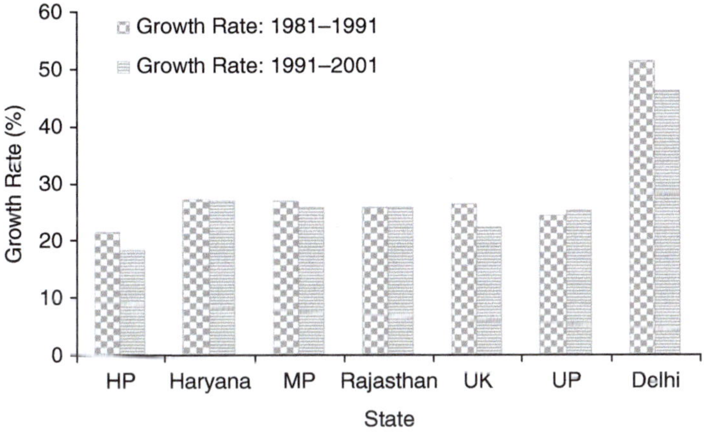

Fig. 7.8 Population growth rate in the Yamuna basin

The spatial decadal growth rate in the Yamuna basin is shown in Fig. 7.9 for the periods of 1981–1991 and 1991–2001, respectively. Comparing Fig. 7.9, a similar remark can be made as for the population density.

7.1.4 The Rate of Urbanization

Urbanization is also defined by the United Nations as the movement of people from rural to urban areas with population growth equating to urban migration. The population growth in urban areas is mainly due to infrastructure development and the lack

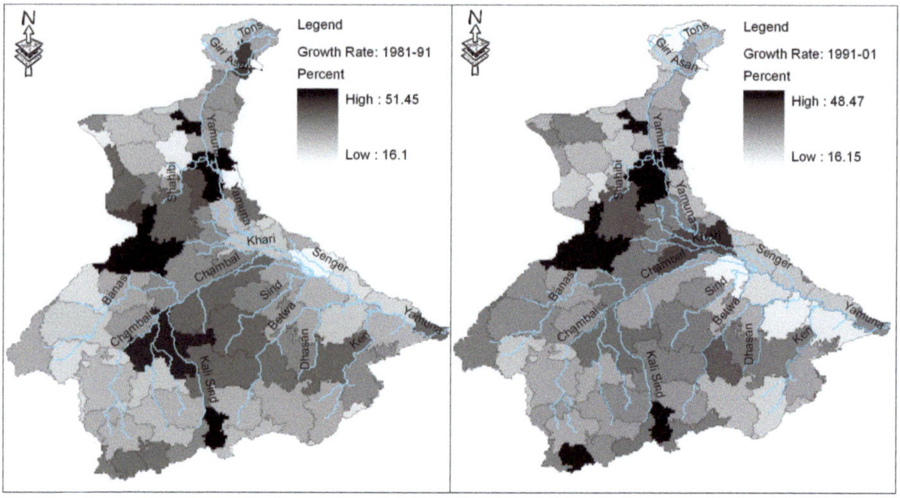

Fig. 7.9 Comparison of population growth rate in the Yamuna River basin

of development and insecurity in agriculture. Urbanization is the physical growth of urban areas.

Urbanization occurs naturally from individual and corporate efforts to reduce time and expense in commuting and transportation while improving opportunities for jobs, education, housing, and transportation. Living in cities permits individuals and families to take advantage of the opportunities of proximity, diversity, and marketplace competition.

In rural areas, often on small family farms, it is difficult to improve one's standard of living beyond basic sustenance. Farm living is dependent on unpredictable environmental and climatic conditions. Urban areas, on the other hand have better basic services as well as other specialist services that are rarely available in rural areas. Similarly, urban centers have better employment opportunities for different qualification levels.

Health is another major factor that causes migration to urban centers. People, especially the elderlies are often forced to move to cities where there are doctors and hospitals that can cater for their health needs. It is also a result of industrialization that farms become more mechanized, putting many laborers out of work. This is currently occurring rapidly in India.

According to the 1981 census, 75% of male workers in non-agricultural activities are required for a settlement to be declared as urban.

The sources of urban growth comprise a natural increase in urban areas, migration, reclassification of rural areas into urban areas and the change in the boundaries of existing towns.

Intra-city migration from smaller to bigger cities is continuing along with the migration from rural to urban areas besides an enormous natural population pressure. Cities are facing a serious shortage of power, water, sewerage, housing, transportation, and communication mixed with dramatic pollution, poor public health, unemployment and poverty. Thus, understanding and monitoring past and current urbanization processes is the basis for future predictions and preparedness, and thus for sustainable urban planning.

Focus on the urban growth of the cities in the Yamuna basin shows interesting features (Fig. 7.9). It shows that population growth is not only in the cities adjacent to main Yamuna River but the development of incipient towns has also started. It might be due the migration of population to adjacent urban centre. This may continue longer, which will be beneficial for metropolitan cities in reducing the population load due to migration.

7.1.5 Migrants

The complexity of movement of population in different parts of the country helps in understanding the dynamics of the society. At this crucial juncture in economic development study on migration assumes special importance. A person is considered as migrant by place of birth if the place in which he is enumerated during the census is other than his place of birth (Census, 2001).

The total migrant population in the basin is 3,90,51788 (Table 7.2) and its state wise distribution is shown in Fig. 7.10. Uttar Pradesh shares the largest population of migrants of 27.7% and Himachal Pradesh lowest as 1.1%.

State wise data shows that the population of Delhi comprises the highest number of migrants as 43.4% followed by Himachal Pradesh and Haryana (Table 7.3). Urban and rural categories of migrants shows the highest urban migrants as 92.46% in Delhi whereas, rural migrants are highest in Rajasthan as 71.4% followed by HP, Uttar Pradesh, Madhya Pradesh, Haryana, Uttarakhand and Delhi.

Urban and rural both type of migration are mainly for better standard of living. In rural areas, a lack of agriculture infrastructure, source of income, poor standard of amenities and monsoon failure mainly lead to migration. However, urban areas face migration because of the lack of job and infrastructure for development.

Table 7.2 Urban and rural migrant population

State	Total migrant	Rural migrant	Urban migrant
Himachal Pradesh	412,274	289,709	122,565
Haryana	5,117,618	3,117,546	2,000,071
Madhya Pradesh	9,229,246	6,035,624	3,193,622
Rajasthan	7,003,496	5,002,008	2,001,488
Uttarakhand	466,216	216,447	249,769
Uttar Pradesh	10,810,933	7,489,107	3,321,826
Delhi	6,012,006	453,243	5,558,763
Total	39,051,788	22,603,684	16,448,105

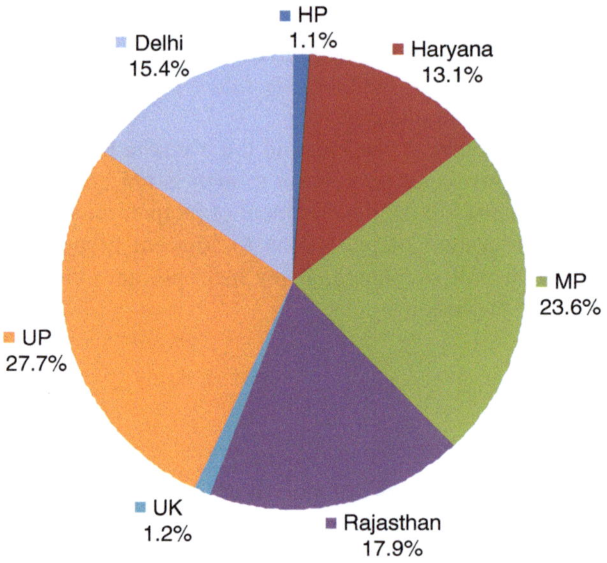

Fig. 7.10 State wise distribution of migrants in the Yamuna basin

Table 7.3 State wise migrants in the Yamuna basin (%)

State	Migrant	Rural migrant	Urban migrant
Himachal Pradesh	37.3	70.3	29.7
Haryana	34.6	60.9	39.1
Madhya Pradesh	30.1	65.4	34.6
Rajasthan	30.1	**71.4**	28.6
Uttarakhand	34.2	46.4	53.6
Uttar Pradesh	25.6	69.3	30.7
Delhi	**43.4**	7.54	92.5

7.2 Sex Ratio

Sex ratio is an important social indicator to measure the extent of prevailing equity between males and females at a given point of time. It is mainly the outcome of the interplay of sex differentials in mortality, sex selective migration, sex ratio at birth and at times the sex differential in population enumeration.

Sex ratio is defined as the number of females per thousand of males. According to the Census of India (2001), the sex ratio of India stands at 933. This is a marginal improvement from the 1991 Census, which had recorded 927 females for every 1000 males. Considering the Census (2001), the sex ratio among the major States ranged between 861 in Haryana to 1058 in Kerala. State wise analysis of sex ratio in the Yamuna basin shows the best status for Uttarakhand as the sex ratio is 938 (Fig. 7.11).

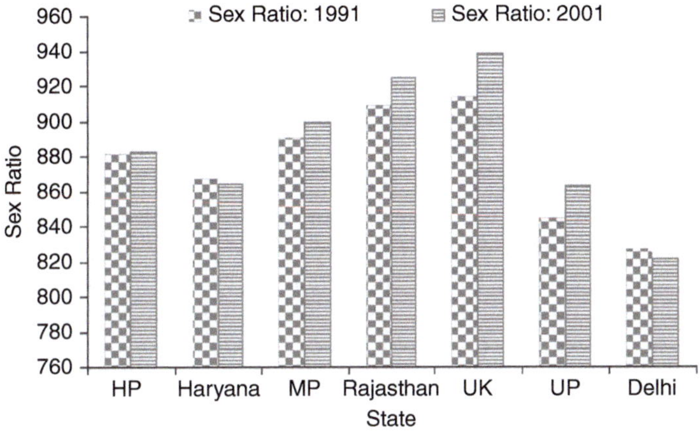

Fig. 7.11 Sex Ratio in the Yamuna basin

In the Yamuna basin, the sex ratio has been considerably increased from census year 1991 to 2001 (Fig. 7.11). The spatial variability in the sex ratio for the Yamuna basin is also shown in Fig. 7.12. It was observed from Fig. 7.12 that the sex ratio is highly dependent on the topography. The sex ratio is generally high in the hilly and forest area as compared to the plains. A gradual reduction in the sex ratio is taking place from hilly region to the plains. It might be due to non-availability of resources and inadequate employment opportunity in the hilly regions. Another region of this variation is the migration of population from hilly areas to plains for employment. In spite of the sex ratio of India (i.e. 933 females per 1000 males), the average sex ratio in the basin is estimated to be 891 with a minimum of 821 in Delhi and a maximum of 1051 in Tehri Garhwal (Census, 2001).

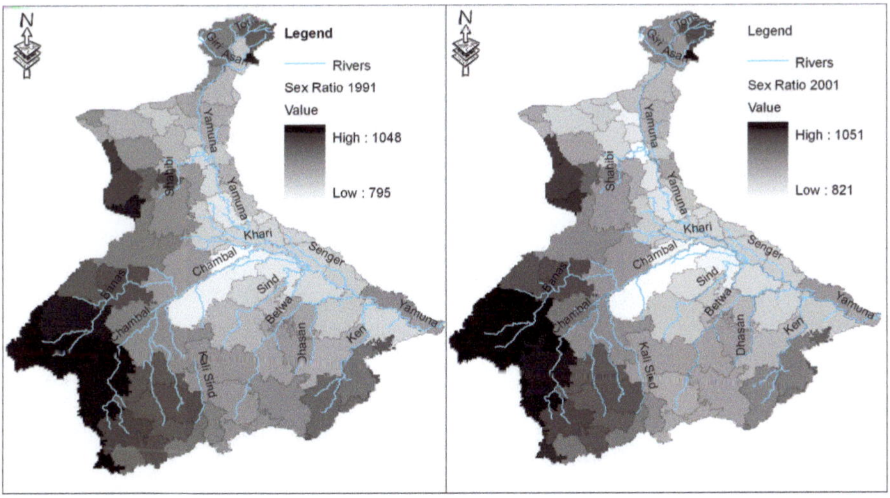

Fig. 7.12 Spatial pattern of sex ratio and comparison for the census years 1991 and 2001

7.3 Literacy

Literacy, in its simplest definition, is the ability to read and write. Literacy is more than just being able to read and write. The literate use reading to learn about a wide range of subjects, share their ideas and tend to have rich imaginations and many interests. Literacy widens a person's world; illiteracy narrows it. Literacy is calculated by Census using the population of the persons of more than 7 years of age having knowledge of reading and writing.

Literacy campaigns have helped spread the knowledge about health care and nutrition, thereby enabling mothers to keep their family in better health and to care better for their children. The National Literacy Mission seeks to achieve 75% literacy goal by imparting functional literacy to illiterates in the 15–35 age group. This age group has been the focus of attention because they are in the productive and reproductive period of life. The basic objective is to create a generation which will ensure that their children are educated, to realize the dream of "education for all."

The literacy rate in the states of the Yamuna basin is the highest in Delhi as 81.67, followed by Himachal Pradesh and Uttarakhand (Table 7.4). It is the lowest in Uttar Pradesh as 56.27. The literacy rate for women is poor in the state of Uttar Pradesh, Rajasthan and Madhya Pradesh (lower than national average). It might be attributed due to gender based inequality, occupation of girl child in domestic tasks, and low enrolment of girls in schools.

Table 7.4 Literacy Rate in State falling under the Yamuna basin

India/state	Category	Person	Male	Female
India	Total	64.84	75.26	53.67
	Rural	58.74	70.7	46.13
	Urban	79.92	86.27	72.86
Himachal Pradesh	Total	76.48	85.35	67.42
	Rural	75.08	84.51	65.68
	Urban	88.95	92.04	85.03
Uttarakhand	Total	71.62	83.28	59.63
	Rural	68.07	81.78	54.7
	Urban	81.44	87.05	74.77
Haryana	Total	67.91	78.49	55.73
	Rural	63.19	75.37	49.27
	Urban	79.16	85.83	71.34
Delhi	Total	81.67	87.33	74.71
	Rural	78.05	86.6	67.39
	Urban	81.93	87.39	75.22
Rajasthan	Total	60.41	75.7	43.85
	Rural	55.34	72.16	37.33
	Urban	76.2	86.45	64.67
Uttar Pradesh	Total	56.27	68.82	42.22
	Rural	52.53	66.59	36.9
	Urban	69.75	76.76	61.73
Madhya Pradesh	Total	63.74	76.06	50.29
	Rural	57.8	71.7	42.76
	Urban	79.39	87.39	70.47

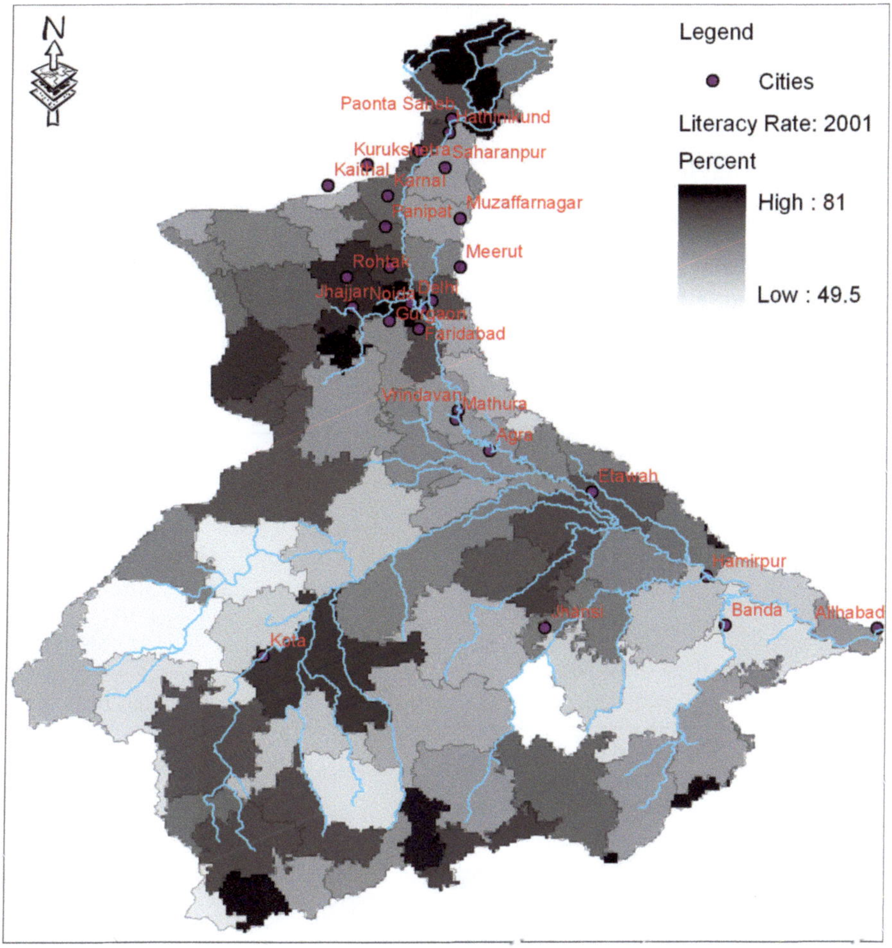

Fig. 7.13 District-wise variation of average literacy rate in the Yamuna basin

The variation of average district wise literacy rate can be seen from Fig. 7.13. The figure shows that the literacy rate is generally high in urban areas. Delhi has the highest literacy rate, among other districts. Shimla (79.1) and Dehradun (79.0) also have very high literacy rate as compared to other districts. District Lalitpur has the lowest literacy rate in the basin as 49.5.

7.4 Main and Marginal Workers

Work is defined as participation in any economically productive activity. All persons engaged in "work" are workers. Persons who are engaged in cultivation or milk production even solely for domestic consumption are also treated as workers. The reference period for determining a person as a worker and a non-worker is one year preceding the date of enumeration (Census, 2001).

All those workers who had worked for a major part of the reference period (i.e. 6 months or more) are termed as Main Workers and workers who had not worked for the major part of the reference period are termed as Marginal Workers.

Yamuna basin has total 47,922,103 workers (Table 7.5). These comprise 36,857,018 Main and 11,065,085 Marginal workers.

State wise distribution of workers shows the highest workers in Uttar Pradesh followed by Madhya Pradesh, Rajasthan, and Delhi as 29.27, 24.82 and 21.75%, respectively (Fig. 7.14). The lowest share of workers is by Uttarakhand as 0.86% in the Yamuna basin.

Table 7.5 Workers in the Yamuna basin

State	Workers	Main workers	Marginal workers
Himachal Pradesh	556,944	448,721	108,224
Haryana	5,856,832	4,298,415	1,558,417
Madhya Pradesh	12,575,951	9,519,908	3,056,043
Rajasthan	10,227,216	7,600,473	2,626,743
Uttarakhand	402,803	334605.1	68,198
Uttar Pradesh	13,758,975	10,339,141	3,419,834
Delhi	4,543,381	4,315,756	227,625
Total	47,922,103	36,857,018	11,065,085

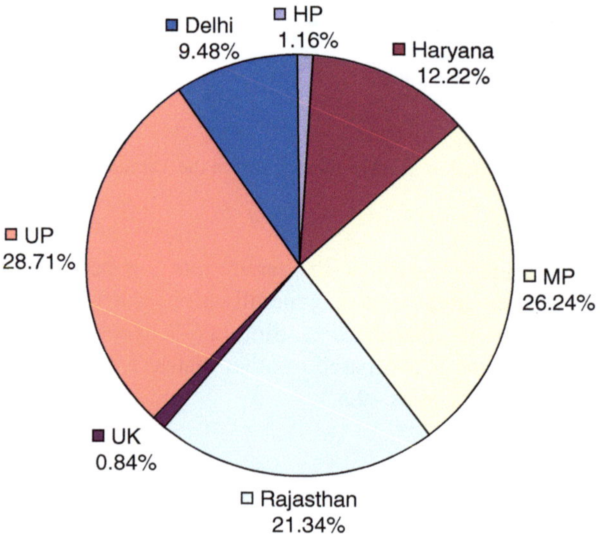

Fig. 7.14 State wise distribution of workers in the Yamuna basin (%)

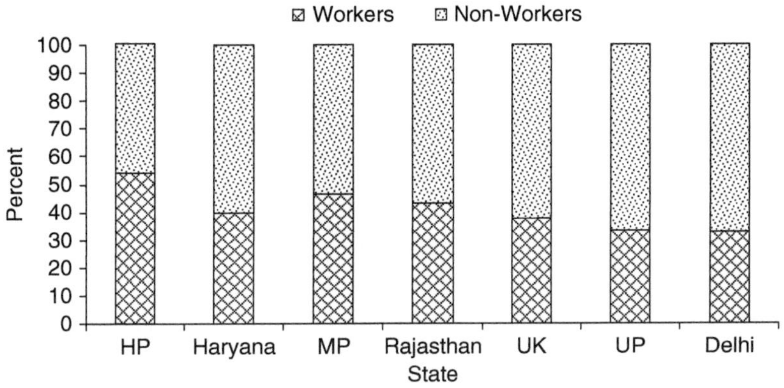

Fig. 7.15 Workers and non workers percentages

A person who does not work at all during the reference period is treated as a non-worker. The non-workers broadly constitute students, those performing household duties, such as daily household chores like cooking, cleaning utensils, looking after children, fetching water, etc. and the unpaid work in the family farm, cultivation or milching, dependents, such as infants or very elderly people are not included in the category of worker. Delhi, Uttar Pradesh and Madhya Pradesh share most of the non-worker population of the basin (Fig. 7.15).

7.4.1 Cultivator and Agricultural Labourer

According to Census, a person is classified as cultivator if he or she is engaged in the cultivation of land for payment in money, kind or share. Cultivation includes effective supervision or direction in cultivation. A person who works on another person's land for wages in money or kind or share is regarded as agricultural laborers. The worker population shows a total 55.5% population of cultivators and agricultural laborers in the Yamuna basin. Cultivator and agriculture laborers are 18,755,795 and 7,855,485, respectively, in the Yamuna basin (Table 7.6).

Table 7.6 Cultivator and agricultural labour in the Yamuna basin

State	Cultivator	Agricultural labor
Himachal Pradesh	3,74,895	14,529
Haryana	21,91,334	8,40,500
Madhya Pradesh	54,19,816	29,83,682
Rajasthan	54,58,481	10,03,340
Uttarakhand	1,14,264	31,744
Uttar Pradesh	51,59,589	29,65,922
Delhi	37,415	15,766
Total	1,87,55,795	78,55,485

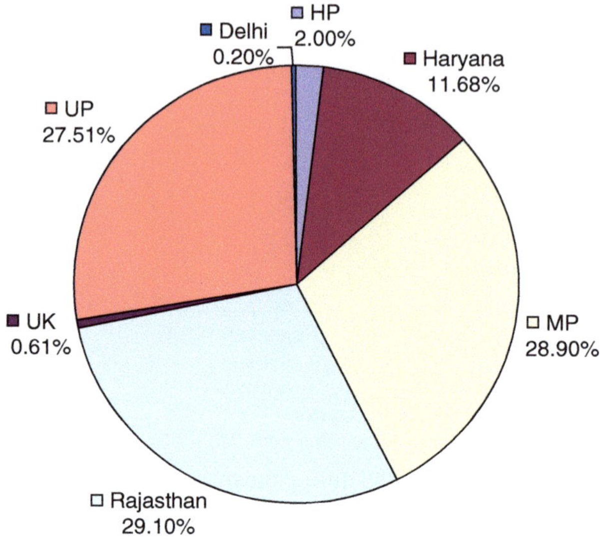

Fig. 7.16 Distribution of cultivators in the Yamuna basin (%)

Cultivator's population is distributed as the highest among workers in Rajasthan which is 29.1% followed by Madhya Pradesh, Uttar Pradesh and Haryana as 28.9, 27.5 and 11.0, 68.0%, respectively (Fig. 7.16). In the Yamuna basin, Delhi has the lowest percentage of cultivators (i.e. 0.2%). Madhya Pradesh and Uttar Pradesh

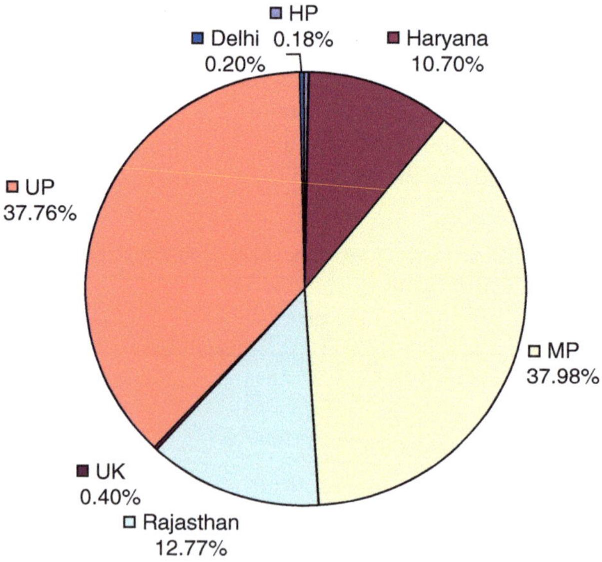

Fig. 7.17 Distribution of agricultural labor in the Yamuna basin (%)

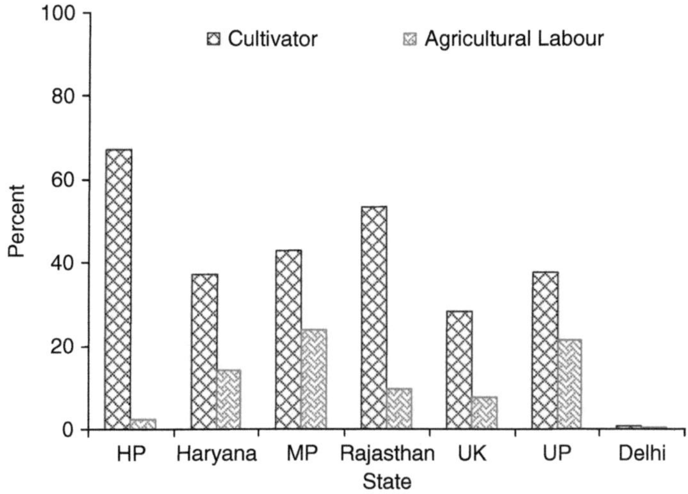

Fig. 7.18 Status of cultivators and agricultural labor among workers in the Yamuna basin

share almost equal percentages of agricultural labor in the worker category in the Yamuna basin (Fig. 7.17).

State wise analysis of workers show Himachal Pradesh has a higher percentage of cultivators followed by Rajasthan, Madhya Pradesh, Uttar Pradesh, Haryana, Uttarakhand, and Delhi (Fig. 7.18).

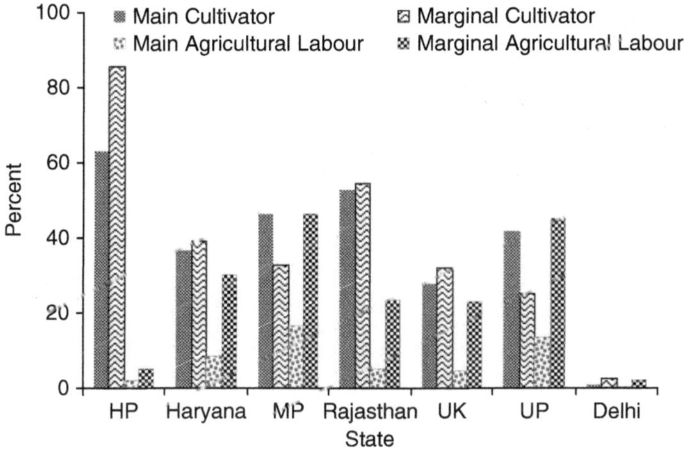

Fig. 7.19 State wise cultivator and agricultural labor (%)

Analysis of main and marginal categories of population shows that Uttar Pradesh and Madhya Pradesh comprise a higher main category of cultivators than marginal category (Fig. 7.19). Main Agricultural laborers are lower in all cases than in marginal category.

7.5 Health: Vital Statistics

***Infant Mortality*:** Infant Mortality is defined as the number of infant deaths that occur per thousand live births in any population in one calendar year, the infant being a child of 1 year or less. The various factors involved in infant mortality are the immediate direct medical causes that lead to the death of infants, economic, socio-cultural, environmental and nutritional factors that play a contributory or predisposing role in the epidemiology of infant mortality. Infant deaths are more frequent among poor families. The poorest live in slum areas where the civic administration spends the least on public amenities, such as safe water supply, sanitation, drainage and personal health services.

The infant mortality rate is the highest in Madhya Pradesh (i.e. 74) followed by Uttar Pradesh and Rajasthan (i.e., 71 and 67, respectively). In these three states it is higher than the national average of 57. Infant mortality is lowest in Delhi (Table 7.7).

***Expectation of Life*:** Expectation of life at birth is the highest for Delhi. Himachal Pradesh and Haryana have almost equal life expectation at birth as 65.9 and 65.2 which are higher than the national average 62.5.

***Birth Rate and Death Rate*:** Comparison of health indicators among the states of the Yamuna basin shows a higher birth rate in Uttar Pradesh, Madhya Pradesh

Table 7.7 Health indicators

	2005			2006			
India/State	Birth rate	Death rate	Infant mortality rate	Birth rate	Death rate	Infant mortality rate	Expectation of life at birth (1998–2002)
India	**23.8**	**7.6**	**58**	**23.5**	**7.5**	**57**	**62.5**
Delhi	18.6	4.6	35	18.4	4.7	37	**70**
Haryana	24.3	6.7	60	23.9	6.5	57	65.2
Himachal Pradesh	20	6.9	14	18.8	6.8	50	65.9
Madhya Pradesh	29.4	9	76	29.1	8.9	**74**	56.9
Rajasthan	28.6	7	68	28.3	6.9	67	61.1
Uttar Pradesh	30.4	8.7	73	30.1	8.6	71	59.1
Uttrakhand	20.9	7.4	42	21	6.7	43	–

Source: SRS Bulletin (2008)

Table 7.8 Index of "A long and healthy life" based on infant mortality rate and life expectancy at age 1

State	2006	1996
Haryana	0.733	0.685
Himachal Pradesh	0.747	0.709
Madhya Pradesh	0.628	0.559
Rajasthan	0.678	0.618
Uttar Pradesh	0.651	0.598
Uttarakhand	0.721	0.643
Delhi	0.759	0.731
India	0.706	0.653

and Haryana as compared to the national average of 23.5. The death rate is higher in Uttar Pradesh and Madhya Pradesh than the national average (SRS, 2006).

Life Expectancy Index: The Life Expectancy Index is a statistical measure used to determine the average lifespan of the population of a certain area. Life expectancy is one of the factors in measuring the Human Development Index (HDI), along with adult literacy, education, and standard of living. Life expectancy is also a factor in finding the physical quality of life of an area. Table 7.8 shows the highest index for Delhi followed by Himachal Pradesh, Haryana, Uttarakhand, Rajasthan, Uttar Pradesh and Madhya Pradesh (UNDP, 2006). Uttar Pradesh, Rajasthan and Madhya Pradesh has lower life expectancy index than national average.

Maternal Mortality: For India, the National Family Health Survey of 1992–1993 was the first to provide a national-level estimate of 437 maternal deaths per 100,000 births for the two-year period preceding the survey (International Institute for Population Sciences, 1995). Bhat (2001) estimated the values of maternal mortality using sex-differentials in adult mortality (SDAM). Results are depicted in Table 7.9. The maternal mortality rate in the Yamuna basin states is higher than the country level status.

Table 7.9 Maternal mortality rate

S. No.	State	MMR	Year
1	Madhya Pradesh	532	1987–1996
2	Haryana	472	1987–1996
3	Rajasthan	399	1987–1996
4	Uttar Pradesh	579	1987–1996
5	India – Total	**410**	1987–1991
6	India – Total	**341**	1992–1996

Source: Bhat (2001)

***Health Facilities*:** Health facilities are required for preventing disease, prolonging life and promoting health through the organized efforts and informed choices of society, organizations, public and private, communities and individuals. Government runs various schemes for prevention of the spread of communicable diseases. The goal of public health is to improve lives through the prevention and treatment of disease. For sustaining the achievement for communicable disease control, particularly HIV/AIDS, tuberculosis, malaria and meningitis; vigilance and surveillance for epidemic-prone diseases and advancing emergency preparedness infrastructure is required. Table 7.10 shows the yearly statistics of the number of hospital beds per lakh of population for the states of Yamuna basin during the period of 1961 to 1998. Tables 7.11 and 7.12 show the total number of hospital beds in rural and urban areas, respectively. Considering these statistics, status of Madhya Pradesh is poor as compared to other states. Based on Table 7.10, the status of the Uttar Pradesh is still poor as the population is very high for this state.

Table 7.10 Total number of hospital beds in Yamuna basin states per 1,00,000 population

	Year									
State	1961	1966	1971	1976	1981	1986	1988	1991	1996	1998
Delhi									168.24	150.5
Haryana	*	52.01	61.28	69.91	62.27	51.33	51.12	45.38	38.45	38.23
Himachal Pradesh	48.32	148.23	132.64	136.62	81.76	89.4	89.7	87.79	83.85	88.51
Madhya Pradesh	30.75	39.39	36.7	39.64	32.24	34.59	36.03	38.27	26.58	26.59
Rajasthan	52.27	63.77	67.38	64.23	52.16	54.73	53.15	44.11	42.33	41.63
Uttar Pradesh	35.76	38	43.87	47.29	45.74	42.47	40.65	38.2	37.83	42.41
All India	52.28	60.75	63.6	72.79	73.64	77.79	78.19	78.7	66.4	71.5

Table 7.11 Total number of beds in hospitals in rural area per 1,00,000 population

	Year					
State	1981	1986	1988	1991	1996	1998
Delhi	*	*	*	*	4.17	*
Haryana	10.05	4.91	4.92	4.57	3.86	3.93
Himachal Pradesh	26.7	17.31	16.51	16.8	9.95	12.46
Madhya Pradesh	3.21	2.99	4.37	43.38	11.79	11.79
Rajasthan	5.93	6.02	5.84	3.11	2.89	2.95
Uttar Pradesh	7.83	7.63	7.34	6.93	2.55	7.62
All India	16.51	18.18	18.49	22.26	17.54	23.25

*Not Available

Table 7.12 Total number of beds in hospitals in urban area per 100,000 population

	Year					
	1981	1986	1988	1991	1996	1998[1]
Delhi	*	*	*	*	186.68	*
Haryana	248.71	202.43	196.99	169.19	144.48	132.87
Himachal Pradesh	749.76	895.7	886.19	832.46	859.41	830.24
Madhya Pradesh	146.28	147.18	145.26	21.37	75.53	75.55
Rajasthan	225.63	226.67	216.86	182.3	174.81	159.55
Uttar Pradesh	219.01	191.55	180.15	165.38	189.2	191.7
All India	261.56	260.27	256.59	241.96	207.64	188.55

Sources: Health Information of India
Notes: Beds includes beds in hospitals and dispensaries
* Not Available, CBHI, MOHFW, GOI, respective years
[1] Figures based on Statistical Abstract India 2000 and Denominator Population used for obtaining Hospital- Population ratio refers to reference period given in the source

7.6 Population Below Poverty Line (BPL)

According to a 2005 World Bank estimate, 42% of India's population falls below the International poverty line of $1.25 a day (Purchasing Power Parity (PPP), in nominal terms Rs. 21.6 a day in urban areas and Rs 14.3 in rural areas). In India after 2002, below poverty line (BPL) estimation is done considering the thirteen socio-economic parameters: (i) size of operational landholding, (ii) type of house, (iii) availability of cloths, (iv) food security, (v) sanitation, (vi) ownership of consumer durables, (vii) literacy status, (viii) status of household labour force, (ix) means of livelihood, (x) status of children (going to school), (xi) type of indebtedness, (xii) reason for migration from household, and (xiii) preference for assistance.

State wise statistics of the BPL population is given in Table 7.13. It evident that the BPL percentage in Madhya Pradesh is high as compared to other states. It

Table 7.13 State wise percentage of population below poverty line

States/UT	1973–1974	1977–1978	1983	1987–1988	1993–1994	1999–2000
Haryana	35.36	29.55	21.37	16.64	25.05	8.74
Himachal Pradesh	26.39	32.45	16.4	15.45	28.44	7.63
Madhya Pradesh	61.78	61.78	49.78	43.07	42.52	37.43
Punjab	28.15	19.27	16.18	13.2	11.77	6.16
Rajasthan	46.14	37.42	34.46	35.15	27.41	15.28
Uttar Pradesh	57.07	49.05	47.07	41.46	40.85	31.15
Delhi	49.61	33.23	26.22	12.41	14.69	8.23
All India	54.88	51.32	44.48	38.86	35.97	26.1

Source: Planning Commission (2005)

forms the order as Madhya Pradesh, Uttar Pradesh, Rajasthan, Haryana, Himachal Pradesh, respectively.

***Rural BPL Household*:** In spite of agricultural, industrial and economic evolution of India, 32.98% of rural families come below the poverty line. In the Yamuna basin the total number of rural households is 2,178,612 (Census-2001). The highest rural households have been reported for Allahabad of the Yamuna basin followed by Udaipur, Alwar and Aligarh; however, BPL households are maximum in Delhi (i.e. 438,015 households) followed by Sagar in Madhya Pradesh and Allahabad in Uttar Pradesh and Udaipur in Rajasthan. Figure 7.20 shows the district wise rural household and BPL percentage in Yamuna basin. Information is based on the database of Ministry of Rural Development, GoI. The estimates are based on the methodology outline in the Report of the Expert Group on the estimation of proportion and number of Poor (Ministry of Rural Development, GoI).

***Share of Poor*:** India still has the world's largest number of poor people in a single country. Of its nearly 1 billion inhabitants, an estimated 350–400 million are below the poverty line, 75% of them in the rural areas. More than 36% of the population is illiterate, with women, tribal and scheduled castes particularly affected. Economic development and urbanization are closely linked. In India, cities contribute over 55% to country's GDP and urbanization has been recognized as an important factor driving economic growth.

It is interesting to note that the ratio of urban poverty in some of the larger states is higher than that of rural poverty leading to the phenomenon of "Urbanization of Poverty". Urban poverty poses the problem of housing and shelter, water, sanitation, health, education, social security and livelihoods along with special needs of vulnerable groups like women, children and aged people. Poor people live in slums which are overcrowded, often polluted and lack basic civic amenities like clean drinking water, sanitation and health facilities.

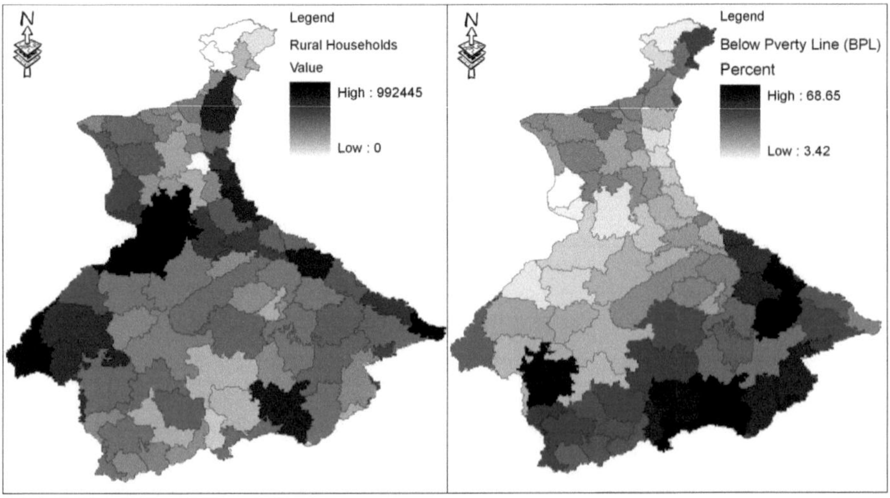

Fig. 7.20 Rural household and BPL percentage in theYamuna River basin

Table 7.14 Trend in percentage (expressed as a percentage of the total population) share of poor in the two groups of States individually and collectively

States	1983–1984	1987–1988	1993–1994	1999–2000
Forward Group				
Andhra Pradesh	5.10	5.22	4.81	4.57
Gujarat	3.65	3.98	3.28	2.61
Haryana	0.92	0.83	1.37	0.67
Karnataka	4.64	5.17	4.88	4.01
Kerala	3.31	2.88	2.39	1.58
Maharashtra	9.01	9.65	9.53	8.76
Punjab	0.89	0.82	0.78	0.56
Tamil Nadu	8.05	7.53	6.31	5.01
Total for forward States	35.57	36.08	33.35	27.77
Backward Group				
Assam	2.41	2.47	3.01	3.63
Bihar	14.31	13.71	15.40	16.36
Madhya Pradesh	8.61	8.61	9.32	11.47
Orissa	5.62	5.40	5.01	6.50
Rajasthan	3.93	4.65	4.01	3.14
Uttar Pradesh	17.24	17.47	18.87	20.36
West Bengal	9.87	9.24	7.95	8.20
Total for Backward States	61.99	61.55	63.57	69.66
All India	100.00	100.00	100.00	100.00

Source: Planning Commission (2005)

Table 7.14 shows the percentage share of poor. Forward and backward groups are on the basis of their economic status. The sharp decline in the share of poor in the forward States since 1987–1988, especially after 1993–1994 is commendable. Table 7.14 shows that the share of poor in backward group has gone up, whereas in forward group has reduced. As the table indicates, each one of the states in this group, except for West Bengal, experiences a considerable increase in the share of the poor. West Bengal's exceptional experience was mainly on account of the fast growth in agricultural production and the associated rural prosperity. It may, however, be mentioned that since the overall poverty in the country has come down substantially in the nineties, an increase in the State share in poverty need not imply an increase in the number of poor. Indeed, between 1993–1994 and 1999–2000, the absolute numbers of poor in all the States have come down.

7.7 Human Development Index

Human Development is a development paradigm that is about much more than the rise or fall of national incomes. It is about creating an environment in which people can develop their full potential and lead productive, creative lives in accord with their needs and interests. People are the real wealth of nations. Dr. Mahbub ul Haq, the Pakistani economist played a key role in formulating the human development paradigm, Prof. Amartya Sen and others provided the conceptual foundation for an alternative and broader human development approach.

***Human Development Index*:** Human Development Index (HDI) introduced a new way of measuring development by combining indicators of life expectancy, educational attainment and income into a composite human development index, the HDI. The breakthrough for the HDI was the creation of a single statistic which was to serve as a frame of reference for both social and economic development. The HDI sets a minimum and a maximum for each dimension, called goalposts, and then shows where each country stands in relation to these goalposts, expressed as a value between 0 and 1.

The *Human Development Index* (HDI) is a simple composite measure that measures the overall achievements of a region in terms of three basic dimensions of human development; health status (measured by longevity), knowledge (measured by literacy and enrolments) and a decent standard of living (measured by per capita incomes). These three dimensions are measured by life expectancy at birth, educational attainment (adult literacy and the combined gross primary, secondary and tertiary enrolment ratio) and GDP per capita (PPP US$). Income enters the HDI as a proxy for a decent standard of living and as a surrogate for all human choices not reflected in the other two dimensions.

A few statistics of the HDI is presented through Table 7.15. It can be seen that HDI has been improved in all the states. HDI is higher in Haryana than the national average. Table 7.16 has HDI values for some nearby Asian countries. India ranks 134th position in HDI (UNDP).

District wise HDI for the Yamuna basin is shown in Fig. 7.21. The average HDI, estimated for the Yamuna basin, is 0.55, with the lowest of 0.45 being in Chhatarpur in Madhya Pradesh and maximum of 0.74 being in Delhi (National Capital of India). For Haryana district wise Human Development Index (HDI) data was not available, therefore, the total HDI average of Haryana has been mapped for all the Haryana districts. However, for Faridabad and Gurgaon this will be higher than the state average.

The socio-economic indicators for Uttar Pradesh, Rajasthan, Madhya Pradesh and Delhi are summarized in Table 7.17. It can be seen that the poverty in the urban

Table 7.15 Trends in human development index (HDI)

State	1981	1991	2001
Haryana	0.36	0.44	0.50
Madhya Pradesh	0.24	0.32	0.39
Rajasthan	0.25	0.34	0.42
Uttar Pradesh	0.25	0.31	0.38
India	0.30	0.38	0.47

Source: Srinivasan (2004)

Table 7.16 Human development index (HDI) (Indicators for 2007)

Country	HDI Value	Rank	Life expectancy at birth (years)	Adult literacy rate (% aged 15 and above)	Gross enrolment ratio in education (%)	GDP per capita (PPP $)
India	0.612	134	63.4	66.0	61.0	2,753
China	0.772	92	72.9	93.3	68.7	5,383
Bhutan	0.619	132	65.7	52.8	54.1	4,837
Pakistan	0.572	141	66.2	54.2	39.3	2,496
Nepal	0.553	144	66.3	56.5	60.8	1,049
Sri Lanka	0.759	102	74.0	90.8	68.7	4,243

Source: UNDP (2007)

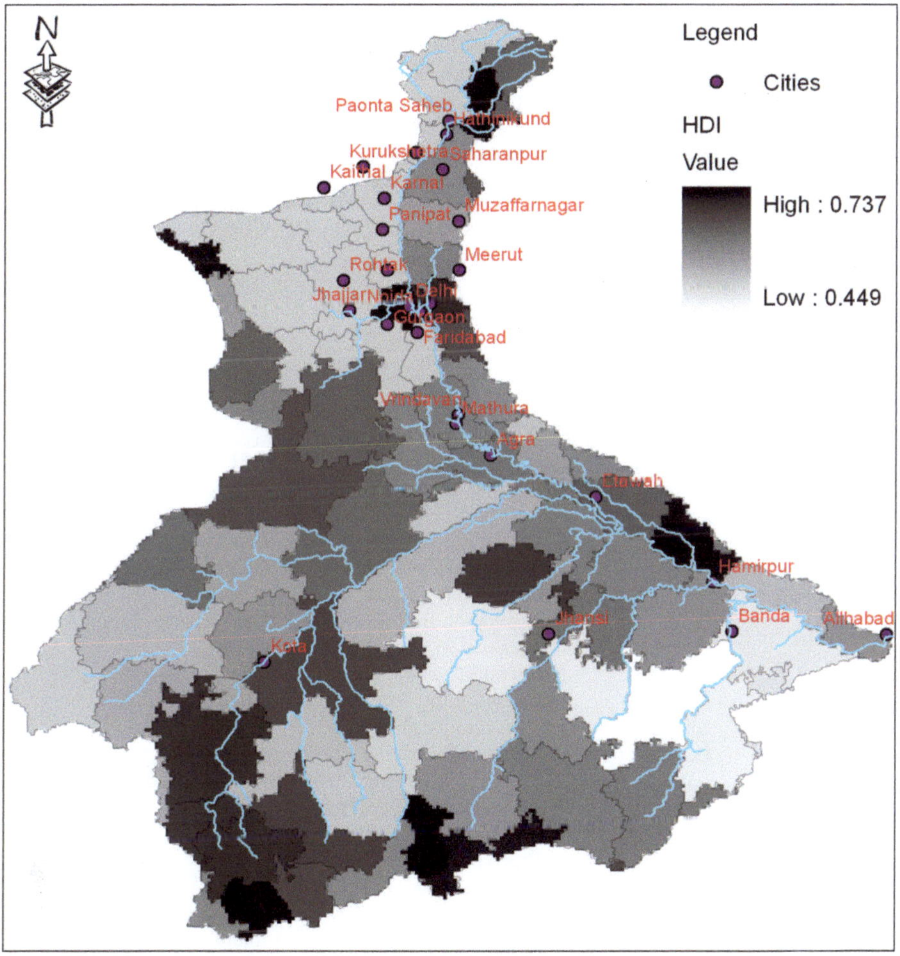

Fig. 7.21 Human development index (HDI) in the Yamuna basin

Table 7.17 Indicators for demography, health and amenities (State wise numbers)

Uttar Pradesh	Demographic Indicators	Total Population (in million)	166
		Urban Population (in million)	34.5
		% Population in Urban Areas	20.80
		Population Growth Rate (1991–2001)	25.80
		Urban Population Growth Rate (1991–2001)	32.90%
		No. of Million Plus Cities (Kanpur, Lucknow, Agra, Varanasi, Meerut, and Allahabad)	6
		No. of Towns with Over 1 Lakh Population	52
		Population Below Poverty Line in Urban Areas (in million)	10.7
		Population Below poverty line in Urban Areas	**30.90%**
	Health Indicators among Urban Poor	Neonatal Mortality Rate	43.7
		Infant Mortality Rate	79
		Child mortality rate	130.6
		% Children Receiving All Immunization	29.7
		% Children Receiving ORS or Recommended Home fluid during Diarrhea	2.1
		% Children Underweight	58.3
		% Pregnant Women Receiving the Recommended 3 ANC Visits	9.1
		% Deliveries Attended by Health Personnel	26.2
		% Using any Modern Method of family Planning	21.3
		% Women Anemic	55.1
		Total Fertility Rate (TFR)	3.74
	Environmental Health Indicators	% Having Access to Piped Water Supply	17.7
		% Having Access to Toilets	33.6
Rajasthan	Demographic Indicators	Total Population (in million)	56.5
		Urban Population (in million)	13.2
		% Population in Urban Areas	23.4
		Population Growth Rate (1991–2001)	28.3
		Urban Population Growth Rate (1991–2001)	31.5
		No. of Million Plus Cities	1
		No. of Towns with Over 1 Lakh Population	17
		Population Below Poverty Line in Urban Areas (in million)	2.61
		% Population Below Poverty Line in Urban Areas	**19.8**
	Health Indicators among Urban Poor	Neonatal Mortality Rate	65.5
		Infant Mortality Rate	98.2
		Child mortality rate	162.3
		% Children Receiving All Immunization	7.4
		% Children Receiving ORS or Recommended Home fluid during Diarrhea	29.0
		% Children Underweight	62.5
		% Pregnant Women Receiving the Recommended 3 ANC Visits	23.6
		% Deliveries Attended by Health Personnel	33.0
		% Using any Modern Method of family Planning	28.5
		% Women Anemic	55.7

Table 7.17 (continued)

	Environmental Health Indicators	% Having Access to Piped Water Supply	12.6
		% Having Access to Toilets	18.9
Madhya Pradesh	Demographic Indicators	Total Population (in million)	60
		Urban Population (in million)	16
		% Population in Urban Areas	27
		Population Growth Rate (1991–2001)	24
		Urban Population Growth Rate (1991–2001)	31
		No. of Million Plus Cities	3
		No. of Towns with Over 1 Lakh Population	23
		Population Below Poverty Line in Urban Areas (in million)	6.14
		% Population Below poverty line in Urban Areas	**38.4**
	Health Indicators among Urban Poor	Neonatal Mortality Rate	69.7
		Infant Mortality Rate	99.4
		Child mortality rate	131.9
		% Children Receiving All Immunization	20.6
		% Children Receiving ORS or Recommended Home fluid during Diarrhea	34.7
		% Children Underweight	72.4
		% Pregnant Women Receiving the Recommended 3 ANC Visits	32.0
		% Deliveries Attended by Health Personnel	38.1
		% Using any Modern Method of family Planning	NA
		% Women Anemic	55.0
	Environmental Health Indicators	% Having Access to Piped Water Supply	56.0
		% Having Access to Toilets	13.3
Delhi	Demographic Indicators	Total Population (in million)	13.78
		Urban Population (in million)	12.82
		% Population in Urban Areas	93.1
		Population Growth Rate (1991–2001)	46.31
		Urban Population Growth Rate (1991–2001)	52.13
		No. of Million Plus Cities	NA
		No. of Towns with Over 1 Lakh Population	NA
		Population Below Poverty Line in Urban Areas (in million)	1.14
		% Population Below poverty line in Urban Areas	**9.42**
	Health Indicators among Urban Poor	Neonatal Mortality Rate	39.3
		Infant Mortality Rate	94.4
		Child mortality rate	135.5
		% Children Receiving All Immunization	24.7
		% Children Receiving ORS or Recommended Home fluid during Diarrhea	44.7
		% Children Underweight	43.4
		% Pregnant Women Receiving the Recommended 3 ANC Visits	35.8
		% Deliveries Attended by Health Personnel	29.1

Table 7.17 (continued)

		% Using any Modern Method of family Planning	31.3
		% Women Anemic	42.7
		Total Fertility Rate (TFR)	NA
	Environmental Health Indicators	% Having Access to Piped Water Supply	84.4
		% Having Access to Toilets	74.6
India	Demographic Indicators	Total Population (in million)	1,027
		Urban Population (in million)	285.4
		% Population in Urban Areas	27.81
		Population Growth Rate (1991–2001)	21.34
		Urban Population Growth Rate (1991–2001)	31.39
		No. of Million Plus Cities (Kanpur, Lucknow, Agra, Varanasi, Meerut, and Allahabad)	35
		No. of Towns with Over 1 Lakh Population	423
		Population Below Poverty Line in Urban Areas (in million)	80.79
		% Population Below poverty line in Urban Areas	25.7
	Health Indicators among Urban	Neonatal Mortality Rate	34.9
		Infant Mortality Rate	54.6
		Child mortality rate	72.7
		% Children Receiving All Immunization	39.9
		% Children Receiving ORS or Recommended Home fluid during Diarrhea	24.9
		% Children Underweight	47.1
		% Pregnant Women Receiving the Recommended 3 ANC Visits	54.3
		% Deliveries Attended by Health Personnel	55.9
		% Using any Modern Method of family Planning	48.7
		% Women Anemic	58.8
		Total Fertility Rate (TFR)	NA
	Environmental Health Indicators	% Having Access to Piped Water Supply	18.5
		% Having Access to Toilets	47.2

Source: Planning Commission (2004)

areas is very high in Uttar Pradesh and Madhya Pradesh. Rajasthan and Madhya Pradesh are very poor in access to toilet as 18.9 and 13.3%, respectively.

Technical Efficiency (TE): Shanmugam and Venkataramani (2006) determined the technical efficiency (TE) using district level data. The main variables used for computation of TE are rural literacy rate, rural infant mortality, villages with pacca road, electrification and landholding size. Based on his study, TE for the districts of Haryana, Uttar Pradesh, Madhya Pradesh, Rajasthan and India are 79.93, 77.27, 77.07, 77.46, and 79.23%, respectively. This study shows that the relative importance of the determinants of TE across districts depends greatly on environmental factors, such as agro-climatic zones, technological factors (such as irrigation regime), and crop mix. The policy implications are clear: interventions to improve the technical efficiency are not "one-size fits- all." Indeed, even districts within the same state would benefit differently from the same set of interventions. Study indicated that health, education, and infrastructure are powerful drivers of efficiency at the district level. In that sense, it might be wise to develop policy interventions at a more localized level.

7.8 Livelihood and Employment

The most crucial need for sustainable livelihoods for an individual or household is employment, or being gainfully employed over a period of time, that ensures a level of remuneration satisfying basic needs and a quality of life. It is difficult to estimate the level of earning/ income/ inflow of money goods or services to a household that ensures a particular standard livelihood. The most widely used measure is one of income poverty. Households subsisting below this level of income (calculated using expenditure estimates) would be the most vulnerable and deprived households. The other measures would be employment figures, employment in different categories.

Livelihoods, in many ways would then be a sum total of both the employment portfolio of a household and what it earns in monetary terms, or as services and goods, and what a household receives as entitlements being a citizen of the state, or gram panchayat.

Agriculture is the main source of employment in the Yamuna basin. Industrial development is higher in Haryana and cities along Yamuna River.

***Special Employment Programmes and Development Schemes*:** Panchayats as institutions and Human Development as the goal of all planned development have the following three dimensions:

(a) as decentralised institutions that play an effective role in furthering the national goal of human development as representative institutions that are able to express the local concerns for human development and undertake consequent action;
(b) eabling institutions that create opportunities for the local human resource to practice their skills for self-governance; and
(c) generating the development and employment.

The following schemes are working for rural areas:

(a) IRD: Integrated Rural Development
(b) NREP: National Rural Employment Programme
(c) DRDA: District Rural Development Agency
(d) TRYSEM: Training of Rural Youth for Self Employment
(e) IRDP: Integrated Rural Development Programme
(f) SLDP: Special Livestock Development Programme
(g) NAREGA: National Rural Employment Guarantee

7.9 Livestock Population

Livestock Census work has been carried out by Department of Animal Husbandry & Dairying, Ministry of Agriculture. Livestock (also cattle) refers to one or more domesticated animals raised in an agricultural setting to produce commodities such as food and work. The term "livestock" does not include poultry or farmed fish; however the inclusion of these, especially poultry, within the meaning of "livestock" is common. Livestock generally are raised for subsistence or for profit. Raising animals (animal husbandry) is an important component of modern agriculture. It has been practiced in many cultures since the transition to farming from hunter-gather lifestyles. Cattle include bovines such as cow, bull, etc.

Table 7.18 shows cattle, livestock and poultry population. Total cattle population in the Yamuna basin is about 19 million. State wise distribution shows highest cattle population in Madhya Pradesh followed by Uttar Pradesh and Rajasthan as 45, 28 and 18%, respectively (Fig. 7.22). Total 38 million livestock population exist in Yamuna basin. Its state wise distribution (Fig. 7.23) is the highest for Rajasthan as 32%. Haryana ranks first with 48% poultry population in the Yamuna basin (Fig. 7.24).

7.10 Water Supply

Globally, 1 billion people have a lack of access to improved water supply and 2.6 billion people have a lack of adequate sanitation (WHO, 2004). Water provision cannot be separated from two other inter-related factors, sanitation and health. The primary cause of contamination of water is the inadequate or improper system for disposal of human (and animal) excreta. This often leads to a cycle of infection (resulting primarily in diarrhoeal diseases) and contamination which remains one of the leading

Table 7.18 Cattle, livestock and poultry population ('000) for year 2003

	Total cattle		Total livestock		Total poultry	
State	State	Basin	State	Basin	State	Basin
Himachal Pradesh	2,236	245.36	2,881	316.13	767	84.16
Haryana	1,540	1,170.88	7,345	5,584.50	13,619	10,354.71
Madhya Pradesh	18,913	8,589.04	16,704	7,585.86	11,705	5,315.64
Rajasthan	10,854	3,459.92	38,284	12,203.78	6,192	1,973.82
Uttarakhand	2,188	221.57	2,755	278.99	1,984	200.91
Uttar Pradesh	18,551	5,460.02	39,980	11,767.12	11,718	3,448.90
Delhi	92	92.25	282	279.71	459	455.28
Total	54,374	19,057.51	108,231	37,933.81	46,444	16,278.12
India	185,181		485,002		489,012	

Source: Ministry of Agriculture, Department of Animal Husbandry Dairying & Fisheries, GoI.

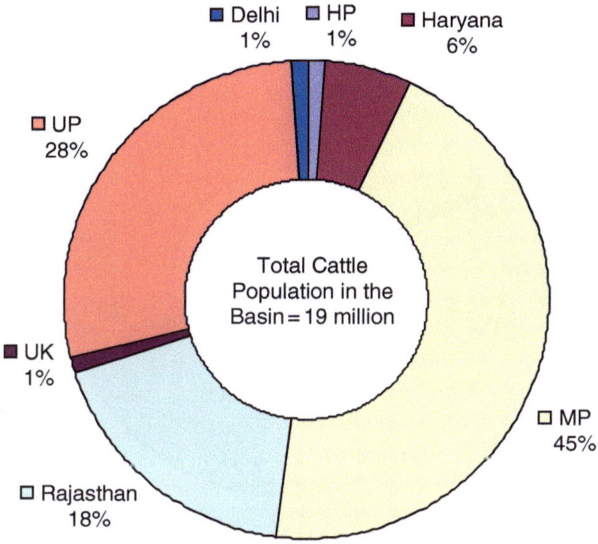

Fig. 7.22 State wise distribution of cattle in the Yamuna basin

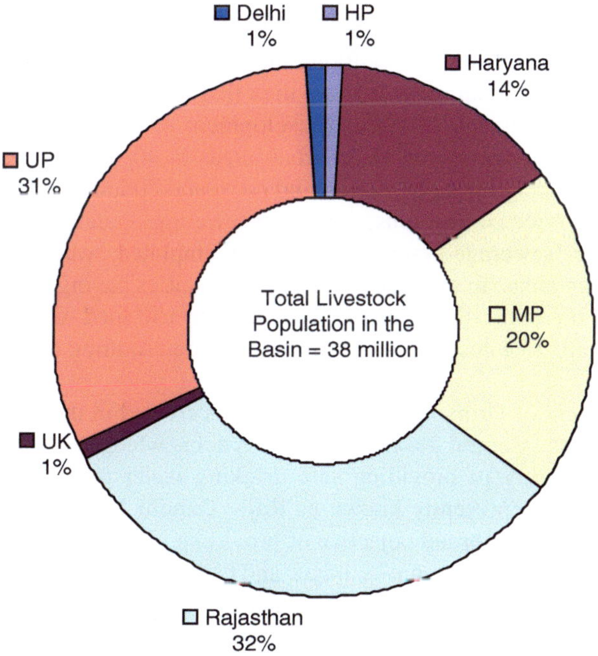

Fig. 7.23 State wise distribution of livestock in the Yamuna basin

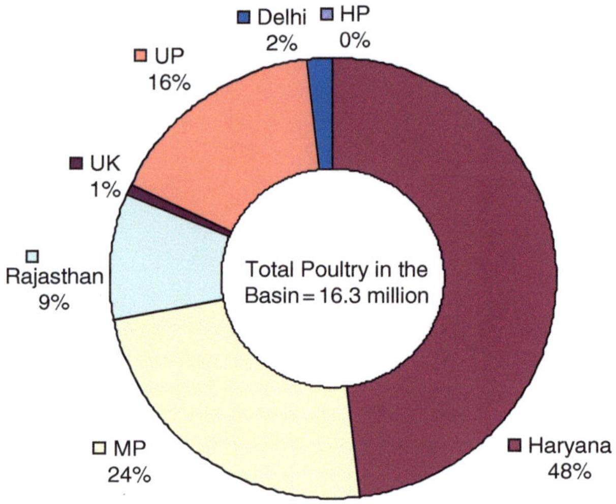

Fig. 7.24 State wise distribution of poultry in the Yamuna basin

causes of illness and death in the developing world. Inadequate water supply is both a cause and an effect of poverty. Invariably adequate and affordable water supplies are poor in the society.

***Per capita Water Requirement*:** Potable drinking water is still not available to various areas. The quantity of water required for domestic purposes depends mainly on habits, social status, climatic conditions and customs of the people. The per capita water requirement in urban areas is more than that in the rural areas. As per yardstick of the Union Ministry of Urban Development & Poverty Alleviation, water requirement for domestic purposes in urban areas is 40 litres per capita per day (lpcd) in case of supply through public stand posts and 70 lpcd in the case of supply through house service connections, where no sewerage system is existing or contemplated. Where sewerage system exists or contemplated, water supply would be 135 lpcd in urban areas. In the case of metropolitan cities having a population more than 1 million, the domestic water supply would be 150 lpcd. Over and above the aforesaid demand, 15% losses may be allowed for determining the quantity of raw water required.

The Department of Drinking Water Supply was created in the Ministry of Rural Development by the Central government, and is acknowledged as the nodal agency with the responsibility of providing safe drinking water to all rural habitations. Technology Mission (presently known as Rajiv Gandhi National Drinking Water Mission) works with the broad objective of providing sustainable and safe drinking water to all uncovered/"no source" villages and creating awareness among the rural people about the hazards of using unsafe water.

***Water Supply Rate*:** The Five Year Plan documents and the National Sample Survey provide official figures about the percentage of population, having access to drinking water in rural and urban areas. The Planning Commission states that these figures are far from the ground realities, which means that the situation is not as projected in the official circles. Traditionally water supply is limited to major towns and cities and that too within the boundaries of state/provincial units. With the spread of the process of urbanization and declining public health standards in both urban and rural areas government has taken a serious initiative in the form of laws and policies. Water supply is done by mainly the government department of respective states. As a matter of policy it is desired as 135 lpcd for urban area and 40 lpcd for rural areas.

A two-way approach is adopted in this regard, viz:

1. Legislations focusing on water supply and at times on water supply and sanitation to be driven by the state agencies
2. Policy initiatives by the central government in order to assist and supplement the states' activities with the overall objective of providing safe drinking water and thereby promoting public health.

Tables 7.19, 7.20, and 7.21 describe the total number of class I and II cities in the states falling under the Yamuna basin as whole and water supply rates. Haryana, Madhya Pradesh, Uttar Pradesh, Rajasthan receive less water supply than the prescribed limit of 135 (Table 7.20).

7.10.1 Water Auditing

In domestic water supply, water audit is considered important, since the treatment of water to bring it to drinking water standard costs a lot of money to the supplier. Water audit helps in determining the amount of water lost from a distribution system due to leakages, etc.

Water audit compares the amount of water supplied with the amount billed and accounts for the water loss. Draft Guidelines for Water Audit and Water Conservation, jointly produced by the Central Water Commission and the Central Ground Water Board are as follows:

Table 7.19 Water supply in class–I cities

State	No. of cities	Water supply in MLD	Per capita supply
Delhi	1	4,346	292.49
Haryana	20	783.39	142.59
Himachal Pradesh	1	36.18	221.3
Madhya Pradesh	25	1,560.91	144.6
Rajasthan	24	1,727.96	179.78
Uttar Pradesh	61	4,382.58	170.12
Uttarakhand	6	221.21	177.06

Source: CPCB (2009)

Table 7.20 Water supply in class–II cities

State	No. of cities	Water supply in MLD	Per capita supply
Haryana	7	49.68	91.32
Madhya Pradesh	23	163.64	93.77
Rajasthan	21	184.76	115.53
Uttar Pradesh	46	432.19	127.77
Uttarakhand	1	11.34	163.26

Source: CPCB (2009)

Table 7.21 Water supply in state wise class–I and II cities (aggregate)

State	No. of cities	Water supply in MLD
Delhi	1	4,346
Haryana	27	833.07
Himachal Pradesh	1	36.18
Madhya Pradesh	48	1,724.55
Rajasthan	45	1,912.72
Uttar Pradesh	107	4,814.77
Uttarakhand	7	232.55

Source: CPCB (2009)

(i) Water audit is an important management tool for effective conservation of water. Broadly water audit should be conducted categorically in two systems, resource audit or supply side audit and the other one as consumption audit on demand side. All efforts should be made for improvement of not only water use efficiency and distribution system, but also on the efficient development and management of the source of water.
(ii) It has been strongly advocated that the water audit system needs to be framed and incorporated in every significant water resources project as a routine exercise during operation and maintenance of the project by the project authorities.
(iii) The periodicity of water audit and its report may be determined in advance at the commencement of commissioning the project by the project authority and the concerned Governments and appropriate provision of fund may be made for its implementation.
(iv) The recommendations in the water audit report for corrective measures of the system may be considered on priority for implementation by the competent authority. All efforts should also be made to provide all technical and financial provisions in a time bound manner.
(v) The irrigation sector utilizes about 83% of water as a major stakeholder. Due to the thrust on account of rapid urbanization and modernization, the demands for domestic and industrial uses are progressively increasing, thus creating a situation of competing of demands from value added sectors of water use and threatening irrigation sector even in maintaining current level of water

use whereas more water is needed for growing more to meet the demand of growing population.

(vi) A systematic comprehensive water audit is useful in bringing out the trend of changes in demand and supply scenario which will help in deciding the methodology for improving the efficiency of the system by adopting conjunctive use of surface and ground water, application of modern irrigation techniques, including drip and sprinkler irrigation wherever feasible and other improvised agricultural devices in addition to development of wasteland and waterlogged areas.

(vii) Due to over-exploitation of ground water, the water table at vulnerable places like thickly populated urban areas are depleting at a very fast rate. Private tube wells are mushrooming without control to meet the growing demand. Industries should be discouraged to exploit ground water on their own.

(viii) As far as possible supplies to industries should be from surface water and if ground water supply is considered essential, it should be managed by a Government Agency. There is general apathy towards conjunctive use of ground water and surface water. Specific water audit needs to be conducted on a regular basis for realistic assessment of ground realities and initiating remedial measures under the umbrella of holistic approach.

(ix) Pollution level of fresh surface water and ground water resources are alarmingly increasing due to the excessive use of pesticides and fertilizers in agriculture and discharge of untreated waste by industries and sewage disposal leading to health hazards and scarcity of fresh water. Water audit from this angle needs to be conducted strategically and periodically. The existing laws regarding pollution control need to be strictly observed by not only imposing penalties but also restricting the polluters.

(x) To prevent wastage of water, pricing of water for irrigation, domestic and industrial uses needs to be revised and updated periodically so that subsidy is phased out as quickly as possible and at least operation and maintenance cost is recovered for sustainability of the system. Further, the pricing of water at flat rate system needs to be replaced gradually by actual cost rate by volume. The differential pricing system should also be suitably introduced keeping in view the socio-economic aspects of the people and the region in addition to their life style and ethnic background.

(xi) Benchmarking system of various suitable parameters for all sectors of water use may be developed and introduced for optimizing and enhancing the efficiency of the system. It is an effective tool for water audit and measurement of relative performance and suggests ameliorative measures for performance improvement.

(xii) To identify the source of water loss due to leakage, the approach of bulk metering system should be installed at various well defined macro and micro systems like various zones, districts, towns, colonies and even large group-consumers to single unit consumers so that water audit can be effectively conducted.

7.10.1.1 Transmission Losses

A study undertaken by the Ministry of Urban Development & Poverty Alleviation through NEERI, Nagpur, has revealed that about 30 to 50% of the water produced and supplied in the cities goes as waste through leakages in the distribution system. About 80% of the aforesaid losses are estimated in the household connections due to worn out pipes, etc.

In view of this, the Ministry of Urban Development & Poverty Alleviation has emphasized the need for control of unaccounted supply of water (non-revenue water) through leak detection programmes for identifying leakages and rectifying the same through suitable replacement of pipelines. A manual containing details of various aspects of operations and maintenance (O&M) of water supply systems has been brought out by the Ministry of Urban Development & Poverty Alleviation recently.

7.10.1.2 Water Conservation

(i) Water Conservation is a prime and challenging concern. There is a gap on the application of the appropriate technologies, which needs to be removed. Due to lack of proper operation and maintenance in irrigation, industry and domestic water distribution system, there is a huge loss of water. Hence, it is emphasized to improve the operation and maintenance (O&M) system.

(ii) For developing water resources, age-old traditional water conservation methods need to be judiciously adopted in conjunction with the latest modern conservation technology. Keeping this in view, rain water harvesting, revival of traditional water storages, check dams and other similar structures need to be adopted. Building byelaws should be suitably modified to introduce mandatory roof top rain water harvesting.

(iii) In order to conserve precious fresh water, recycling of waste water may be incorporated wherever feasible. The dual water supply system, one for treated wastewater and the other for fresh water may be introduced so that treated waste water can be used for secondary purposes, such as toilets flushing, gardening, agriculture and selective industries, etc. New urban colonies, big hotel industries and other similar establishments should have mandatory dual water supply systems.

(iv) Cropping patterns and crops water requirement varies from time to time due to the dynamic socio-economic condition of the people and the region in addition to geo-morphological, climatic and metrological changes. Hence, for effective management, appropriate base line data for water demand under different situations need to be brought out for optimum crop water management and field activities considering effective rainfall in different physiological stages.

(v) Night irrigation practice may be introduced to minimize evaporation loss thus conserving irrigation water. Timely and need based irrigation should be done to minimize loss of water. Further, for boosting productivity, rotational

Table 7.22 Norms for providing potable drinking water

Purpose quantity	lpcd
Drinking	3
Cooking	5
Bathing	15
Washing utensils and house	7
Ablution	10

cropping pattern may be introduced for balancing fertility of soil and natural pest control.

(vi) Various water saving devices are being developed under various ongoing R&D programmes. These devices may be suitably adopted in the system.

(vii) Strategic mass awareness campaign should be conducted regularly to cover all stakeholders, including service providers and consumers, for water conservation in irrigation, domestic and industrial sectors. Special attention must be given so that the fruits of the campaign must reach children, housewives and farmers effectively.

Drinking water supply is a state subject. *Accelerated rural water supply programme (ARWSP) certain guidelines and norms* that have been adopted since the inception of the programme (1972) for providing potable drinking water to the rural population based on basic minimum need is as follows: 40 lpcd for humans to meet the following requirements based on basic minimum need as defined under the ARWSP guideline (Table 7.22). The implementing agencies for the programme may be decided by the State Government. The implementation may be through the PHED or Rural Development Department/Panchayati Raj Department/Board, Corporation or Authority. In addition, provision should be allowed at 30 lpcd for animals in hot and cold desert ecosystems.

References

Bhat (2001) Maternal mortality in India: an update. www.iussp.org/Brazil2001/S10/S1601Bhat.pdf

Census (1991) Census of India – 1991, Ministry of Home Affairs, Government of India

Census (2001) Census of India – 2001, Ministry of Home Affairs, Government of India

CPCB (2009) Status of Water supply, Wastewater generation and Treatment in class-I cities and Class-II towns of India, Central Pollution Control Board of India

International Institute for Population Sciences (1995) National family health survey (MCH and Family Planning), India 1992–93. IIPS, Mumbai, India

Planning Commission (2004) Government of India, New Delhi

Planning Commission (2005) Government of India, New Delhi

Shanmugam KR, Venkataramani A (2006) Technical efficiency in agricultural production and its determinants: an exploratory study at the district level. Working paper: Madras School of Economics, Chennai

Srinivasan K (2004) Population and development in India since independence: an overview. The J Family Welfare 50 Special Issue

SRS Bulletin (2008) Registrar General of India

Statistical Abstract of India (2000) Central Statistical Organization, Ministry of Statistics and Programme Implementation, Government of India

UNDP (2006, 2007) Human Development Report, United Nations Development Programme

WHO (2004) Water, sanitation and hygiene links to health: facts and figures. World Health Organization, Geneva

WSSD (2002) World Summit on Sustainable Development, Johannesburg

Chapter 8
Water Resources and Water Budgeting

Abstract This chapter presents a catchment-wise water resource analysis of the Yamuna River basin. Both surface and ground water resources are included. Over-exploitation of ground water resources is causing persistent lowering of the water table in the basin. This major issue is discussed using the indicator of the stage of ground water development. In order to know the exact status of the basin in terms of water resources, surplus and deficit analyses are also presented. Furthermore, catchment wise flood estimation techniques, based on the synthetic unit hydrograph and geomorphological unit hydrograph approaches, are also presented.

Water resources are sources of water that is vital for life. Uses of water include agricultural, industrial, household, recreational and environmental activities. Virtually all these human uses require fresh water. Globally, nearly 97% of the water on the earth is salt water, and only about 3% of the water exists as fresh water of which nearly two third is frozen in glaciers and polar ice caps. The remaining unfrozen fresh water is mainly found as groundwater, with a small fraction (or percent) above ground or in the air.

Although fresh water is a renewable resource, but the supply of the clean and fresh water is steadily decreasing as a result of exceeding water demand and increasing pollution in many parts of the world, and Yamuna River basin is no exception. Besides, the availability of water greatly varies in time and space. Therefore, it becomes vital to estimate the extent and availability of surface and ground water for economic and optimum utilization, planning, design and operation of water resources.

This chapter discusses the availability of water in the Yamuna basin, including catchment wise water budgeting for analysing surplus and deficit catchments in terms of water resources needed for integrated water resources management, and methodology for flood estimation for hydrological design.

8.1 Water Resources Assessment

Water resources of a basin are equal to the volume of average annual flow of surface water generated from precipitation and groundwater. However, if human intervention in the catchment has been done in the form of surface storages, then the water resource of a catchment can be estimated using the following formula.

R.K. Rai et al., *The Yamuna River Basin*, Water Science and Technology Library 66,
DOI 10.1007/978-94-007-2001-5_8,

$$\mathrm{WR} = Q_\mathrm{S} + \mathrm{GW}_\mathrm{rec} - (Q_\mathrm{out} - Q_\mathrm{in}) + S_\mathrm{s} + S_\mathrm{sm} \tag{8.1}$$

where WR is the water resources availability in the basin, Q_s is the surface runoff volume, i.e., the total volume of the long term average annual flow of surface runoff due to precipitation, GW_rec is the groundwater recharge generated from precipitation, Q_out is the groundwater contribution to the base flow, and Q_in is the seepage from river or canal to the aquifer, S_s is the surface storages, S_sm is the soil moisture storage generally filled up during the monsoon period; however, it supplies water to meet agricultural demands.

Introduction of the term net recharge simplifies eq. (8.1) as

$$\mathrm{WR} = Q + \mathrm{GW}_\mathrm{Net} + S_\mathrm{s} + S_\mathrm{sm} \tag{8.2}$$

where Q is the total volume of long term average annual flow in the river, which includes surface runoff and base flow in the river (i.e. $Q_\mathrm{s} + Q_\mathrm{out}$), and GW_Net is the net annual groundwater recharge, which includes GW_rec, Q_in and Q_out. All the variables appearing in eqs. (8.1) and (8.2) are in volumetric unit.

8.1.1 Surface Water Resources

The surface water resources of the basin are assessed based on stream flow measured at a terminal site (i.e. site either close to the sea or confluence with the other higher order stream). The average annual flow at the terminal point of river is normally denoted as the water resources of the basin.

When water resources in the basin have been developed, the surface water resources availability would be estimated as the sum of Q and S_s. The long term average annual runoff volume, Q is estimated using the discharge data at different sites. The locations of the gauge-discharge sites maintained by Central Water Commission are shown in Fig. 8.1, and their salient features are given in Table 8.1.

The steps involved in the estimation of surface water are to first derive the flow duration curve (FDC) followed by estimation of 50, 75 and 90% dependable flow. The FDCs for the discharge sites of the Yamuna River are derived using the Weibull's plotting position formula which is described in Chapter 3. The derived FDCs for various sites of the Yamuna River are shown in Figs. 8.2, 8.3, and 8.4; and the estimated values of dependable flows are given in Table 8.2. The table also includes dependable water availability (in volumetric terms) used to estimate the surface water availability for different dependabilities, and mean annual discharge with the corresponding values of surface water availability at different locations in the basin and sub-basins. For Yamuna basin, an attempt is to establish the relationship between catchment area and mean discharge (Fig. 8.5), catchment area and water availability (Fig. 8.6), and catchment area and dependable flows (Fig. 8.7). These relationships can be used as ready reference for the Yamuna basin.

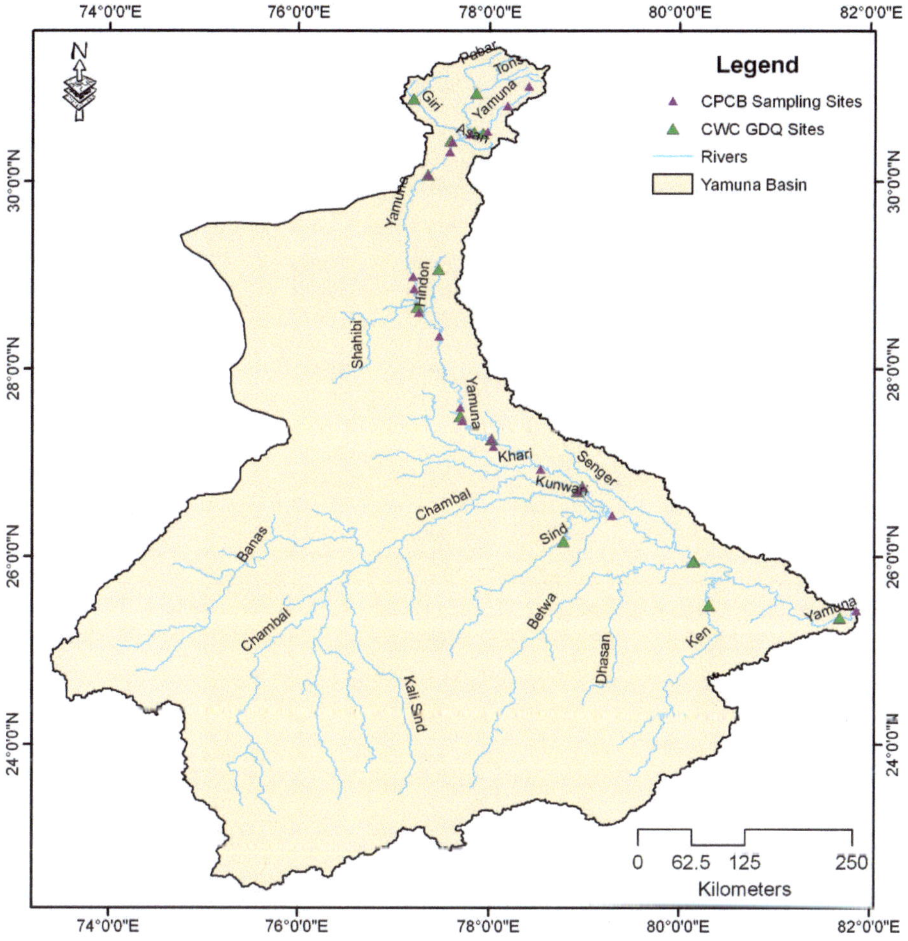

Fig. 8.1 CWC GD and CPCB WQ sites of Yamuna River and its tributaries

8.1.1.1 Surface Water Storage

There are large number of reservoirs and dams constructed in the Yamuna River basin. The list of water storages along with salient features of dams and reservoirs in the basin are given in Appendix A. The location of important dams in the basin is shown in Fig. 8.8. Therefore, the catchment wise surface water availability in the basin estimated by summing up the mean annual surface runoff in the river (Table 8.2), and live storages in the catchments (Appendix A) is given in Table 8.3.

Based on the above analysis (Table 8.3), the total estimated surface water resources availability in the Yamuna basin is 61,219.2 MCM, of which Chambal River basin 43.67% contribution has followed by Betwa (19.28%), Ken (13.21%) and Sind (3.08%) catchments.

Table 8.1 Salient features of CWC GD sites

S. No.	Site	River	Latitude	Longitude	Type	Catchment Area (km^2)[a]	% Area
1	Paonta	Yamuna	30.4306	77.5911	GDSQ	10,875.50	3.27
2	Kalanaur	Yamuna	30.0696	77.3528	GDSQF	12,720.50	3.83
3	Delhi Rly. Bridge	Yamuna	28.6624	77.2459	GDSQWF	30,793.84	9.27
4	Mathura	Yamuna	27.4984	77.6956	GDSQ	41,703.84	12.55
5	Agra	Yamuna	27.2556	78.0226	GDSQ	42,572.24	12.81
6	Etawah	Yamuna	26.7500	78.9833	GDSQWF	71,999.24	21.67
7	Hamirpur	Yamuna	25.9572	80.1601	GDQWF	247,432.20	74.47
8	Pratappur	Yamuna	25.3556	81.6841	GDSQ	331,318.80	99.72
9	Naini Bridge	Yamuna	25.4244	81.8616	GF	332,264.8	100.00
10	Galeta	Hindon	26.0667	77.4667	GDQ		
11	Udi	Chambal	26.7000	78.9333	GDSQ	134,679	99.99
12	Seondha	Sind	26.1667	78.7833	GDSQ	16,701	59.77
13	Sahjina	Betwa	25.9455	80.1507	GDSQF	43,651.0	99.99
14	Banda	Ken	25.4864	80.3122	GDSQWF	27,616	96.65

G Gauge, *S* Silt, *W* Wireless, *D* Discharge, *Q* Water Quality, *F* Flood Forecasting
[a] Catchment area does not include the WYC command area, but it is based on flow direction

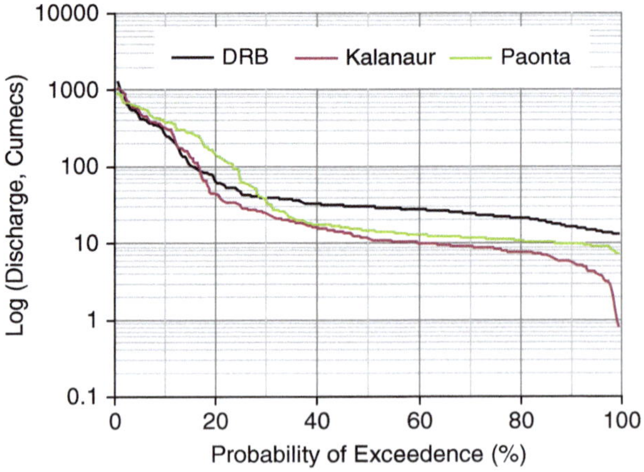

Fig. 8.2 Flow duration curve for different locations in the Yamuna basin

8.1.2 Groundwater Resources

Groundwater resources are important for water resources planning as they supplement surface water resources. In the Yamuna basin more than 65% of water supply is dependent on the groundwater potential. The total annual groundwater potential for a unit refers to the available annual recharge after allowing for the natural

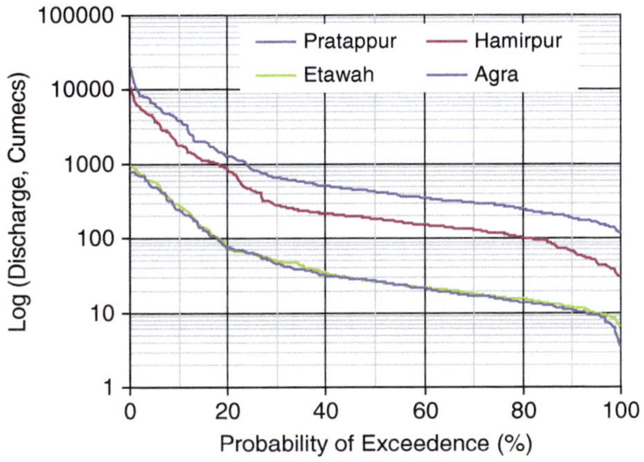

Fig. 8.3 Flow duration curve for different locations in the Yamuna basin

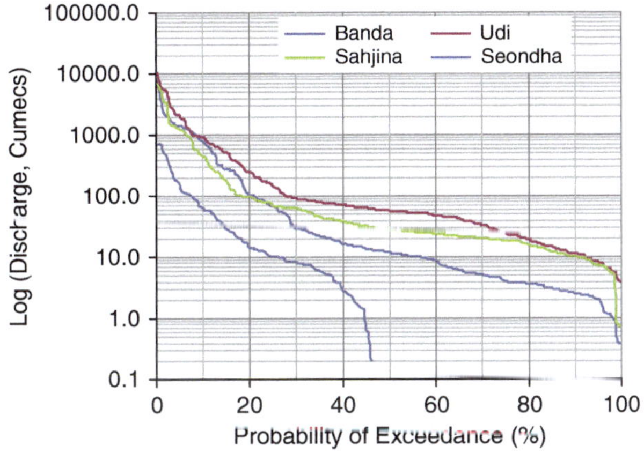

Fig. 8.4 Flow duration curve for different locations in the Yamuna basin

discharge during the monsoon and non-monsoon season through base flow and subsurface inflow/outflow. In the study, the groundwater potential is estimated using the methodology presented in GEC-97 of Central Ground Water Board (CGWB). The district wise groundwater potential is shown in Fig. 8.9 which reveals that the groundwater availability is high along the course of the Yamuna due to alluvial geological formation. The district-wise groundwater information is summarized in Appendix B which also includes total recharge, groundwater flow, net availability, total draft and stage groundwater development.

A large fraction of water demand that is being fulfilled by ground water is resulting in ground water depletion. Further, due to the over-exploitation of groundwater for intensive agriculture, a large part of the basin has fallen under dark zone, and it can be assessed through the stage of ground water development.

Table 8.2 Discharge and runoff water availability at different locations in the Yamuna basin

Site	River	Area (km^2)	Dependable flow (m^3/s)			Dependable water availability (MCM)			Mean annual flow (m^3/s)	Mean annual availability (MCM)
			50%	75%	90%	50%	75%	90%		
Paonta	Yamuna	10,875.5	14.95	11.50	10.12	471.46	362.66	319.14	114.32	3,605.20
Kalanaur	Yamuna	12,720.5	11.73	8.53	5.76	369.92	269.00	181.65	80.35	2,533.92
DRB	Yamuna	30,793.8	30.34	22.52	16.53	956.80	710.19	521.29	88.71	2,797.56
Mathura	Yamuna	41,703.8	–	–	–	–	–	–	–	–
Agra	Yamuna	42,572.2	27.01	15.68	10.79	851.90	494.44	340.27	85.26	2,688.76
Etawah	Yamuna	71,999.2	26.78	16.56	11.91	844.58	522.08	375.47	87.31	2,753.41
Hamirpur	Yamuna	247,432.2	183.84	119.14	66.98	5,797.50	3,757.07	2,112.38	644.87	20,336.62
Pratappur	Yamuna	331,318.8	428.41	283.43	188.84	13,510.47	8,938.30	5,955.31	1,249.53	39,405.18
Allahabad	Yamuna	332,264.8	–	–	–	–	–	–	–	–
Udi	Chambal	134,679	57.70	24.00	11.65	1,819.63	756.86	367.39	353.71	11,154.60
Seondha	Sind	16,701	0.00	0.00	0.00	0.00	0.00	0.00	22.91	722.49
Shahjina	Betwa	43,651	29.50	19.15	10.34	930.31	603.91	326.08	235.57	7,428.94
Banda	Ken	27,616	12.03	4.13	2.60	379.38	130.24	81.99	234.57	7,397.40

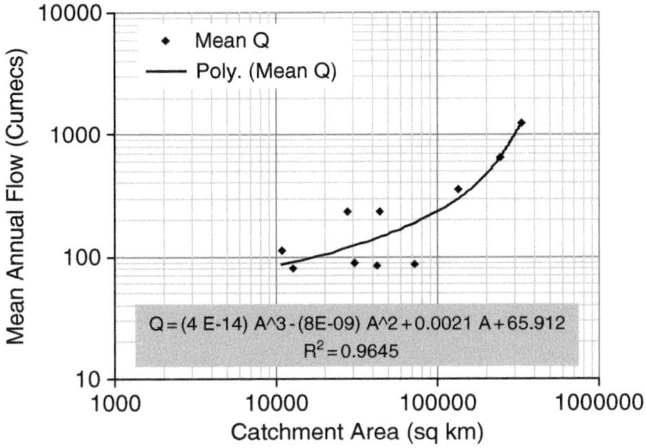

Fig. 8.5 Catchment area vs mean annual flow relation for the Yamuna basin

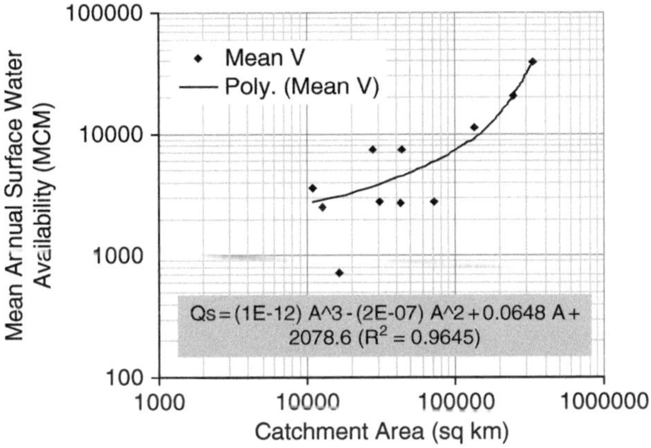

Fig. 8.6 Catchment area vs water availability relation for the Yamuna basin

The stage groundwater development of the unit area can be estimated using the following relationship:

$$\text{Stage GW Development } (\%) = \frac{\text{Annual GW Draft}}{\text{Annual Net GW Recharge}} \times 100\% \qquad (8.3)$$

The district wise stage groundwater development for the Yamuna basin is shown in Fig. 8.10 from which it is observed that most part of the basin in the southwestern and central parts are under dark zone. Most of the Haryana area is either under grey zone or dark zone, though it is the richest in terms of canal networks in the basin. It is also reported that 90% of groundwater draft is used for irrigation; only 10% is being used for domestic and other purposes (GEC, 1997).

Further, the catchment wise groundwater resource availability is summarized in Table 8.4 which also includes the status of GW development in the basin, and it is observed that the Chambal basin is critical in terms of the ground water exploitation followed by the whole Yamuna basin (82.3%) and the Betwa catchment (51.2%).

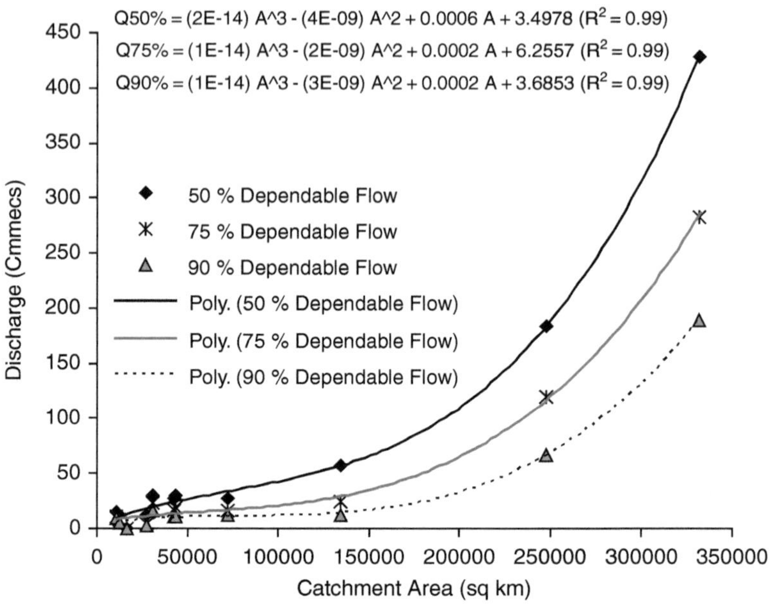

Fig. 8.7 Area-Discharge relations for Yamuna basin at 50-, 75-, and 90% probability

8.1.3 Soil Moisture Storage

In the Yamuna basin annual soil moisture storage varies between 100 and 250 mm (http://www.iwmi.cgiar.org/). For the Yamuna River basin, 18% of mean annual rainfall of the catchment is considered in the analysis.

Considering individual estimates of water resources (i.e., surface water, ground water and soil moisture storage) described under section 8.1, the total water resources of the catchments and the basin that have been estimated are presented in Table 8.5.

8.2 Water Demand Estimation

The water uses in the basin can be broadly categorized as: (i) domestic, (iii) agriculture, (iii) industrial, (iv) power, (v) livestock, etc. In the Yamuna basin, domestic, agriculture, industrial and livestock water use are important.

The domestic water demand is estimated from the population data. The standard water use of 135 lpcd in urban, 175 lpcd for Delhi (average), and 40 lpcd in rural area is considered for the domestic water requirement. According to the water supply of Delhi master Plan: (a) domestic = 135 lpcd, (b) industrial and commercial = 70 lpcd, (c) public use = 10 lpcd, (d) losses and thefts = 55 lpcd,

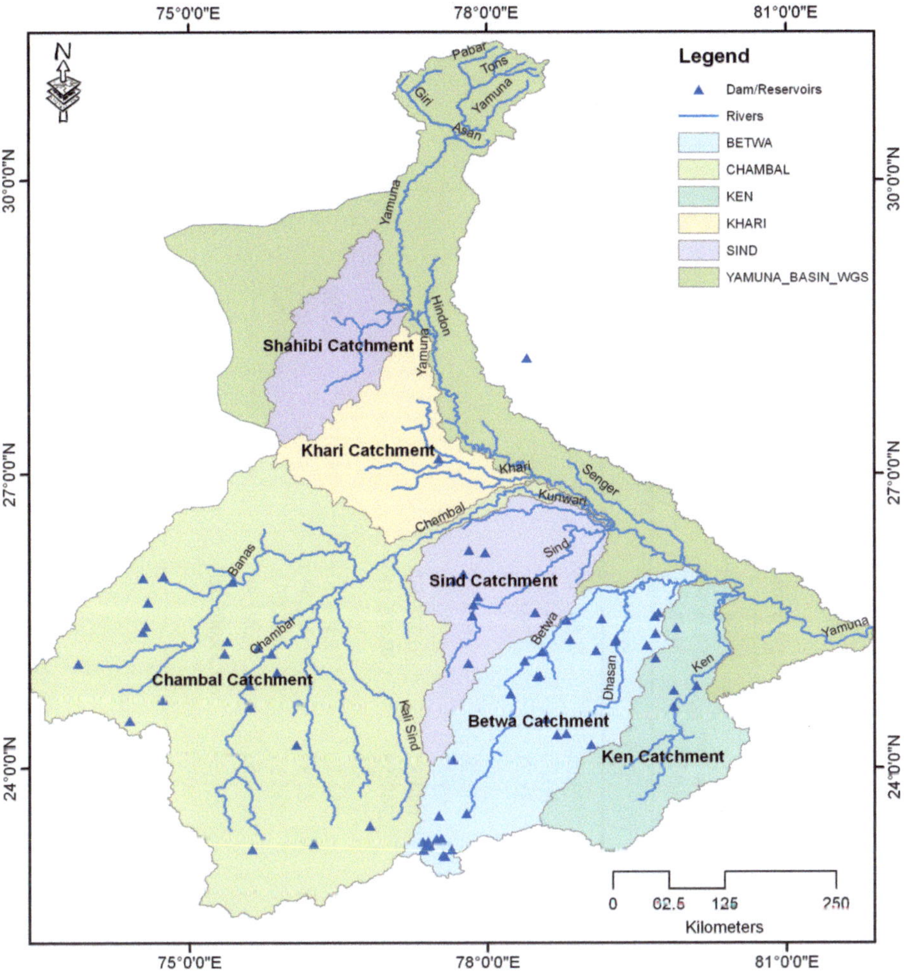

Fig. 8.8 Location of important dams in the Yamuna basin

(e) total = 270 lpcd. The district wise estimated domestic water demand is given in Appendix B. The catchment wise estimates of domestic demand are summarized in Table 8.6. The spatial distribution of domestic water demand in the basin is shown in Fig. 8.11 based on which it can be stated that the domestic water demand is high along the Yamuna Riveras compared to the other area. The domestic water demand is the largest in the Delhi state and NCR region.

For industrial water demand, a thumb rule of nearly 20–30% of municipal water is being utilized to meet industrial demand. Since the whole basin cannot be set at the same level of industrialization like Delhi, the 20% of municipal water demand can be taken as the industrial demand (Table 8.6).

Table 8.3 Surface water resources of the Yamuna River basin

S. No.	Site	Catchment	Area (km^2)	Mean annual flow (m^3/s)	Mean annual availability (MCM)	Gross storage (MCM)	Live storage (MCM)	Surface water resources (MCM)
(i)	(ii)	(iii)	(iv)	(v)	(vi)	(vii)	(viii)	(ix)
1	Paonta	Yamuna	10,875.5	114.32	3,605.20	3,550	1,556[a]	3,605.2 (5,162.2)
2	Kalanaur	Yamuna	12,720.5	80.35	2,533.92			
3	DRB	Yamuna	30,793.8	88.71	2,797.56			
4	Mathura	Yamuna	41,703.8	–	–			
5	Agra	Yamuna	42,572.2	85.26	2,688.76			
6	Etawah	Yamuna	71,999.2	87.31	2,753.41			
7	Hamirpur	Yamuna	247,432.2	644.87	20,336.62			
8	Pratappur	Yamuna	331,318.8	1,249.53	**39,405.18**			
9	Allahabad	Yamuna	332,264.8	–	–			
10	Udi	**Chambal**	134,679	353.71	11,154.60	19,613	15,582	**26,736.6**
11	Seondha	**Sind**	16,701	22.91	722.49	1,248	1,166	**1,888.5**
12	Shahjina	**Betwa**	43,651	235.57	7,428.94	4,923	4,374	**11,802.9**
13	Banda	**Ken**	27,616	234.57	7,397.40	812	692	**8,089.4**
		Yamuna Basin			**39,405.18**		**21,814** (23,371)	**61,219.2** (62,776.0)

[a] Refers to the proposed

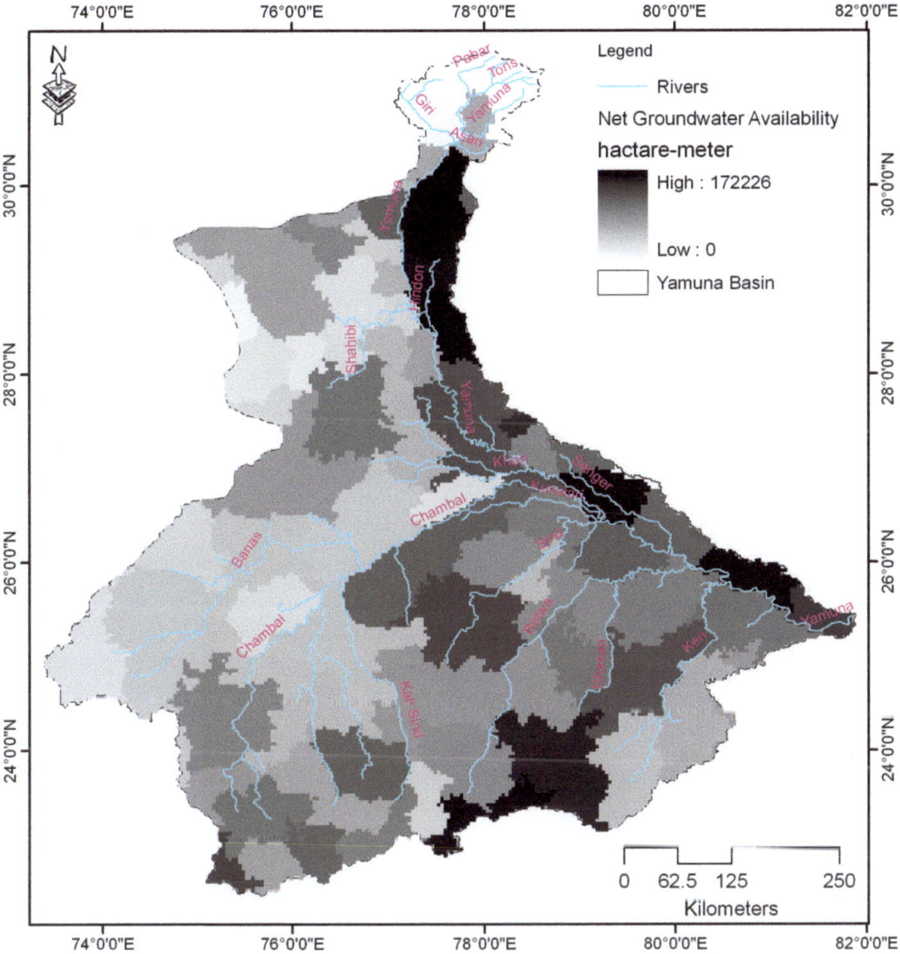

Fig. 8.9 District-wise net groundwater availability in the basin (ha-m)

8.2.1 Agricultural Water Demand

Agricultural demand is estimated using the following relationship (FAO, 1998) [however, a detailed procedure of crop water demand is presented in Chapter 13]:

$$\mathrm{CWR} = \sum_{i=1}^{N} \sum_{j=1}^{M} (K_{\mathrm{c}})_{i,j} \times (ET_0)_j \tag{8.4}$$

where CWR is the crop water requirement (mm), $(K_{\mathrm{c}})_{i,j}$ is the crop coefficient [dimensionless] for the jth day of the ith crop, $(ET_0)_j$ is the reference crop evapotranspiration (mm) of the jth day, N is the number of crops, and M is the

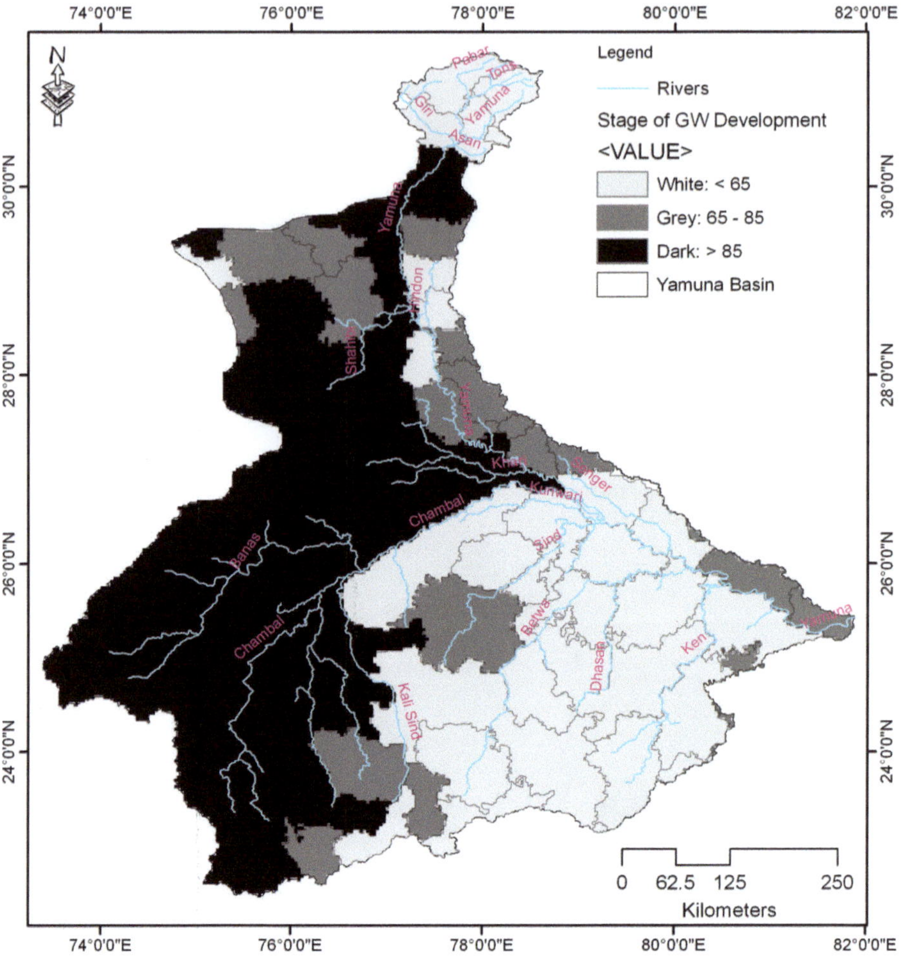

Fig. 8.10 District-wise stage groundwater developments in the basin (%)

crop duration. Approximate date of sowing, growth stage and crop coefficients for important crops of Yamuna basin are given in Table 8.7. Considering the cropping patter of the district, the crop water demand is estimated using eq. (8.4).

The spatial variability of crop water demand in the basin is shown in Fig. 8.12 [detailed estimates are presented in Chapter 13]. It is observed that the crop water demand is high along the alluvial plains of the basin. Based on Figs. 8.11 and 8.12, it can be stated that the demand is very high along the main river course and alluvial system of major tributaries, resulting high ground water abstraction and induced pollution. Further to this, crop water demand is less in the Ken catchment followed by Betwa catchment. The catchment wise summary of the crop water demand is presented in Table 8.6.

Table 8.4 Sub-basin-wise groundwater resources availability

S. No.	Catchment	Area (km^2)	Total recharge (MCM)	GW flow (MCM)	Net GW availability (MCM)	Total draft (MCM)	Stage GW development (%)
1	Upper Yamuna	11,058.4	487.0	30.1	457.0	40.5	12.1
2	Chambal	134,967.6	10,075.4	692.5	9,382.9	9,133.9	101.2
3	Sind	27,940.0	3,630.7	189.0	3,441.7	1,413.8	40.9
4	Betwa	43,835.9	6,050.3	349.2	5,701.0	2,837.3	51.2
5	Ken	28,806.5	3,293.7	180.7	3,113.0	1,497.9	50.2
6	Yamuna	365,870.7	48,907.2	3,478.4	45,428.8	34,669.6	82.3

Table 8.5 Sub-basin-wise estimates of demand and groundwater resources

S. No.	Catchment	Area (km^2)	Soil moisture storage (MCM)	Net GW availability (MCM)	Surface water resources (MCM)	Total water resources (MCM)
(i)	(ii)	(iii)	(iv)	(v)	(vi)	(vii) = (iv)+(v)+(vi)
1	Upper Yamuna	11,058.4	1,935.0	457.0	3,605.2 (5,162.2)	5,997.0 (7,554.0)
2	Chambal	134,967.6	23,619.0	9,382.9	26,736.6	59,739.0
3	Sind	27,940.0	4,890.0	3,441.7	1,888.5	10,220.0
4	Betwa	43,835.9	7,671.0	5,701.0	11,802.9	25,175.0
5	Ken	28,806.5	5,041.0	3,113.0	8,089.4	16,244.0
6	Yamuna	365,870.7	**640,27.0**	**45,428.8**	**61,219.2 (62,776)**	170,675.0 (172,231.0)

Values in the parenthesis are the estimates after functioning of three dams in the upper Himalayan segment

8.2.2 Surplus and Deficit Analysis

In the Sections 8.1 and 8.2, the water availability and water demand has been estimated; and used to estimate the surplus and deficits in the basin/catchments. The estimated gross water demand in the basin is 205,216 MCM, of which 98.2% water is used for agriculture in the basin; whereas the water availability in the basin is 170,675 MCM. Therefore, total deficit in the basin is of the order of 34,541 MCM (Table 8.6).

From analysis it may be concluded that all catchments of the basins are experiencing water deficits, except the upper Himalayan and Ken catchments as the crop water demand is quite low in these catchments due to their prevailing geographic conditions.

Table 8.6 Catchment wise estimates of demand and groundwater resources

S. No.	Catchment/ basin	Domestic water demand (MCM)	Crop water demand (MCM)	Industrial water demand (MCM)	Gross water demand (MCM)	Soil moisture storage (MCM)	Net GW availability (MCM)	Surface water resources (MCM)	Total water resources (MCM)	Surplus/deficit
(i)	(ii)	(iii)	(iv)	(v)	(vi) = (iii)+(iv)+(v)	(vii)	(viii)	(ix)	(x) = (vii)+(viii)+(ix)	(xi) = (x)–(vi)
1	Upper Yamuna	54.3	1,332	10.86	1,397.16	1,935	457	3,605.2 (5,162.2)	5,997 (7,554)	4,599.8 (6,157.0)
2	Chambal	721	62,771	144.2	63,636.2	23,619	9,382.9	26,736.6	59,739	−3,897.2
3	Sind	166.8	13,786	33.36	13,986.16	4,890	3,441.7	1,888.5	10,220	−3,766.1
4	Betwa	287.5	26,456	57.5	26,801	7,671	5,701	11,802.9	25,175	−1,626.0
5	Ken	137.2	12,808	27.44	12,972.64	5,041	3,113	8,089.4	16,244	3,271.4
6	Yamuna	3,712.9	200,761	742.58	205,216.48	64,027	45,428.8	61,219.2 (62,776)	170,675 (172,231)	–34,541.5 (–32,984.7)

Values in the parenthesis are the estimates when proposed dams are considered

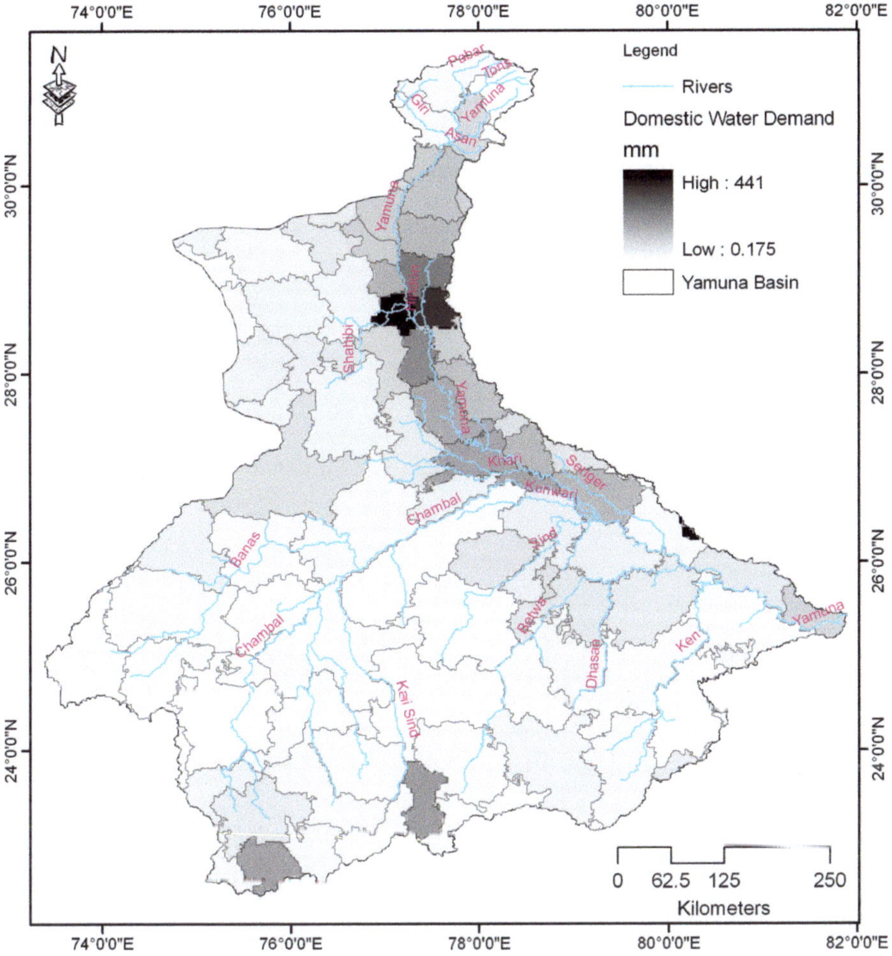

Fig. 8.11 Spatial distribution of domestic water demand in the Yamuna basin

8.3 Flood Hydrograph Estimation for the Catchments

Accurate estimation of flood hydrograph is important for hydrological design, such as design of culverts, design of the spillway capacity of reservoirs, hydropower design, etc. It requires a variety of data, including climatic, topographic, soils, land use information along with event based hydrologic data for model calibration. However, practically most of the small streams are ungauged in India; and it becomes difficult to estimate the flood hydrograph using sophisticated models (Singh and Frevert, 2002a, b). Under such circumstances, regional flood estimation becomes more useful for the derivation of flood hydrograph. Apart from this, the geomorphology based conceptual models have become popular for the estimation of synthetic hydrographs.

Table 8.7 Approximate date of sowing, growth stage and crop coefficients for important crops of the Yamuna basin

Crop	Sowing date (approx)	Growth stage (days) Initial	Development	Middle	Late	Total duration (Days)
Wheat	1st November	20	60	20	50	150
K_c		0.40	1.00	1.20	0.60	
Barley	1st November	15	25	50	30	120
K_c		0.20	1.00	1.15	0.25	
Gram	20th October	20	30	30	20	100
K_c		0.35	0.75	1.05	0.50	
Potato	25th October	25	30	30	30	115
K_c		0.52	0.81	1.10	0.70	
Mustard	1st November	25	35	35	40	135
K_c		0.38	0.38	1.00	0.56	
Bajra	10th June	20	35	25	30	110
K_c		0.41	1.00	1.20	0.40	
Paddy	20th June	30	30	60	30	150
K_c		1.05	1.20	1.20	0.75	
Jowar	20th June	20	40	45	20	125
K_c		0.42	1.00	1.20	0.40	
Maize	15th May	20	35	40	30	125
K_c		0.42	0.95	1.00	0.50	
Arhar	15th June	20	35	40	30	125
K_c		0.63	0.63	0.63	0.63	
Soybean	15th May	20	35	60	25	140
K_c		0.40	1.00	1.15	0.50	
Groundnut	15th May	35	45	35	25	140
K_c		0.52	0.85	1.00	0.55	
Cotton	10th April	30	50	60	55	195
K_c		0.43	1.00	1.15	0.60	
Sugarcane	15th November	30	50	180	60	320
K_c		0.47	1.00	1.25	0.70	

Source: FAO (1998)

This section, therefore, discusses the application of a regional flood estimation technique given by Central Water Commission (CWC, 1983) as well as the geomorphology based conceptual modelling technique for the generation of synthetic hydrograph (Rai et al., 2009; Singh, 1988). Both of these techniques follow the assumptions of Unit Hydrograph.

8.3.1 Regional Synthetic Unit Hydrograph Technique

The CWC (1983) method is well accepted by field engineers and hydrologists for the derivation of unit hydrograph (UH) for ungauged catchments in India for the estimation of design floods for installing waterway of bridges/cross drainage

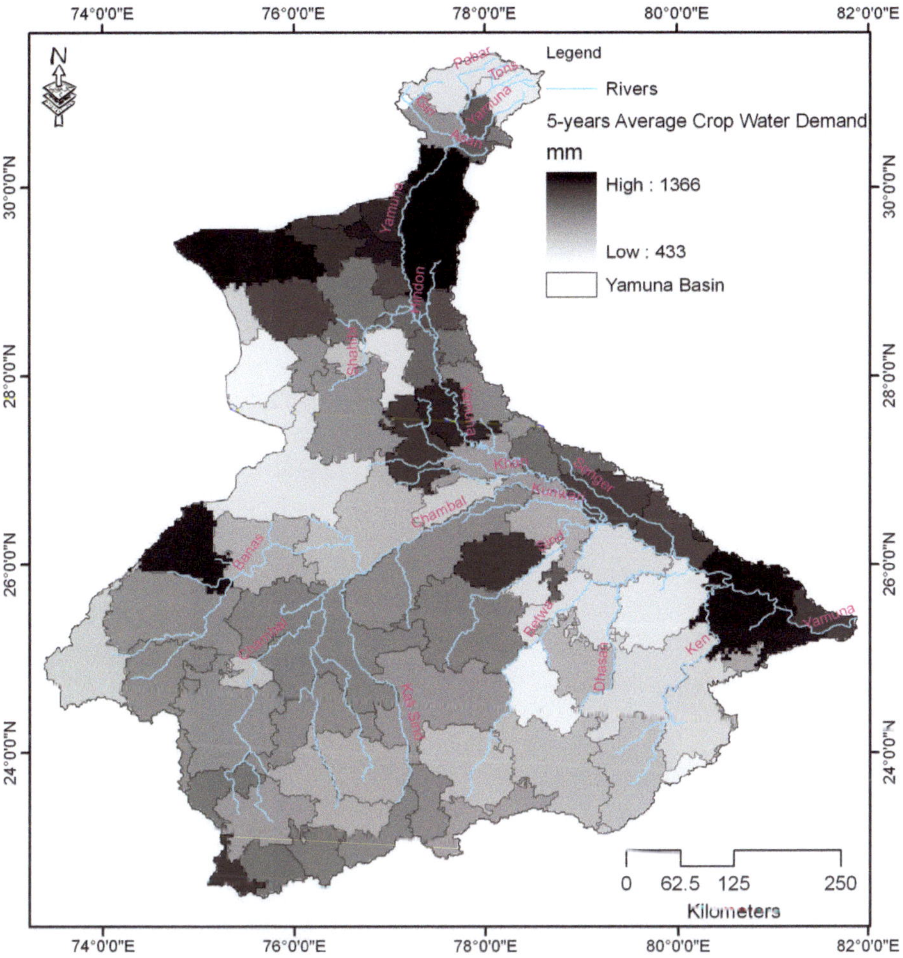

Fig. 8.12 Spatial distribution of crop water demand in the basin

structures across small and medium streams. In this method a set of equations for UH parameters (viz., peak flow rate, time to peak flow rate and time base of the hydrograph) as a function of slope, length of the longest stream and drainage area are developed for the regions, based on the regional analysis of observed data from 42 representative sub-watersheds (40 sub-watersheds data were collected by Northern and North-eastern Railways, India; and 2 sub-watersheds data were collected by Ministry of Transport, India).

Considering the hydro-climate and geography and soil characteristics of the area, India has been divided into seven hydro-meteorologically homogeneous zones, and is further divided into twenty six hydro-meteorological sub-zones [i.e. sub-zones

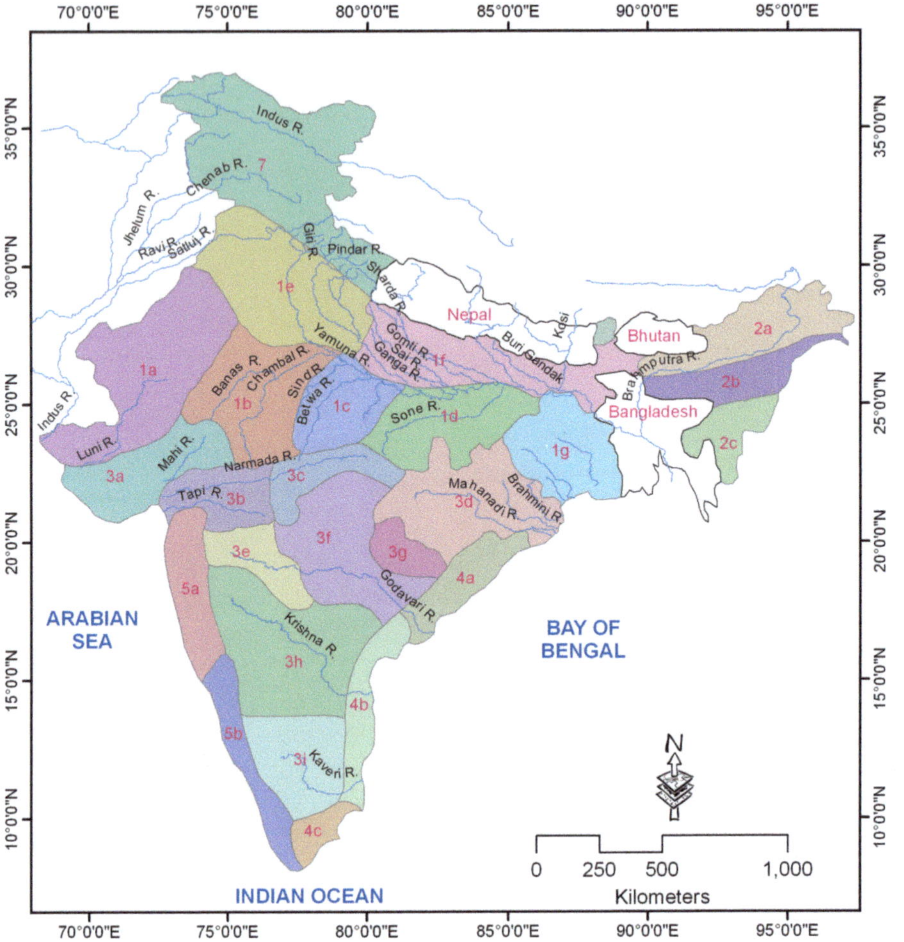

Fig. 8.13 Hydro-meteorological sub-zones of India

1(a) to 1(g), 2(a) to 2(c), 3(a) to 3(i), 4(a) to 4(c), 5(a), 5(b), 6 and 7] (Jain et al., 2007). The hydro-meteorological sub-zones of India are shown in Fig. 8.13.

Due to varying topography and hydro-climatic conditions, the Yamuna River basin comprises five sub-zones, viz. 7, 1b, 1c, 1e and 1f (Fig. 8.13). Detailed computational steps for computing the synthetic unit hydrograph (SUH) for these sub-zones are presented in Tables 8.8, 8.9, 8.10, 8.11, and 8.12. Parameters of the synthetic unit hydrograph (SUH) are shown in Fig. 8.14. The UH parameters so derived using the procedure are used to generate the synthetic unit hydrograph by smoothing the rest of the ordinates. It should be kept in mind that the total volumetric assumption of unit depth should be preserved. If the unit depth is not achieved then adjustments can be made in the ordinates of the UH.

Table 8.8 Relationships of 1 h SUH of ungauged catchments in hydro-meteorological sub-zone 7 (CWC, 1994)

Steps	Equations	Remarks
Step-1: Analysis of "catchment physiography"	$S = \frac{\sum L_i (D_{i-1} - D_i)}{L^2}$ where L_i is the length of the i th segment in km, D_{i-1}, D_i are the depths of the river at the point of intersection of $(i-1)$ and i th contours respectively from the base line (datum)	A is the catchment area, L is the length of the longest stream and S is the equivalent stream slope
Step-2: Determination of "time from center of effective rainfall duration to the peak value for 1 h SUH"	time to peak $t_p = 2.498 \times (L \times L_c/S)^{0.156}$ (h) L_c is the length of the longest stream from a point opposite to the center of gravity of the catchment to the point of study.	
Step-3: Determination of "peak value for 1 h SUH"	unit hydrograph peak per unit catchment area $q_p = 1.048 \times (t_p)^{-0.178}$ (cumec/km^2)	Peak discharge $Q_p = q_p \times A$
Step-4: Estimation of the "time to peak (T_m) for 1 h SUH"	time to peak $T_m = t_p + t_r/2$	
Step-5: Estimation of "time base for 1 h SUH"	time base $T_B = 7.845(t_p)^{0.453}$	
Step-6: Estimation of "time widths for different % values of peak discharge of 1 h SUH"	time width parameters (in h) $W_{50} = T_{50} + T\,T_{50} = 1.954 \times (L \times L_c/S)^{0.099}$ $W_{75} = T_{75} + T\,T_{75} = 0.972 \times (L \times L_c/S)^{0.124}$ $W_{R50} = T_{50} = 0.189 \times (W_{50})^{1.769}$ $W_{R75} = T_{75} = 0.419 \times (W_{75})^{1.246}$	

Table 8.9 Relationships of 1 h SUH of ungauged catchments in hydro-meteorological sub-zone 1b (CWC)

Steps	Equations	Remarks
Step-1: Analysis of "catchment physiography"	$S = \frac{\sum L_i (D_{i-1} - D_i)}{L^2}$ where L_i is the length of the i th segment in km, D_{i-1}, D_i are the depths of the river at the point of intersection of $(i-1)$ and i th contours respectively from the base line (datum)	A is the catchment area, L is the length of the longest stream and S is the equivalent stream slope
Step-2: Determination of "time from center of effective rainfall duration to the peak value for 1 h SUH"	time to peak $t_p = 0.339 \times \left(L/\sqrt{S}\right)^{0.829}$ (h)	
Step-3: Determination of "peak value for 1 h SUH"	unit hydrograph peak per unit catchment area $q_p = 1.215/(t_p)^{0.610}$ (cumec/km^2)	Peak discharge $Q_p = q_p \times A$
Step-4: Estimation of the "time to peak (T_m) for 1 h SUH"	time to peak $T_m = t_p + t_r/2$	
Step-5: Estimation of "time base for 1 h SUH"	time base $T_B = 6.662(t_p)^{0.613}$	
Step-6: Estimation of "time widths for different % values of peak discharge of 1 h SUH"	time width parameters (in h) $W_{50} = T_{50} + TT_{50} = 2.215/(q_p)^{1.034}$ $W_{75} = T_{75} + TT_{75} = 1.191/(q_p)^{1.057}$ $W_{R50} = T_{50} = 0.834/(q_p)^{1.077}$ $W_{R75} = T_{75} = 0.502/(q_p)^{1.065}$	

Table 8.10 Relationships of 1 h SUH of ungauged catchments in hydro-meteorological sub-zone 1c (CWC)

Steps	Equations	Remarks
Step-1: Analysis of "catchment physiography"	$S = \frac{\sum L_i (D_{i-1} - D_i)}{L^2}$ where L_i is the length of the i th segment in km, D_{i-1}, D_i are the depths of the river at the point of intersection of $(i - 1)$ and i th contours respectively from the base line (datum)	A is the catchment area, L is the length of the longest stream and S is the equivalent stream slope
Step-2: Determination of "peak value for 1 h SUH"	unit hydrograph peak per unit catchment area $q_p = 1.331 \times (L/S)^{-0.492}$ (cumec/km^2)	Peak discharge $Q_p = q_p \times A$
Step-3: Determination of "time from center of effective rainfall duration to the peak value for 1 h SUH"	time to peak $t_p = 2.195 \times (q_p)^{-0.944}$ (h)	
Step-4: Estimation of the "time to peak (T_m) for 1 h SUH"	time to peak $T_m = t_p + t_r/2$	
Step-5: Estimation of "time base for 1 h SUH"	time base $T_B = 3.917(t_p)^{0.990}$	
Step-6: Estimation of "time widths for different % values of peak discharge of 1 h SUH"	time width parameters (in h) $W_{50} = T_{50} + T\,T_{50} = 2.04 \times (q_p)^{-1.026}$ $W_{75} = T_{75} + T\,T_{75} = 1.25 \times (q_p)^{-0.864}$ $W_{R50} = T_{50} = 0.739 \times (q_p)^{-0.968}$ $W_{R75} = T_{75} = 0.500 \times (q_p)^{-0.813}$	

Table 8.11 Relationships of 2 h SUH of ungauged catchments in hydro-meteorological sub-zone 1e (CWC, 1984)

Steps	Equations	Remarks
Step-1: Analysis of "catchment physiography"	$S = \frac{\sum L_i (D_{i-1} - D_i)}{L^2}$ Where L_i is the length of the i th segment in km, D_{i-1}, D_i are the depths of the river at the point of intersection of $(i - 1)$ and i th contours respectively from the base line (datum)	A is the catchment area, L is the length of the longest stream and S is the equivalent stream slope
Step-2: Determination of "peak value for 2 h SUH"	unit hydrograph peak per unit catchment area $q_p = 2.030/\left(L/\sqrt{S}\right)^{0.649}$ in (cumec/km^2)	Peak discharge is $Q_p = q_p \times A$
Step-3: Estimation of the "time to peak (T_m) for 2 h SUH"	time lag $t_p = 1.858/\left(q_p\right)^{1.038}$ The time to peak $T_m = t_p + t_r/2$	
Step-4: Estimation of "time base for 2 h SUH"	time base $T_B = 7.744\left(t_p\right)^{0.779}$	
Step-5: Estimation of "time widths for different % values of peak discharge of 2 h SUH"	time width parameters (in h) $W_{50} = T_{50} + T\,T_{50} = 2.217/\left(q_p\right)^{0.990}$ $W_{75} = T_{75} + T\,T_{75} = 1.477/\left(q_p\right)^{0.876}$ $W_{R50} = T_{50} = 0.812/\left(q_p\right)^{0.907}$ $W_{R75} = T_{75} = 0.606/\left(q_p\right)^{0.791}$	

Table 8.12 Relationships of 6 h SUH of ungauged catchments in hydro-meteorological sub-zone 1f (CWC)

Steps	Equations	Remarks
Step-1: Analysis of "catchment physiography"	$S = \frac{\sum L_i (D_{i-1} - D_i)}{L^2}$ where L_i is the length of the i th segment (km), D_{i-1}, D_i are the depths of the river at the point of intersection of (i–1) and i th contours respectively from the base line (datum) [m]	A is the catchment area (km^2), L is the length of the longest stream (km) and S is the equivalent stream slope (m/km)
Step-2: Determination of "time from center of effective rainfall duration to the peak value for 6 h SUH"	time to peak $t_p = 3.065 \times \left(L/\sqrt{S}\right)^{0.471}$ (h) $t_p = 1.217/(q_p)^{1.034}$	
Step-3: Determination of "peak value for 6 h SUH"	unit hydrograph peak per unit catchment area $q_p = 0.409/\left(L/\sqrt{S}\right)^{0.456}$ (cumec/km^2) $q_p = 0.138/(t_p)^{0.968}$	Peak discharge $Q_p = q_p \times A$
Step-4: Estimation of the "time to peak (T_m) for 6 h SUH"	time to peak $T_m = t_p + t_r/2$	
Step-5: Estimation of "time base for 6 h SUH"	time base $T_B = 16.432(t_p)^{0.646}$	
Step-6: Estimation of "time widths for different % values of peak discharge of 6 h SUH"	time width parameters (in h) $W_{50} = T_{50} + T\,T_{50} = 1.743/(q_p)^{1.104}$ $W_{75} = T_{75} + T\,T_{75} = 0.902/(q_p)^{1.108}$ $W_{R50} = T_{50} = 0.736/(q_p)^{0.928}$ $W_{R75} = T_{75} = 0.478/(q_p)^{0.902}$	

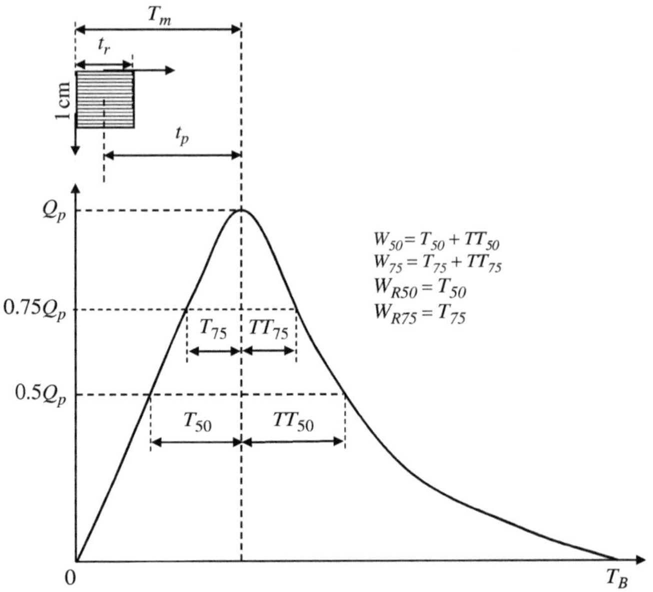

Fig. 8.14 Components of synthetic unit hydrograph

8.3.2 Geomorphology Based Synthetic Unit Hydrograph Technique

Geomorphology of a river basin describes the status of topographic features of surfaces and streams, and its relationship with hydrology provides the geomorphologic control on the basin hydrology (Jain and Sinha, 2003). Geomorphology reflects the topographic and geometric properties of the watershed and its drainage channel network. It controls the hydrologic processes from rainfall to runoff, and the subsequent flow routing through the drainage network. The role of basin geomorphology in controlling the hydrological response of a river basin is known for a long time. Earlier works have provided an understanding of basin geomorphology-hydrology relationship through empirical relationships (Snyder, 1938; Horton, 1945; Taylor and Schwartz, 1952; Singh, 1988, 1989, 1992; Jain and Sinha, 2003). Snyder (1938) proposed a synthetic unit hydrograph approach (SUH) for ungauged basins as a function of catchment area, basin shape, topography, channel slope, stream density and channel storage; and derived the basin coefficient by averaging out other parameters. Therefore, geomorphology based approach becomes one of the most popular modeling tools for the computation of runoff hydrographs (Rodriguez-Iturbe and Valdes, 1979; Rodriguez-Iturbe et al., 1982; Yen and Lee, 1997; Rai et al., 2009). The geomorphologic instantaneous unit hydrograph (GIUH) (Rodriguez-Iturbe and Valdes, 1979) and its simplification by Gupta et al. (1980) are hydrological models that relate the geomorphological features of a basin to its response to rainfall. They can be applied to ungauged basins having scarce hydrologic data (Al-Wagdany and Rao, 1998). The derivation of the GIUH uses the assumption that a stream of a certain order has a known linear response function of the familiar or complex probability distribution type (Rodriguez-Iturbe and Valdes, 1979; Fleurant et al., 2006). Thus, the GIUH based transfer function approach is applicable where rainfall data is available but runoff data is not, and it is a more powerful technique for the flood estimation than the commonly used parametric models (Yen and Lee, 1997; Bhaskar et al., 1997; Jain et al., 2000; Lohani et al., 2001; Kumar et al., 2002, 2007; Sarangi et al., 2007; Rai et al., 2009; etc.).

Another advantage of the GIUH technique is its potential for deriving the unit hydrograph (UH) using the geomorphologic characteristics obtainable from topographic map/remote sensing, possibly linked with geographic information system (GIS) and digital elevation model (DEM) (Rodriguez-Iturbe and Valdes, 1979; Rai et al., 2009).

8.3.2.1 Derivation of Geomorphologic Unit Hydrograph (GUH)

The GIUH theory was introduced by Rodriguez-Iturbe and Valdes (1979) by relating the peak discharge and time to peak discharge with the geomorphologic characteristics of the catchment and a dynamic velocity parameter. This work explicitly integrates the geomorphologic details and the climatologic characteristics of the basin, in a framework of travel time distribution for stream flow synthesis in the basin having no or scanty information of flow data. This formulation of GIUH is based on the probability density function (*pdf*) of the time history of a randomly chosen drop of effective rainfall reaching the trapping state of a hypothetical basin, treated as a continuous Markovian process, where the state is the order of the stream in which the raindrop is located at any time. The value at the mode of this *pdf* produces the main characteristics of GIUH. Rodriguez-Iturbe and Valdes (1979) derived the peak and time to peak characteristics of the IUH as a function of Horton's order ratios (Horton, 1945), and are expressed as follows (Kumar et al., 2002):

$$q_{\mathrm{p}} = 1.31\, R_{\mathrm{l}}^{0.43}\; V/L_{\Omega} \tag{8.5}$$

$$t_{\mathrm{p}} = 0.44\; (L_{\Omega}/V)\; (R_{\mathrm{b}}/R_{\mathrm{a}})^{0.55} \cdot R_{l}^{-0.38} \tag{8.6}$$

where q_{p} is the peak flow (h^{-1}), t_{p} is the time to peak (h), L_{Ω} is the length of the highest order stream (km), V is the dynamic velocity parameter ($\mathrm{m\ s}^{-1}$), R_{b} is the bifurcation ratio, R_{a} is the stream area ratio, and R_{l} is the stream length ratio. Multiplying eqs. (8.5) and (8.6) a non-dimensional term $q_{\mathrm{p}}^{*}t_{\mathrm{p}}$ is derived at.

$$q_{\mathrm{p}}^{*}t_{\mathrm{p}} = 0.5764\; \left[R_{\mathrm{b}}/R_{\mathrm{a}}\right]^{0.55} R_{\mathrm{l}}^{0.05} \tag{8.7}$$

Term $q_{\mathrm{p}}^{*}t_{\mathrm{p}}$ is not dependent on velocity and thereby on storm characteristics and hence, it is a function of only the geomorphologic characteristics of the basin. The dynamic velocity parameter in the formulation of GIUH incorporates the effect of climatic variation. Rodriguez-Iturbe and Valdes (1979) showed that the dynamic velocity parameter of GIUH can be taken as the velocity at the peak discharge for a given rainfall-runoff event in the catchment. Rodriguez-Iturbe and Valdes (1979) compared the GIUHs for some real world basins with the IUHs derived from the discharge hydrograph produced by a physically based rainfall-runoff model of the same basin and found them to be remarkably similar. For deriving the SUH for the catchments of the Yamuna basin, the Nash model was linked with the geomorphologic characteristics of the catchments. A brief description of the Nash model and its linkage with geomorphologic parameters are now given.

8.3.2.2 The Nash Model

The Nash model (Nash, 1957) is based on the concept of routing of the instantaneous inflow through a cascade of linear reservoirs with equal storage coefficient. The Nash model can be expressed as:

$$u(t) = \frac{1}{k\,\Gamma(n)}\,(t/k)^{n-1}\,\exp(-t/k) \tag{8.8}$$

where $u(t)$ is the ordinates of IUH (hour^{-1}), t is the sampling time interval (hour), n and k are the parameters of the Nash model, in which n is the number of linear reservoirs, and k is the storage coefficient (hour).

A unit hydrograph (UH) of desired duration (D) may be derived by using the following expression:

$$U(D, t) = \frac{1}{D}\left[I(n, t/k) - I\,(n, (t-D)/k)\right] \tag{8.9}$$

where $U(D, t)$ denotes the ordinates of D hour UH (h^{-1}), t is the sampling time interval (h), $I\,(n, t/k)$ is the incomplete gamma function of order n at (t/k), and D is the duration of UH (h).

8.3.2.3 Geomorphology Based Parameter Estimation of Nash's GIUH

The complete shape of the GIUH can be obtained by linking q_p and t_p of the GIUH with scale (k) and shape (n) parameters of the Nash model. Equating the first derivative with respect to t of eq (8.8) to zero, t becomes the time to peak discharge, t_p. Thus, taking the natural logarithm of both sides of eq. (8.8), differentiating with respect to t and by simplification eq. (8.10) is derived.

$$\frac{\partial}{\partial t}\ln\left[u(t)\right] = \left[-\frac{1}{k} + \frac{(n-1)}{t}\right] \tag{8.10}$$

Equating eq. (8.10) to zero and replacing t with t_p result in:

$$t = t_\text{p} = k\,(n-1) \tag{8.11}$$

Simplifying the value of t_p from eq. (8.11) in eq. (8.8) and simplifying yields

$$q_p = \frac{1}{k\Gamma(n)} \exp\left[-(n-1)\right] \cdot (n-1)^{n-1} \tag{8.12}$$

From eqs. (8.11) and (8.12), the result is:

$$q_p \cdot t_p = \frac{(n-1)}{\Gamma(n)} \exp\left[-(n-1)\right] \cdot (n-1)^{n-1} \tag{8.13}$$

Equating eq. (8.13) to eq. (8.7) one obtains:

$$\frac{(n-1)}{\Gamma(n)} \cdot \exp\left[-(n-1)\right] \cdot (n-1)^{n-1} = 0.5764\left[R_b/R_a\right]^{0.55} \times R_l^{0.05} \tag{8.14}$$

The Nash parameter n can be obtained by solving eq. (8.14) using the Newton-Raphson method.

The Nash's parameter k for the given velocity V is obtained using eqs. (8.6) and (8.11) and the known value of parameter n as:

$$k = \frac{0.44\, L_{\Omega}}{V} \cdot \left[\frac{R_b}{R_a}\right]^{0.55} \cdot R_l^{-0.38} \cdot \frac{1}{(n-1)} \tag{8.15}$$

The derived values of n and k are used to determine the complete shape of the Nash based GIUH using eq. (8.8). Subsequently, the D-hour UH can be derived from eq. (8.9). The direct runoff hydrograph (DRH) is estimated by convoluting the excess rainfall hyetograph with the UH.

8.3.2.4 Application of the GIUH Methodology

To implement the described technique requires a geomorphologic investigation of the catchments of the Yamuna basin, and is presented in Chapter 5. However, the geomorphologic parameters required to derive the GSUH are summarized in Table 8.13.

Using Horton's ratio and Ω, parameter n of the Nash model is estimated using eq. (8.14) and is given in Table 8.14 for the catchments of the basin. Using the value of n and range of average velocities, the values of storage coefficients, k are computed using eq. (8.15). The values of k are depicted in Table 8.14. Using the values of n and k, the GIUH is derived (eq. (8.8)) for the catchments and subsequently the 1-h SUH is generated (eq. (8.9)). The resulting SUHs are shown in Figs. 8.15, 8.16, 8.17, 8.18, and 8.19 for the Upper Himalayan Yamuna, Chambal, Sind, Betwa and Ken catchments, respectively.

The above described approach becomes difficult for field engineers because of the involvement of large computational steps. Therefore, an attempt is made to

Table 8.13 Horton's ratio of the catchments of the Yamuna basin

Catchment	Horton's ratio: Stream-area ratio, R_a	Bifurcation ratio, R_b	Length ratio, R_l	Length of highest order stream, Ω (km)
Upper Yamuna	6.622	5.703	2.945	120.9
Chambal	5.21	4.76	1.361	389.1
Sind	4.01	3.55	1.86	51.4
Betwa	4.33	3.91	2.15	116.8
Ken	4.00	3.58	2.304	273.4

Table 8.14 The Nash parameters for the catchments of the Yamuna basin

S. No.	Velocity (m/s)	Upper Yamuna *n*	Upper Yamuna *k* (h)	Chambal *n*	Chambal *k* (h)	Sind *n*	Sind *k* (h)	Betwa *n*	Betwa *k* (h)	Ken *n*	Ken *k* (h)
1	0.25	3.1324	60.97	3.1084	274.90	3.1017	30.93	3.1737	66.84	3.1679	152.06
2	0.50		30.48		137.45		15.46		33.42		76.03
3	0.75		20.32		91.63		10.31		22.28		50.69
4	1.00		15.24		68.72		7.73		16.71		38.02
5	1.50		10.16		45.82		5.15		11.14		25.34
6	2.00		7.62		34.36		3.87		8.35		19.01
7	2.50		6.09		27.49		3.09		6.68		15.21
8	3.00		5.08		22.91		2.58		5.57		12.67
9	4.00		3.81		17.18		1.93		4.18		9.50
10	5.00		3.05		13.74		1.55		3.34		7.60

establish relationships to avoid a detailed analysis. For this purpose, first the relationships between storage coefficients k and average flow velocities are developed (Fig. 8.20). Numerical expressions for the catchments are given through eqs. (8.16) through (8.20) in Table 8.15.

Coefficients (k_{coeff}) of the above relationships [eqs. (8.16) to (8.20)] have been further fitted for the length of the highest order stream of the catchment (Ω) using the second order polynomial (Fig. 8.21). The derived relationship can be mathematically expressed as:

$$K_{\text{coeff}} = 0.0003 \times \Omega^2 + 0.0323 \times \Omega + 6.4608 \quad (R^2 = 0.9966) \tag{8.22}$$

Using eq. (8.22), the value of k_{coeff} can be estimated and substituted in eq. (8.21) to estimate the value of storage coefficient k corresponding to the reliable value of velocity of flow. The value of n and k for the catchment are then used to derive the IUH and subsequently the UH using eqs. (8.8) and (8.9).

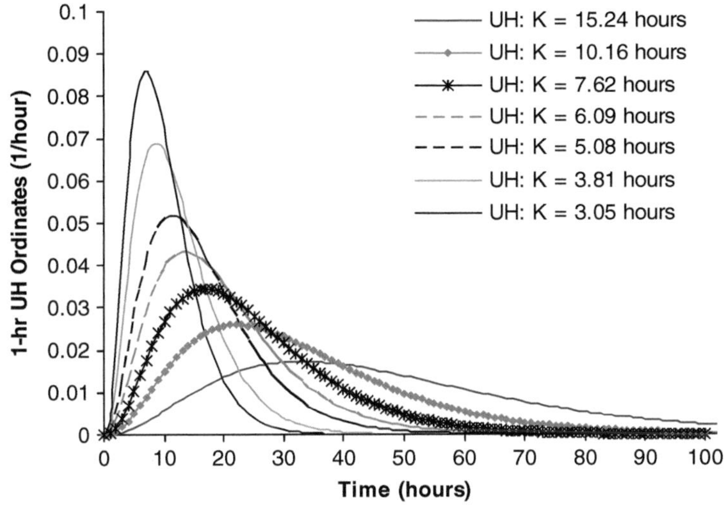

Fig. 8.15 One-hour Unit hydrograph for the Upper Himalayan Yamuna catchment

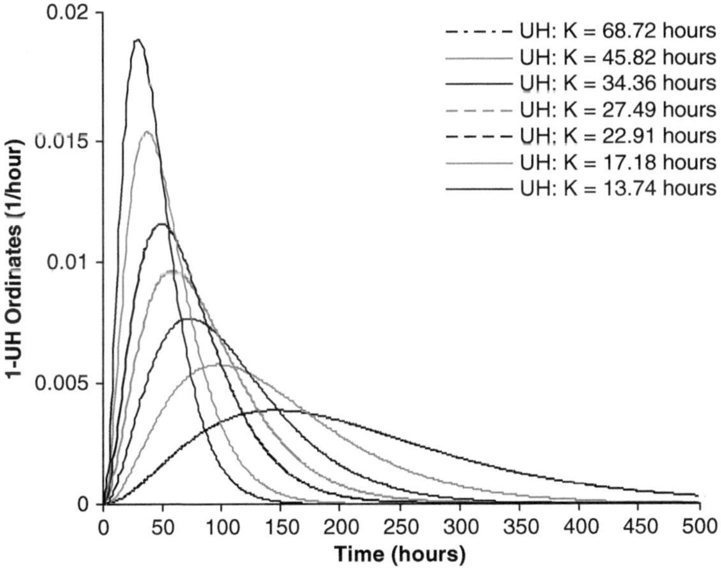

Fig. 8.16 One-hour Unit hydrograph for the Chambal catchment

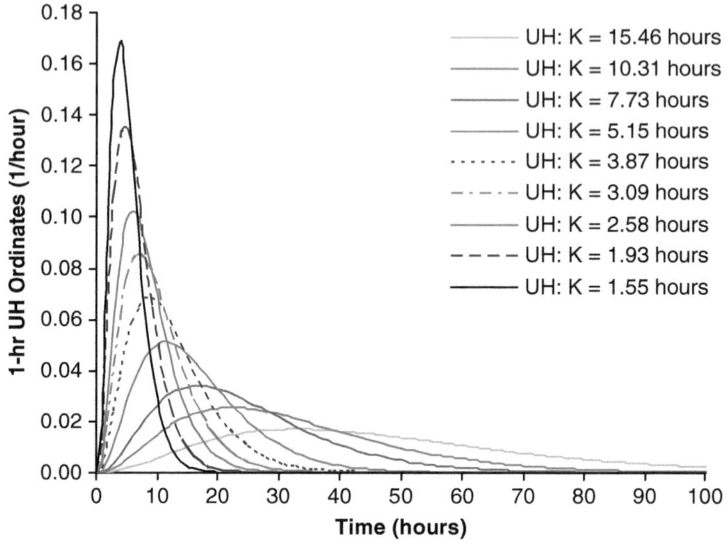

Fig. 8.17 One-hour Unit hydrograph for the Sind catchment

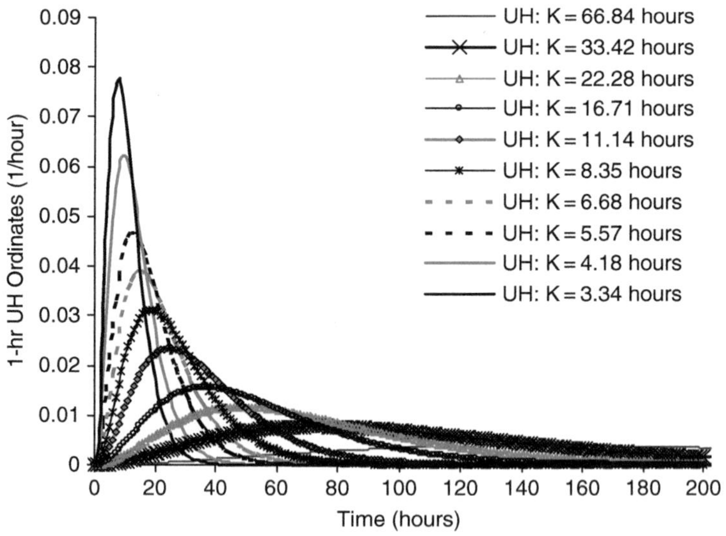

Fig. 8.18 One-hour Unit hydrograph for the Betwa catchment

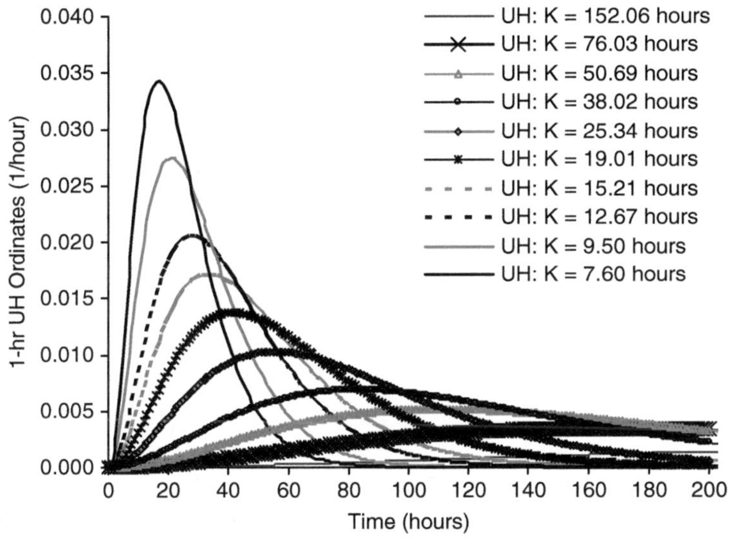

Fig. 8.19 One-hour Unit hydrograph for the Ken catchment

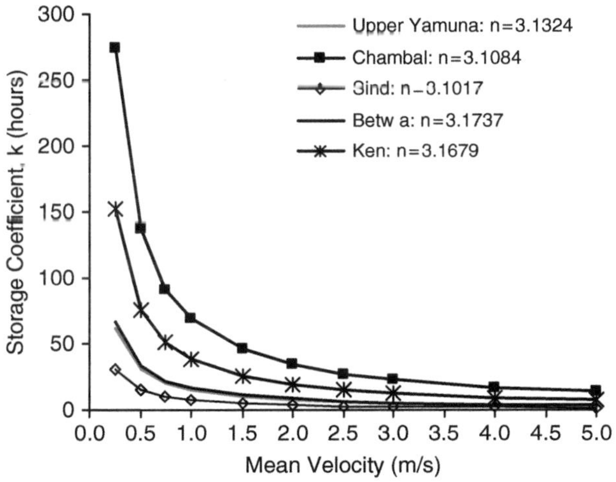

Fig. 8.20 Relationship between the mean flow velocity and k

Table 8.15 Relationship between flow velocity and k for different catchments

Catchment	n	Relation between k and $\bar{u}$	Eq.
Upper Himalayan Yamuna	3.1324	$k = 15.243/\bar{u}$	(8.16)
Chambal	3.1084	$k = 68.726/\bar{u}$	(8.17)
Sind	3.1017	$k = 7.7328/\bar{u}$	(8.18)
Betwa	3.1737	$k = 16.711/\bar{u}$	(8.19)
Ken	3.1679	$k = 38.016/\bar{u}$	(8.20)
General equation		$k = k_{\text{coeff}}/\bar{u}$	(8.21)

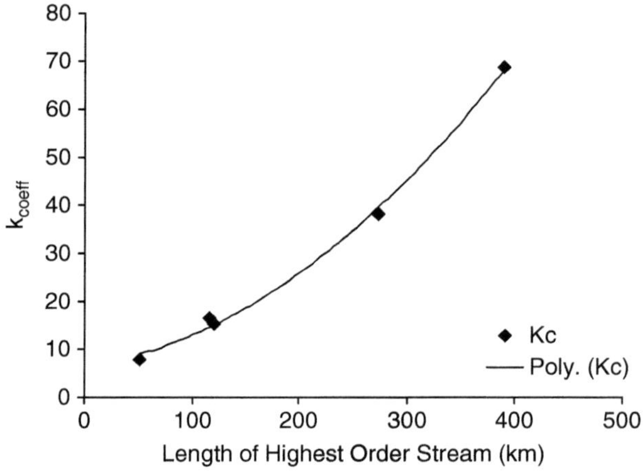

Fig. 8.21 Relationship between k_{coeff} and Ω

References

Al-Wagdany AS, Rao AR (1998) Correlation of the velocity parameter of three geomorphological instantaneous unit hydrograph models. Hydrol Process 12:651–659

Bhaskar NR, Parida BP, Nayak AK (1997) Flood estimation for ungauged catchments using the GIUH. J Water Resour Plann Manage 12(4):228–238

Central Water Commission Research Designs (CWC) (Directorate of Hydrology [Small Catchment]) & Standards Organisation (Ministry of Railways) & India Meteorological Deptt. & Ministry of Transport (1984) Flood Estimation Report for Upper Indo-Ganga Plains (Sub zone – 1e), A method based on Unit Hydrograph Principle. Report No. UGP/9/1984, CWC, New Delhi

Central Water Commission Research Designs (CWC) (Directorate of Hydrology [Regional Studies]) & India Meteorological Dept & Ministry of Surface Transport (1994) Flood Estimation Report for Western Himalayas – Zone 7, Design Office Report No WH/22/1994, CWC, New Delhi

CWC (1983) Flood estimation report. Directorate of Hydrology (Small Catchments), CWC, New Delhi

FAO (1998) FAO Irrigation and Drainage Paper No 56: Crop Evapotranspitration (Guidelines for computing crop water requirements). Food and Agriculture Organisation, Land and Water, Rome, Italy

Fleurant C, Kartiwa B, Roland B (2006) Analytical model for a geomorphological instantaneous unit hydrograph. Hydrol Process 20:3879–3895

GEC (1997) Ground Water Estimation Methodology, Report of the Groundwater Resources Estimation Committee, Ministry of Water Resources, Government of India

Gupta VK, Waymire E, Wang CT (1980) A representation of an instantaneous unit hydrograph from geomorphology. Water Resour Res 16(5):855–862

Horton RE (1945) Erosional development of streams and their drainage basins: hydrophysical approach to quantitative morphology. Bull Geol Soc Am 56:275–370

Jain SK, Agarwal PK, Singh VP (2007) Hydrology and water resources of India. Water Science and Technology Library, Springer

Jain SK, Singh RD, Seth SM (2000) Design flood estimation using GIS supported GIUH approach. Water Resour Manage 14:369–376

Jain V, Sinha R (2003) Derivation of unit hydrograph from GIUH analysis for a Himalayan river. Water Resour Manage 17:355–375

Kumar R, Chatterjee C, Lohani AK, Kumar S, Singh RD (2002) Sensitivity analysis of the GIUH based Clark model for a catchment. Water Resour Manage 16:263–278

Kumar R, Chatterjee C, Singh RD, Lohani AK, Kumar S (2007) Runoff estimation for an ungauged catchment using geomorphological instantaneous unit hydrograph (GIUH) model. Hydrol Process 21:1829–1840

Lohani AK, Singh RD, Nema RK (2001) Comparison of geomorphological based rainfall-runoff models. CS(AR) Technical Report, National Institute of Hydrology, Roorkee

Nash JE (1957) The forms of instantaneous unit hydrograph. Int Ass Sci Hydrol Pub 1, 45(3):114–121

Rai RK, Upadhyay Alka, Sarkar S, Upadhyay A, Singh VP (2009) GIUH based transfer function for Gomti River basin of India. J Spat Hydrol 9(2):29–50

Rodriguez-Iturbe I, Valdes JB (1979) The geomorphologic structure of hydrologic response. Water Resour Res 15(6):1409–1420

Rodriguez-Iturbe I, Gonzalas SM, Bras RC (1982) The geomorpho-climatic theory of the instantaneous unit hydrograph. Water Resour Res 18(4):877–886

Sarangi A, Madramootoo CA, Enright P, Prasher SO (2007) Evaluation of three unit hydrograph models to predict the surface runoff from a Canadian watershed. Water Resour Manage 21:1127–1143

Singh VP (1988) Hydrologic systems: vol I, rainfall-runoff modeling. Prentice-Hall, Englewood Cliffs, NJ

Singh VP (1989) Hydrologic systems: vol II, watershed modeling, Prentice-Hall, Englewood Cliffs, NJ

Singh VP (1992) Elementary hydrology. Prentice-Hall, Englewood Cliffs, NJ

Singh VP, Frevert DK (eds) (2002a) Mathematical models of large watershed hydrology. Water Resources Publications, Highlands Ranch, CO

Singh VP, Frevert DK (eds) (2002b) Mathematical models of small watershed hydrology and applications. Water Resources Publications, Highlands Ranch, CO

Snyder FF (1938) Synthetic unitgraphs. Transactions American Geophysical Union, 19th Annual Meeting, Part 2, p 447

Taylor AB, Schwartz HE (1952) Unit-hydrograph lag and peak flow related to basin characteristics. Trans Am Geophys Union 33:235–246

Yen BC, Lee KT (1997) Unit hydrograph derivation for ungauged watersheds by stream-order laws. J Hydrol Eng 2(1):1–9

Chapter 9
Water Pollution

Abstract This chapter discusses the issues related to water quality in the Yamuna River basin, including the sources of pollution. The discussion includes different types of pollution sources, which are categorized into point- and non-point sources. The role of barrages on water quality and their impact on river segmentation are adequately addressed. Further, water quality monitoring sites on the river and monitoring agencies, followed by the importance of physico-chemical characteristics of the water on human and river health, are described.

9.1 Sources of Pollution

The entire stretch of Yamuna River from origin to confluence with Ganga River is highly used for various human activities. The accumulated results of these activities are the low flow availability in the river during the lean period and water pollution. During the course of travel, Yamuna River receives a lot of pollutants of various kinds.

Water pollution can be broadly categorized into three main classes based on the pollution sources: (i) domestic, (ii) industrial, and (iii) agricultural. Among these categories, the city sewage and industrial waste are important. These pollutants enter groundwater, rivers, and other water bodies and finally back to the households and carries disease-causing microbes. Agricultural runoff, or the water from the fields that drains into rivers, is another major water pollutant as it contains fertilizers and pesticides. Domestic sewage refers to the waste water that is discarded from households. Also referred to as sanitary sewage, the water contains a wide variety of dissolved and suspended impurities (CPCB, 1978; CPCB, 1981; CPCB, 1982–1983; CPCB, 1984–2008; CPCB, 1986; CPCB, 1994; CPCB, 2006; CPCB, 2009).

The use of land for agriculture and the practices followed in cultivation greatly affect the quality of groundwater. Intensive cultivation of crops causes chemicals from fertilizers (e.g. nitrate) and pesticides to seep into the groundwater, a process commonly known as leaching. Routine applications of fertilizers and pesticides for agriculture and indiscriminate disposal of industrial and domestic wastes are increasingly being recognized as significant sources of water pollution. These pollutants also affect surface water bodies when agricultural runoff is produced during monsoon. Generally, agricultural pollution is treated as non-point source pollution, as it is generated from diffuse sources. The high nitrate content in groundwater

R.K. Rai et al., *The Yamuna River Basin*, Water Science and Technology Library 66,
DOI 10.1007/978-94-007-2001-5_9,

is mainly from irrigation runoff from agricultural fields where chemical fertilizers have been used indiscriminately.

Waste water from manufacturing or chemical processes in industries contributes to water pollution. Industrial waste water usually contains specific and readily identifiable chemical compounds. During the last fifty years, the number of industries has grown rapidly, but water pollution is concentrated within a few sub-sectors, mainly in the form of toxic wastes and organic pollutants. Out of this a large portion can be traced to the processing of industrial chemicals and to the food products industry. In fact, a number of large- and medium-sized industries in the region covered by the Ganga Action Plan do not have adequate effluent treatment facilities. Most of these defaulting industries are sugar mills, distilleries, leather processing industries, and thermal power stations. Most major industries have treatment facilities for industrial effluents, but this is not the case with small-scale industries, which cannot afford enormous investments in pollution control equipment as their profit margin is very slender.

9.1.1 Point Sources of Pollution

When the source of pollution is single, well specified and generates a significant amount of pollutants, such a source is known as point source. Urban centers located along or near the banks of Yamuna River are the major pollution sources. The point sources of pollution cover two major categories: domestic and industrial pollution.

9.1.1.1 Domestic Pollution

The domestic pollution is the major source of pollution in Yamuna River. About 85% of the total pollution in the river is caused by domestic sources. The domestic pollution is mainly caused by urban centers. The major urban centres dumping domestic waste into Yamuna River are Panipat, Sonepat, Delhi, Ghaziabad, Mathura-Vrindavan, Agra, Etawah and Allahabad. The impact of domestic pollution on river depends on the efficiency of wastewater collection systems, and type and length of waste transportation systems. If wastewater gets more retention time within urban premises before reaching receiving water bodies, the pollution load will reduce due to biodegradation and settling. The organic matters and microorganisms are the main constituents of domestic waste. Further, total salts, chlorides, nutrients, detergents, oil and grease, etc. are also contributed by the domestic sources. There are numerous unauthorized colonies existing in various urban centres. Due to the non-availability of sewerage system in these colonies, the night soil is collected, transported and dumped either in drains, tributaries or directly into river without any treatment. During the last few years because of proliferation of Jhuggi Jhompri settlement this activity has increased significantly and has now become a major non-point source of river water pollution.

9.1.1.2 Industrial Pollution

After independence, rapid industrialization occurred in the Yamuna River basin. There are large clusters of industries established in Kota, Gwalior, Indore, Nagda, Khetri, Yamuna Nagar, Panipat, Sonepat, Delhi, Baghpat, Ghaziabad, Gautam Budha Nagar, Faridabad, Mathura and other places. The categories of industries discharging wastewater into Yamuna River includes pulp and paper, sugar, distilleries, textiles, leather, chemical, pharmaceuticals, oil refineries, thermal power plants, food, etc. In order to comply with environment laws, it is compulsory for these industries to treat the effluent to achieve prescribed standards before discharging into the environment.

9.1.2 Non Point or Diffused Sources of Pollution

Just opposite to the point sources the diffused sources are unspecified; numerous in numbers and contribution of each is of less significance, though, in combination the resulting contribution is significant. This pollution is original in the catchment area of the river and transported regularly or occasionally by leaching, drainage and surface water runoff during monsoon. The pollutants originated from diffused sources are topsoil, organic matter, plant residues, nutrients, organic chemicals, toxicants, microorganisms etc. The important diffused pollution sources contributing to Yamuna River are: (i) agricultural pollution sources; (ii) dumping of garbage and dead bodies; (iii) immersion of idols; (iv) pollution due to in-stream uses of water.

9.1.2.1 Agricultural Pollution

There are four major sources that contribute agricultural pollution in the river: (i) agricultural residues; (ii) fertilizer and pesticides; (iii) animal husbandry; and (iv) excess salts from applied irrigation water.

The various types of agricultural residues generated in the river basin are degraded naturally as they are part of saprophytic food chain and thus not contributing much to the river pollution. In the Yamuna basin the application rate of fertilizer is still low, moreover, due to marginal irrigation in the basin there is little chance of leftover nutrients from fertilizer application to leach, drain or wash away except during monsoon. The nutrients and pesticides are generally adsorbed by the sediment particles and reach the river along with run off sediments particularly during early floods. These chemicals geo-accumulate in riverbeds, which are not stable due to the sandy nature of river bottom. Hence, with the flushing of riverbed by water current these chemicals mix slowly with the supernatant river water.

The solid waste generated from animal husbandry is generally collected dry and rarely washed. A major part of this solid waste is used as organic manure. In rural areas the animal husbandry solid waste particularly animal dung is formed into cakes and dried for burning as dung cakes. Because of this, there is no significant water pollution caused from animal husbandry. However, with the change in cooking style with more and more use of cooking gas, urbanization of rural areas, modernized cattle farming, illegal and unorganized dairy farming in urban centers, etc. increase the organic pollution from these sources.

Plants consume water only through transpiration and the salts present in the applied irrigation water accumulate in the soil and ultimately leach or are washed off to adjoining water bodies. As a result, the build-up of salinity is frequently observed in parts of Haryana.

In almost the entire stretch of Yamuna River that lies in the plain area, the riverbed is extensively used for farming during the lean period when the river stream shrinks to minimum. The riverbed farming, which mainly includes vegetables, melons, and cucumbers, starts after the monsoon season and continues till the end of summer season. During farming, there is frequent use of pesticide, washing of vegetables, disposal of farming residues, etc., which provides a direct impact on river water quality.

9.1.3 Dumping of Garbage and Dead Bodies

A large portion of solid waste generated by unauthorized inhabitants all along the banks of Yamuna River or its tributaries finds its way into the river. The waste includes domestic waste, waste from dairies, unauthorized slaughtering, flowers and other material used during worships carcasses of animals, etc. The dumping of human and animal dead bodies are also sometimes observed in the Yamuna River. Disposal of infant dead bodies in the river water is practiced in the entire Yamuna stretch. Floating of human dead bodies partially eaten by animals and in rotten state are generally observed in the lower part of the river. Such disposal of dead bodies increases the risk of pathogenic contamination besides other negative impact.

9.1.4 Immersion of Idols

Immersion of idols, especially during Durga Puja, and Ganesh Puja, takes place all along the banks of rivers. Flowers, Straw, Bamboo, Clay/Plaster of Paris, harmful chemicals used for paints, and plastic bags finds their way into the river. As the Durga Puja festival is also getting popularity in Northern India the rate of idols immersion is also increasing every year.

9.1.5 Pollution Due to In-Stream Use of Water

The various sources of pollution caused by the stream use of water are: bathing and clothe washing; cattle wading; and open defecation. Because of spiritual faith, the bathing in Yamuna River is very common. Bathing, especially mass bathing, significantly contributes disease causing pathogens in the river water and enhance the bacterial load. The religious activities, e.g., offering flowers, milk, sweets, etc. into the river water further increase organic loading in the river, since the food items are not consumed by aquatic animals due to their limited availability. The other activities associated with bathing are cloth washing. This activity contributes inorganic, organic and biological contaminants in the river water besides detergents. Excessive presence of detergents causes significant foaming at the sites of turbulence. Foaming not only hampers the oxygen diffusion rate in the river water, essential for self-purification but also affects various biological activities.

The Yamuna basin is one of the densely populated river basins in the country. In the basin, due to the non-existence of sanitary facilities in rural areas and urban areas, especially in slum clusters, a large section of population uses either the catchment area or directly the river for open defecation. The activity contributes organic pollution and pathogens in the river water.

9.2 Water Quality Issues in Yamuna River

Most of the rivers, including Yamuna River, are spiritually regarded as mother. People from all over the country visit various stretches of this river especially at Yamunotri, Paonta Sahib, Mathura-Vrindavan and Bateshwar to take holy dip in river water to purge away their sins. Thus, the river portrays Indian culture and traditions. Deteriorating water quality and decreasing quantity of Yamuna River water hurt the sentiments of Indian masses besides having several adverse impacts on life processes in the river. The issues related with water quality of Yamuna River are described in what follows (CPCB, 1978; CPCB, 1981; CPCB, 1982–1983; CPCB, 1984–2008; CPCB, 1986; CPCB, 1994; CPCB, 2006; CPCB, 2009):

9.2.1 High Organic Contents

Yamuna River receives significantly high amounts of organic matter, which generally originate from domestic sources. For biodegradation, this organic waste requires oxygen, causing a significant depletion of dissolved oxygen in the river water. The oxygen depletion not only affects biotic community of the river but also affects its self-purification capacity. This problem is critical in the river stretch between Delhi and its confluence with Chambal River. In the Delhi stretch, the load of organic matter is so high that it consumes the entire dissolved oxygen available in river water.

9.2.2 High Nutrients

The organic matter after biodegradation releases nutrients in the water. High nutrient concentration leads to eutrophication, a condition characterized by significant diurnal variation in dissolved oxygen concentration and excessive algal growth.

9.2.3 Excessive Presence of Pathogens

Continuous flow of sewage waste, dumping of animal dead bodies, etc. and instream uses of water like bathing, cattle wading, etc. contribute significant load of pathogens in the river water, making it unsuitable for drinking and bathing purposes.

9.2.4 Accumulation of Pollutants in the Catchment Area

Organic, inorganic and toxic pollutants generated from agricultural and industrial sources are accumulated near the source during dry seasons and get mixed with river water, threatening aquatic life during monsoon or percolated to ground water and making water unfit for human consumption.

9.2.5 Aesthetic Value

Yamuna River loses its aesthetic value and glory due to severe odour that releases to the surrounding environment from the anaerobic activities occurring in the river strata and the ugly surface look contributed by blackish water, floating of garbage, plastic bags, dead bodies of animals. The religious activities and tourism are greatly affected because these adversely transform the characteristics of river water.

9.2.6 Deforestation in the Catchment Area

Forest cover in the catchment area of the Yamuna is vanishing rapidly. This leads to soil erosion with rainfall, and results in mixing of high amounts of silt, mud etc. in the river water, which in turn increases the turbidity. The turbidity of river water is also increased due to the direct influx of domestic and industrial wastewater. Increased turbidity has an impact on the productivity of water body besides affecting biotic life of aquatic system.

9.2.7 Reduction in the Quantity of Water

The fresh water of Yamuna River is over exploited for irrigation use, drinking and industrial uses resulting in very little or sometimes no water in the river at certain locations during summer season. The water scarce condition is so severe that to avoid percolation and evaporation losses, the Delhi's share of Yamuna water transported through WJC is added back into the river through Drain No. 2. All this leads to the stagnation of water and formation of dry zones in the drainage area of the river. The non-availability of fresh water hampers the purification capacity of the river and causes an increase in the concentration of pollutants in the river water.

9.2.8 Discharges from Sewage Treatment Plants

Sewage treatment plants (STP's) have been constructed at various urban centers to conserve the water quality of Yamuna River. The treated, untreated or partially treated sewage from these STPs are generally discharged directly or through carrier drain into the river. Prior to installation of STPs the sewage of urban centers was discharged and got mixed with river water at various locations in the wide stretch of river through long and slow transportation system. After installation of STPs along with swift collection and transport systems, the sewage from urban centers concentrated at a few places, where STP's are located. The connection of STP with the river sometimes poses great threat to water quality during non-operation of STP due to unavoidable reasons, e.g., power failure, mechanical problems or maintenance of plants. In such cases the collected sewage is generally bypassed and eventually discharged into the river at a few locations without any treatment. Such problem is very significant in those stretches of river where the STPs are located upstream of the river e.g. Mathura-Vrindavan and Agra. The discharges from these STP's located upstream from the water abstraction point have an impact on the water quality making it unsuitable for various human activities occurring downstream of these STP's.

9.2.9 Role of Barrages

Presently there are six barrages in the Yamuna River and some others are in planning stage. The barrages have an impact on the characteristics of Yamuna River:

- Discontinue the river flow.
- Less demand for irrigation water or rainfall in the catchment area leads to intermittent release of water from the barrages and thus affects the river water quality.
- Sludge containing inorganic, organic, and toxic matters are generally deposited upstream of barrages. This settled material flushes downstream along with a sudden release of water from the barrages. Thus, pollutants mix further with the river water at downstream, deteriorating its quality.
- The water is generally released from the barrages during monsoon after a gap of 6–9 months. The water releases from the barrage after a considerable gap, significant amounts of deposited sludge in the reaches downstream, which is dominantly organic in nature also flushes with it and flow in the river downstream. This sludge, after mixing with the water downstream, consumes the available dissolved oxygen rapidly resulting into fish mortality and killing of other fauna of the river.
- Besides the negative impact of barrages on river characteristics, barrages also have one positive impact. Barrage forms a sort of reservoir upstream. This reservoir acts as oxidation pond to treat the river water.

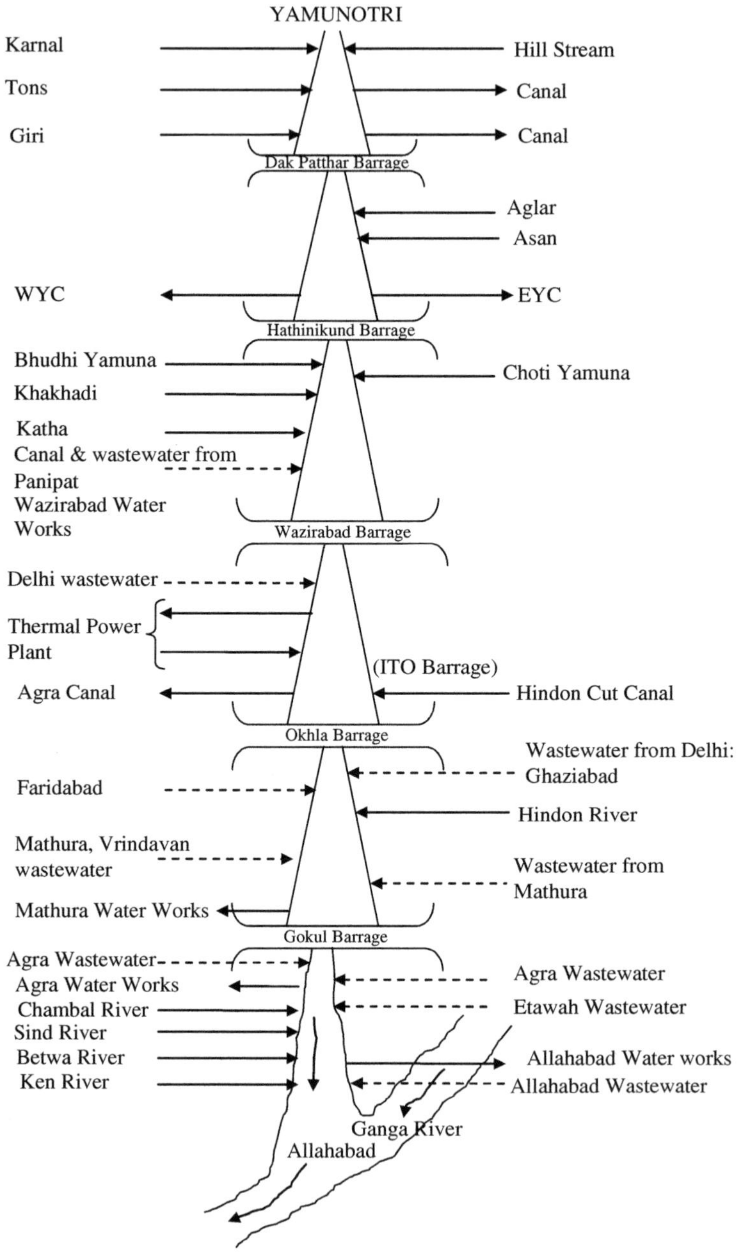

Fig. 9.1 Segmentation of the Yamuna River system

The water pollution in the river varies with the reaches. Due to the water quality status in the river and changes in flow regime due to human intervention, during the lean period, the Yamuna River is divided into four independent segments (Fig. 9.1). The characteristics of these segments are now discussed.

9.3 River Segmentation

The four independent segments are as follows (CPCB, 1994):

Segment I: This segment (length 172 km) is identified from Yamunotri and terminates at Hathnikund/Tajewala barrage. The major source of water in this segment is the melting of glaciers. The water flow in this segment terminates into Western Yamuna canal (WJC) and Eastern Yamuna Canal (EJC) for irrigation and drinking water purposes in command areas.

Segment II: This segment (about 224 km) lies between Hathnikund barrage and Wazirabad barrage. The main source of water in this segment is ground water accrual. A few small tributaries also contribute water in this segment. The water is diverted in this segment from WJC through drain No. 2 to fulfil the raw water demand for drinking water supply to Delhi. The water segment is terminated into Wazirabad reservoir formed due to stagnation of water at Wazirabad barrage. The reservoir water is pumped to the various water works as raw water for treatment to meet the drinking water demand of the capital city. No or very little water is allowed to flow downstream of the Wazirabad barrage during lean seasons.

Segment III: This 22 km segment of Yamuna River is located between the Wazirabad barrage and Okhla barrage. This segment receives water from seventeen sewage drains of Delhi and also from WJC and Upper Ganga Canal via Najafgarh drain and Hindon cut canal, respectively.

Little contribution of water is also made in this segment by Surghat, where the Ganga and Yamuna water are provided for bathing purposes. This river segment terminates into Agra Canal, which is used to augment its flow for irrigation in the states of Haryana and Uttar Pradesh.

Segment IV: This Segment of Yamuna River is about 958 km long initiate immediately downstream of the Okhla barrage and extends up to the confluence with Ganga River at Allahabad. The source of water in this segment are ground water accrual, its tributaries like Hindon, Chambal, Sind, Ken, Betwa, etc. and waste water carrying drains of Delhi, Mathura-Vrindavan, Agra and Etawah. The water of this segment is used for drinking and industrial uses in Mathura and Agra.

In Mathura, recently the Gokul barrage has been constructed to trap the Yamuna River water for drinking purposes. Due to low drinking water demand, only part of water is pumped out and the rest flows downstream. As the water demand will increase in future. It is likely that no water will be allowed to flow downstream like the Wazirabad and Okhla barrages. This may create further segmentation of segment IV into two segments of 154 and 804 km. With the construction of another barrage near Sikandara in Agra the river would be further segmented.

9.4 Water Quality Monitoring

Water quality is one of the fundamental constraints important for drinking, bathing and irrigation and is regularly monitored by the Central and State Government Agencies for major river systems and tributaries of major rivers, respectively. The importance of the water quality parameters and their standards has been summarized in the next sections. The responsible departments are: Central Pollution Control Board (CPCB), State Pollution Control Boards, Central Water Commission, etc having different objectives. For example, CPCB is responsible for disseminating information in respect of matters relating to water and air pollution and their prevention and control, whereas CWC collect, maintain and publish the data relating to water resources and its utilization, including the quality of water throughout India and to act as the Central Bureau of Information relating to water resources. However, the ground water quality is being monitored by the combined efforts of Central Ground Water Board (CGWB), State Ground Water Departments (SGWDs), etc. The major responsible water quality agencies are summarized in Table 9.1.

Data collection primarily falls under the Ministry of Environment and Forests in the Central Pollution Control Board and the various State Pollution Control Boards, though the Central Ground Water Board collects some groundwater data as well. The data collection program is under the National Water Quality Monitoring program and within that the sites are divided into three subsets: Global Environmental Monitoring Program, Monitoring of Indian National Aquatic Resources, and the various river action plans. The majority of data is stored on the Environmental Data Bank maintained by the CPCB. The Global Environmental Monitoring System (GEMS), WISDOM, and (Environmental Information System) ENVIS databases also provide some water quality data. Water quality data is collected primarily, and in many places solely by governmental organizations, through central and state pollution control boards, under the umbrella of the Ministry of Environment & Forests and various departments belonging to the Ministry of Water Resources. The water quality monitoring locations of CPCB and CWC for the Yamuna River System from its origin to the confluence with Ganga River are depicted in Fig. 8.1.

The important and frequent water quality monitoring parameters are: temperature, pH, electrical conductivity (EC), dissolved oxygen (DO), biological oxygen demand (BOD), chemical oxygen demand (COD), nitrate (NO_3^-), nitrite (NO_2^-), total coliform (TC), feacal coliform (FC), etc.

Other measured physio-chemical parameters include Calcium, Magnesium, Sodium, Potassium, Chloride, Sulphate, Boron, Fluoride, Sodium Percent, Total Alkalinity, Hardness, Phosphate, Turbidity, Total Kjeldahl-N, Total Dissolved Solids, and Total Solids. Other biological parameters include S.A.R, Saprobity Index and Diversity Index etc. he type of monitoring station along with the frequency and parameters are given in Table 9.2.

Table 9.1 Water quality monitoring agencies

Data base	Objectives	Type of data	Supporting sources	Frequency of monitoring	SW/GW/Both	Purpose
CPCB	Prevention and control of pollution	Water quality	SPCBs	Monthly, quarterly, yearly	Both	Baseline, trend and flux stations
EDB	To facilitate quick and easy retrieval of data base	Air, Water, urban pollution, industrial pollution, hazardous waste	CPCB, SPCBs, etc.	Derived data (annual min., max., mean)	Both	Baseline indicator for environment
ENVIS	Disseminating information speedily to the user, to gear up the modern technologies of acquisition, processing, storage, retrieval, and dissemination of information of environmental nature.	Air, Water, Noise	CPCB	Regular updating of website (Derived data)	Both	Baseline information
GEMS (UNEP)	State and trends of regional and global water quality; global water quality assessment, prevention and control and strategic planning	Water quality and hydro-meteorological data	UNEP monitoring stations (peninsular states); CPCB	Annual, monthly and diurnal	Both	Baseline, trend and flux stations
WISDOM	Collection, storage, analysis, publication	Water quality and hydro-meteorological data	CWC, CGWB	Derived data	Both	Baseline data
GEMS and GWIS	Data storage in HIS system	GW quality	CGWB	Primary and derived data	Geospatial GW data	Baseline and Trend

Table 9.2 Type of station, frequency of observations and parameters used in water quality monitoring

Type of station	Frequency	Parameter
Baseline	Perennial rivers and Lakes: Four times a year Seasonal rivers: 3–4 times (at equal spacing) during flow period. Lake: 4 times a year	(A) Pre-monsoon: Once a year Analyse 25 parameters as listed below: (a) General: Colour, odour, temp, pH, EC, DO, turbidity, TDS (b) Nutrients: NH_3-N, NO_2 + NO_3, Total P (c) Organic Matter: BOD, COD (d) Major ions: K, Na, Ca, Mg, CO_3, HCO_3, Cl, SO_4, (e) Other inorganic: F, B and other location- specific parameter, if any (f) Microbiological: Total and Faecal Coliforms (B) Rest of the year (after the pre-monsoon sampling) at every three months' interval: Analyse 10 parameters: Colour, Odour, Temp., pH, EC, DO, NO_2 + NO_3, BOD, Total and Faecal Coliforms.
Trend	Once every month starting April-May (pre-monsoon), i.e. 12 times a year	(A) Pre-monsoon: Analyse 25 parameters as listed for baseline monitoring (B) Other months: Analyse 15 parameters as listed below (a) General: Colour, Odour, Temp, pH, EC, DO and Turbidity (b) Nutrients: NH_3-N, NO_2 + NO_3, Total P (c) Organic Matter: BOD, COD (d) Major ions: Cl (e) Microbiological: Total and Faecal coliforms (C) Micropollutant: Once in a year in monsoon season (i) Pesticides-Alpha BHC, Beta BHC, Gama BHC (Lindane), OP-DDT, PP-DDT, Alpha Endosulphan, Beta Endosulphan, Aldrin, Dieldrin, 2,4-D, Carboryl (Carbamate), Malathian, Methyl Parathian, Anilophos, Chloropyriphos (ii) Toxic Metals-As,Cd,Hg,Zn,Cr,Pb,Ni,Fe (Pesticides and Toxic metals may be analysed once a year)

9.5 Physico-Chemical Characteristics of Water Quality Parameters

A basic understanding of the parameters described above is important not only for creating the awareness of water use, but also in protecting the environment. A description of each parameter is given in Table 9.3.

Table 9.3 Importance of water quality parameters

S. No.	Parameter	Feature
1	Temperature	The temperature of water body varies with seasons. Temperature of the water body has an impact on the biota and its habitat. Fish and most aquatic organisms are cold-blooded. Consequently, their metabolism increases as the water warms and decreases as it cools. Each species of aquatic organism has its own optimum water temperature. If the water temperature shifts too far from the optimum, the organism suffers.
2	pH	Potentia hydrogeni is negative logarithm of H+ions. It is measured in scale of 0–14. As the pH goes on increasing from 1 to 6.9, the hydrogen ion concentration goes on decreasing and the solution is acidic. pH 7 represents the neutral condition and beyond 7 is alkaline. Most fish can tolerate pH values of about 5.5–9.0, but serious anglers look for waters between pH 6.5 and 8.2.
3	Electrical Conductance	EC is the ability of an aqueous solution to conduct the electric current. In water it is the property caused by the presence of various ionic species. Conductivity therefore measures the concentration of substance in solution and does not indicate the nature of these substances. The rock composition determines the chemistry of the watershed soil and ultimately the water body. For example, limestone leads to higher EC because of the dissolution of carbonate minerals in the basin.
4	Turbidity	When light is passed through a sample having suspended turbidity, some of the light is scattered by the particles. The scattering of light is generally proportional to turbidity. Pollution tends to reduce water clarity. Watershed development and poor land use practices cause increase in erosion, organic matter, and nutrients, all of which cause increases in suspended particulates and algae growth. High concentrations of particulate matter can modify light penetration, cause shallow lakes and bays to fill in faster, and repress benthic habitats impacting both organisms and eggs. Fine particulate material can clog or damage sensitive gill structures, prevent proper egg and larval development, and potentially interfere with particle feeding activities. Very high levels of turbidity for a short period of time may not be significant and may even be less of a problem than a lower level that persists longer.

Table 9.3 (continued)

S. No.	Parameter	Feature
5	Total Dissolved Solids	Total dissolved solids (TDS) denote mainly dissolved solids such as minerals present in the water. TDS comprises inorganic salts and small amounts of organic matter that are dissolved in water. The principal constituents are usually the cations calcium, magnesium, sodium and potassium and the anions carbonate, bicarbonate, chloride, sulphate and, particularly in groundwater, nitrate from agricultural use.
		The presence of dissolved solids in water may affect its taste. Water with extremely low TDS concentrations may also be unacceptable because of insipid taste. In addition to palatability, certain components of TDS, such as chlorides, sulphates, magnesium, calcium and carbonates also affect corrosion or encrustation in water distribution systems.
6	Total suspended solids	Total suspended solids (TSS) denote the suspended impurities present in the water. In most of the cases they are of organic nature and pose severe problems of water pollution. Also, TSS is a parameter that directly relates to land uses in the watershed and is a key parameter used for modelling efforts and for assessing the success of mitigation and restoration efforts.
		High TSS can block light from reaching submerged vegetation. As the amount of light passing through the water is reduced, photosynthesis slows down. Reduced rates of photosynthesis causes less dissolved oxygen to be released into the water by plants. If light is completely blocked from bottom dwelling plants, the plants will stop producing oxygen and will die. As the plants are decomposed, bacteria will use up even more oxygen from the water. Low dissolved oxygen can lead to fish kills.
		High TSS can also cause an increase in surface water temperature, because the suspended particles absorb heat from sunlight. This can cause dissolved oxygen levels to fall even further because warmer waters can hold less dissolved oxygen, and can harm aquatic life in many other ways.
7	Hardness	Hardness is generally imparted by the calcium and magnesium ions present in water. Polyvalent ions of some other metals like strontium, iron, aluminium, zinc and manganese etc. are also capable of precipitating the soap; concentration of these ions is very low in natural waters.
		Therefore, hardness is generally measured as concentration of calcium and magnesium which are far higher in quantities over other hardness producing ions. Hardness can be a nuisance due to the mineral buildup on plumbing fixtures and poor soap and detergent performance. It often causes aesthetic problems, such as an alkali taste to the water that makes coffee taste bitter; build-up of scale on pipes and fixtures than can lead to lower water pressure; build-up of deposits on dishes, utensils and laundry basins; difficulty in getting soap and detergent to foam; and lowered efficiency of electric water heaters.

Table 9.3 (continued)

S. No.	Parameter	Feature
8	Alkalinity	It is the measure of a solution's capacity to react with a strong acid (usually sulfuric acid H_2SO_4) to a predetermined pH. The alkalinity of a solution is usually made up of carbonate, bicarbonate, and hydroxides. Alkalinity is exhibited in solution by alkalies such as sodium hydroxide and by alkaline salts such as sodium carbonate. Soap and soap-based products are alkaline, since soap is a moderately alkaline salt and performs well only in an alkaline medium.
		For protection of aquatic life the buffering capacity should be at least 20 mg/l. Alkalinity is important for fish and aquatic life because it protects rapid pH changes.
9	Dissolved Oxygen (DO)	The presence of dissolved oxygen (DO) is essential to maintain the higher forms of biological life and to keep proper balance of various populations thus making the water bodies healthy. The chemical and biochemical processes undergoing in a water body are largely dependent upon the presence of oxygen. The main sources of dissolved oxygen are from the atmosphere and the photosynthetic processes of the green plants. Estimation of dissolved oxygen is a key test in water pollution and waste treatment process control. The solubility of oxygen in water depends upon the partial pressure of oxygen in air, temperature of water and mineral content of water.
		Oxygen is produced during photosynthesis and consumed during respiration and decomposition. Because it requires light, photosynthesis occurs only during daylight hours. Respiration and decomposition, on the other hand, occur 24 h a day. This difference alone can account for large daily variations in DO concentrations.
		Other sources of oxygen include the air and inflowing streams. Oxygen concentrations are much higher in air, which is about 21% oxygen, than in water, which is a tiny fraction of 1% oxygen. Where the air and water meet, this tremendous difference in concentration causes oxygen molecules in the air to dissolve into the water. More oxygen dissolves into water when wind stirs the water; as the waves create more surface area, more diffusion can occur.
		Another physical process that affects DO concentrations is the relationship between water temperature and gas saturation. Cold water can hold more oxygen, than warmer water. Warmer water becomes "saturated" more easily with oxygen. As water becomes warmer it can hold DO. So, during the summer months in the warmer top portion of a water body, the total amount of oxygen present may be limited by temperature. If the water becomes too warm, even if 100% saturated, O_2 levels may be suboptimal for many species.

Table 9.3 (continued)

S. No.	Parameter	Feature
10	Biochemical Oxygen Demand (BOD)	Biochemical oxygen demand (BOD) is the oxygen required by the microorganisms in depleting the organic matter. Microorganisms' breakdown the carbohydrate and use the energy thus released, but the aquatic system is depleted in oxygen in the process. An estimate of the oxygen consumed in the unit volume of water over a period of time is called BOD.
		The BOD values are very useful in process design, treatment plant efficiency, and stream pollution control management and in evaluating the self purification capacities. BOD5 determination is the measurement of dissolved oxygen content of the sample before and after 5 day incubation.
11	Chemical Oxygen Demand (COD)	Chemical oxygen demand (COD) is a measure of oxygen consumed during the oxidation of the oxidizable organic matter by a strong oxidizing agent. The determination of COD value is of great importance where BOD values cannot be determined accurately due to the presence of toxins and other unfavourable conditions for growth of microorganisms. The COD gives no indication of whether or not the waste is degradable biologically and nor does it indicates the rate at which biological oxygen would proceed and hence the rate at which the oxygen would be required in a biological system.
		Despite all the limitations, the COD test continues to remain a very important parameter in management and design of treatment plants because of its rapidity in determination.
12	Total organic carbon (TOC)	Total organic carbon (TOC) provides a speedy and convenient way of determining the degree of organic contamination. Domestic and industrial wastewaters mainly contribute organic contaminants in fresh water bodies. A high organic content means an increase in the growth of micro-organisms which contribute to the depletion of oxygen supplies.
13	Ammonia	Ammonium of mineral origin is rare in natural waters. The most important source of ammonia is the ammonification of organic matter. Sewage has large quantities of nitrogenous matter, thus its disposal tends to increase the ammonia content of the water. Occurrence of ammonia in water can be accepted as the chemical evidence or organic pollution. If only ammonia is present, pollution by sewage must be very recent. Occurrence of nitrite with ammonia indicates a long time has been passed after nitrogenous matter has been oxidized.

Table 9.3 (continued)

S. No.	Parameter	Feature
14	Nitrate	Most surface water is deficient in nitrate. The most important source of nitrate is biological oxidation of organic nitrogenous substances which come in sewage and industrial wastes. Domestic waste water contains very high amount of nitrogenous compounds. Runoff from agricultural fields is also high in nitrate, which is the highest oxidized form of nitrogen. Atmospheric nitrogen fixed into nitrates by the nitrogen fixing organisms is also a significant contributor to nitrates in the waters.
		It is essential nutrient for photosynthetic autotrophs and in some cases has been identified as the growth limiting nutrient. Nitrates cause reactions with haemoglobin in blood, causing the oxygen carrying capacity of the blood to decrease.
15	Phosphate	Phosphorus is the eleventh most abundant element on the surface of the earth and is most commonly found as phosphate. It plays an important role in biochemical processes and is a key factor in the eutrophication of surface water. Increased phosphate concentrations are linked with increasing rates of plant growth.
		Phosphorus is necessary for plant and animal growth. Nearly all fertilizers contain phosphates (chemical compounds containing the element, phosphorous). When it rains, varying amounts of phosphates wash from farm soils into nearby waterways. Phosphates stimulate the growth of plankton and water plants that provide food for fish. This may increase the fish population and improve the waterway's quality of life. If too much phosphate is present, algae and water weeds grow wildly, choke the waterway, and use up large amounts of oxygen.
16	Chloride	Chloride as free ions is one of the major inorganic anions in water and wastewater. The salty taste produced in potable water is due to the presence of chloride ions. High amount of chloride increases corrosive nature.
		Chlorine has been used as an effective disinfectant in drinking water supplies for nearly 100 years. Chlorine is considered necessary to destroy many of the bacteria in your drinking water. Health officials are concerned with the chlorinating by-products, also known as "chlorinated hydrocarbons" or trihalomethanes. Most trihalomethanes are formed in drinking water when chlorine reacts with naturally occurring substances such as decomposing plant and animal materials. Risks for certain types of cancer are now being correlated to the consumption of chlorinated drinking water.

Table 9.3 (continued)

S. No.	Parameter	Feature
17	Sulphate	Sulphate occurs in water naturally as result of leaching from gypsum and other common minerals. In addition sulphate may be added to water system, such as in municipal water supply sulphate content increases during clarification of alum. Sulphate content also increases corrosive nature.
		The damaging effects of sulfur with animals are mostly brain damage, through malfunctioning of the hypothalamus, and damage to the nervous system. Laboratory tests with test animals have indicated that sulfur can cause serious vascular damage in veins of the brains, the heart and the kidneys. These tests have also indicated that certain forms of sulfur can cause foetal damage and congenital effects. Mothers can even carry sulfur poisoning over to their children through mother milk. It can damage the internal enzyme systems of animals.
18	Calcium	Calcium is one of the most abundant substances in natural water. It is present in higher quantities in the rocks, from where it is leached to contaminate water. Sewage and industrial wastes are also important source of calcium. Concentration of calcium is reduced at higher pH due to its precipitation as $CaCO_3$. It is one of the important nutrients required by the organisms.
		High concentration of calcium in water suppresses the formation of lather with soap. It coagulates with soap and makes dirty layers on sinks and tubes etc.
		Calcium may have beneficial effects when ingested. It may block the absorption of heavy metals in the body and is thought to increase bone mass and prevent certain types of cancer. Very high concentrations of calcium may adversely affect the absorption of other essential minerals in the body.
19	Magnesium	Magnesium is common constituent of natural water and occurs with calcium. Its concentration is lower than calcium. Magnesium salts breakdown when heated, forming scale in boilers.
		Magnesium may contribute undesirable tastes to drinking water.
		Sensitive people may find the taste unpleasant at 100 mg/l. The average person finds the taste unpleasant at about 500 mg/l. These levels are well above the magnesium concentrations found in most water. Magnesium in drinking water may have a laxative effect, particularly with magnesium sulphate concentrations above 700 mg/l. However, the human body tends to adapt to this laxative effect with time.
20	Carbonate and bicarbonate	These are responsible for hardness and also an important parameter for drinking and irrigation water. Bicarbonate and carbonate ions combined with calcium and magnesium precipitate as calcium carbonate ($CaCO_3$) and magnesium carbonate ($MgCO_3$) respectively when the soil solution concentrates in drying conditions.

Table 9.3 (continued)

S. No.	Parameter	Feature
21	Fluoride	Fluoride contamination in water has geogenic as well as anthropogenic origin. It is consideredas the main contaminant for drinking water as it causes deformity in bones. Fluoride is commonly added to the water supply in developing countries because at certain concentrations it can have dental hygiene benefits. However, naturally-occurring Fluoride in the groundwater is sometimes present at very high concentrations and this can cause dental and skeletal problems.
22	Oil and Grease	The determination of oil and grease includes all the substances that are extractable by the specified solvent. Generally, the substances extractable are oils, fats and waxes. The result obtained indicates only the non volatile fraction of these materials. Oil and grease has the natural tendency to float on the water surface under quiescent conditions, as the density of oil and grease is usually less than one. Not all the oil and grease is in liquid or solid form. Appreciable amounts remain in a finely divided emulsified form. The high oil and grease concentration observed in effluent receiving water bodies is responsible for the depletion of fish and other aquatic life. The petroleum films flow on the surface of the water cutting out sunlight for aquatic photosynthesis and reducing oxygen concentration for aquatic life in the water, thereby negatively affecting these aquatic life forms. Some marine animals and reptiles such as turtles are predominantly vulnerable to adverse effects from oil contamination because of their need to breath from the water surface. Petrochemicals such as benzene can cause cancer even at low exposure levels.
23	Total coli and faecal coli	Total coliforms include bacteria that are found in the soil, in water that has been influenced by surface water, and in human or animal waste. These are indicator parameters for detecting microbiological quality of water. Fecal coliforms are the group of the total coliforms that are considered to be present specifically in the gut and feces of warm-blooded animals. The coliform group comprises all aerobic and facultative anaerobic gram negative, non sporulating, rod shaped bacteria that ferment lactose with gas formation within 48 h at 35°C. Coliform includes bacteria of most common intestinal organism Klebsiella pneumonia and of organism rare association in intestine Enterobactor aerogenes. Other common organisms included in this class are Salmonella shigella, Streptococci, Aeromonas etc. Coliform densities can be determined by multiple tube fermentation technique. This technique provides the most probable number (MPN).

Table 9.3 (continued)

S. No.	Parameter	Feature
24	Chlorophyll	Surface waters that have high chlorophyll conditions are typically high in nutrients, generally phosphorus and nitrogen. These nutrients cause the algae to grow or bloom. High levels of nitrogen and phosphorus can be indicators of pollution from man-made sources, such as septic system leakage, poorly functioning wastewater treatment plants, or fertilizer runoff. Thus, chlorophyll measurement can be utilized as an indirect indicator of nutrient levels. Chlorophyll a is most common and predominant in all oxygen-evolving photosynthetic organisms such as higher plants, red and green algae.
		Chlorophyll b differs from chlorophyll a only in one of the functional groups bonded to the porphyrin (a -CHO group in place of a -CH_3 group). Chlorophyll b is an accessory pigment and acts indirectly in photosynthesis by transferring the light it absorbs to chlorophyll a. Chlorophyll c occurs only in algae, specifically the diatoms, dinoflagellates and brown algae. Its role is to pass on the light excitation to chlorophyll a.
25	Primary Productivity	Primary production in water body can be determined by light and dark bottle oxygen method. By increasing the production of phytoplankton, fertilization may increase fish production considerably, though much remains to be learned about the optimum amount and frequency of fertilization. Primary production determinations can be used to evaluate the effect of different fertilizer treatments on organic production.
		Net primary productivity (NPP) is the rate at which all the plants in an ecosystem produce net useful chemical energy; it is equal to the difference between the rate at which the plants in an ecosystem produce useful chemical energy and the rate at which they use some of that energy during respiration. The total energy fixed by plants in a community through photosynthesis is referred to as gross primary productivity (GPP).
26	Heavy metals: in general	Heavy metals are natural components of the earth's crust. To a small extent they enter our bodies via food, drinking water and air. As trace elements, some heavy metals (e.g. copper, selenium, zinc) are essential to maintain the metabolism of the human body. However, at higher concentrations they can lead to poisoning. Heavy metal poisoning could result, for instance, from drinking-water contamination (e.g. lead pipes), high ambient air concentrations near emission sources.
		Heavy metals are dangerous because they tend to bioaccumulate. Bioaccumulation means an increase in the concentration of a chemical in a biological organism over time, compared to the chemical's concentration in the environment.
		Heavy metals can enter a water supply by industrial and consumer waste, or even from acidic rain breaking down soils and releasing heavy metals into streams, lakes, rivers, and groundwater.

Table 9.3 (continued)

S. No.	Parameter	Feature
27	Zinc	Zinc is a very common substance that occurs naturally. Many foodstuffs contain certain concentrations of zinc. Drinking water also contains certain amounts of zinc, which may be higher when it is stored in metal tanks. Industrial sources or toxic waste sites may cause the zinc amounts in drinking water to reach levels that can cause health problems. Most zinc is added during industrial activities, such as mining, coal and waste combustion and steel processing. Some soils are heavily contaminated with zinc, and these are to be found in areas where zinc has to be mined or refined, or were sewage sludge from industrial areas has been used as fertilizer. Zinc is the 23rd most abundant element in the Earth's crust. Water is polluted with zinc, due to the presence of large quantities of zinc in the wastewater of industrial plants. Some fish can accumulate zinc in their bodies, when they live in zinc-contaminated waterways. When zinc enters the bodies of these fish it is able to bio magnify up the food chain.
28	Manganese	Manganese compounds may be present in the atmosphere as suspended particulates resulting from industrial emissions, soil erosion, volcanic emissions and the burning of Methylcyclopentadienyl manganese tricarbonyl (MMT) containing petrol. In surface waters, manganese occurs in both dissolved and suspended forms, depending on such factors as pH, anions present and oxidation–reduction potential. It can bioaccumulate in lower organisms (e.g., phytoplankton, algae, molluscs and some fish) but not in higher organisms; biomagnifications in food chains is not expected to be very significant.
29	Iron	Iron in an essential element in both plant and animal metabolism. Iron can cause the water to look cloudy, taste poor, and can stain laundry or discolour rice. The drinking water quality standard for Iron is 0.3 mg/l based on aesthetic effects, but this is often exceeded in ground waters. Contamination sources are mainly mining, auto body rust, engine parts etc.
30	Lead	Major sources of lead contamination are waste from industrial, vehicular emissions, paints and burning of plastics, papers, etc. Lead has impact on blood lead levels below 10 μg/dl of blood. Its chronic exposure effects impairment of neurological development, suppression of the hematological system and kidney failure.

Table 9.3 (continued)

S. No.	Parameter	Feature
31	Mercury	Mercury is a compound that can be found naturally in the environment. It can be found in metal form, as mercury salts or as organic mercury compounds. Mining, electronics and plastic waste, pesticides, pharmaceutical and dental waste are major source of mercury pollution. It causes gastro-intestinal disorders, respiratory tract irritation, renal failure and neurotoxicity. Mercury also causes damage to the brain and the central nervous system. Mercury enters the environment as a result of normal breakdown of minerals in rocks and soil through exposure to wind and water. Most of the mercury released from human activities is released into air, through fossil fuel combustion, mining, smelting and solid waste combustion. Some forms of human activity release mercury directly into soil or water, for instance the application of agricultural fertilizers and industrial wastewater disposal. All mercury that is released in the environment will eventually end up in soils or surface waters.
32	Nickel	Nickel contamination is mainly due to diesel fuel and gasoline, lubricating oil, brake emissions. High nickel concentrations on sandy soils can clearly damage plants and high nickel concentrations in surface waters can diminish the growth rates of algae. Microrganisms can also suffer from growth decline due to the presence of nickel, but they usually develop resistance to nickel after a while. For animals nickel is an essential foodstuff in small amounts. But nickel is not only favourable as an essential element; it can also be dangerous when the maximum tolerable amounts are exceeded. This can cause various kinds of cancer on different sites within the bodies of animals, mainly of those that live near refineries.
33	Cadmium	Sources of cadmium contamination are mainly electronics, plastics, batteries and contaminated water. Cadmium causes irritation of the lungs and gastrointestinal tract, kidney damage, abnormalities of the skeletal system and cancer of the lungs and prostate. Cadmium in humans, long-term exposure is associated with renal disfunction. High exposure can lead to obstructive lung disease and has been linked to lung cancer, and damage to human's respiratory systems.
34	Chromium	Chromium is used in metal alloys and pigments for paints, cement, paper, rubber, and other materials. Low-level exposure can irritate the skin and cause ulceration. Long-term exposure can cause kidney and liver damage, and damage to circulatory and nerve tissue. Chromium often accumulates in aquatic life, adding to the danger of eating fish that may have been exposed to high levels of chromium.

Table 9.3 (continued)

S. No.	Parameter	Feature
35	Copper	Copper is an essential substance to human life, but in high doses it can cause anaemia, liver and kidney damage, and stomach and intestinal irritation.
		In surface water copper can travel great distances, either suspended on sludge particles or as free ions. Copper does not break down in the environment and because of that it can accumulate in plants and animals when it is found in soils. On copper-rich soils only a limited number of plants have a chance of survival. That is why there is not much plant diversity near copper-disposing factories. Due to the effects upon plants copper is a serious threat to the productions of farmlands. Copper can seriously influence the proceedings of certain farmlands, depending upon the acidity of the soil and the presence of organic matter. Copper can interrupt the activity in soils, as it negatively influences the activity of microorganisms and earthworms. The decomposition of organic matter may seriously slow down because of this.
		When the soils of farmland are polluted with copper, animals will absorb concentrations that are damaging to their health. Mainly, sheep and goat suffer great from copper poisoning, because the effects of copper are manifest at fairly low concentrations.
36	Flow velocity	Velocity depends on the physical characteristics of streams, such as slope, roughness and cross section. Flow of streams affects everything from temperature of the water and concentration of various substances in the water to the distribution of habitats and organisms. It has impact on physico-chemical characteristics of water. Stream flow, acting together with the gradient, and the geology of the bottom substrate, determines the types of habitats present, the shape of the channel, and the composition of the stream bottom.
		Low flow periods in summer allow the stream to heat up rapidly in warm weather while in the fall and winter temperatures may fall rapidly when flow is low. Flow directly affects the amount of oxygen dissolved in the water.
37	Flow rate	The flow rate of the water body is a primary factor in TSS concentrations. Fast running water can carry more particles and larger-sized sediment. Heavy rains can pick up sand, silt, clay, and organic particles such as leaves, soil, sheared particles from the land and carry it to surface water. A change in flow rate can also affect TSS; if the speed or direction of the water current increases, particulate matter from bottom sediments may be suspended.

9.6 Water Quality Standards

The Bureau of Indian Standard (BIS) has developed standards for drinking water quality. The CPCB has provided best use criteria for surface water bodies. River Action Plans led by Ministry of Environment & Forests are for reviving the river water up to B-class i.e. bathing category. In the Tables 9.4, 9.5, and 9.6 standards of water quality for drinking (BIS, 1991), CPCB classification of designated use, and effluent disposal limits are presented, respectively.

Table 9.4 Bureau of Indian standard/specification for drinking water (BIS: 10,500, 1991): maximum limit

Parameter	Requirement desirable limit	Undesirable effect outside the desirable	Permissible limit in the absence of alternate source	Remarks
Essential characteristic				
Colour Hazen Units	5	Above 5, consumer acceptance decreases	25	Extended to 25 only if toxic substance are not suspect in absence of alternate sources
Odour	Un-objectionable	–	–	Test cold and when heated test are several dilutions
Taste	Agreeable	–	–	Test to be conducted only after safely has been established
Turbidity (NTU)	5	Above 5, consumer acceptance decreases	10	–
pH	6.5–8.5	Beyond this range the water will affect the mucous membrane and/or water supply system	No relaxation	–
Total Hardness, $CaCO_3$ (mg/l)	300	Encrustation in water supply structure and adverse effects on domestic use	600	–

Table 9.4 (continued)

Parameter	Requirement desirable limit	Undesirable effect outside the desirable	Permissible limit in the absence of alternate source	Remarks
Iron, Fe (mg/l)	0.3	Beyond this limit taste/appearance are affected; has adverse effects on domestic uses and water supply structure and promotes iron bacteria	1.0	–
Chlorides, Cl (mg/l)	250	Beyond effects outside the desirable limit	1,000	–
Residual free Chlorine (mg/l)	0.2	–	–	To be applicable only when water is chlorinated. Tested at customer end. When protection against viral infection is required, it should be min. 0.5 mg/l.
Desirable characteristics				
Dissolved solids (mg/l)	500	Beyond this, palatability decreases and may cause gastrointestinal irritation.	2,000	–
Calcium, Ca (mg/l)	75	Encrustation in water supply structure and adverse effects on domestic use.	200	–
Magnesium, Mg (mg/l)	30	Encrustation in water supply structure and adverse effects on domestic use.	100	–
Copper, Cu (mg/l)	0.05	Astringent taste dis-coloration and corrosion of pipes fittings and utensils will be caused beyond this.	1.5	–

Table 9.4 (continued)

Parameter	Requirement desirable limit	Undesirable effect outside the desirable	Permissible limit in the absence of alternate source	Remarks
Manganese, Mn (mg/l)	0.1	Beyond this limit taste/appearance are affected, has adverse effect on domestic use and water supply structure	0.3	–
Sulphate, SO_4 (mg/l)	200	Beyond this causes gastro intestinal irritation when magnesium or sodium are present	400	May be extended upto 400 provided magnesium, Mg does not exceed 30
Nitrate, NO_3 (mg/l)	45	Beyond this methaemoglobinemia takes place.	100	–
Fluoride, F (mg/l)	1.0	Fluoride may be kept as low as possible. High fluoride may cause fluorosis.	1.5	–
Phenolic Compounds, C_6H_5OH (mg/l)	0.001	Beyond this, it may cause objectionable taste and odour	0.002	–
Mercury, Hg (mg/l)	0.001	Beyond this the water becomes toxic	No Relaxation	To be tested when pollution is suspected
Cadmium, Cd (mg/l)	0.01	Beyond this the water becomes toxic	No Relaxation	To be tested when pollution is suspected
Selenium, Se (mg/l)	0.01	Beyond this the water becomes toxic.	No Relaxation	To be tested when pollution is suspected
Arsenic, As (mg/l)	0.05	Beyond this the water becomes toxic	No Relaxation	To be tested when pollution is suspected
Cyanide, CN (mg/l)	0.05	Beyond this the water becomes toxic	No Relaxation	To be tested when pollution is suspected

Table 9.4 (continued)

Parameter	Requirement desirable limit	Undesirable effect outside the desirable	Permissible limit in the absence of alternate source	Remarks
Lead, Pb (mg/l)	0.05	Beyond this the water becomes toxic	No Relaxation	To be tested when pollution is suspected
Zinc, Zn (mg/l)	5	Beyond this limit it can cause astringent taste and an opalescence in water	15	To be tested when pollution is suspected
Anionic detergents, MBAS (mg/l)	0.2	Beyond this limit it can cause a light froth in water	1.0	To be tested when pollution is suspected
Chromium, Cr+6 (mg/l)	0.05	May be carcinogenic above this limit	–	–
Poly-nuclear Aromatic Hydrocarbons, PAH (mg/l)	–	May be carcinogenic	–	–
Mineral oil (mg/l)	0.01	Beyond this limit, undesirable taste and odour after chlorination takes place	0.03	To be tested when pollution is suspected
Pesticides (mg/l)	Absent	Toxic	0.001	–
Alpha emitters (Bq/l)	–	–	0.1	–
Beta emitters (Pci/l)	–	–	1.0	–
Alkalinity (mg/l)	200	Beyond this limit, taste becomes unpleasant	600	–
Aluminium, Al (mg/l)	0.03	Cumulative effect is reported to cause dementia	0.2	
Boron, B (mg/l)	1.0	–	5.0	–

Table 9.5 Water quality criteria for best designated use (CPCB)

Designated-best-use	Class of water	Criteria
Drinking Water source without conventional treatment but after disinfection	A	T. Coli: ≤ 50 MPN/100 ml pH: 6.5–8.5 DO: ≥ 6 mg/l 5 days BOD at 20°C: ≤ 2 mg/l
Outdoor bathing (Organized)	B	T. Coli: ≤ 500 MPN/100 ml pH: 6.5–8.5 DO: ≥ 5 mg/l 5 days BOD at 20°C: ≤ 3 mg/l
Drinking water source after conventional treatment and disinfection	C	T. Coli: ≤ 5,000 MPN/100 ml pH: 6.0–9.0 DO: ≥ 4 mg/l 5 days BOD at 20°C: ≤ 3 mg/l
Propagation of Wild life and Fisheries	D	pH: 6.5–8.5 DO: ≥ 4 mg/l Free Ammonia, N: ≤ 1.2 mg/l
Irrigation, Industrial Cooling, Controlled Waste disposal	E	pH: 6.0–8.5 EC at 25°C: ≤ 2,250 micro mhos/cm Sodium absorption Ratio Max. 26 Boron: ≤ 2 mg/l
No use	Below-E	Not Meeting A, B, C, D and E Criteria

Table 9.6 General standards for industrial effluents (maximum limit)

Parameter	Inland surface water	Public sewers	Land for irrigation	Marine/coastal areas
Suspended solids (mg/l)	100	600	200	(a) For process waste water (b) For cooling water effluent 10% above total suspended matter of influent.
Particle size of suspended solids	≤850 micron IS Sieve	–	–	(a) Floatable solids, solids maximum 3 mm (b) Settleable solids, maximum 856 microns
pH value	5.5 – 9.0	5.5 – 9.0	5.5 – 9.0	5.5 – 9.0
Temperature above the receiving water	≤5°C	–	–	≤5°C
Oil and grease (mg/l)	10	20	10	20
Total residual chlorine (mg/l)	1.0	–	–	1.0
Ammoniacal Nitrogen, N (mg/l)	50	50	–	50
Total Kjeldahl Nitrogen, N (mg/l)	100	–	–	100
Free Ammonia, NH_3 (mg/l)	5.0	–	–	5.0
3 days BOD at 27°C (mg/l)	30	350	100	100
COD (mg/l)	250	–	–	250
Arsenic, As (mg/l)	0.2	0.2	0.2	0.2
Mercury, Hg (mg/l)	0.01	0.01	–	0.01
Lead, Pb (mg/l)	0.1	1.0	–	2.0
Cadmium, Cd (mg/l)	2.0	1.0	–	2.0
Hexavalent Chromium, Cr+6 (mg/l)	0.1	2.0	–	1.0
Total chromium, Cr (mg/l)	2.0	2.0	–	2.0

Table 9.6 (continued)

Parameter	Inland surface water	Public sewers	Land for irrigation	Marine/coastal areas
Copper, Cu (mg/l)	3.0	3.0	–	3.0
Zinc, Zn (mg/l)	5.0	15	–	15
Selenium, Se (mg/l)	0.05	0.05	–	0.05
Nickel, Ni (mg/l)	3.0	3.0	–	5.0
Cyanide, CN (mg/l)	0.2	2.0	0.2	0.2
Fluoride, F (mg/l)	2.0	15	–	15
Dissolved phosphates, P (mg/l)	5.0	–	–	–
Sulphide, S (mg/l)	2.0	–	–	5.0
Phenolic compounds, C_6H_5OH (mg/l)	1.0	5.0	–	5.0
Radioactive materials: Alpha emitter's micro curie	10–7	10–7	10–8	10–7
Beta emitters micro curie	10–6	10–6	10–7	10–6
Bio-assay test	90% survival of fish after 96 hours in 100% effluent	90% survival of fish after 96 h in 100% effluent	90% survival of fish after 96 h in 100% effluent	90% survival of fish after 96 h in 100% effluent
Manganese (mg/l)	2.0	2.0	–	2.0
Iron, Fe (mg/l)	3.0	3.0	–	3.0
Vanadium, V (mg/l)	0.2	0.2	–	0.2
Nitrate nitrogen (mg/l)	10.0	–	–	20.0

These standards shall be applicable for industries, operations or processes other than those industries, operations or process for which standards have been specified in Schedule of the Environment Protection Rules (MoEF, 1989). CPCB has separate effluent disposal standard for individual industries.

References

BIS (1991) Indian standard drinking water specification. Bureau of Indian Standards, IS 10500: 1991, New Delhi

CPCB (1978) Union Territory of Delhi (Abridged), Control of Urban Pollution Series, CUPS/1/1978-79, Central Pollution Control Board of India

CPCB (1981) Basin Sub Basin Inventory of Water Pollution – The Ganga Basin Part One: The Yamuna sub – basin, Assessment and Development study of River Basin Series, ADSORBS/2/1980-81, Central Pollution Control Board of India

CPCB (1982–1983) Quality and Trend of Yamuna River (1977–1982), Assessment and Development Study of River Basin Series, ADSORBS/10/1982-83, Central Pollution Control Board of India

CPCB (1984–2008) Water Quality Statistics of India, Monitoring of Indian Notional Aquatic Resources Series (MINARS/1 to 30/1984 to 2008), Central Pollution Control Board of India

CPCB (1986) Yamuna Ek Paristhiki Vishneshan, Assessment and Development Study of River Basin Series, ADSORBS/14/1986-87, Central Pollution Control Board of India

CPCB (1994) Water Quality Status of Yamuna River, Assessment and Development Study of River Basin Series, ADSORBS/29/1993-94, Central Pollution Control Board of India

CPCB (2006) Status of Sewage Treatment in India, Control of Urban Pollution Series: CUPS/61/2005–2006, Central Pollution Control Board of India

CPCB (2009) Status of Water supply, Wastewater generation and Treatment in class-I cities and Class-II towns of India, Central Pollution Control Board of India

MoEF (1989) Hazardous wastes (Management and Handling) Rules: 1989. Environment (Protection) Act, 1906 (29 of 1986), Ministry of Environment and Forests, Government of India, New Delhi

Chapter 10
Wastewater Generation and Treatment

Abstract The chapter deals with wastewater generation and treatment, including the inventory of wastewater treatment mechanisms in the Yamuna River basin and available gaps in the treatment infrastructure and generation. The trend analysis is also performed to determine the future impact of wastewater generation on the river health. Detailed analysis of city-wise and state-wise wastewater generation are also included in the chapter for determining the critical issues of water quality management of the Yamuna River.

In the Yamuna basin there are two important sources of wastewater, i.e., domestic sewage from urban centres and industrial effluents from different categories of industries. As per the studies carried out by the Central Pollution Control Board (CPCB, 1982–1983, 1994), the domestic wastewater is the predominant source of pollution of the Yamuna River.

Urbanization in the sense of more and more people living in urban areas is by far the most important social change that has taken place in India in recent times. From a modest base of 25.8 million persons in 1901, the number of urban dwellers has risen to 285 million in 2001, signalling a phenomenal eleven-fold increase in urban population over a period of 100 years. During the past three decades of 1971–1901 itself, India's urban population has grown almost three fold, increasing from 109 million to 285 million. The proportion of population living in urban areas (level of urbanization) increased from 17.3% in 1951 to 28.5% in 2001 and the number of urban centres (town and urban agglomerations) swelled from 2843 in 1951 to 5162 in 2001. The earlier trend of migration from rural areas to urban areas gained further momentum due to fast economic growth in urban areas creating large employment opportunities and declining resources availability per capita and shrinking economic opportunities in rural areas (Chapter 7). A large number of rural-urban migrants get absorbed in the informal sector, and several among them took shelter in the slums, with the result that today nearly one half of India's urban employment sector is informal and unorganized. Thus rapid urbanisation places enormous strain on authorities to provide infrastructure necessary to support the population. This is also posing serious environmental problems related to water supply and waste water generation and their disposal.

The annual rate of growth of urban population in India (3.09%) is distinctly higher than that of the high-income industrial market economies (1.4%) and also of the East European economies (1.8%). The latter groups of countries, with around

R.K. Rai et al., *The Yamuna River Basin*, Water Science and Technology Library 66,
DOI 10.1007/978-94-007-2001-5_10, © Springer Science+Business Media B.V. 2012

two thirds of their population living in urban places, are already highly urbanised. The low-income economies, as a group with a 4.4% growth rate and the middle-income economies with 4.1%, have an edge over India in this regard. As many as sixty, out of a total of ninety-four countries, belonging to low and middle income categories outrank India in urban growth rate. The annual urban growth rate is 4.2% in Mexico, 4.8% in Nigeria and 6.5% in Bangladesh. The corresponding rate in high-income, oil-exporting countries is still higher, being about 8.2%.

In the Indian context, urban population is usually presented in as a six-part classification in Indian censuses. While the first size-class comprises cities i.e., urban places having 100 thousand and more population, the last category consists of the tiny towns, i.e., places having population less than 5,000. In order to provide a meaningful analysis of the changes in size-class composition of urban population, the last three categories namely towns having a population 10,000–19,999; 5,000–9,999; and less than 5,000 are grouped together and termed as small towns. Medium towns are defined to have a population between 20,000 and 49,999; and large towns constitute a population in the range of 50,000–99,999. The places having more than 100 thousand population are named as Class-I cities. On the other hand, cities having population 1 million and more are termed as million plus cities or Metro-cities (Census, 2001).

The importance of different size classes of cities in total urban population has shifted in favour of large cities. In 1,901, 26% of urban inhabitants lived in settlements with over 1,00,000 population. By 2001 this percentage has risen to 60.3% heralding what would seem to be one of the most significant changes in the pattern of habitation in the country. With this change, India is no longer a country that lives in villages and small towns; rather it has acquired a complexion of a country that has an extensive network of large urban settlements. Population growth does have a significant impact on the environment and resources and will heighten demand for food, energy, water, health care, sanitation and housing.

Rapid urbanization and economic growth has led to a steep increase in waste generation. Unregulated growth of urban areas, particularly over the last two decades, without infrastructural services for proper collection, transportation, treatment and disposal of domestic waste water led to increased pollution and health hazards. The municipalities and such other civic authorities who are responsible for management of wastes have not been able to cope with this massive task. This could be attributed to various reasons, including the lack of resources, inability to raise revenues, erosion of authority and inadequate managerial capabilities.

10.1 Urbanization in the Yamuna Basin

As per the Census of India, 2001, the total population of the Yamuna River basin is about 130 million. There are 6 metro-cities (population of about 20 million), 61 Class-I cities (Population 11.4 mllion), 61 Class-II towns (Population 4.27 million), 201 Class-III towns (Population 5.94 million), 452 Class-IV towns (Population

Table 10.1 State-wise urban and rural population in the Yamuna basin

	Population in Yamuna basin			
States	Urban	Rural	Total	No. of villages
Himachal Pradesh	149,738	715,698	865,436	2,284
Uttarakhand	168,269	247,804	416,073	619
Haryana	4,723,131	9,317,117	14,040,248	4,210
Uttar Pradesh	9,233,115	25,032,382	34,265,497	18,639
Delhi	12,819,761	963,215	13,782,976	158
Madhya Pradesh	10,906,554	19,448,621	30,355,175	22,889
Rajasthan	8,479,742	28,303,047	36,782,789	26,003
Total	46,480,310	84,027,884	130,508,194	74,802

3.63 million) 148 Class-V towns (Population 1.09 million) and 74,802 villages (Population 84 million). It is observed that urbanization in the Yamuna basin is 35.6%, which is higher than the national average of 28.5% (Census, 2001). Table 10.1 depicts state-wise summary of urban and rural population in the Yamuna basin.

As clear from Table 10.1, Delhi has the highest urban population followed by Madhya Pradesh, Uttar Pradesh and Rajasthan. The State of Himachal Pradesh and Uttarakhand are not very significant from the population point of view as their population put together is less than 1% of the total population in the basin. In the Yamuna basin, the population share under different categories of urban centers is summarized in Table 10.2.

10.2 Water Supply and Wastewater Generation

Wastewater generation is a function of water supply. In India piped water supply is largely restricted to urban areas. Rural population and smaller towns still depend on community wells, hand-pumps or even rivers/springs/ponds for their drinking water requirements. As per the CPCB study the water supply facilities are more advanced and effective in large urban centers than smaller ones (CPCB, 2009). The per capita water supply is also higher in larger cities than smaller ones. In the Yamuna basin, it is observed that the per capita water supply varies from 82 lpcd in Rajasthan to 352 lpcd in New Delhi. In smaller towns and rural areas the water supply is just around 45 lpcd. This results in meager wastewater generation from smaller urban centers or rural areas. As per the CPCB's study the Class-I cities in the country generate about 90% of the wastewater, whereas the Class-II towns only 10%. However, in the Yamuna basin, the share of sewage generation from class-I and II cities are 94 and 6%, respectively. If we go further down to Class-III or lower categories of urban centers or villages, practically insignificant or no wastewater is generated there. Hence, in the present study, the wastewater estimation is restricted to only Class-I cities and Class-II towns. The total water supply in the class-I cities and Class-II towns in

Table 10.2 Distribution of urban population in different categories of urban centers in the Yamuna basin

State	Class-I		Class-II		Class-III		Class-IV		Class-V		Total	
	No	Pop	No	Pop	No	Pop	No	Pop	No	Pop	No	Pop
Uttarakhand	0	0	0	0	2	50,956	0	0	2	6,747	4	57,707
Himachal Pradesh	0	0	0	0	3	82,763	1	19,087	7	29,648	11	131,509
Uttar Pradesh	15	5,295,761	20	1,357,105	42	1,233,719	170	1,004,874	44	341,658	291	9,233,393
Haryana	14	3,328,821	5	299,861	18	555,291	29	422,121	15	117,037	81	4,723,198
Delhi	10	11,256,671	9	635,925	20	698,194	12	156,547	11	72,424	62	12,819,813
Madhya Pradesh	16	6,248,127	16	1,153,375	60	1,709,801	200	1,400,124	51	395,127	343	10,906,881
Rajasthan	12	5,300,518	11	821,819	56	1,605,840	40	626,329	18	125,236	137	8,479,867
Total	67	31,429,898	61	4,268,085	201	5,936,564	452	3,629,082	148	1,087,877	929	46,352,368

the Yamuna basin is about 11,555 MLD. It is also observed that due to the lack of proper collection facilities in most of the cities and towns the wastewater either flows in storm water drains often unlined or accumulates in the vacant plots, low-lying areas, percolates into the ground and pollutes groundwater. The wastewater, which flows in storm water drains ultimately leads to receiving water bodies and pollutes them. Thus, surface and groundwater in and around urban centres especially large urban centres is seriously affected by the discharge of untreated urban wastewater. However, the state-wise expected urban wastewater generation in the Yamuna basin is shown in Fig. 10.1, and it appears that Delhi has the highest wastewater generation followed by Uttar Pradesh, Madhya Pradesh and Rajasthan. Beside this, the city-wise wastewater generation is given in Tables 10.3, 10.4, 10.5, 10.6, and 10.7.

It is also important to consider that those cities/towns located on the banks of a river are most important from pollution point of view. Contribution of pollution from those located away from a river is generally less due to percolation, evaporation losses and lifting by farmers for irrigation en-route. Thus, all those towns, which are located along different rivers in the basin, are very important from a pollution control point of view.

It is observed that a total of 8,444 MLD of wastewater is generated from all the Class-I cities and Class-II towns in the basin out of which the treatment capacity is available for only 3,447 MLD. Thus, there is a large gap of about 5,000 MLD between generation and treatment of wastewater (Table 10.8).

It is also estimated by CPCB (2006) that out of 8,444 MLD of wastewater generated in the basin, about 4,458 MLD is discharged directly into the Yamuna River and about 1,200 MLD is discharged into its tributaries, the remaining 2,786 MLD is either disposed of on land or used for irrigation. Delhi alone generates about 3,743 MLD of wastewater, which is 44% of the entire sewage generated in the basin and 84% of the sewage being discharged into the Yamuna along its entire course. Uttar Pradesh, Madhya Pradesh and Rajasthan share almost equal magnitudes of sewage

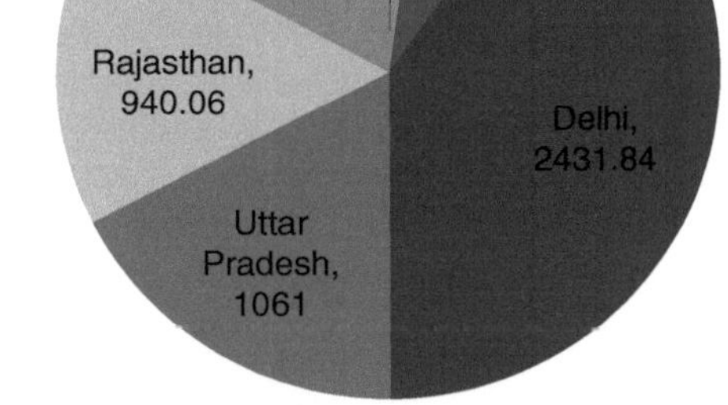

Fig. 10.1 State-wise wastewater generation (in MLD) in the Yamuna basin

Table 10.3 Wastewater generation in the cities of Haryana

City	Population	Water supply rate (lpcd)	Water supply (mld)	Wastewater generation (mld)
Faridabad	1,054,981	140	148	118
Rohtak	286,773	123	35	28
Panipat	261,665	128	33	27
Hisar	256,810	123	32	25
Sonipat	216,213	123	27	21
Karnal	210,476	170	36	29
Yamunanagar	189,587	180	34	27
Gurgaon	173,542	240	42	33
Sirsa	160,129	130	21	17
Kurukshetra	120,072	123	15	12
Jhajjar	119,839	123	15	12
Kaithal	117,226	123	14	12
Yamunanagar	101,300	123	12	10
Rewari	100,946	123	12	10
Faridabad	100,528	123	12	10
Hisar	75,730	123	9	7
Mahendragarh	62,091	90	6	4
Sirsa	53,812	90	5	4
Jind	50,659	90	5	4
Sonipat	48,518	90	4	3
Jhajjar	39,004	90	4	3
Faridabad	38,306	90	3	3
Kurukshetra	37,130	90	3	3
Panipat	35,150	90	3	3
Kurukshetra	33,547	90	3	2
Hisar	33,130	90	3	2
Sirsa	32,786	90	3	2

generation in this basin. On the sewage treatment front Delhi has the highest sewage treatment capacity of 2,330 MLD which is about 68% of the total sewage treatment capacity in the basin (Table 10.8). However, there is still a large gap between generation and treatment in Delhi itself.

10.3 Trends in Sewage Generation and Treatment

The CPCB (2006, 2009) has been regularly updating information on Class-I cities and Class-II towns in the country for water supply and wastewater generation, treatment and disposal. As per the available information the water supply and wastewater generation have increased much faster than population growth in these urban centres, which is clearly depicted in Fig. 10.2. This figure also includes trends in the number of cities, population, water supply, wastewater generation and treatment capacity. However, the treatment capacity has been almost stagnant for one and half decades.

Table 10.4 Wastewater generations in Delhi

City	Population	Water supply rate (lpcd)	Water supply (mld)	Wastewater generation (mld)
In all 9 districts	9,817,439	270	2,651	2,121
New Delhi, Central, South West and South	294,783	352	104	83
North West	163,716	140	23	18
North West	153,874	140	22	17
North West	151,427	140	21	17
West	150,371	140	21	17
North East	148,549	140	21	17
East	132,628	120	16	13
South West	124,452	130	16	13
South	119,432	140	17	13
North East	90,564	120	11	9
North East	89,117	120	11	9
West	85,848	120	10	8
North	69,182	120	8	7
East	68,978	120	8	7
East	65,969	120	8	6
South	58,220	130	8	6
North East	57,460	120	7	6
North West	50,587	120	6	5
North East	48,028	120	6	5
South	47,336	120	6	5
North East	45,090	120	5	4
South	44,895	120	5	4
West	43,898	120	5	4
North East	43,364	120	5	4
North East	42,564	120	5	4
South	41,243	120	5	4
South	39,267	120	5	4
South West	38,580	120	5	4

Table 10.5 Wastewater generations in Uttar Pradesh

City	Population	Water supply rate (lpcd)	Water supply (mld)	Wastewater generation (mld)
Agra	634,622	170	108	86
Jhansi	198,135	44	9	7
Mathura	140,150	80	11	9
Firozabad	130,863	71	9	7
Muzaffarnagar	114,783	68	8	6
Saharanpur	225,396	84	19	15
Ghaziabad	118,836	213	25	20
Allahabad	990,298	158	156	125

Table 10.6 Wastewater generations in Rajasthan

City	Population	Water supply rate (lpcd)	Water supply (mld)	Wastewater generation (mld)
Jaipur	636,768	156	99	79
Ajmer	264,291	78	21	16
Kota	212,991	149	32	25
Udaipur	161,278	125	20	16
Alwar	100,378	77	8	6

Table 10.7 Wastewater generations in Madhya Pradesh

City	Population	Water supply rate (lpcd)	Water supply (mld)	Wastewater generation (mld)
Indore	560,936	149	84	67
Gwalior	406,140	93	38	30
Bhopal	284,859	180	51	41
Sagar	154,785	55	9	7
Ratlam	119,247	50	6	5
Ujjain	208,561	95	20	16

Table 10.8 Water supply and wastewater generation and treatment in the class-I and II cities/towns in the Yamuna basin

	Class-I cities			Class-II towns		
States	Water supply (MLD)	WW gen (MLD)	WW treat (MLD)	Water supply (MLD)	WW gen (MLD)	WW Treat (MLD)
Himachal Pradesh	0	0	0	0	0	0
Uttarakhand	0	0	0	0	0	0
Haryana	819	655	333	70.4	56.3	3.5
Delhi	4,679	3,743	2,330	100.0	80.0	0
Uttar Pradesh	1,349	1,079	393.5	251.2	201.0	4.5
Madhya Pradesh	1,519	1,215	319	151.9	121.5	9
Rajasthan	1,537	1,230	54	78.3	62.6	0
Total	9,903	7,922	3,430	651.9	521.5	17

10.4 Inventory of Water Treatment and Sewage Treatment Plants

Sewage treatment plants (STPs) play an important role in keeping rivers clean. Though several STPs have been installed along the river course with a designed capacity of the order of 2,332.25 MLD for Delhi, 327 MLD for Haryana, and 403.25

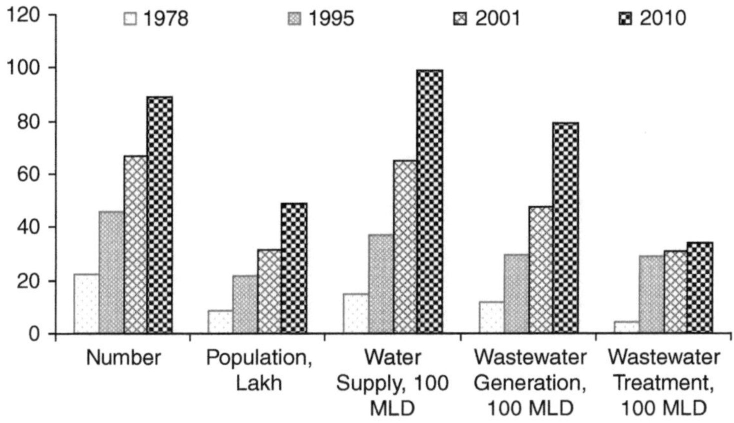

Fig. 10.2 Change in number, population, water supply, WW generation and treatment in class-I cities of the Yamuna basin

MLD for Uttar Pradesh (CPCB, 1978; CPCB, 2006; CPCB, 2009). However, the river water quality has not been improved significantly. Detailed state-wise inventory of the STPs and common effluent treatment plant (CETPs) in the basin are summarized in Tables 10.9, 10.10, 10.11, and 10.12. These tables also include the designed capacity and running status of the STPs along with influent and effluent BOD values. The main cause of the river quality not yet improved may be: (i) STPs are in general practically not meeting their compliance, (ii) under running of most of the STPs due to lack of sewer connection, (iii) improper drainage system, etc. The density of STPs in Delhi is the highest and is mostly based on the activated sludge process (ASP). However, on the other part, type belongs to either Oxidation Pond (OP) or Up-flow Anaerobic Sludge Blanket (UASB) type. A descriptive map showing the location and type of STPs in the basin and NCT Delhi are depicted in Figs. 10.3 and 10.4, respectively.

The overall inventory of water treatment plants (WTPs) in the Yamuna River basin and Delhi is shown in Figs. 10.5 and 10.6, respectively, and the capacities of these WTPs have been summarized in Table 10.13.

10.5 Industrial Wastewater Generation, Treatment and Disposal

The Indian economy has never been in such a good shape as it is now with the liberalization measures initiated in 1991 becoming effective. The continued impressive performance of the agricultural sector and a significant improvement in the industrial production since September 1993 have enabled the rate of growth of the Gross Domestic Product (GDP) to rise from 0.8% in 1991–1992 to 6–8% in 2008–2009. This has resulted in rapid industrialization in the Yamuna basin also. There are large clusters of industries established in Kota, Gwalior, Indore, Nagda, Khetri, Yamuna Nagar, Panipat, Sonepat, Delhi, Baghpat, Ghaziabad, Gautam

Table 10.9 Sewage treatment plants in Delhi

S. No	Name of the STP's	Design capacity (MLD)	Type of STP	Running Status	Under YAP	*BOD Influent/ effluent
1	Coronation Pillar	181.84	Activated sludge process (ASP) and Ttrickling filter	Under utilized		170/26
2	Delhi Gate	10.00	High Rate Biofilters (HBR) Densadeg technology	Running on designed capacity	Yes	140/22
3	Ghitorni	22.73	ASP	Not in operation		
4	Keshopur	327.31	ASP	Under utilized		216/34
5	Kondli	204.57	ASP	Under utilized		110/28
6	Mehrauli	22.73	Extended aeration	Under utilized		226/22
7	Najafgarh	22.73	ASP	Under utilized		208/28
8	Nilothi	181.84	ASP	Under utilized		264/64
9	Narela	45.46	ASP	Under utilized		160/42
10	Okhla	636.43	ASP	Under utilized		242/24
11	Papankalan	90.92	ASP	Under utilized		168/24
12	Rithala	363.68	ASP and HBR	Under utilized		218/20
13	Rohini	68.19	Activated sludge process	Not in operation		
14	Sen N.H.	10.0	HBR	Running on designed capacity	Yes	155/12
		2	Ultra-violet radiation			
15	Timarpur	27.27	Oxidation ponds (OP)	Under utilized		106/4
16	Yamuna Vihar	90.92	ASP	Under utilized		256/32
17	Vasant Kunj	23.63	ASP and Extended aeration	Under utilized		260/30
Total STP = 30		2,332.25				

Buddha Nagar, Faridabad, Mathura and other places. The categories of industries discharging wastewater into Yamuna River includes pulp and paper, sugar, distilleries, textiles, leather, chemical, pharmaceuticals, oil refineries, thermal power plants, food processing, etc. In order to comply with environment laws, it is compulsory for these industries to treat their effluents to achieve prescribed standards before discharging effluent into the environment.

The rapid industrialization has exerted extra pressure on infrastructure facilities of human settlements manifesting in environmental problems, in terms of increased incidences of air, water, noise pollution and land degradation. It is unfortunate

Table 10.10 Sewage treatment plants in Haryana

S. No.	Name of the STP's		Design capacity (MLD)	Type of STP	Running status	Constructed under YAP	BOD influent/ effluent[a]
1	Yamuna Nagar	STP-I	10	UASB	Under utilized	Yes	188/38
	Jagadhari	STP-II	25	UASB	Over the capacity	Yes	144/26
	Chhchhrauli	STP	1	OP	Under utilized	Yes	
2	Karnal	STP-I	40	UASB	Full capacity	Yes	166/30
			1	Down flow hanging sponge (DHS-Bio tower)	Full capacity		154/29
		STP-II	8	OP	Under utilized	Yes	
3	Panipat	STP-I	10	UASB	Under utilized	Yes	176/54
		STP-II	35	UASB	Under utilized	Yes	168/28
4	Sonepat	STP-I	30	UASB	Under utilized	Yes	166/29
5	Gurgaon	STP-I	30	UASB	Full capacity	Yes	
6	Faridabad	STP-I	20	UASB	Under utilized	Yes	77/39
			2	Ultra-violet radiation			
			2	Solar radiation			
		STP-II	45	UASB	Under utilized	Yes	81/33
		STP-III	50	UASB	Under utilized	Yes	
7	Gharaunda	STP	3	OP	Under utilized	Yes	
8	Gohana	STP	3.5	OP	Under utilized	Yes	
9	Indri	STP	1.5	OP	Under utilized	Yes	
10	Palwaal	STP	9	OP	Under utilized	Yes	
11	Radaur	STP	1	OP	Under utilized	Yes	
	Total STP = 17		**327**				

[a] Limit for disposal of effluent in natural streams <30 mg/l

Table 10.11 Sewage treatment plants in Uttar Pradesh

S. No	Name of the STP's		Design capacity (MLD)	Type of STP	Running status	Under YAP	BOD influent/effluent[a]
1	Sharanpur	STP	38	UASB	Full capacity	Yes	67/8
2	Muzaffar Nagar	STP	32.5	OP	Over the capacity	Yes	156/32.8
3	Ghaziabad	STP-I	70	UASB	Full capacity	Yes	209/50
			3	karnal technology (KT)	Full capacity	Yes	
		STP-II	56	UASB	Full capacity	Yes	140/37
4	Noida	STP-I	34	UASB	Full capacity	Yes	92/35
		STP-II	27	UASB	Full capacity	Yes	139/30
			2	Chlorination			
		STP-III	9	OP	Full capacity	Yes	161/39
5	Vrindavan	STP-I	4	OP	Under utilized	Yes	
		STP-II	0.5	OP	Under utilized	Yes	118/57
6	Mathura	STP-I	14.5	OP	Under utilized	Yes	176/37
		STP-II	12.5	OP	Under utilized	Yes	
7	Agra	STP-I	78	UASB	Under utilized	Yes	NA/21
		STP-II	10	OP	Under utilized	Yes	79/23
		STP-III	2.25	OP	Under utilized	Yes	94/47
8	Etawah	STP	10	OP	Under utilized	Yes	135/20
Total STP = 15			**403.25**				

[a] Limit for disposal of effluent in natural streams <30 mg/l

that environmental considerations tend to get a back seat in economic planning, especially while planning for industrial development. The industrial development inevitably results in the deterioration of environment. Unchecked industrialization processes have disturbed the environmental balance in the Yamuna basin. Most of the urban centres are not only facing problems of domestic pollution loads generated from the urban population but also from pollution loads from large numbers of industries in the residential areas in these urban centres. With the fast growth

Table 10.12 Common effluent treatment plants in Delhi

S. No.	Name of the CETP	Design capacity (MLD)	Type	Running Status	Under YAP	BOD influent/effluent[a]
1	Badli CETP	12	ASP	Under utilized		185/24
2	GTK CETP	6	ASP	Under utilized		160/29
3	Jhilmil CETP	16.8	ASP	Under utilized		170/27
4	Lawrence Road CETP	12	ASP	–		
5	Mangolpuri CETP	2.4	ASP	Full capacity		150/21
6	Nangloi CETP	12	ASP	Under utilized		190/27
7	Narela CETP	4	ASP	Full capacity		210/34
8	Okhla CETP	12	ASP	Under utilized		180/36
9	Okhla CETP	24	ASP	Under utilized		95/27
10	SMA CETP	12	ASP	Under utilized		140/28
11	Wazirpur CETP	24	ASP	Under utilized		130/25

[a] Limit for disposal of effluent in natural streams <30 mg/l

of industries a large number of environmental problems arise. A most important environmental problem associated with industrialization is disposal of their wastes. A large number of industrial activities generate solid, liquid or gaseous wastes. Many of these wastes, when released into the environment, cause serious problems of pollution. Normally, control of pollution is "un-economical" and therefore the industries are reluctant to invest their resources on this aspect. Thus, the industrialization has resulted in serious environmental problems in many parts of India.

For rational planning of pollution control strategies it is important to have a complete knowledge of the nature and magnitude of pollution loads generated from different categories of industries. The Central Pollution Control Board has been assessing pollution loads generated from different sources. However, due to the lack of reliable data, the industrial pollution quantification cannot be done by the CPCB. Since the information on the quantification of industrial pollution is of crucial importance, it is decided by CPCB to quantify the pollution load from the industrial production data. The industrial production data are fairly well documented by Central Statistical Organisation, Ministry of Planning, Government of India, and are being regularly updated. The pollution load generated in terms of important pollutants per unit production is also fairly well established by the CPCB under its elaborate studies called "Comprehensive Industries Document Series" on different categories of industries in India. Taking the help of this existing information an attempt is made to estimate the pollution load that is being generated in terms of important pollutants in the Yamuna basin.

As per the information available from different SPCBs there are about 400 large and medium industries belonging to 13 categories of highly polluting industries in the Yamuna basin. The state-wise distribution of these industries is given in

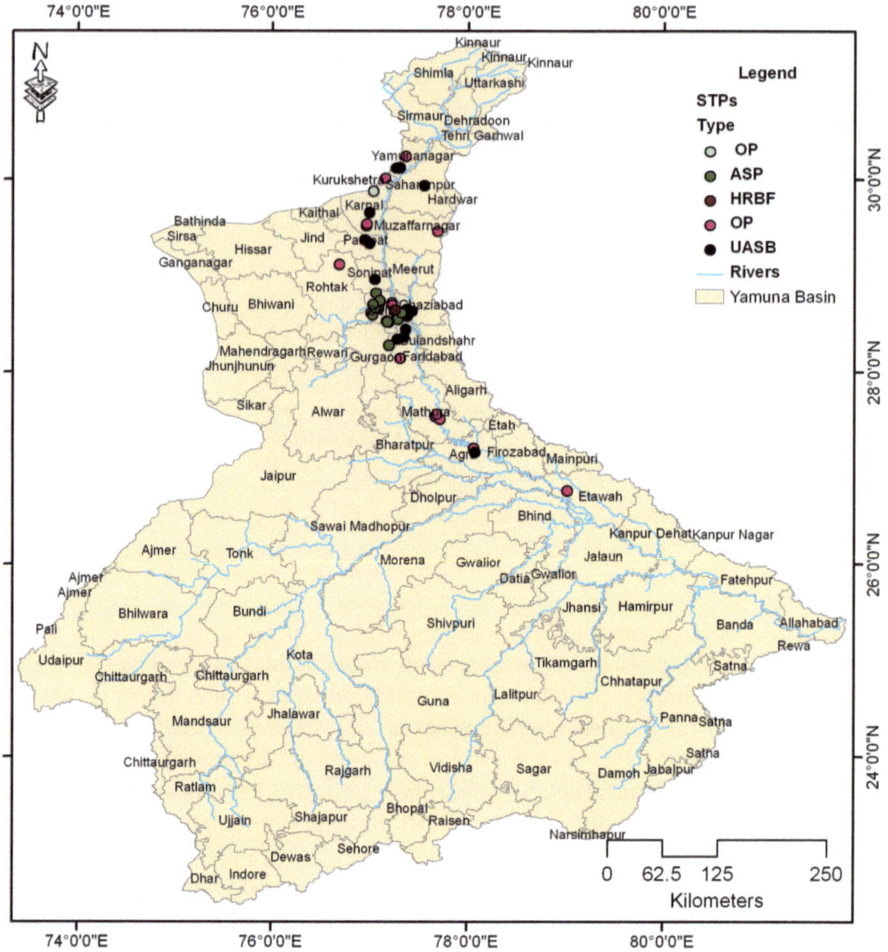

Fig. 10.3 Sewage treatment plants in the basin

Table 10.14. From this table, it is clear that Uttar Pradesh have the highest number of large and medium industries in the basin followed by Haryana, Rajasthan and Madhya Pradesh. Among different categories, paper mills has highest number followed by pharmaceutical industries, cement, sugar and distilleries, fertilizer plants, thermal power plants and tanneries.

10.6 Pollution Load Generation from Industrial Sources

Information on industry-wise pollution load generation is either not available or not reliable due to various reasons. Thus, an approach adopted to estimate pollution contribution from industrial sources is based on industrial production. The

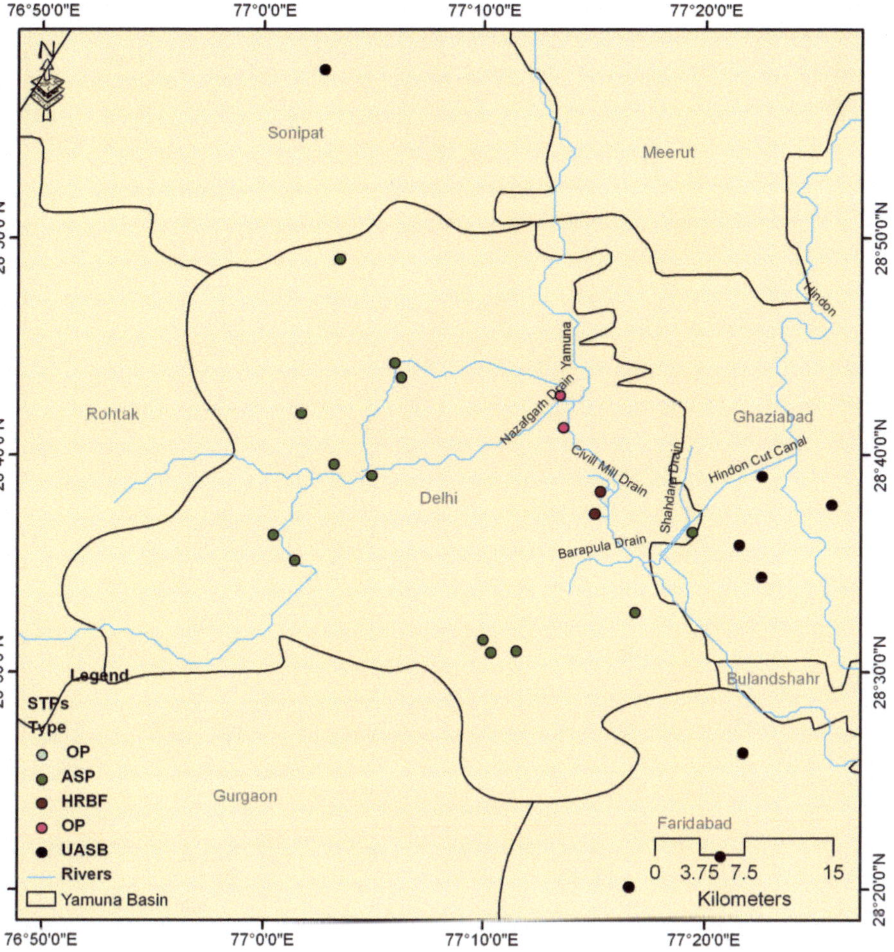

Fig. 10.4 Sewage treatment plants in NCT Delhi

pollution load in terms of volume of wastewater and BOD load state-wise and industrial category-wise is presented in Figs. 10.7, 10.8, 10.9, 10.10, 10.11, and 10.12.

As revealed from Figs. 10.7, 10.8, 10.9, 10.10, 10.11, and 10.12, UP contributes about 55% of the industrial effluents by volume, followed by Delhi 18%, MP 14%, Haryana 7%, Rajasthan 4% and Himachal Pradesh 2%. From the BOD load generation point of view Uttar Pradesh is contributing 64% of BOD in the basin (Fig. 10.7), followed by Madhya Pradesh, Haryana, Rajasthan, Delhi and Himachal Pradesh. Their contributions are 15%, 10%, 6%, 3% and 2%, respectively. Similarly, a look into the category-wise generation of the volume of wastewater reveals that paper mills contributes the highest volume of wastewater in the basin i.e. 52% followed by sugar industries 15%, textiles 6%, distilleries 3% and others 24%. From the BOD

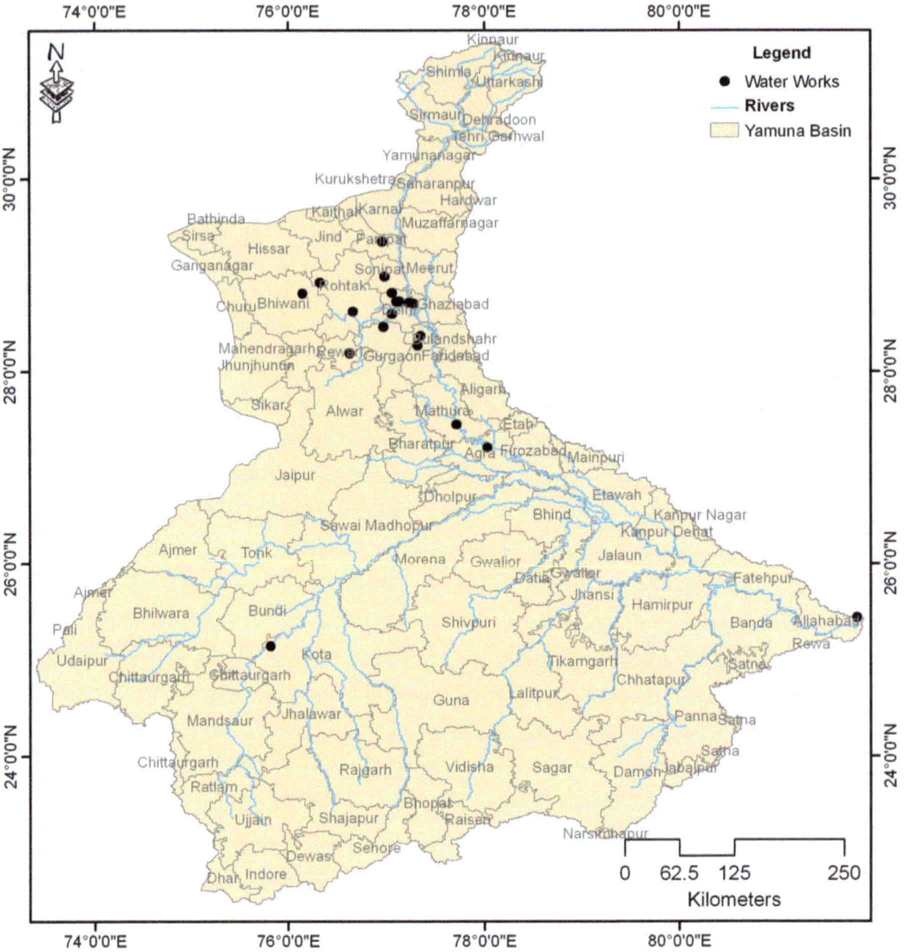

Fig. 10.5 Location map of water treatment plants in Yamuna basin

load generation point of view it is the distilleries dominates the whole industrial pollution load scenario in the basin, which is 60% (Fig. 6.8) followed by paper mills 27%, sugar 8% and others 5%. The BOD load contributed after treatment into the environment is the highest from paper mills, i.e., 72% (Fig. 10.11) followed by distilleries 22%, sugar 2%, tanneries 3% and others 1%. State-wise estimation of the BOD load contribution into the environment revealed that Uttar Pradesh contributes 70% of the pollution load into the environment (Fig. 10.12) followed by Madhya Pradesh 12%, Haryana 7%, Delhi 5%, Himachal Pradesh and Rajasthan 3% each.

Since Uttar Pradesh dominates in generation and contribution of industrial pollution load, it is important to look into the disposal points. The pollution contribution from Uttar Pradesh is mostly either directly into the Yamuna River or in the Krishni,

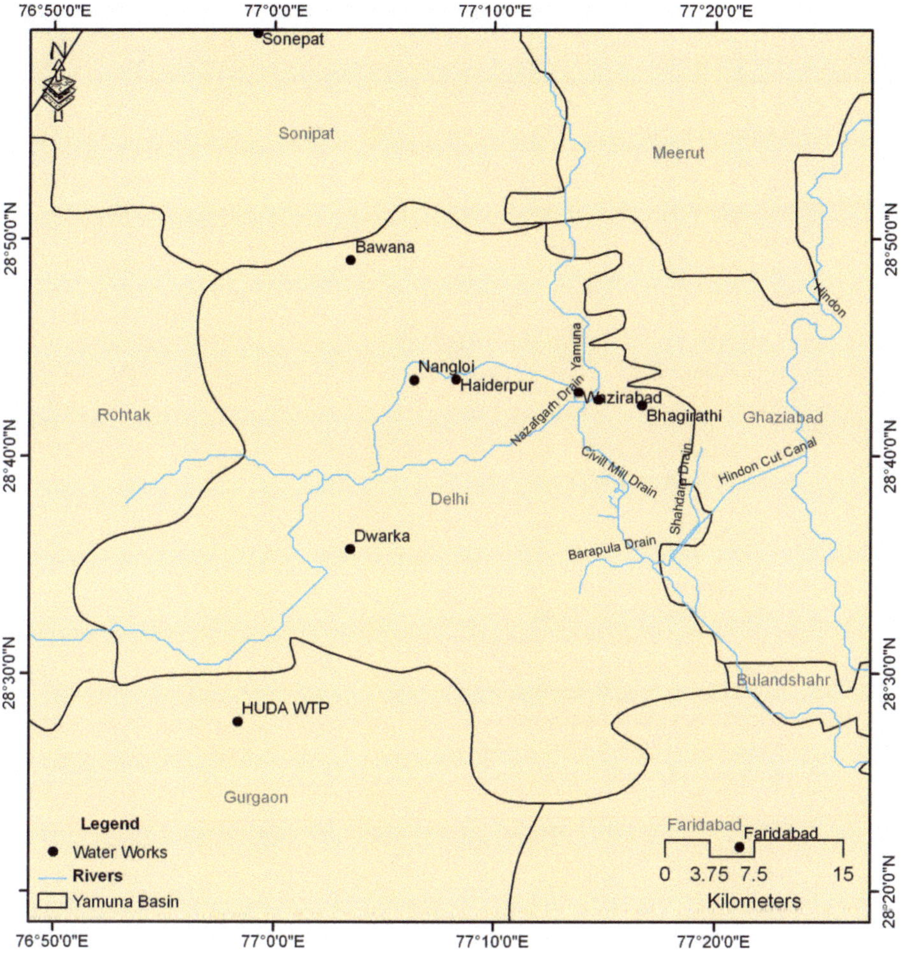

Fig. 10.6 Location map of water treatment plants in Delhi

the Kali and the Hindon. Most of the industries contributing pollution are agro-based and hence, it is the organic pollution which is most important. The other important state is Haryana, which contributes its industrial pollution either directly into the Yamuna River through, drains No. 2, 6 and 8. These drains are carrying the effluent from the towns located along the river. There is one very important industrial town located on the Western Yamuna Canal (WYC) which discharges its effluent episodically in the WYC, affecting the drinking water quality of Delhi very often. In MP the pollution loads are contributed mostly in Chambal River, Kshipra, Khan, Betwa, Kalisindh, Parwati and Ken rivers. Since these rivers are not perennial and flow for 3–4 months in a year, their contribution of pollution to the Yamuna is not very significant. Similarly, the industrial pollution load contribution from Rajasthan is mostly received by the Chambal and the Banas, which are again not perennial

Table 10.13 Inventory of water treatment plants

S. No.	Name/location	Capacity
Delhi		
1	Wazirabad WTP	120 MGD
2	Bhagirathi WTP	100 MGD
3	Sonia Vihar WTP	140 MGD
4	Hyderpur WTP	200 MGD
5	Nangloi WTP	40 MGD
6	Chandrawal WTP	90 MGD
7	Bawana	20 MGD
8	Dwarka	40 MGD
9	Okhla	–
10	Okhla Ranny Well	100 MGD
11	Palla and GW	≈30 MGD
Uttar Pradesh		
1	Agra Waterworks WTP	410 MLD
2	Agra Jeevanimandi WTP	
3	Mathura	25 MLD
4	Allahabad	
Haryana		
1	Jind – Kaithal Canal based water works	
2	Jind – Narwana Canal based water works	
3	Panipat Canal based water works	
4	Palwal Canal based water works	
5	Jhajjar Canal based water works	
6	Sonepat Canal based water works	
7	Gurgaon Canal based water works	
8	Faridabad Canal based water works	
Kota		
1	Kota	64 MLD
2	Jodhpur	60 MLD
3	Udaipur (2)	36.7 MLD
4	Nagaur	UC
5	Jaipur	
6	Sawai Madhopur	
7	Karauli	UC

and hence the pollution load is either assimilated in those rivers or percolates in the ground and may pollute the groundwater.

10.7 Wastewater Discharged into the Yamuna

As stated above, about 8,444 MLD of wastewater is generated from all the Class-I cities and Class-II towns in the basin out of which about 5,000 MLD is directly discharged into the Yamuna River from 16 cities/towns located on its banks, out of which the treatment capacity exists for about 3,100 MLD. The growth of urban centres exerts pressure on water resources in two ways. The increasing need for

Table 10.14 State-wise and category-wise number of large and medium industries under 13 categories of polluting industries

State	Distilleries	Cement	Chloralkalis	Dyes and Dye Int	Fertilizer	Pharmaceuticals	Paper mills	Rayon	Sugar	TPP	Tanneries	Engineering	Misc.	Total
Madhya Pradesh	9	8	1	2	5	17	1	1	5	1	0	0	0	**50**
Uttar Pradesh	6	0	0	5	0	5	47	0	14	3	10	18	69	**177**
Rajasthan	3	28	2	2	14	10	1	1	2	8	0	3	4	**78**
Delhi	0	3	0	0	0	0	0	0	0	5	0	0	0	**8**
Haryana	7	10	0	2	4	18	8	0	15	3	7	13	3	**90**
Himachal Pradesh	3	2	0	0	0	1	0	0	0	0	0	0	0	**6**
Total	**28**	**51**	**3**	**11**	**23**	**51**	**57**	**2**	**36**	**20**	**17**	**34**	**76**	**409**

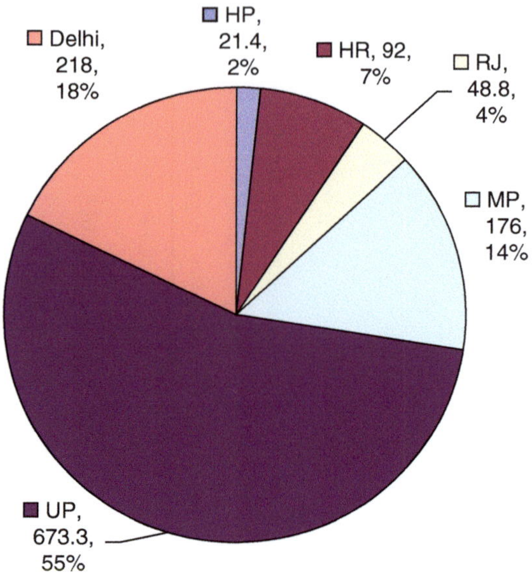

Fig. 10.7 State-wise industrial effluent generation in the Yamuna basin

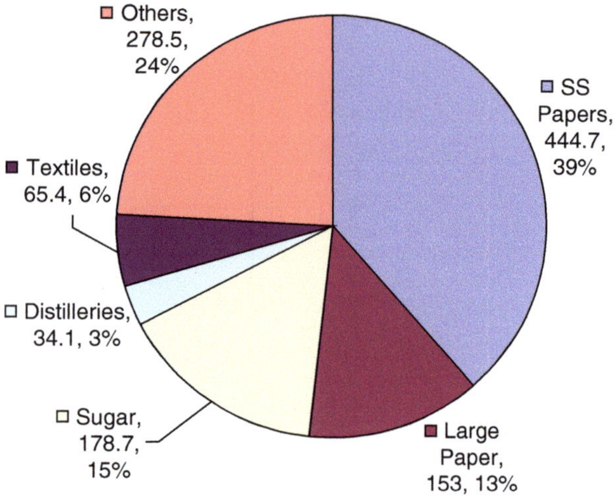

Fig. 10.8 Category-wise volume of wastewater generated, MLD

water to meet the domestic requirement and the impact of resultant waste water discharge on the receiving waters has the cumulative effect of deteriorating the quality of receiving waters. All the 16 cities along its banks have grown steeply in the past 2–3 decades. The wastewater added from these urban centres has increased manifold and the fresh water availability in many stretches of the river has gone down to zero. This has resulted in the Yamuna to become the main sewer in the Delhi, Mathura and Agra stretches. Delhi, being the capital of India and large business and

Fig. 10.9 State-wise industrial BOD load generation in the Yamuna basin, t/d

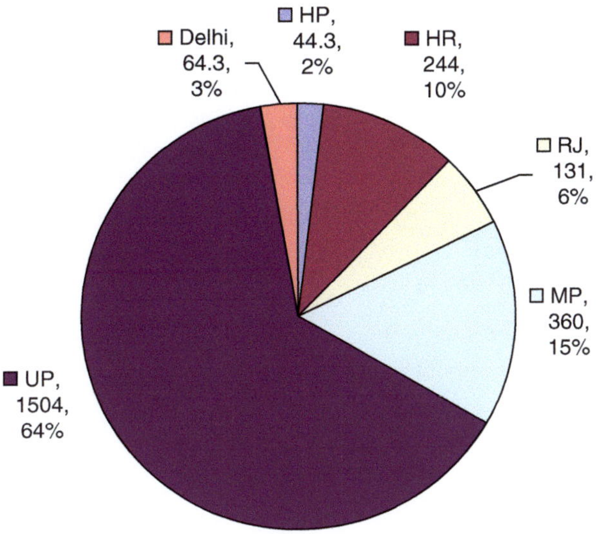

Fig. 10.10 Category-wise industrial BOD load generation in the Yamuna basin, t/d

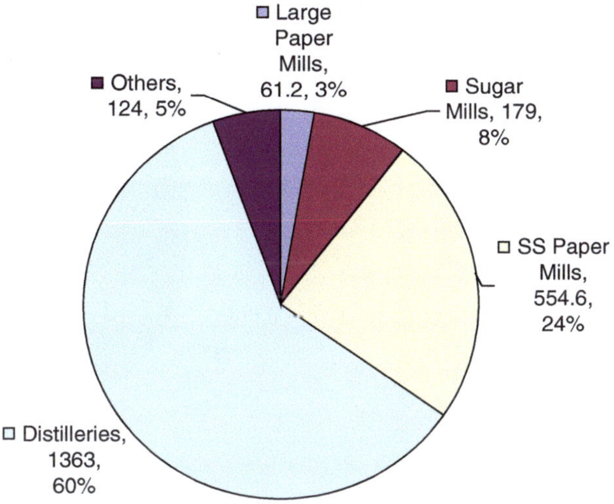

commercial centre in North India, provides good employment opportunities. This has resulted in the migration of large population from all over the country to settle in this city and around, leading to heavy pressure on the Yamuna.

The pollution loads estimated in the towns along the river by CPCB are presented in Table 10.15 and in Fig. 10.13.

It is revealed that out of the 16 towns along the river, 10 including Delhi and downstream towns discharge their wastewater directly into the river. Three towns upstream of Delhi in Haryana, i.e., Karnal, Panipat and Sonepat discharge

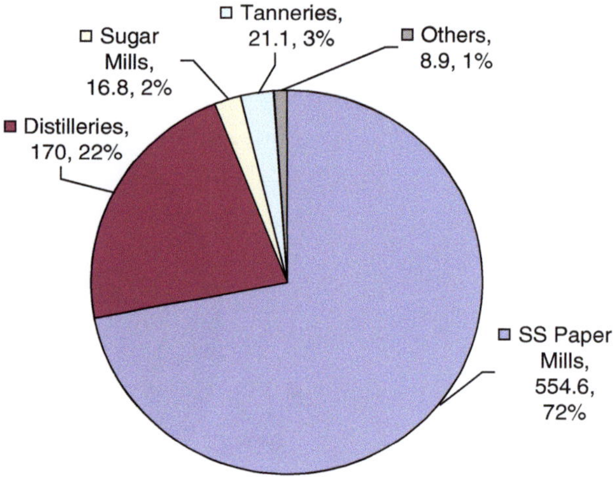

Fig. 10.11 Category-wise industrial BOD load discharged, t/d

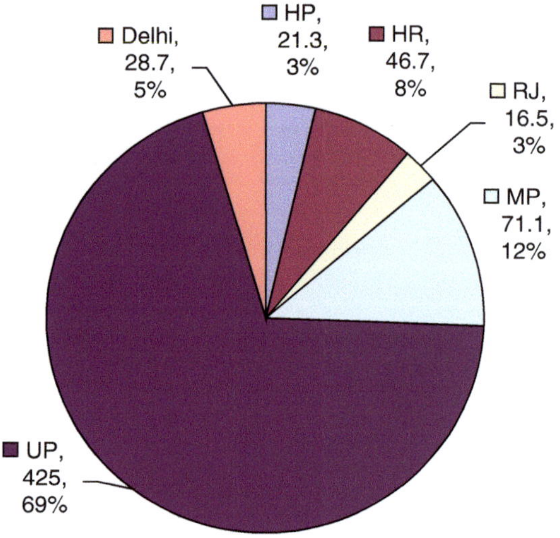

Fig. 10.12 State-wise Industrial BOD load discharged, t/d

their wastewater into drains leading to the Yamuna after traversing a distance of 20–45 km, resulting in most of the wastewater being either used-up by the farmers for irrigation, evaporated or percolated and not reaching the river in dry season. Yamunanagar discharges its wastewater into the misused Western Yamuna Canal. The two towns in U.P. upstream of Delhi, i.e., Saharanpur and Muzaffarnagar discharge their wastewater into Hindon and Kali Rivers respectively. Delhi contributes highest load into the river. It generates about discharges about 80% of its wastewater into the Yamuna River and about 20% into the Agra Canal. Out of the 80% wastewater discharged into the Yamuna nearly 25% is discharged downstream

Table 10.15 Major polluting sources of Yamuna River

Cities/towns	Bank	Volume of WW, MLD	Treatment capacity MLD
Yamunanagar	R	45	35
Saharanpur	L	45	38
Muzaffarnagar	L	40	32.5
Karnal	R	60	48
Panipat	R	60	45
Sonepat	R	45	30
Delhi	R/L	3, 800	2,330
Gurgaon	R	45	30
Faridabad	R	140	115
Ghaziabad	L	150	126
Noida	L	90	70
Vrindavan	R	5	4.5
Mathura	R	35	28
Agra	R	190	90
Etawah	R	13	10
Allahabad	L	223	89
Total		4, 986	3,121

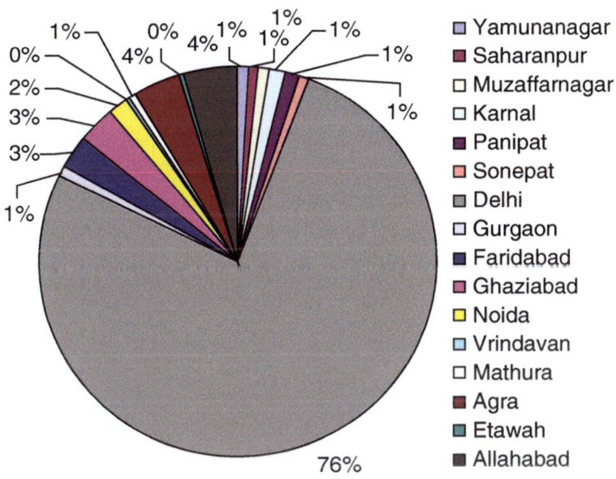

Fig. 10.13 Town share of wastewater generation along the Yamuna River

of the Okhla Barrage. The wastewater from Faridabad is partly (about 30%) discharged into the Agra Canal and partly discharged into the Yamuna River through a drain being used by farmers for irrigation and partly into the Gaunchi Drain (about 30%) leading to the Yamuna after traversing a distance of about 70 km. During the course of 70 km there is biodegradation and lifting by farmers for irrigation and hence contribution in the river through this drain is almost insignificant.

10.7.1 Wastewater Discharged from Delhi

Yamuna River, which is the main source of water supply for Delhi, plays a crucial role in its growth. The perennial increase in the population and urban activity in Delhi is placing exigent pressures and demands upon this natural resource. As a result, today, there is a heavy pressure on the civic amenities, water supply, and sanitation in particular, leading to a severe impact on the river water quality. Therefore, in order to assess the water quality status of Yamuna River, it becomes imperative to study the growth of the Union Territory (UT) of Delhi in detail.

To understand the city growth the spatial growth trend of Delhi is studied over the years. For this, all the settlements falling within the confines of Union Territory of Delhi are taken and their expansion is noticed in the time series (1901–2001). In order to ascertain the real picture of growth, the decadal growth has been studied. Table 10.16 shows the decadal population growth rate from 1901 to 2001.

The urban area of Delhi has been growing continuously since 1901. As evident from Table 10.16, from 1921 to 1931, there was an enormous rise of 288%, mainly due to the developments coming up after the transfer of the Imperial Capital from Calcutta to Delhi. After the completion of construction of New Delhi in 1931, the expansion became almost stagnant. Therefore, the significant change in the urban area has been experienced only in the post-colonial period. After independence, Delhi has been growing more or less steadily with the urban area expanding to 918.7 km^2 during 2001 census, which is as much as 62.5% of the total area of UT of Delhi.

Population has been an integral part of the urban sprawl. The ever increasing population of the country manifests itself by way of the metropolitan cities becoming increasingly congested. This is reflected in the change in population of Delhi since 1901–2001, as shown in Table 10.17.

From Tables 10.16 and 10.17, it can be inferred that Delhi as well as its urban component have followed an upward trend all through this century. However, the

Table 10.16 Growth of urban area in Delhi

Year	Area km^2	Net addition km^2	Percentage change
1901	43.3	–	–
1911	43.3	0	0
1921	168.1	124.8	288.22
1931	169.6	1.5	0.89
1941	170.1	0.5	0.29
1951	195.8	25.7	15.11
1961	326.3	130.5	66.65
1971	451.4	125.1	38.34
1981	591.9	140.5	31.13
1991	685.3	93.4	15.78
2001	918.7	233.4	74.59

Source: Directorate of Census Operations, Delhi

Table 10.17 Population growth of Delhi during 1901–2001

Year	Total population	Percentage increase	Urban population	Percentage to total	Percentage increase
1901	405,809	–	214,112	51	–
1911	413,851	2	237,944	56	11
1921	488,452	18	304,420	62	28
1931	636,246	30	447,442	70	47
1941	917,939	44	695,686	76	55
1951	1,744,072	90	1,437,134	82	107
1961	2,658,612	52	2,359,408	88	64
1971	4,065,698	53	3,647,023	90	55
1981	6,220,406	53	5,768,200	93	58
1991	9,370,475	51	8,427,083	90	46
2001	13,782,976	47	12,819,761	93	52
2010	20,498,805	52	19,316,842	94	51 (Projected)

Source: Directorate of Census Operations, Delhi

biggest jump was witnessed in the decade 1941–1951, when the number really doubled following the influx of refugees from Pakistan. During the last four decades, the growth rate is more or less constant at about 50% per decade.

The density of population in the UT of Delhi in 1981 was 4,231 persons per sq km. This increased to 6,374 persons per sq. km in 1991, registering an increase of as much as 51% in the decade 1981–1991. During the year 2001 the population density has increased to 9,376 persons per km^2, which is a further increase of 47% in the decade 1991–2001.

The rate of growth of urban population of Delhi is the highest in the country. The urban population has increased to about 60 times in 100 years. Since the Delhi Master Plan 1962 came into being, DDA has developed more than 20,000 ha of land for various uses. In spite of such huge developmental activities in Delhi, about 70% of population of Delhi still lives in sub-standard areas with a break up of 1.0 million in *jhuggi* clusters, 1.2 million in unauthorised colonies and unauthorised regularised colonies, 1.4 million in slum designated areas, 0.5 million in urban villages, 0.5 million in rural villages and 1.2 million in resettlement colonies.

10.7.1.1 Pollution Load Generation from Domestic Sources in Delhi

Every human activity results in the generation of waste though the nature and quantities of such wastes may vary in a wide range from activity to activity. Rural and urban areas are very much distinct from each other in terms of activities as well as the provision of infrastructure, particularly water supply and sanitation.

The Delhi Jal Board (DJB) suggests that an average of about 240 lpcd of domestic sewage is generated in the sewered areas of Delhi. The entire sewered areas of Delhi can be divided into five sewerage zones, namely, Okhla, Keshopur, Rithala, Coronation Pillar and Shahdra. These sewerage zones accommodate populations of

Table 10.18 Sewerage zone wise generation of domestic pollution loads

Sewerage zone	Population	Sewage Generated (MLD)
1. Okhla	3,981,035	955.4
2. Keshopur	2,044,928	490.8
3. Rithala	2,461,719	590.8
4. Coronation Pillar	579,238	139.0
5. Shahdara	2,384,520	572.3
Total	11,451,442	2,748.3

varying densities. As a result, the distribution of population in these zones is not uniform. In order to ascertain the total population in each sewerage zone, an intersection of boundaries of sewerage zones and planning zones is obtained. The areas of resulting sub zones are then multiplied with the corresponding Planning Zone densities to find the population of each sub zone. The total population of all the sub zones belonging to one particular sewerage zone is finally obtained. From the total population, total volume of domestic sewage generated can be calculated. Table 10.18 gives the particulars regarding the generation of domestic pollution loads in various sewerage zones.

10.7.1.2 Pollution Load Generation from Industrial Sources

Industries, unlike other human activity, can generate water pollution loads that are highly concentrated in terms of space and time, highly varied in nature (some being refractory and very difficult to either destroy or remove once they have entered a stream), and being highly toxic, pose much more serious hazards, thereby, deserving a much more careful consideration.

It is not possible here to go into the specifics of the wide variety of polluting chemicals generated and discharged by industries. For illustration, only BOD is discussed to indicate the severity of the problem.

There are a large number of large, medium and small scale industries in the U.T. of Delhi. A major fraction of these industrial units are located within the 28 major industrial areas and thus, act as another source of concentrated generation of pollutants. Most of the large and many small and medium scale industries have their own effluent treatment systems, but in the absence of relevant information regarding the extent of treatment provided by the industries, it is difficult to consider pollution reduction due to treatment. The name of industrial areas with their respective sewerage zone is provided in Table 10.19. There are 15 common effluent treatment plants planned for the 28 industrial areas as shown in Table 10.20. Some of them are installed and working. However, the performance of these treatment plants is not satisfactory due to various reasons, including the nature of wastes, non-existence of pre-treatment needed in many industries, etc.

Table 10.19 Sewerage zone-wise generation of industrial pollution loads

Sewerage zone	Industrial area	Industrial discharge (MLD)
1. Okhla	Okhla Industrial Area	12.0
	Okhla Industrial Estate	10.0
	Okhla Flatted Factories	7.5
	Friends Colony Industrial Estate	3.0
	Mohan Co-operative Industrial Estate	13.8
	Sub Total	**46.3**
2. Keshopur	Rohtak Road	14.00
	Karampura Industrial Estate	9.09
	Mayapuri Industrial Estate	12.0
	Najafgarh Road Industrial Estate (60%)	7.20
	Kirti Nagar Industrial Estate	8.0
	Tilak Nagar Industrial Estate	8.0
	Naraina Industrial Estate	10.56
	Mangolpuri Industrial Estate (40%)	1.5
	Sub Total	**70.35**
3. Rithala	Mangolpuri Industrial Estate (60%)	2.4
	Najafgarh Road Industrial Estate (40%)	4.80
	Anand Parbat Industrial Estate	8.0
	Lawrence Road Estate	2.0
	Badli Indusrial Area	1.53
	Udyognagar Industrial Estate	3.21
	Rajasthan Udyognagar Industrial Estate	10.0
	SMA Industrial Area	12.0
	Narela Industrial Estate (90%)	10.0
	Wazirpur Industrial Estate	20.0
	GT Karnal Road Industrial Estate	4.0
	SSI Indusrial Area	3.5
	DSIDC Nangloi Industrial Area	4.0
	Sub Total	**85.44**
4. Coronation Pillar	Azadpur Industrial Estate	1.94
	Narela Industrial Estate (10%)	0.54
	Sub Total	**2.48**
5. Shahdara	Jhilmil Industrial Estate	7.0
	Patparganj Industrial Estate	3.0
	Shahdara Industrial Estate	3.5
	Sub Total	**13.5**
	Grand Total	**218.07**

Note: Narela (30%) lies outside the Sewerage Zones
Source: Central Pollution Control Board, Delhi

10.7.1.3 Pollution Loads Contributed to Yamuna and Agra Canal Through Drains

About 60% of the Delhi population is covered under sewerage facilities; a large part of sewage generated from unsewered areas and also sewered areas where sewer line is defunct due to chocking, silting or sinking the sewage flows in storm water

Table 10.20 Common effluent treatment plants with design capacity in Delhi

S. No.	Name of CETP	Capacity (MLD)
1	Anand Parbat Industrial Estate	24.0
2	Badli Industrial Estate	12.0
3	Nangloi Industrial Area	12.0
4	G.T. Karnal Road	6.0
5	Jhilmil Industrial Estate	16.8
6	Lawrence Road Industrial Estate	12.0
7	Mangolpur Industrial Estate	2.4
8	Mayapuri Industrial Estate	12.0
9	Mohan Co-operative Industrial Estate	16.8
10	Najafgarh Road Industrial Estate	7.2
11	Naraina Industrial Estate	24.0
12	Okhla Industrial Area	24.0
13	Okhla Industrial Estate	12.0
14	S.M.A. Industrial Area	12.0
15	Wazirpur Industrial Estate	24.0
Total		**218.2**

Table 10.21 Discharge and BOD loads in different drains of Delhi

Drains	BOD Load (T/d)	Discharge (m^3/s)	Ultimate disposal
Najafgarh Drain	79.68	24.14	Yamuna u/s Okhla Barrage
Shahdara Drain	50.53	7.65	Yamuna d/s Okhla Barrage
Drain Near Sarita Vihar Bridge	22.85	6.24	Agra Canal
Sen Nursing Home Drain	17.61	1.24	Yamuna u/s Okhla Barrage
Power House Drain	15.09	0.63	Yamuna u/s Okhla Barrage
Maharani Bagh Drain	9.28	0.48	Yamuna u/s Okhla Barrage
Civil Mill Drain	8.78	0.40	Yamuna u/s Okhla Barrage
Sarita Vihar Drain	8.39	0.41	Yamuna d/s Okhla Barrage
Barapulla Drain	4.18	0.75	Yamuna u/s Okhla Barrage
Drain At LPG Bottling Plant	4.05	0.74	Agra Canal
ISBT Drain	3.77	0.41	Yamuna u/s Okhla Barrage
Tuglakabad Drain	3.35	0.34	Agra Canal
Tehkhand Drain	2.97	0.26	Agra Canal
Magazine Road Drain	1.42	0.05	Yamuna u/s Okhla Barrage
Sweeper Colony Drain	1.08	0.09	Yamuna u/s Okhla Barrage
Tonga Stand Drain	0.76	0.05	Yamuna u/s Okhla Barrage
Metcalf house Drain	0.32	0.07	Yamuna u/s Okhla Barrage
Drain No.14	0.21	0.13	Yamuna u/s Okhla Barrage
Kalkaji Drain	0.15	0.03	Yamuna u/s Okhla Barrage
Khyber Pass Drain	0.05	0.09	Yamuna u/s Okhla Barrage
Total	234.52	44.2	

drains. The CPCB has been regularly monitoring the drains in Delhi. The average values of monitoring results of drains are presented in Table 10.21. It is clear from the table that the Najafgarh Drain along with the supplementary drain, carries maximum pollution load to the Yamuna River followed by the Shahdara drain. The two drains together contribute about 55% by load and 72% by volume of waste water.

10.8 Concluding Remarks

Based on the discussions and data, the following is concluded:

- It is estimated that out of 8,444 MLD of wastewater generated in the basin, about 4,458 MLD is discharged directly into the Yamuna River.
- About 1,200 MLD is discharged into tributaries of Yamuna River.
- Approximately 2,786 MLD is either disposed of on land or used for irrigation.
- Delhi generates about 3,743 MLD of wastewater, which is 44% of the entire sewage generated in the basin and 84% of the sewage being discharged into the Yamuna along its entire course.
- Uttar Pradesh, Madhya Pradesh and Rajasthan share almost equal amounts of sewage generation in this basin.
- Delhi has the highest sewage treatment capacity of 2,330 MLD which is about 68% of the total sewage treatment capacity in the basin. However, there is still a large gap of 1,413 MLD between generation and treatment capacity.
- Though several STPs have been installed along the river course with a designed capacity of the order of 2,332.25 MLD for Delhi, 327 MLD for Haryana, and 403.25 MLD for Uttar Pradesh, but the river water quality is still the worst. The main causes of the inefficient functioning of STPs are (i) STPs are in general practically not meeting their compliance, (ii) under running of most of the STPs due to the lack of connection, (iii) improper drainage system, (iv) excess BOD concentration coming to the plant due to inadequate water supply, etc.
- The density of STPs in Delhi is the highest and is mostly based on the Activated Sludge Process (ASP). However, on the other part, the type belongs to either Oxidation Pond (OP) or Up-flow Anaerobic Sludge Blanket (UASB) type.

References

Census (2001) Census of India – 2001, Ministry of Home Affairs, Government of India

CPCB (1978) Union Territory of Delhi (Abridged), Control of Urban Pollution Series, CUPS/1/1978-79, Central Pollution Control Board of India

CPCB (1982–1983) Quality and Trend of Yamuna River (1977–1982), Assessment and Development Study of River Basin Series, ADSORBS/10/1982-83, Central Pollution Control Board of India

CPCB (1994) Water Quality Status of Yamuna River, Assessment and Development Study of River Basin Series, ADSORBS/29/1993-94, Central Pollution Control Board of India

CPCB (2006) Status of Sewage Treatment in India, Control of Urban Pollution Series: CUPS/61/2005-2006, Central Pollution Control Board of India

CPCB (2009) Status of Water supply, Wastewater generation and Treatment in class-I cities and Class-II towns of India, Central Pollution Control Board of India

Chapter 11
Water Quality Index and Status

Abstract This chapter includes a methodology for determining the water quality status of a river. The water quality of the Yamuna River is described using indirect and direct methods. The indirect method is based on various water quality indices (WQI), whereas the direct method uses water quality data at various locations on the river. Detailed mathematical formulations of the various WQIs, such as National Sanitation Foundation (NSF), Nonlinear NSF, Canadian Council method, Nasirian (2007) method, Universal method, and European Classification Scheme, are presented. Besides the WQI-based approach, a detailed analysis of water quality data is presented in order to determine water quality trends, profile, and monthly variations in the river. The bio-mapping technique is also used to detect the gradual change in the water quality status of the Yamuna River.

11.1 Water Quality Index

Continuous assessment of physical, chemical and biological parameters of water is an essential part of water quality control programmes. These efforts lead to the accumulation of considerable information which cannot usually produce direct qualitative determination of water quality. Various methods of data manipulations have been used to condense these records into a more suitable form, and promote effective communication to all concerned. Moreover, these methods can produce a proper means for classification and comparison of different water bodies. The water quality index (WQI) system is a well-known method of expressing water quality that offers a simple, stable and reproducible unit of measure which responds to changes in the principal characteristics of water (Brown et al., 1972).

WQI can be defined as a single numerical expression which reflects the composite influence of significant physical, chemical and microbiological parameters of water quality. A number of indices have been developed to summarize water quality data in an easily expressible and easily understood format (Couillard and Lefebvre, 1985). Since Horton (1965) proposed the first WQI, a great deal of consideration has been given to the development of index methods. The basic differences among these indices are the way their sub-indices were developed. Walski and Parker (1974) used an exponential function to represents the sub indices of various quality variables. Landwehr (1979) suggested the Pearson type 3-distribution function to represent the sub-indices of all quality variables. Bhargava (1987) modified the

R.K. Rai et al., *The Yamuna River Basin*, Water Science and Technology Library 66, DOI 10.1007/978-94-007-2001-5_11, © Springer Science+Business Media B.V. 2012

exponential formula; Dinius (1987) used a power function for the majority of sub indices. Various modifications have also been done in the recent years (Nives, 1999; Swamee and Tyagi, 2000; Harrison et al., 2000; Faisal et al., 2003; Ahmad et al., 2004; Shiow-Mey et al., 2004; etc.). However, Nasirian (2007) compared various WQI systems currently in use and showed that none of them describes quality of water from mining effluent because most of sub-indices in current WQIs are not relevant to indicating changes in water quality brought about by mining activities. Looking into the importance of WQI, this chapter addresses the application and comparison of different WQIs. The WQI techniques described herein are: (i) National Sanitation Foundation's WQI (NSF-WQI) (1970); (ii) Multiplicative NSF-WQI technique (1974); (iii) CCME WQI (2001); (iv) WQI (Nasirian, 2007); and (v) Universal Water Quality Index based on the European classification scheme (Boyacioglu, 2007).

11.1.1 National Sanitation Foundation's WQI (NSF-WQI)

During the year 1970, a Water Quality Index (WQI) was developed at the National Sanitation Foundation (NSF). It appears to be the most comprehensive form of WQI (Brown et al., 1972; Landwehr and Deininger, 1976; Sharifi, 1990; etc.). While formulating, the NSF adopted opinions from more than 140 selected water quality experts. On the basis of questionnaire, the NSF was able to draw up a list of valid parameters which had been rated on a scale of importance. They also established the relation of water quality to values in the form of rating curves. In developing rating curves the experts were asked to attribute values for variation in the level of water quality caused by different levels of each of the selected parameters.

WQI is a 100 point scale that summarizes results from a total of nine different parameters. The selected water quality parameters and their weights are given in Table 11.1. The WQI can be mathematically expressed as follows:

$$\mathrm{WQI} = \sum_{i=1}^{n} W_i Q_i \tag{11.1}$$

Table 11.1 Water quality parameters and their weight (NSF-WQI)

S. No.	Parameters	Unit	Weight
1	Dissolved Oxygen (DO)	% Saturation	0.17
2	Fecal Coliform (FC)	Count/100 ml	0.16
3	pH	Standard unit	0.11
4	Biological Oxygen Demand (BOD)	mg/l	0.11
5	Temperature change	(°C)	0.10
6	Total Phosphate (TP)	mg/l	0.10
7	Nitrate (TN)	mg/l	0.10
8	Turbidity	NTU	0.08
9	Total Suspended Solids (TSS)	mg/l	0.07
Total Weight			1.00

where WQI is the index number between 0 and 100; Q_i is the water quality score for ith parameter (number between 0 and 100); W_i is the weight of the ith parameter; and n is the number of parameters ($n = 9$).

The water quality score of each parameter is obtained from the related rating curve or equation. The rating curve rapidly transforms the concentration or quotient of a parameter into a quality score. For all the nine parameters, the rating curve can be developed, which is discussed as follows.

11.1.1.1 Rating Curve for Fecal Coliform

Water quality rating curve for the fecal coliform (FC) established by NSF (Fig. 11.1) has been fitted for computational ease.

$$Q_{\mathrm{FC}} = -12.12\ \ln(\mathrm{FC}) + 99.231;\ \text{if FC} \leq 500\,(R^2 = 0.9996) \tag{11.2}$$

$$Q_{\mathrm{FC}} = 224.55\ (\mathrm{FC})^{-0.343};\ \text{if } 500 < \mathrm{FC} \leq 45{,}000\,(R^2 = 0.995) \tag{11.3}$$

$$Q_{\mathrm{FC}} = -3.233\ \ln(\mathrm{FC}) + 40.585;\ \text{if } 45{,}000 < \mathrm{FC} < 100{,}000\,(R^2 = 0.994) \tag{11.4}$$

$$Q_{\mathrm{FC}} = 2.0;\ \text{if FC} > 100{,}000 \tag{11.5}$$

11.1.1.2 Rating Curve for Total Solids

The fitted rating curve (Fig. 11.2) can be expressed through the following set of numerical expressions:

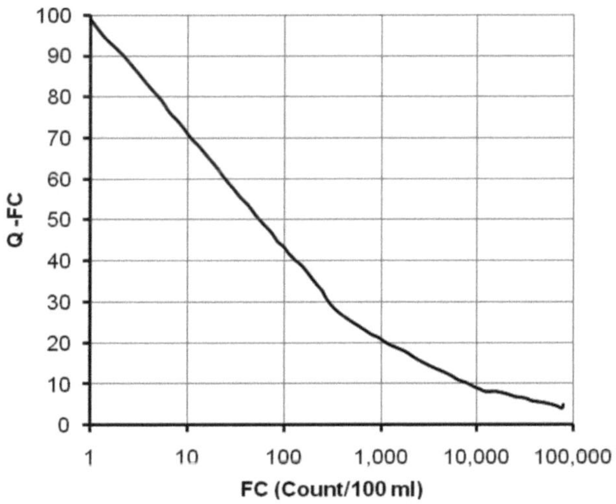

Fig. 11.1 Rating curve for fecal coliform

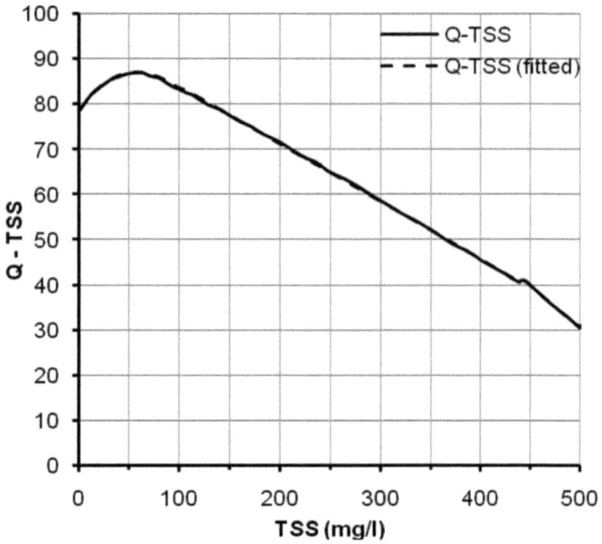

Fig. 11.2 Rating curve for TSS

$$Q_{\mathrm{TSS}} = 0.000007(\mathrm{TSS})^3 - 0.0034(\mathrm{TSS})^2 + 0.3209(\mathrm{TS}) + 78.557;\ \text{if } 0 < \mathrm{TSS} \leq 100\,(R^2 = 0.996) \tag{11.6}$$

$$Q_{\mathrm{TSS}} = -0.1277(\mathrm{TSS}) + 96.866;\ \text{if } 100 < \mathrm{TSS} \leq 440\,(R^2 = 0.998) \tag{11.7}$$

$$Q_{\mathrm{TSS}} = -0.1889(\mathrm{TSS}) + 124.92;\ \text{if } 440 < \mathrm{TSS} \leq 500\,(R^2 = 0.999) \tag{11.8}$$

$$Q_{\mathrm{TSS}} = 20;\ \text{if TSS} > 500 \tag{11.9}$$

11.1.1.3 Rating Curve for Dissolved Oxygen

The dissolved oxygen is expressed in terms of the percentage saturation. The DO_{sat} can be estimated from the following relationship:

$$\%\,\mathrm{DO_{sat}} = \frac{\mathrm{DO}}{\mathrm{DO_{sat}}} \times 100 \tag{11.10}$$

where DO_{sat} is the saturated DO at T°C for water. The DO_{sat} is estimated using the following relationship:

$$\mathrm{DO_{sat}} = \frac{1}{(0.001639\,T + 0.06282)} - 1.279 \tag{11.11}$$

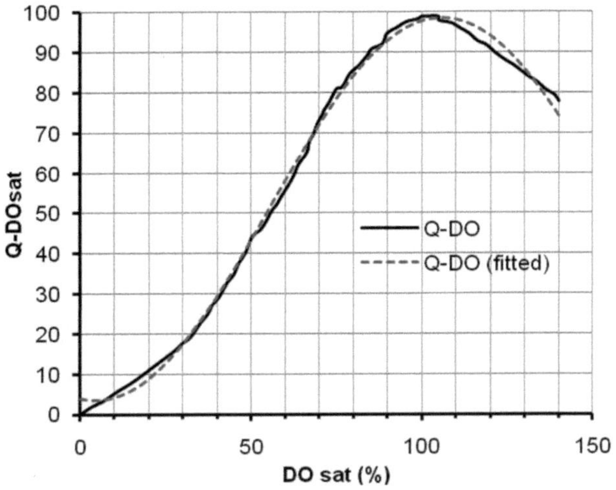

Fig. 11.3 Rating curve for DO

where T is the water temperature (°C). The rating curve based on Fig. 11.3 can numerically be expressed as:

$$Q_{\mathrm{DO}} = 0.0; \text{ if } \%\mathrm{DO}_{\mathrm{sat}} \leq 0.0 \tag{11.12}$$

$$Q_{\mathrm{DO}} = 47.54 \cos[0.03103(\%\mathrm{DO}_{\mathrm{sat}}) - 3.282] + 50.96; \quad \text{if } 0 < \%\mathrm{DO}_{\mathrm{sat}} \leq 140 \tag{11.13}$$

$$Q_{\mathrm{DO}} = 50.0; \text{ if } \%\mathrm{DO}_{\mathrm{sat}} > 140 \tag{11.14}$$

11.1.1.4 Rating Curve for Turbidity

Turbidity is measured in terms of NTU. The rating curve developed by NSF is shown in Fig. 11.4. For computational ease the rating curve has been best fitted with appropriate techniques and is shown in Fig. 11.4.

The established fitted rating curve is expressed with the following set of equations.

$$Q_{\mathrm{Turb}} = \left[8424 + 18.88 \times (\mathrm{Turb} + 8.527)^2\right]^{0.5} - 4.628\,\mathrm{Turb}; \text{ if } 0 < \mathrm{Turb} \leq 100\,\mathrm{NTU} \tag{11.15}$$

$$Q_{\mathrm{Turb}} = 5; \text{ if Turb} > 100\,\mathrm{NTU} \tag{11.16}$$

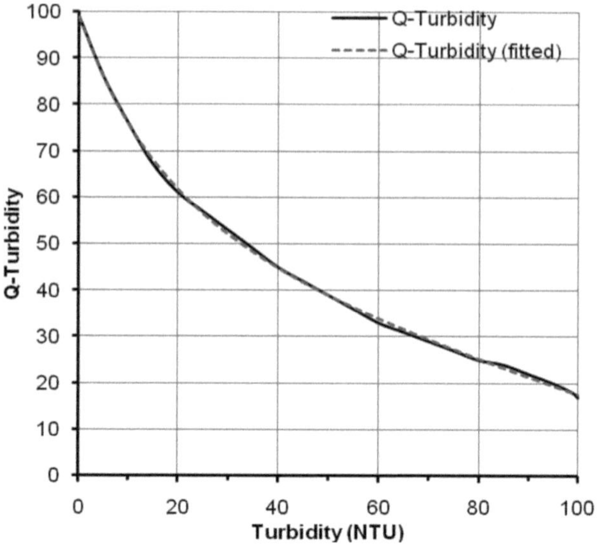

Fig. 11.4 Rating curve for turbidity

11.1.1.5 Rating Curve for pH

The rating curve developed by NSF was fitted and the comparison is shown in Fig. 11.5. The rating curve can be mathematically expressed as follows:

$$Q_{\text{pH}} = 0; \text{ if pH} < 2 \tag{11.17}$$

$$\begin{aligned} Q_{\text{pH}} = &-0.0799\,\text{pH}^5 + 1.3838\,\text{pH}^4 - 8.115\,\text{pH}^3 + 21.389\,\text{pH}^2 \\ &- 23.761\,\text{pH} + 9.209; \text{ if } 2 \le \text{pH} < 7.5\,(R^2 = 0.9997) \end{aligned} \tag{11.18}$$

$$\begin{aligned} Q_{\text{pH}} = &-0.9647\,\text{pH}^4 + 38.69838\,\text{pH}^3 - 572.38\,\text{pH}^2 \\ &+ 3676.2\,\text{pH} - 8555.0; \text{ if } 7.5 \le \text{pH} \le 12.0\,(R^2 = 0.9999) \end{aligned} \tag{11.19}$$

$$Q_{\text{pH}} = 0; \text{ if pH} > 12 \tag{11.20}$$

11.1.1.6 Rating Curve for BOD

The fitted rating curve developed from NSF for biological oxygen demand (BOD) is shown in Fig. 11.6. A list of mathematical expressions for computing the parameter score is:

$$\begin{aligned} Q_{\text{BOD}} = &-0.2311\,\text{BOD}^4 + 3.0227\,\text{BOD}^3 - 11.754\,\text{BOD}^2 \\ &+ 3.3994\,\text{BOD} + 100; \text{ if BOD} < 5\,(R^2 = 0.996) \end{aligned} \tag{11.21}$$

$$\begin{aligned} Q_{\text{BOD}} = &-0.0027\,\text{BOD}^3 + 0.236\,\text{BOD}^2 - 7.4486\,\text{BOD} + 87.674; \\ &\text{if } 5 \le \text{BOD} < 30\,(R^2 = 0.9996) \end{aligned} \tag{11.22}$$

$$Q_{\text{BOD}} = 2; \text{ if BOD} \ge 30 \tag{11.23}$$

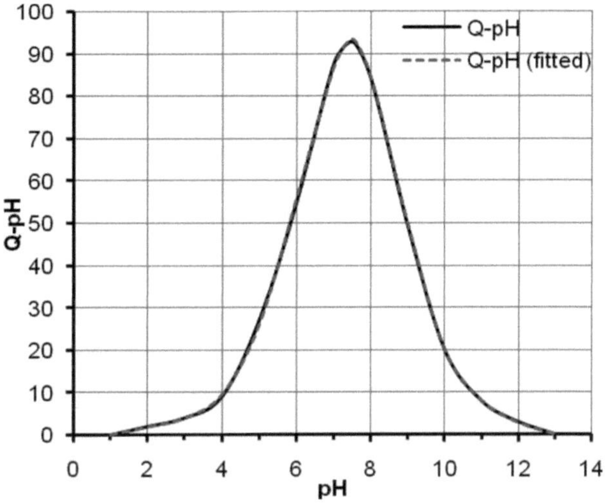

Fig. 11.5 Rating curve for pH

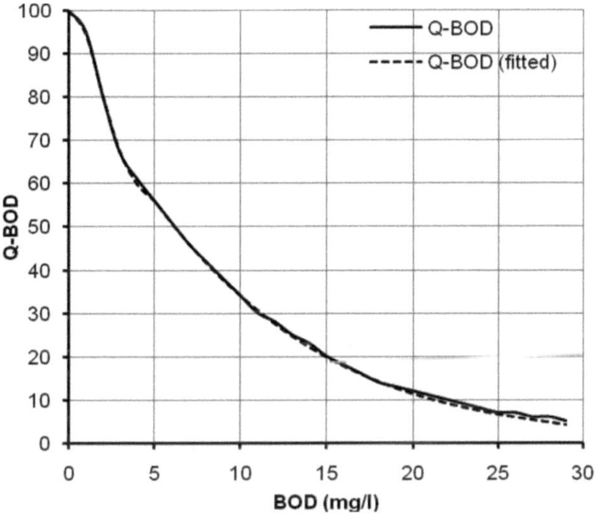

Fig. 11.6 Rating curve for BOD

11.1.1.7 Rating Curve for Nitrate

The rating curve of nitrate was exponentially fitted (Fig. 11.7) and expressed as follows:

$$Q_{NO_3} = 28.0\exp[-0.270(NO_3)] + 72.0\exp[-0.035(NO_3)]\,; \quad \text{if } 0 \le NO_3 \le 100 \tag{11.24}$$

$$Q_{NO_3} = 1.0;\ \text{if } NO_3 > 100 \tag{11.25}$$

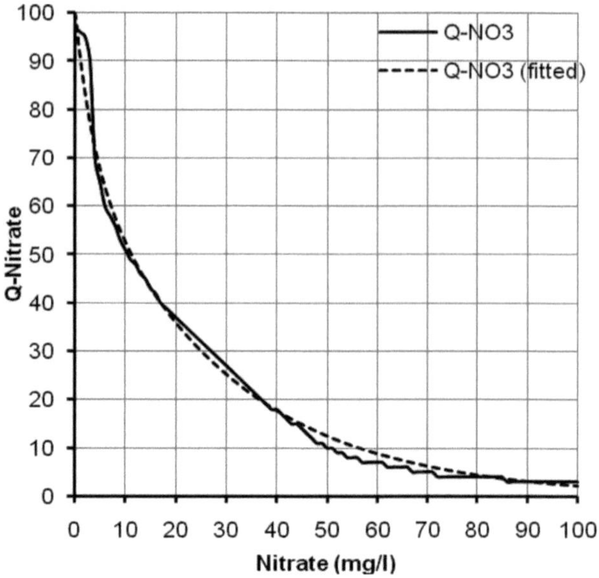

Fig. 11.7 Rating curve for nitrate

11.1.1.8 Rating Curve for Phosphate

The rating curve of phosphate is suitably fitted (Fig. 11.8) and expressed as follows:

$$Q_{\mathrm{TP}} = 98.37 - 316.5\,\mathrm{TP}^2 + \frac{0.01637}{\mathrm{TP}};\quad \text{if } 0.01 < \mathrm{TP} \leq 0.09 \tag{11.26}$$

$$Q_{\mathrm{TP}} = 35.22\ \exp\left\{\frac{0.3312}{\mathrm{TP}} - 0.2135\,\mathrm{TP}\right\};\quad \text{if } 1.0 \leq \mathrm{TP} \leq 7.5 \tag{11.27}$$

$$Q_{\mathrm{TP}} = 7;\ \text{if } 7.5 < \mathrm{TP} \leq 10.0 \tag{11.28}$$

$$Q_{\mathrm{TP}} = 2;\ \text{if } \mathrm{TP} > 10.0 \tag{11.29}$$

11.1.1.9 Rating Curve for Temperature Change

Temperature is a critical water quality and environmental parameter because it governs the kinds and types of aquatic life, regulates the maximum dissolved oxygen concentration of the water, and influences the rate of chemical and biological reactions. The organisms within the ecosystem have preferred temperature regimes

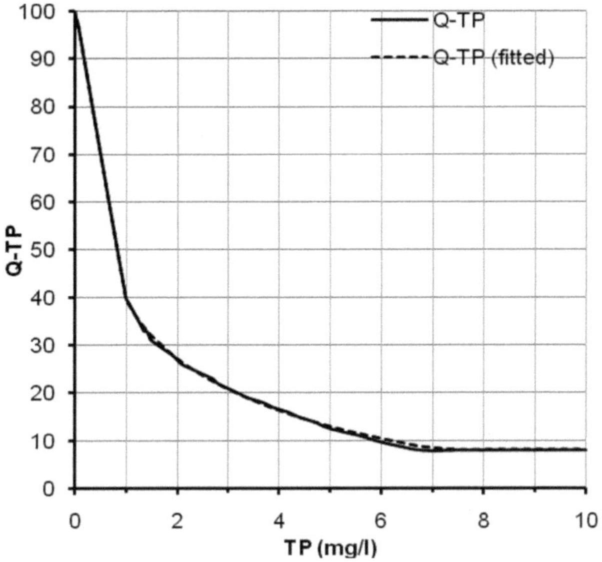

Fig. 11.8 Rating curve for total phosphate

that change as a function of season, organism age or life stage, and other environmental factors. With respect to chemical and biological reactions, the higher the water temperature the higher the rate of chemical and metabolic reactions.

Seasonal variations in stream temperature may be caused by changing air temperature, solar angle, meteorological events, and a number of physical aspects related to the stream and watershed. These physical features include stream origin, velocity, vegetation types and coverage, stream configuration, land-use, and percentage of impervious area. For example, a narrow, deep well-shaded shoreline reduces the impact of warming by the sun; whereas, a wide shallow stream would be more impacted by solar heating.

The change in temperature (ΔT) refers to the change in temperature between upstream and downstream control points. The rating curve of temperature change developed by NSF is suitably fitted for quick computational purposes (Fig. 11.9) and is expressed with following set of equations.

$$Q_{\Delta T} = 3.8\,\Delta T + 93;\ \text{if}\ -10 \leq \Delta T \leq 0 \tag{11.30}$$

$$Q_{\Delta T} = -5.6\,\Delta T + 101;\ \text{if}\ 0 < \Delta T \leq 10 \tag{11.31}$$

$$Q_{\Delta T} = -2.8\,\Delta T + 73;\ \text{if}\ 10 < \Delta T \leq 15 \tag{11.32}$$

$$Q_{\Delta T} = -1.8\,\Delta T + 58;\ \text{if}\ 15 < \Delta T \leq 20 \tag{11.33}$$

$$Q_{\Delta T} = -1.2\,\Delta T + 46;\ \text{if}\ 20 < \Delta T \leq 30 \tag{11.34}$$

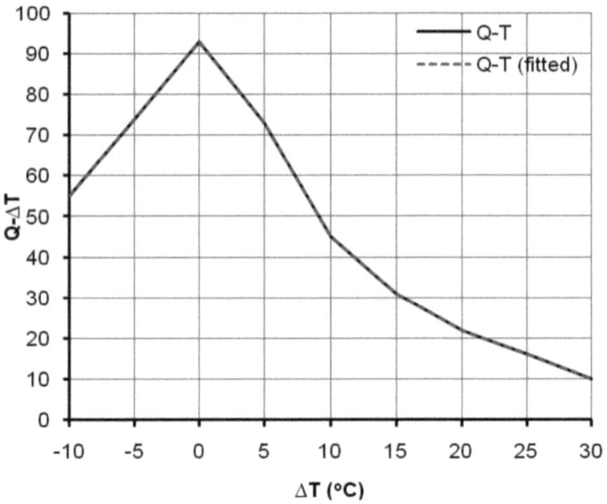

Fig. 11.9 Rating curve for temperature change

Table 11.2 Water class based on NSF-WQI

WQI	Water Class
91–100	Excellent
71–90	Good
51–70	Medium to average
26–50	Fair
0–25	Poor

Once the quality score of the water quality parameters is estimated using the rating curves, the WQI can be estimated using eq. (11.1). Based on the WQI, the water quality can be classified as in Table 11.2.

The computational step for using the NSF-WQI is shown in Example 11.1.

11.1.2 Multiplicative NSF-WQI

Despite the apparent responsiveness of WQI to changes in water quality conditions, analysis of data from the various field studies suggests that the additive WQI (eq. (11.1)) lacks sensitivity in adequately reflecting the effect of a single low value parameter on the overall water quality. O'Connor (1971) noted that the additive model is a good choice when all the parameters are within a reasonable range, but a multiplicative form is more sensitive to discontinuities in the overall quality which may result in zero or poor quality in any parametric dimension. WQI (M) is significantly more sensitive to the effect of a single bad parameter. As a result, a multiplicative form of WQI (WQI-M) can be numerically expressed as follows:

$$\mathrm{WQI}(M) = \prod_{i=1}^{n} Q_i^{w_i} \tag{11.35}$$

where WQI(M) is the multiplicative water quality index, a number between 0 and 100; Q_i is the score of the ith parameter, a number between 0 and 100; w_i is the unit weight of the parameter; and n is the number of parameter considered.

In the determination of WQI(M), Q_i, W_i, and n are similar to the NSF-WQI. When all the parameters are not measured (i.e. only 5 parameters out of 9) then the weights are adjusted, instead of WQI, as follows:

$$w_i^{\text{new}} = w_i + w_i \times \left(\frac{1 - \sum_i w_i}{\sum_i w_i} \right) \tag{11.35a}$$

The calculation procedure of WQI (M) is given in Example 11.2.

Example 11.1 During January, 2008 the water quality parameters measured at Palla site (i.e. upstream to the Wazirabad barrage) in Yamuna River are: Temperature at Sonepat (*d/s*) = 13°C, Temperature at Palla = 12°C, pH = 7.67, DO = 10.5, BOD = 2.0, FC = 1700. Calculate the WQI and water quality class using NSF WQI technique for Palla.

S. No.	Parameters	Measured value	Unit	Weight	Sub-index	Weight used	WQI (A)
(a)	DO	10.5	mg/l				
(b)	DO_{sat}	10.84	mg/l				
(c)	T (u/s)	13	°C				
(d)	T (site)	12	°C				
1	$\%DO_{sat}$	96.83	%	0.17	96.68	0.17	16.44
2	FC	1,700.0	count/100 ml	0.16	17.51	0.16	2.80
3	pH	7.67	std unit	0.11	91.69	0.11	10.09
4	BOD	2	mg/l	0.11	80.27	0.11	8.83
5	Delta T	1	°C	0.10	95.40	0.10	9.54
6	TP	*	mg/l	0.10	2.00	0.00	0
7	TN	*	mg/l	0.10	1.00	0.00	0
8	Turbidity	*	NTU	0.08	5.00	0.00	0
9	TSS	*	mg/l	0.07	20.00	0.00	0
Total				**1.00**		**0.65**	**47.69**

* represents the values which were not measured

When all the nine parameters are not measured then WQI will be adjusted as:
$$WQI_{\text{adjusted}} = \frac{WQI_{\text{calculated}}}{\sum_{i=1}^{n} w_i}$$

The water quality index is calculated as: WQI = 47.69/0.65 = 73.37; and the water class can be categorized as "GOOD".

Example 11.2 During January, 2008 the water quality parameters measured at Palla site (i.e. upstream to the Wazirabad barrage) in Yamuna River are: Temperature at Sonepat (d/s) = 13°C, Temperature at Palla = 12°C, pH = 7.67, DO = 10.5, BOD = 2.0, FC = 1,700. Calculate the WQI and water quality class using WQI (M) technique for Palla.

S. No.	Parameters	Measured value	Unit	Weight	Sub index	Weight used	Adjusted weight	WQI (M)
(a)	DO	10.5	mg/l					
(b)	DO_{sat}	10.84	mg/l					
(c)	T (u/s)	13	°C					
(d)	T (site)	12	°C					
1	$\%DO_{sat}$	96.83	%	0.17	96.68	0.17	0.26	3.31
2	FC	1,700	Counts/100 ml	0.16	17.51	0.16	0.25	2.02
3	pH	7.67	Std unit	0.11	91.69	0.11	0.17	2.15
4	BOD	2	mg/l	0.11	80.27	0.11	0.17	2.10
5	Delta T	1	°C	0.10	95.40	0.10	0.15	2.02
6	TP	*	mg/l	0.10	2.00	0.00	0.00	1.00
7	TN	*	mg/l	0.10	1.00	0.00	0.00	1.00
8	Turbidity	*	NTU	0.08	5.00	0.00	0.00	1.00
9	TSS	*	mg/l	0.07	20.00	0.00	0.00	1.00
Total				**1.00**		**0.65**	**1.00**	**60.84**

* represents the values which were not measured

Therefore the Multiplicative WQI [WQI(*M*)] is 60.84 and the corresponding water class can be categorized as "MEDIUM to AVERAGE". The lesser value as compared to the WQI(*A*) shows the importance of monitored parameter.

11.1.3 CCME WQI

The Water Quality Index (WQI) of the Canadian Council of Ministers of the Environment (CCME) is based on a formula developed by the British Columbia Ministry of Environment, Lands and Parks and modified by Alberta Environment. The index incorporates three elements: (a) *scope*: the number of variables not meeting water quality objectives; (b) *frequency*: the number of times these objectives are not met; and (c) *amplitude*: the amount by which the objectives are not met. The index produces a number between 0 (worst water quality) and 100 (best water quality). These numbers are divided into five descriptive categories to simplify presentation.

11.1.3.1 Data Requirement

The time period chosen will depend on the amount of data available and the reporting requirements of the user. A minimum period of one year is often used because data are usually collected to reflect this period (monthly or quarterly monitoring data). Data from different years may also be clubbed, especially when monitoring in certain years is incomplete, but as with combining stations some degree of variability will be lost.

The calculation of CCME-WQI requires at least four variables, sampled a minimum of four times, be used. However, a maximum number of variables or samples are not specified. The selection of appropriate water quality variables for a particular region is necessary for the index to yield meaningful results. Clearly, choosing a small number of variables for which the objectives are not met will provide a different picture than if a large number of variables are considered, only some of which do not meet objectives.

11.1.3.2 Calculation of Index

The CCME WQI can be expressed through the three elements as follows:

$$\text{CCME} : \text{WQI} = 100 - \left(\frac{\sqrt{F_1^2 + F_2^2 + F_3^2}}{1.732} \right) \tag{11.36}$$

The divisor 1.732 normalizes the resultant values within the range between 0 and 100.

F_1 (Scope) represents the percentage of variables that do not meet their objectives a least once during the time period under consideration ("failed variables"), relative to that number of variables measured:

$$F_1 = \left(\frac{\text{Number of failed variables}}{\text{Total number of variables}} \right) \times 100 \tag{11.37}$$

F_2 (Frequency) represents the percentage of individual tests that do not meet the objectives ("failed test").

$$F_2 = \left(\frac{\text{Number of failed tests}}{\text{Total number of tests}} \right) \times 100 \tag{11.38}$$

F_3 (Amplitude) represents the amount by which failed test values do not meet their objectives. F_3 is calculated using the following three steps:

(a) The number of times by which an individual concentration is greater than (or less than, when the objective is a minimum) the objective is termed an "excursion" and is expressed as follows. When the test value must not exceed the objective (for example, BOD):

$$\text{Excursion}_i = \left(\frac{\text{Failed test value}_i}{\text{Objective}_i} \right) - 1 \tag{11.39}$$

For the cases in which the test value must not fall below the objectives (for example DO):

$$\text{Excursion}_i = \left(\frac{\text{Objective}_i}{\text{Failed test value}_i}\right) - 1 \tag{11.40}$$

(b) The collective amount by which individual tests are out of compliance is calculated by summing the excursions of individual tests from their objectives and dividing by the total number of tests (both those meeting objectives and those not meeting objectives). This variable, referred to as the normalized sum of excursions, or *nse*, is calculated as:

$$\text{nse} = \frac{\sum_{i=1}^{n} \text{Excursion}_i}{\text{Number of tests}} \tag{11.41}$$

F_3 is then calculated by an asymptotic function that scales the normalized sum of the excursions from objectives (*nse*) to yield a range between 0 and 100.

$$F_3 = \frac{\text{nse}}{0.01 \times \text{nse} + 0.01} \tag{11.42}$$

Once the factors have been obtained, the CCME WQI can be calculated using eq. (11.36). Based on the CCME WQI value, water quality is ranked by relating it to a descriptive category as follows (Table 11.3):

Application of the CCME-WQI is presented in Example 11.3.

Example 11.3 Estimate the CCME-WQI from the monthly data collected from the Palla site of Yamuna River. The water quality parameters measured by CPCB along with objectives are listed in the following Table 11.4.

Table 11.3 Water class based on CCME-WQI

Category	CCME WQI	Characteristics
Excellent	95–100	Water quality is protected with a virtual absence of threat or impairment; conditions very close to natural or pristine levels.
Good	80–94	Water quality is protected with only a minor degree of threat or impairment; conditions rarely depart from natural or desirable levels.
Fair	65–79	Water quality is usually protected but occasionally threatened or impaired; conditions sometimes depart from natural or desirable levels.
Marginal	45–64	Water quality is frequently threatened or impaired; conditions often depart from natural or desirable levels.
Poor	0–44	Water quality is almost always threatened or impaired; conditions usually depart from natural or desirable levels.

Table 11.4 Monthly water quality data of Palla site in Yamuna River for the year of 2005

Date	DO	pH	AMM.	TKN	BOD	EC	TC	FC
Jan, 05	8.4	8.08	**1.36**		1	475	**8,000**	**900**
Feb, 05	8.1	8.11	0.69	1.65	2	574	**20,000**	**600**
Mar, 05	8.8	8.15	0.48	1.95	4	446	**15,000**	**470**
Apr, 05	8.8	8.61	0.41	1.12	2	518	**18,200**	**1,650**
May, 05	7.2	7.75	BDL	0.56	1	465	**42,000**	**1,950**
Jun, 05	7.9	7.09	BDL	BDL	1	496	**24,000**	**7,400**
Jul, 05	6.3	8.33	**1.63**	4.43	2	398	**79,000**	380
Aug, 05	7.3	7.25	0.31	0.84	1	190	**61,000**	300
Sep, 05	10.9	7.26	BDL		3	485	**17,900**	310
Oct, 05	6.1	7.3	0.14		2	265	**26,000**	**12,100**
Nov, 05	11.2	7.41	0.7		2	**850**	**49,000**	**2,500**
Dec, 05	9.5	7.42	0.81		2	662	**130,000**	**6,600**
Objective	**>4**	**6.5–8.5**	**<1.2**	**<10**	**<3**	**<700**	**<5,000**	**<4,500**

Calculation:

Case I: When all the listed parameters are considered for the analysis

Date	DO	pH	AMM.	TKN	BOD	EC	TC	FC	
Jan, 05	8.4	8.08	**1.36**		1	475	**8,000**	900	
Feb, 05	8.1	8.11	0.69	1.65	2	574	**20,000**	600	
Mar, 05	8.8	8.15	0.48	1.95	**4**	446	**15,000**	470	
Apr, 05	8.8	**8.61**	0.41	1.12	2	518	**18,200**	1,650	
May, 05	7.2	7.75	BDL	0.56	1	465	**42,000**	1,950	
Jun, 05	7.9	7.09	BDL	BDL	1	496	**24,000**	**7,400**	
Jul, 05	6.3	8.33	**1.63**	4.43	2	398	**79,000**	380	
Aug, 05	7.3	7.25	0.31	0.84	1	190	**61,000**	300	
Sep, 05	10.9	7.26	BDL		3	485	**17,900**	310	
Oct, 05	6.1	7.3	0.14		2	265	**26,000**	**12,100**	
Nov, 05	11.2	7.41	0.7		2	**850**	**49,000**	2,500	
Dec, 05	9.5	7.42	0.81		2	662	**130,000**	**6,600**	
Objective	**>4**	**6.5–8.5**	**<1.2**	**<10**	**<3**	**<700**	**<5,000**	**<4,500**	**Total**
Number of failed tests	0	1	2	0	1	1	12	3	**20**
Number of failed variables	0	1	1	0	1	1	1	1	**6**

No. of Variables = 8

No. of Tested Samples = 91

$$F_1 = \left(\frac{\text{Number of failed variables}}{\text{Total number of variables}}\right) \times 100 = 100 \times 6/8 = 75.00$$

$$F_2 = \left(\frac{\text{Number of failed tests}}{\text{Total number of tests}}\right) \times 100 = 100 \times 20/91 = 21.98$$

Estimation of Excursions:

Date	DO	Excursion	pH	Excursion	AMM.	Excursion	TKN	Excursion	BOD	Excursion	EC	Excursion	TC	Excursion	FC	Excursion
Jan, 05	8.4	0	8.08	0.000	1.36	0.13		0.00	1	0.00	475	0.00	8,000	0.60	900	0.00
Feb, 05	8.1	0	8.11	0.000	0.69	0.00	1.65	0.00	2	0.00	574	0.00	20,000	3.00	600	0.00
Mar, 05	8.8	0	8.15	0.000	0.48	0.00	1.95	0.00	4	0.33	446	0.00	15,000	2.00	470	0.00
Apr, 05	8.8	0	8.61	0.013	0.41	0.00	1.12	0.00	2	0.00	518	0.00	18,200	2.64	1,650	0.00
May, 05	7.2	0	7.75	0.000	0.00	0.00	0.56	0.00	1	0.00	465	0.00	42,000	7.40	1,950	0.00
Jun, 05	7.9	0	7.09	0.000	0	0.00	0	0.00	1	0.00	496	0.00	24,000	3.80	7,400	0.64
Jul, 05	6.3	0	8.33	0.000	1.63	0.36	4.43	0.00	2	0.00	398	0.00	79,000	14.80	380	0.00
Aug, 05	7.3	0	7.25	0.000	0.31	0.00	0.84	0.00	1	0.00	190	0.00	61,000	11.20	300	0.00
Sep, 05	10.9	0	7.26	0.000	0	0.00		0.00	3	0.00	485	0.00	17,900	2.58	310	0.00
Oct, 05	6.1	0	7.3	0.000	0.14	0.00		0.00	2	0.00	265	0.00	26,000	4.20	12,100	1.69
Nov, 05	11.2	0	7.41	0.000	0.7	0.00		0.00	2	0.00	850	0.21	49,000	8.80	2,500	0.00
Dec, 05	9.5	0	7.42	0.000	0.81	0.00		0.00	2	0.00	662	0.00	130,000	25.00	6,600	0.47
Objective	>4	0	6.5–8.5	0.013	<1.2	0.49	<10	0.00	<3	0.33	<700	0.21	<5,000	86.02	<4,500	**2.80**

The total excursion is computed using the following equations:

$$\text{Excursion}_i = \left(\frac{\text{Failed test value}_i}{\text{Objective}_i}\right) - 1;$$
$$\text{Excursion}_i = \left(\frac{\text{Objective}_i}{\text{Failed test value}_i}\right) - 1$$

$$\sum_{i=1}^{n} \text{Excursion}_i = 89.87$$

$$\text{nse} = \frac{\sum_{i=1}^{n} \text{Excursion}_i}{\text{Number of tests}} = 89.87/91 = 0.99$$

$$F_3 = \frac{\text{nse}}{0.01 \times \text{nse} + 0.01} = \frac{0.99}{0.01 \times 0.99 + 0.01} = 49.68$$

The WQI using CCME approach is:

$$\text{CCME} : \text{WQI} = 100 - \left(\frac{\sqrt{F_1^2 + F_2^2 + F_3^2}}{1.732}\right)$$

$$= 100 - \left(\frac{\sqrt{75.0_1^2 + 21.98^2 + 49.68^2}}{1.732}\right) = 46.52$$

Based on the CCME WQI, the water quality class can be categorized as "MARGINAL", i.e., the conditions often depart from natural or desirable levels.

Case II: When TC and FC is not considered (i.e. for aquatic lives)

Date	DO	pH	AMM.	TKN	BOD	EC	
Jan, 05	8.4	8.08	**1.36**		1	475	
Feb, 05	8.1	8.11	0.69	1.65	2	574	
Mar, 05	8.8	8.15	0.48	1.95	4	446	
Apr, 05	8.8	8.61	0.41	1.12	2	518	
May, 05	7.2	7.75	BDL	0.56	1	465	
Jun, 05	7.9	7.09	BDL	BDL	1	496	
Jul, 05	6.3	8.33	**1.63**	4.43	2	398	
Aug, 05	7.3	7.25	0.31	0.84	1	190	
Sep, 05	10.9	7.26	BDL		3	485	
Oct, 05	6.1	7.3	0.14		2	265	
Nov, 05	11.2	7.41	0.7		2	**850**	
Dec, 05	9.5	7.42	0.81		2	662	
Objective	**>4**	**6.5–8.5**	**<1.2**	**<10**	**<3**	**<700**	**Total**
Number of failed tests	0	1	2	0	1	1	5
Number of failed variables	0	1	1	0	1	1	4

Estimation of Excursions:

Date	DO	Excursion	pH	Excursion	AMM.	Excursion	TKN	Excursion	BOD	Excursion	EC	Excursion
Jan, 05	8.4	0	8.08	0.000	**1.36**	0.13		0.00	1	0.00	475	0.00
Feb, 05	8.1	0	8.11	0.000	0.69	0.00	1.65	0.00	2	0.00	574	0.00
Mar, 05	8.8	0	8.15	0.000	0.48	0.00	1.95	0.00	4	0.33	446	0.00
Apr, 05	8.8	0	8.61	0.013	0.41	0.00	1.12	0.00	2	0.00	518	0.00
May, 05	7.2	0	7.75	0.000	0.00	0.00	0.56	0.00	1	0.00	465	0.00
Jun, 05	7.9	0	7.09	0.000	0	0.00	0	0.00	1	0.00	496	0.00
Jul, 05	6.3	0	8.33	0.000	**1.63**	0.36	4.43	0.00	2	0.00	398	0.00
Aug, 05	7.3	0	7.25	0.000	0.31	0.00	0.84	0.00	1	0.00	190	0.00
Sep, 05	10.9	0	7.26	0.000	0	0.00		0.00	3	0.00	485	0.00
Oct, 05	6.1	0	7.3	0.000	0.14	0.00		0.00	2	0.00	265	0.00
Nov, 05	11.2	0	7.41	0.000	0.7	0.00		0.00	2	0.00	**850**	0.21
Dec, 05	9.5	0	7.42	0.000	0.81	0.00		0.00	2	0.00	662	0.00
Objective	**>4**	**0**	**6.5–8.5**	**0.013**	**<1.2**	**0.49**	**<10**	**0.00**	**<3**	**0.33**	**<700**	**0.21**

No. of Variables = 6

No. of Tested Samples = 67

$$F_1 = \left(\frac{\text{Number of failed variables}}{\text{Total number of variables}}\right) \times 100 = 100 \times 4/6 = 66.67$$

$$F_2 = \left(\frac{\text{Number of failed tests}}{\text{Total number of tests}}\right) \times 100 = 100 \times 5/67 = 7.46$$

The total Excursion is computed by using the following equations:

$$\text{Excursion}_i = \left(\frac{\text{Failed test value}_i}{\text{Objective}_i}\right) - 1;$$

$$\text{Excursion}_i = \left(\frac{\text{Objective}_i}{\text{Failed test value}_i}\right) - 1$$

$$\sum_{i=1}^{n} \text{Excursion}_i = 1.05$$

$$\text{nse} = \frac{\sum_{i=1}^{n} \text{Excursion}_i}{\text{Number of tests}} = 1.05/67 = 0.02$$

$$F_3 = \frac{\text{nse}}{0.01 \times \text{nse} + 0.01} = \frac{0.02}{0.01 \times 0.02 + 0.01} = 1.546$$

The WQI using CCME approach is:

$$\text{CCME : WQI} = 100 - \left(\frac{\sqrt{F_1^2 + F_2^2 + F_3^2}}{1.732}\right)$$

$$= 100 - \left(\frac{\sqrt{66.67_1^2 + 7.46^2 + 1.546^2}}{1.732}\right) = 61.3$$

Based on CCME WQI, the water quality class can be categorized as "MARGINAL BUT TENDS TO FAIR", i.e., water quality is usually protected but occasionally threatened or impaired; conditions sometimes depart from natural or desirable levels.

11.1.4 WQI

WQI discussed are either highly specialized or very simple in terms of the number of variables considered. None seem to be geared to the protection of multiple water uses or to encompass the variety of measurements of water quality. Nasirian (2007) introduced the new WQI that provides an indication of the quality of water influenced by domestic and some general industrial effluents along with the water impacted by mining. The parameters included in the estimation of WQI are pH, NO_3, radionuclides, dissolved oxygen, PO_4, heavy metals, electrical conductivity (EC), SO_4, and suspended solid. A WQI should be made up sub-indices of parameters that can explain water quality comprehensively. The survey showed that both total dissolved solids (TDS) and electrical conductivity (EC) were selected in the top 10 preferred parameters.

11.1.4.1 Formulation of WQI

The new WQI (Nasirian, 2007) proposed is in accordance with Nives (1999). The WQI was calculated by summing up individual quality rating (q_i) and weighting factor (w_i) as follows:

$$\mathrm{WQI} = \sum_{i=1}^{n} w_i q_i \tag{11.43}$$

where q_i is the water quality score of the ith parameter, w_i is the weighting factor for the ith parameter, and n is the number of parameters. Parameters and their weights are listed in Table 11.5. The water quality standards are given Table 11.6.

Table 11.5 Parameters and weights for Nasirian's WQI

S. No.	Parameters	Unit	Weight
1	pH	Standard Unit	0.18
2	Nitrate (NO_3^-)	mg/l	0.13
3	Radionuclides (^{238}U, ^{232}Th)	Bq/l	0.12
4	Dissolved Oxygen (DO)	mg/l	0.11
5	Phosphate (PO_4^-)	mg/l	0.11
6	Heavy Metals (Cd, Pb, As, Ni, Mn, Fe, Hg)	mg/l	0.11
7	Electrical Conductivity (EC)	μs/cm	0.08
8	Sulphate (SO_4^-)	mg/l	0.08
9	Suspended Solids (SS)	mg/l	0.08

Table 11.6 List of water quality standard

Parameters	DL[a] (USEPA)	MPL[b] (USEPA)	Canadian quality	BIS[c] for drinking water
pH	6.5–8.5			6.5–8.5
TDS (mg/l)				500 (DL) 2,000 (MPL)
PO_4 (mg/l)	0.1			
Turbidity (NTU)				
DO (mg/l)	5	3		3–5
BOD (mg/l)				3
NO_3 (mg/l)				45 (DL) 100 (MPL)
Cl (mg/l)				250 (DL) 1,000 (MPL)
F (mg/l)				1.0 (DL) 1.5 (MPL)
Fecal Coliform (count/100 ml)				0 (DL)
Total Coliform (count/100 ml)				10 (DL)

Table 11.6 (continued)

Parameters	DL[a] (USEPA)	MPL[b] (USEPA)	Canadian quality	BIS[c] for drinking water
Cd (mg/l)	0.005	0.005	0.005: Drinking 0.00002: Aquatic 0.005: Agriculture	0.01 (DL) No relaxation (MPL)
Cr (mg/l)				0.05 (DL) No relaxation (MPL)
Cu (mg/l)				0.05 (DL) 1.5 (MPL)
Cyanide (mg/l)				0.05 (DL) No relaxation (MPL)
Zn (mg/l)				5 (DL) 15 (MPL)
Pb (mg/l)	0	0.015	0.01: Drinking 0.001: Aquatic 0.2: Agriculture	0.05 (DL) No relaxation (MPL)
As (mg/l)	0	0.010	0.025: Drinking 0.005: Aquatic 0.1: Agriculture	0.05 (DL) No relaxation (MPL)
Hg (mg/l)	0.002	0.002	0.0001: Drinking 0.0001: Aquatic 0.003: Agriculture	0.001 (DL) No relaxation (MPL)
Ni (mg/l)	–	0.100	0.2: Drinking 0.025: Aquatic 0.02: Agriculture	
Mn (mg/l)	–	0.050	0.05: Drinking 0.1: Aquatic 0.2: Agriculture	0.10 (DL) 0.30 (MPL)
Total Fe (mg/l)	–	0.300	0.3: Drinking 0.3: Aquatic 5.0: Agriculture	
SO_4^{2-} (mg/l)			500: Drinking NA: Aquatic 1000: Agriculture	200 (DL) 400 (MPL)
Suspended Solids (mg/l)				
EC (μ mhos/cm)			NA: Drinking NA: Aquatic 700: Agriculture	300 (DL) 700 (MPL)
^{238}U (Bq/l)	0	0.37	–	
^{232}Th (Bq/l)	0	3.40	–	

USEPA US Environmental Protection Agency

[a] *DL* Desirable limit – The level of contamination in drinking water below which there is no known or expected risk to health.

[b] *MPL* Maximum Permissible Limit – The highest level of a contaminant that is allowed in drinking water.

[c] *BIS* Bureau of Indian Standards

11.1.4.2 Water Quality Rating Curve

The rating curve developed by the National Sanitation Foundation (NSF) for pH, nitrate (NO_3), dissolved oxygen (DO) and phosphate (PO_4) is used. For other parameter viz., radionuclides, heavy metals, electrical conductivity (EC), sulphate (SO_4^{2-}), and suspended solids (SS), the rating curve proposed by Nasirian (2007) is adequate to use.

11.1.4.3 Rating Curve for Radionuclides

The water quality score for radionuclides parameters (^{238}U, ^{232}Th) is estimated from the following rating curve formulae:

$$Q_{\text{radio}} = -100 \times \bar{r} + 100; \quad \text{if } \bar{r} < 1 \tag{11.44}$$

$$Q_{\text{radio}} = 0; \quad \text{if } \bar{r} \geq 1 \tag{11.45}$$

in which $\bar{r}$ is estimated from following equation:

$$\bar{r} = \frac{1}{n} \sum_{i=1}^{n} r_i \tag{11.46}$$

where, r_i is the activity concentration quotient of radionuclides i (eq. (11.47)), while n is number of radionuclides considered.

$$r_i = C_i / C_\text{s} \tag{11.47}$$

In eq. (11.47), C_i is the concentration of the ith radionuclide, and C_s is the maximum permissible activity concentration of radionuclide. The maximum permissible activity concentration is given in Table 11.6. Due to the toxicity of radionuclides, a limit for r_i is introduced. The water quality score (Q_{radio}) becomes zero if anyone of the radionuclide quotient is greater or equal to one ($r_i \geq 1$).

11.1.4.4 Rating Curve for Heavy Metals

Similar to the procedure of radionuclides, the water quality score for the heavy metal can be estimated through the rating curve proposed by Nasirian (2007):

$$Q_{\text{heavy metal}} = -100 \times \bar{h} + 100; \quad \text{if } \bar{h} < 1 \tag{11.48}$$

$$Q_{\text{heavy metal}} = 0; \quad \text{if } \bar{h} \geq 1 \tag{11.49}$$

in which $\bar{h}$ can be estimated from following equation:

$$\bar{h} = \frac{1}{n} \sum_{i=1}^{n} h_i \tag{11.50}$$

where, h_i is the activity concentration quotient of heavy metals i (eq. (11.47)), while n is number of heavy metal parameters considered:

$$h_i = C_i/C_s \tag{11.51}$$

In eq. (11.51), C_i is the concentration of the ith heavy metal element, and C_s is the maximum permissible concentration of element (Table 11.6). Due to the toxicity of heavy metal presence, a limit for h_i is introduced. The water quality score ($Q_{\text{heavy metal}}$) becomes zero if anyone of the element quotients is greater than or equal to one ($h_i \geq 1$) [i.e. $C_i \geq C_s$].

11.1.4.5 Rating Curve for EC

The best fit equation developed by Nasirian (2007) for electrical conductivity to compute the water quality score from measured data (μS/cm = μ mhos/cm) are as follows:

$$Q_{\text{EC}} = -0.0714 \times \text{EC} + 100; \text{ if EC} \leq 700\,\mu\text{S/cm} \tag{11.52}$$

$$Q_{\text{EC}} = -0.00943 \times \text{EC} + 56.60; \text{ if } 700 < \text{EC} \leq 6000\,\mu\text{S/cm} \tag{11.53}$$

$$Q_{\text{EC}} = 0; \text{ if EC} > 6000\,\mu\text{S/cm} \tag{11.54}$$

[Note: Unit conversion: 1 dS/m = 1 mmho/cm = 1,000 μS/cm; 1,000 m mho/cm = 1 μ mho/cm].

11.1.4.6 Rating Curve for SO_4

The best fit equation developed for sulphate to compute the water quality score from measured data is:

$$Q_{SO_4} = -0.1 \times (SO_4) + 100; \text{ if } SO_4 < 1000\,\text{mg/l} \tag{11.55}$$

$$Q_{SO_4} = 0; \text{ if } SO_4 \geq 1000\,\text{mg/l} \tag{11.56}$$

11.1.4.7 Rating Curve for Suspended Solids (SS)

The best fit equation developed for suspended solids to compute the water quality score from measured data is:

$$Q_{\text{SS}} = 100; \text{ if SS} < 20\,\text{mg/l} \tag{11.57}$$

$$Q_{\text{SS}} = -0.77 \times \text{SS} + 115.3; \text{ if } 20 \leq \text{SS} \leq 150\,\text{mg/l} \tag{11.58}$$

$$Q_{\text{SS}} = 0; \text{ if SS} > 150\,\text{mg/l} \tag{11.59}$$

11.1.4.8 Water Quality Classification

Based on the new WQI proposed by Nasirian (2007) water quality can be classified as follows (Table 11.7):

The application of the methodology is presented in example 11.4.

Example 11.4 Estimate the water quality index and class of Nizamuddin Bridge site (Quarter stream) of Yamuna River using the CPCB data. The average water quality parameter measured during the year 2003 by CPCB is given in Table 11.8.

Table 11.7 Water class based on Nasirian's WQI

Quality	WQI range	Class
Clean	80–100	90–100: Class I 80–90: Class II
Slightly polluted	60–80	50–80: Class III
Very polluted	0–60	40–50: Class IV 0–40: Class V

Table 11.8 Water quality parameters of Nizamuddin Bridge site of Yamuna River and computational step for Nasirian's WQI

Parameters	Unit	Perm. limit	Measured	Quotient (r_i)	$\bar{r}$	WQ score	Weights	Weight used	WQI
Temp	°C		26						
DO_{sat}	mg/l		8.21						
DO	mg/l	5	0.9						
pH	Std unit	6.5–8.5	7			87.12	0.18	0.18	15.68
Nitrate	mg/l	100	18.6			37.73	0.13	0.13	4.91
Phosphate	mg/l	0.1	NM			2.00	0.11	0	0.00
$\%DO_{sat}$	%		10.97			4.37	0.11	0.11	0.48
Sulphate	mg/l	400	NM			0.00	0.08	0	0.00
SS	mg/l		NM			0.00	0.08	0	0.00
EC	µS/cm	700	894			48.2	0.08	0.08	3.85
^{238}U	Bq/l	0.37	NM	0.00					
^{232}Th	Bq/l	3.4	NM	0.00	0.0	100.	0.12	0	0.00
Cd	mg/l	0.01	NM						
Pb	mg/l	0.05	NM						
As	mg/l	0.05	NM						
Ni	mg/l	0.025	0.02	0.80					
Mn	mg/l	0.3	NM						
Fe	mg/l	0.3	5.05	16.83					
Hg	mg/l	0.001	NM						
Cu	mg/l	1.5	0.02	0.01					
Zn	mg/l	15	0.08	0.01	4.4	0.00	0.11	0.11	0.00
Total								**0.61**	**24.9**

NM refers to the parameters not measured; µS/cm = micro mhos/cm; Bq/l = Becquerel/litre

In the Yamuna River, radionuclides are not reported so far, therefore, the activity concentration is assumed zero. For heavy metals, four parameters, viz., Ni, Fe, Cu, and Zn are considered.

The WQI is estimated as:

$$\mathrm{WQI}_{\mathrm{adjusted}} = \frac{\mathrm{WQI}}{\text{Weight used}} = \frac{26.62}{0.61} = 40.58$$

Based on the classification, the water is "VERY POLLUTED" and class is very close to "CLASS-V"

11.1.5 Universal WQI Based on European Classification Scheme (UWQI)

The Universal Water Quality Index (UWQI) was developed to provide a simpler method for describing the quality of surface water used for drinking water supply. UWQI has an advantage over pre-existing indices by reflecting appropriateness of water for specific use, particularly for drinking water supply rather than general supply and has been developed by studying the super-national standard. The UWQI was developed on the basis of the following water quality standards (Boyacioglu, 2007): (a) The quality required of surface water intended for the abstraction of drinking water in the Member States 75/440/EEC set by the Council of the European Communities (EC, 1991); (b) The classification of inland waters according to quality – Turkish water pollution control regulation – WPCR (Official Gazette, 1988); and (c) other reported scientific information. Water quality standard for drinking water supply for the development of UWQI is given in Table 11.9.

According to the EC legislation (75/440/EEC), water quality of surface waters intended for the abstraction of drinking water is classified into three groups. For each class the treatment level required to transform surface water into drinking water is different and can be summarized as follows:

- Class I: Simple physical treatment and disinfection, e.g., rapid filtration and disinfection
- Class II: Normal physical treatment, chemical treatment and disinfection, e.g., pre-chlorination, coagulation, flocculation, decantation, filtration, disinfection (final chlorination)
- Class III: Intensive physical and chemical treatment, extended treatment and disinfection, e.g., chlorination to break-point, coagulation, flocculation, decantation, filtration, adsorption (activated carbon), disinfection (ozone, final chlorination) (EC, 1991).

Table 11.9 Classification of water quality for the development of UWQI

Parameter	Unit	Class I (excellent)	Class II (acceptable)	Class III (polluted)	Remarks
Total coliform	Count/ 100 ml	50	5,000	50,000	It is used to indicate whether other potentially harmful bacteria may be present
Cadmium	mg/l	0.003	0.005	0.010	Chemicals from industrial and domestic discharges
Cyanide	mg/l	0.010	0.050	0.100	
Mercury	mg/l	0.0001	0.0005	0.002	
Selenium	mg/l	0.010	0.010	0.020	Naturally occurring chemicals
Arsenic	mg/l	0.020	0.050	0.100	
Fluoride	mg/l	1	1.5	2	
Nitrate-nitrogen	mg/l	5	10	20	Chemicals from agricultural activities
DO	mg/l	8	6	3	Operational monitoring parameters
pH		6.5–8.5	5.5–6.4 8.6–9.0	<5.5 >9.0	
BOD	mg/l	<3	<5	<7	Indicator of organic pollution
Total phosphorous-PO_4-P	mg/l	0.02	0.16	0.65	It is included to satisfy the ecological requirement of certain types of environment

11.1.5.1 Aggregation Function of UWQI

The aggregation function of Universal Water Quality Index (NWQI) is defined as follows:

$$\text{UWQI} = \sum_{i=1}^{n} w_i q_i \tag{11.60}$$

where UWQI is the universal water quality index, w_i is the weight of the ith parameter, and q_i is the sub-index for ith parameter (Table 11.10).

The weights of the parameters, w_i corresponding to the rating is given in Table 11.11. Based on the value of UWQI, the water quality can be categorized using the Table 11.12.

Example 11.5 Pumping station at Wazirabad for Delhi water supply receives water from Palla via drain no 8. The water quality parameters published by CPCB for the year 2005 is shown in the table below. Calculated the suitability of water for drinking using the Universal Water Quality Index (UWQI).

Table 11.10 Mathematical equations formulated for UWQI

Variable	Range	Sub-index function
BOD	$BOD < 3$	$q_{BOD} = 100$
	$3 \leq BOD < 6$	$q_{BOD} = -25 \times BOD + 175$
	$6 \leq BOD < 7$	$q_{BOD} = -22.5 \times BOD + 162.5$
	$BOD \geq 7$	$q_{BOD} = 0$
Nitrate	$NO_3 \leq 5$	$q_{NO_3} = 100$
	$5 < NO_3 \leq 10$	$q_{NO_3} = -10 \times NO_3 + 150$
	$10 < NO_3 \leq 20$	$q_{NO_3} = -4.5 \times NO_3 + 95$
	$NO_3 > 20$	$q_{NO_3} = 0$
Arsenic	$As \leq 0.02$	$q_{As} = 100$
	$0.02 < As \leq 0.05$	$q_{As} = -1666.7 \times As + 133.33$
	$0.05 < As \leq 0.1$	$q_{As} = -900 \times As + 95$
	$As > 0.1$	$q_{As} = 0$
DO	$DO \geq 8$	$q_{DO} = 100$
	$8 < DO \leq 6$	$q_{DO} = 25 \times DO - 100$
	$6 < DO \leq 3$	$q_{DO} = 15 \times DO - 40$
	$DO < 3$	$q_{DO} = 0$
Fluoride	$F \leq 1$	$q_F = 100$
	$1 < F \leq 2$	$q_F = -95 \times F + 194.17$
	$F > 2$	$q_F = 0$
Total phosphorous	$TP \leq 0.02$	$q_{TP} = 100$
	$0.02 < TP \leq 0.16$	$q_{TP} = -357.14 \times TP + 107.14$
	$0.16 < TP \leq 0.65$	$q_{TP} = -91.83 \times TP + 64.694$
	$TP > 0.65$	$q_{TP} = 0$
Mercury	$Hg \leq 0.0001$	$q_{Hg} = 100$
	$0.0001 < Hg \leq 0.0005$	$q_{Hg} = -125000 \times Hg + 112.5$
	$0.0005 < Hg \leq 0.002$	$q_{Hg} = -30000 \times Hg + 65$
	$Hg > 0.002$	$q_{Hg} = 0$
Selenium	$Se \leq 0.01$	$q_{Se} = 100$
	$0.01 < Se \leq 0.02$	$q_{Se} = 4500 \times Se + 95$
	$Se > 0.02$	$q_{Se} = 0$
Cyanide	$CN \leq 0.01$	$q_{CN} = 100$
	$0.01 < CN \leq 0.05$	$q_{CN} = -1250 \times CN + 112.5$
	$0.05 < CN \leq 0.10$	$q_{CN} = -900 \times CN + 95$
	$CN > 0.10$	$q_{CN} = 0$
Cadmium	$Cd \leq 0.003$	$q_{Cd} = 100$
	$0.003 < Cd \leq 0.005$	$q_{Cd} = -25000 \times Cd + 175$
	$0.005 < Cd \leq 0.010$	$q_{Cd} = -9000 \times Cd + 95$
	$Cd > 0.010$	$q_{Cd} = 0$
Total coliform	$TC \leq 50$	$q_{TC} = 100$
	$50 < TC \leq 5000$	$q_{TC} = -10.857 \ln(TC) + 142.47$
	$5000 < TC \leq 50000$	$q_{TC} = -21.715 \ln(TC) + 284.95$
	$TC > 50000$	$q_{TC} = 0$
pH	$6.5 \leq pH \leq 8.5$	$q_{pH} = 100$
	$5.5 \leq pH \leq 6.4$ and $8.6 \leq pH \leq 9.0$	$q_{pH} = 50$
	$pH < 5.5$ and $pH > 9.0$	$q_{pH} = 0$

Table 11.11 Significance ratings and weights for UWQI

Category	Variable	Rating	Weight factor
Health hazard	Total coliform	4	0.114
	Cadmium	3	0.086
	Cyanide	3	0.086
	Mercury	3	0.086
	Selenium	3	0.086
	Arsenic	4	0.113
	Fluoride	3	0.086
	Nitrate-nitrogen	3	0.086
Operational monitoring	DO	4	0.114
	pH	1	0.029
Oxygen depletion	BOD	2	0.057
	Total phosphorous	2	0.057

Table 11.12 Ranking of water quality based on UWQI

Rank	UWQI	Rank	UWQI
Excellent	95–100	Marginal	25–49
Good	75–94	Poor	0–24
Fair	50–74		

Calculation:

Category	Parameter	Unit	Weight	Measured	Sub-index	Weight used	WQI
Health hazard	TC	count/ 100 ml	0.114	40,842	54.39	0.114	6.20
	Cd	mg/l	0.086	NM	0.00	0.000	0.00
	CN	mg/l	0.086	NM	0.00	0.000	0.00
	Hg	mg/l	0.086	NM	0.00	0.000	0.00
	Se	mg/l	0.086	NM	0.00	0.000	0.00
	As	mg/l	0.113	NM	0.00	0.000	0.00
	F	mg/l	0.086	NM	0.00	0.000	0.00
	Nitrate	mg/l	0.086	1.51	100.0	0.086	8.60
Operational monitoring	DO	mg/l	0.114	7	75.00	0.114	8.55
	pH		0.029	7.7	100.0	0.029	2.90
Oxygen depletion	BOD	mg/l	0.057	2	100.0	0.057	5.70
	TP	mg/l	0.057	NM	0.00	0.000	0.00
Total						**0.400**	**31.95**

NM refers to the parameters not measured

$$\text{WQI}_{\text{adjusted}} = \frac{\text{WQI}_{\text{calculated}}}{\text{Weight used}} = \frac{31.95}{0.40} = 79.88$$

Therefore the water quality can be ranked "GOOD". However, for drinking purposes all the health hazard parameters need to be collected for better decision making.

11.1.6 Concluding Remarks on Water Quality Index

In this section the estimation of water quality class based on the numeric system (i.e. water quality index) has been presented in detail. Various techniques have been discussed with their advantages and disadvantages. To understand the applicability of these techniques, suitable examples are also presented with the case of the Yamuna River system using the data of CPCB. It has been observed that Indian Water Quality Monitoring System does not satisfy the requirements of the parameters used in the discussed WQIs.

Therefore, finally the Water Quality Index (WQI) proposed by the National Sanitation Foundation (NSF) is applied to determine the water quality status of the Yamuna River based on the water quality parameters reported by CPCB for the year 2005. The analysis is presented in Example 11.6. From the analysis, it can be revealed that the water quality is steeply polluted within the Palla to the Nizamuddin Bridge stretch, whereas the water quality has gradually improved from Mazawali to Allahabad near the confluence with Ganga River, though the quality is still in the category of medium to average (i.e., NSF-WQI = 70) (Fig. 11.10).

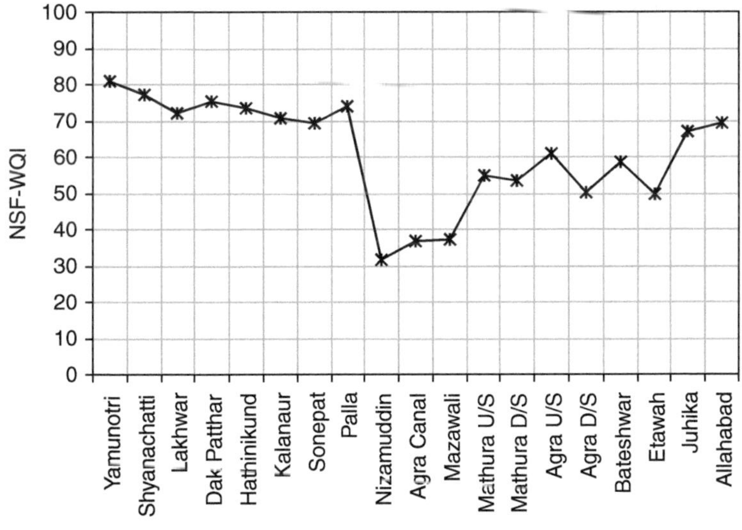

Fig. 11.10 NSF-WQI profile of the Yamuna River for the year 2005

Table 11.13 Average value of water quality parameters of CPCB sites of Yamuna River for the year of 2005

Location/Site	pH	COD (mg/l)	BOD (mg/l)	AMM (mg/l)	TKN (mg/l)	DO (mg/l)	WT (°C)	TC (count/100 ml)	FC (count/100 ml)
Yamunotri		1	1	0.57	0.84	8.4	6.0	80	18
Shyanachatti		2	1	0.66	0.84	10.2	9.5	1,650	360
Lakhwar		2	1	0.63	1.12	8.5	16.5	88,500	1,584
Dak Patthar		2	1	0.71	1.40	9.3	19.3	108,393	2,023
Hathinikund	7.7	8	2	0.29	1.46	9.2	19.7	569,617	8,379
Kalanaur	7.8	10	2	0.41	1.49	8.1	22.5	616,125	12,881
Sonepat	7.7	16	3	0.67	1.87	7.5	23.2	910,783	14,309
Palla	7.7	11	2	0.55	1.51	8.4	24.1	40,842	2,930
Nizamuddin Bridge	7.4	67	23	17.42	24.24	0.5	25.7	37,010,000	2,952,100
Agra Canal	7.4	47	15	9.78	13.94	0.8	25.0	13,500,000	981,333
Mazawali	7.6	57	17	19.62	26.33	2.4	24.6	4,512,727	755,727
Mathura U/S	7.7	37	10	10.65	15.22	6.3	26.4	1,700,250	57,708
Mathura D/S	7.7	32	8	10.89	16.66	5.6	27.0	3,152,500	328,850
Agra U/S	7.9	32	9	5.49	10.18	8.3	26.3	5,421,333	229,083
Agra D/S	7.7	46	13	7.44	12.34	5.1	26.8	35,317,500	1,728,333
Bateshwar	8.0	44	11	4.19	6.95	10.2	27.5	16,838,333	23,858
Etawah	8.2	46	14	2.49	4.91	11.6	27.0	14,492,750	17,983
Juhika	8.1	25	5	1.04	3.62	8.31	25.75	2,36,500	9,206
Allahabad	8.0	7	2	0.64	2.96	7.3	25.1	1,021,000	31,350

Example 11.6 Calculate the NSF-WQI and assess the average water class for the entire monitoring sites of Yamuna River located by CPCB for the year 2005 (Table 11.13).

Calculation:
Results can be summarized as follows:

Location/Site	WQI	Class	Location/Site	WQI	Class
Yamunotri	81.13	Good	Mazawali	36.98	Fair
Shyanachatti	76.98	Good	Mathura U/S	55.05	Average
Lakhwar	72.32	Good	Mathura D/S	53.62	Average
Dak Patthar	75.58	Good	Agra U/S	60.88	Average
Hathinikund	73.28	Good	Agra D/S	50.12	Average
Kalanaur	70.76	Average to Good	Bateshwar	58.47	Average
Sonepat	69.22	Average	Etawah	49.93	Fair[a]
Palla	73.86	Good	Juhika	66.86	Average
Nizamuddin Bridge	31.72	Fair	Allahabad	69.49	Average
Agra Canal	36.79	Fair			

[a] Average BOD and DO reported by CPCB for the year 2005 are 14 and 11.6 mg/l with average temperature of 27°C, respectively. The saturated DO at 27°C can be estimated to be 8.06 mg/l, resulting in the percentage DO saturation of nearly 144%, which reduced the overall WQI number.

The water quality status of the Yamuna River based on NSF-WQI is shown in Fig. 11.10, which reflects that the stretch between Palla to Agra D/S is critical and need to be protected.

11.2 Water Quality Status Based on the Data

The water quality of the river is highly dependent on the availability of freshwater in the river, which is greatly varied in time and space. Precipitation is confined to only three months in a year and varies greatly. Most of the water flows in the Yamuna (nearly 80%) in the monsoon period (July, August and September) only. Whatever water flows during the non-monsoon period (October to June) is extensively used for irrigation and drinking leaving very little or no water in the river to flow.

It is observed that about 500 km long stretch of the river is in bad shape, having poor water quality most of the time, which is below the desired limit (i.e. designated best use). In the dry season four distinct gradients of pollution load can be distinguished in the river stretch between Wazirabad and Etawah. The stretch between Wazirabad and Okhla is the most heavily polluted one, carrying the massive input of wastewater from Delhi (CPCB, 2006). This input sets off a progressive series of chemical and biological events in the downstream water. This stretch is characterized by high bacterial population, cloudy appearance, high BOD and strong disagreeable odour – all indicating general depletion of oxygen. Masses of gaseous sludge rising from the bottom are often noticed floating near the surface of the water. During monsoon, due to flood the sludge deposited in this stretch is flushed and stays in suspension which causes a rise in the oxygen uptake downstream. This causes heavy fish mortality every year during the first flushing after the onset of monsoon. Even in this short stretch, remarkable purification takes place due to high temperature and long retention time due to the two barrages one at Okhla and the other at ITO (nearly 10 km upstream of the Okhla Barrage). The ITO Barrage is used to divert the Yamuna water for cooling of the two Thermal Power Plants located near ITO. In the stretch between Okhla and Agra (225 km) the same status of water quality can be observed, which is due to the wastewater inflow from the Okhla barrage, and several wastewater lateral drains between downstream of Okhla up to Agra. After a few kilometers the repeated additions of sewage are mainly noticeable by a higher state of eutrophication leading to the formation of algal mats in the river. Excessive algal cause problems associated with the oxygen balance in the water (daytime super saturation and night-time oxygen depletion). The water quality from DO, BOD, and bacterial points of view is not fit for designated best uses in this stretch. The Agra Water Works is drawing its raw water from this only. The next 210 km long stretch from Agra to the confluence with the Chambal River at Etawah is characterized by self- purification processes of the Agra effluents. The confluence with relatively clean Chambal River is of great value in diluting the pollution load of Yamuna River before it joins the Ganga at Allahabad. During the monsoon period due to great discharge in the river, most of the barrages are opened, which makes the river as continuous system. However, during this period (i.e. monsoon) entire waste deposits in the river comes in suspension and flowed downstream, which leads extensive damage to the river's ecosystem. More or less, the Yamuna River hardly fulfil the criteria of designated best uses (CPCB, 1978; CPCB, 1981; CPCB, 1982–1983; CPCB, 1984–2008; CPCB, 1994).

11.2.1 Water Quality Trends in the Yamuna

The water quality data acquired by National River Conservation Directorate (NRCD) during the past 10 years were compiled for all monitoring locations. For all the locations, temporal variations in DO, BOD, total coliform and fecal coliform have been analysed and depicted in Fig. 11.11. It is clear from Fig. 11.11 that the water quality of the river has gradually shown similar trends of deterioration throughout the 10 years period. The year 2006 was worst for the river in terms of water quality. This can be attributed to the drought in the catchment of the river resulting in low flow conditions. In general, water quality is gradually deteriorating due to the decrease in the discharge.

11.2.2 Water Quality Profile of the Yamuna

The water quality of the Yamuna River in terms of organic pollution and coliform count has been satisfactory from origin till Palla as evident from Figs. 11.12, 11.13, 11.14, and 11.15.

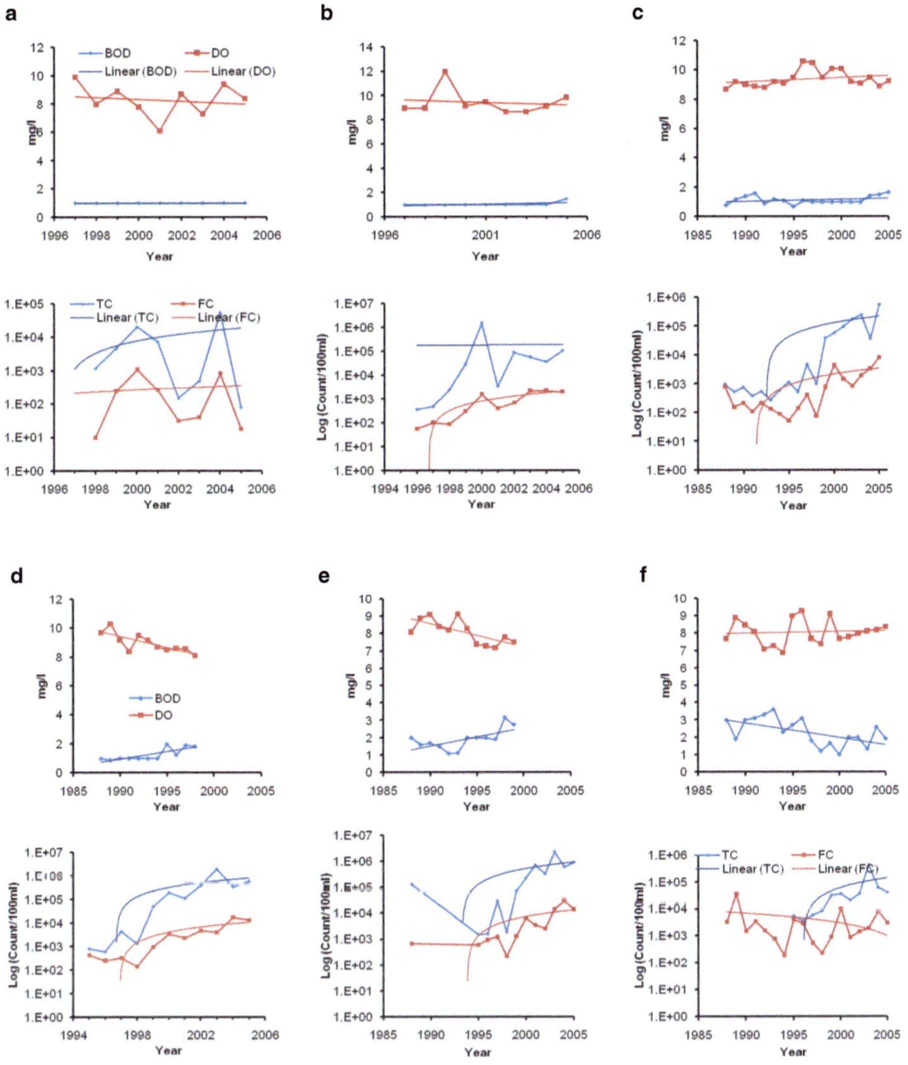

Fig. 11.11 Water quality trend at different locations in the Yamuna River (**a**) Water quality trends at Yamunotri site; (**b**) Water quality trends at Dak Patthar site; (**c**) Water quality trends at Hathinikund site; (**d**) Water quality trends at Kalanaur site; (**e**) Water quality trends at Sonepat site; (**f**) Water quality trends at Palla (U/S Wazirabad) site; (**g**) Water quality trends at Nizamuddin Bridge site; (**h**) Water quality trends at Agra canal site; (**i**) Water quality trends at Mazawali site; (**j**) Water quality trends at Mathura U/S site; (**k**) Water quality trends at Mathura D/S site; (**l**) Water quality trends at Agra U/S site; (**m**) Water quality trends at Agra D/S site; (**n**) Water quality trends at Bateshwar site; (**o**) Water quality trends at Etawah site; (**p**) Water quality trends at Juhika site; (**q**) Water quality trends at Allahabad site

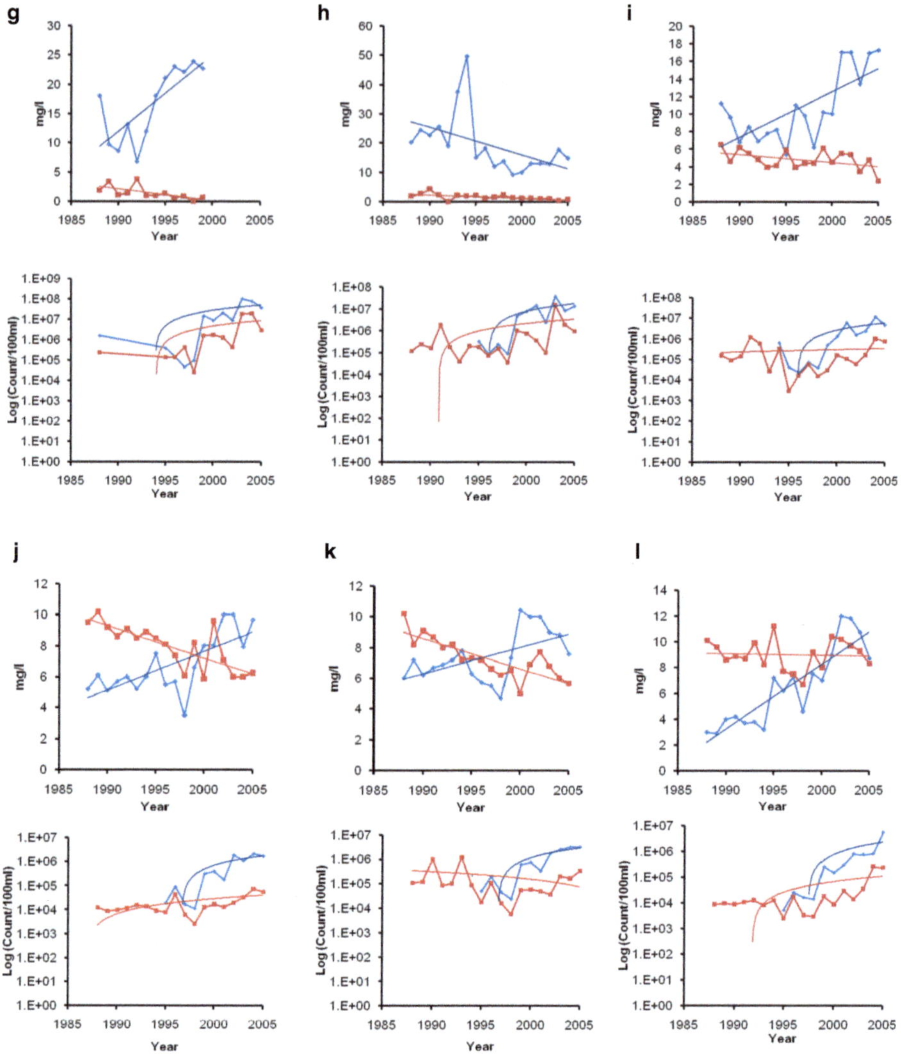

Fig. 11.11 (continued)

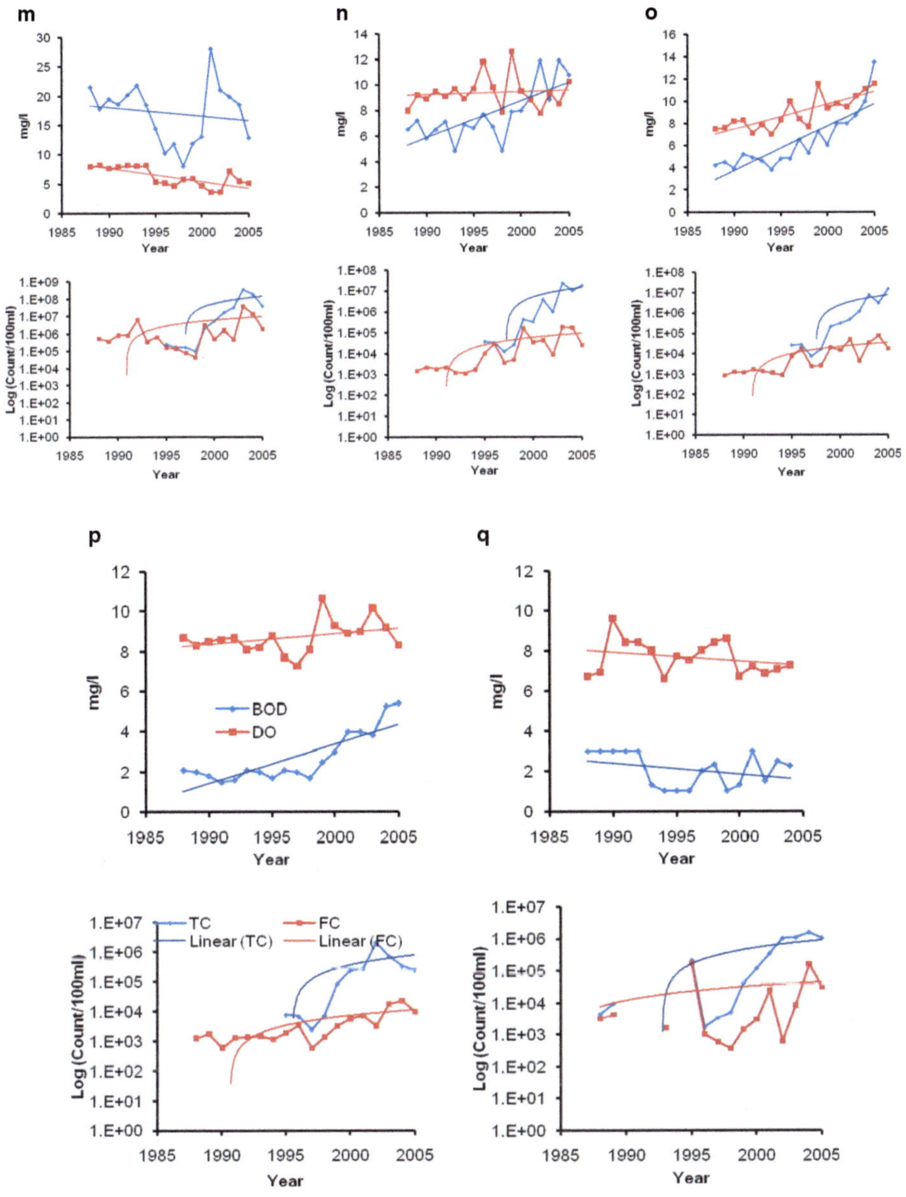

Fig. 11.11 (continued)

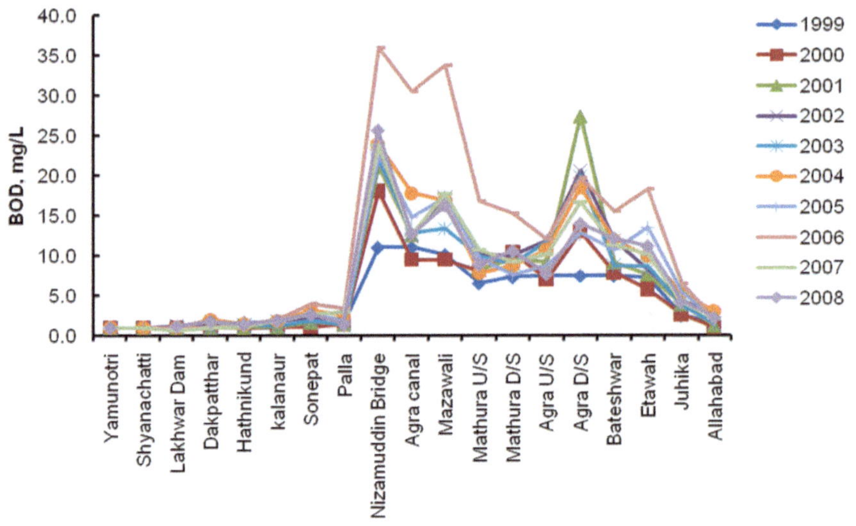

Fig. 11.12 Ten year water quality profile of Yamuna River – BOD

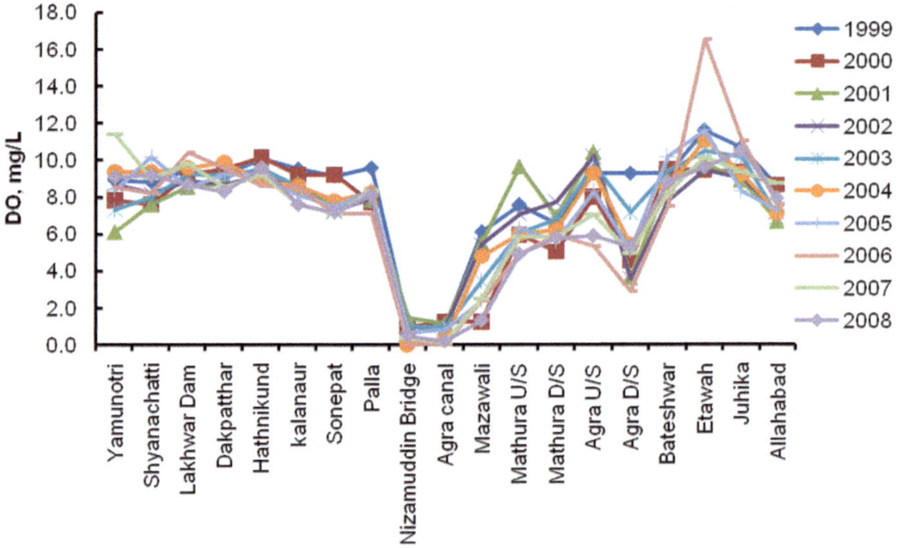

Fig. 11.13 Ten year water quality profile of Yamuna River – DO

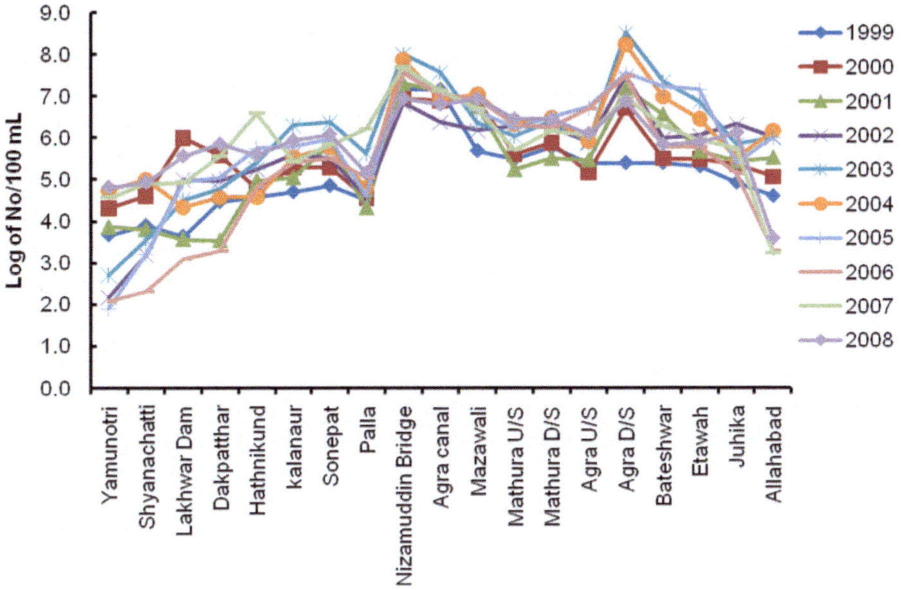

Fig. 11.14 Ten year water quality profile of Yamuna River – Total Coliform

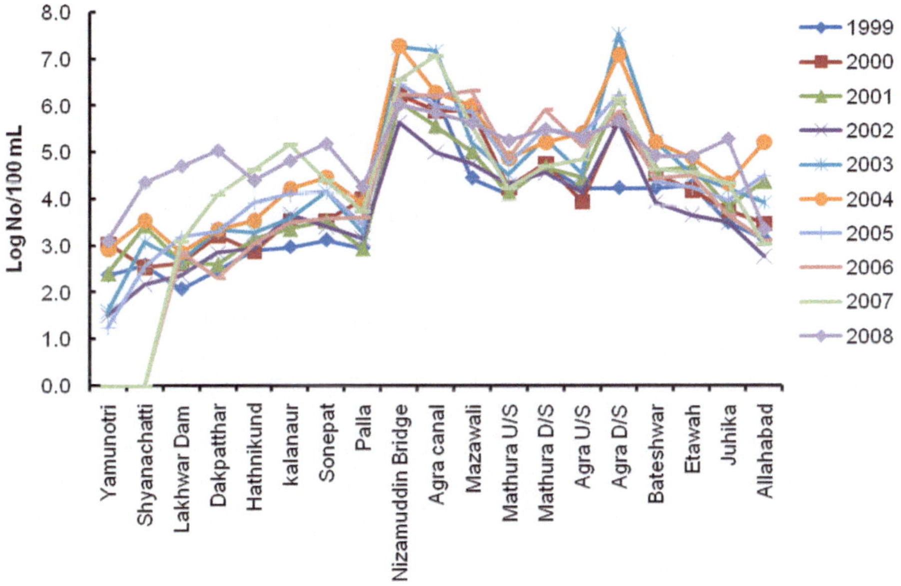

Fig. 11.15 Ten year water quality profile of Yamuna River – Fecal Coliform

Though there has been a gradual increase in BOD from origin to the upstream of Delhi. However, the average BOD values have been well below the designated best use criteria in this stretch. The BOD level has increased significantly after the Wazirabad barrage. This is mainly due to the discharge of partially treated wastewater through several drains in Delhi. The average BOD values are not confirming the desired water quality criteria till the confluence of Chambal River. In the Delhi stretch (Wazirabad barrage to Okhla barrage), several bathing Ghats (i.e. place of spiritual bathing along the river banks) are located though the water quality is not up to the bathing standards.

11.2.2.1 Bio-mapping of the Yamuna River

Bio-mapping is a good tool for water quality analysis of the river, and is commonly used to visualize the gradual changes in the river. Here, dissolved oxygen (DO) has been used as an indicator for the analysis of water quality status, and therefore, the DO-maps have been prepared for various years for comparison. Using the average DO values at various locations in the river Yamuna for the years 2003, 2008 and 2009, the DO map of the river has been prepared for the entire length. The DO maps for these years are depicted in Figs. 11.16, 11.17, and 11.18. Comparison of the DO levels (Figs. 11.16, 11.17, and 11.18) reveals that the water quality of the river is continuously degrading. It can be seen from these figures that the stretch of the red colour (DO < 4 mg/l) is elongating and the stretch presented by green colour (DO > 6 mg/l) has been significantly decreased, showing further deterioration of water quality of the river.

For detailed water quality analysis of the river Yamuna, samples at various locations in the river and wastewater drains have been collected in the year 2009; and simultaneously the photographs were taken. The water colour can be easily seen in the photographs (Plate 11.1) which themselves describe the water quality. Along with photographs the water quality parameters are also listed for visual correlation.

11.2.3 Monthly Variation in Water Quality Parameters

Monthly variation in the water quality parameter is important for management of the river system. Based on the data collected from CPCB during the period 1999–2008, the monthly average has been estimated for important sites, and is shown in Figs. 11.19, 11.20, 11.21, 11.22, 11.23, 11.24, 11.25, and 11.26. It is clear from the figures that in general, DO is reducing and BOD is increasing during monsoon. This is due to the inflow of sediment-laden runoff into the river. However, for the Delhi segment, monthly variation is almost reverse as compared to other stretches. In this stretch, during monsoon, water quality is significantly improving because of high discharges. After monsoon period, water quality of the river gradually deteriorated due to wastewater influence.

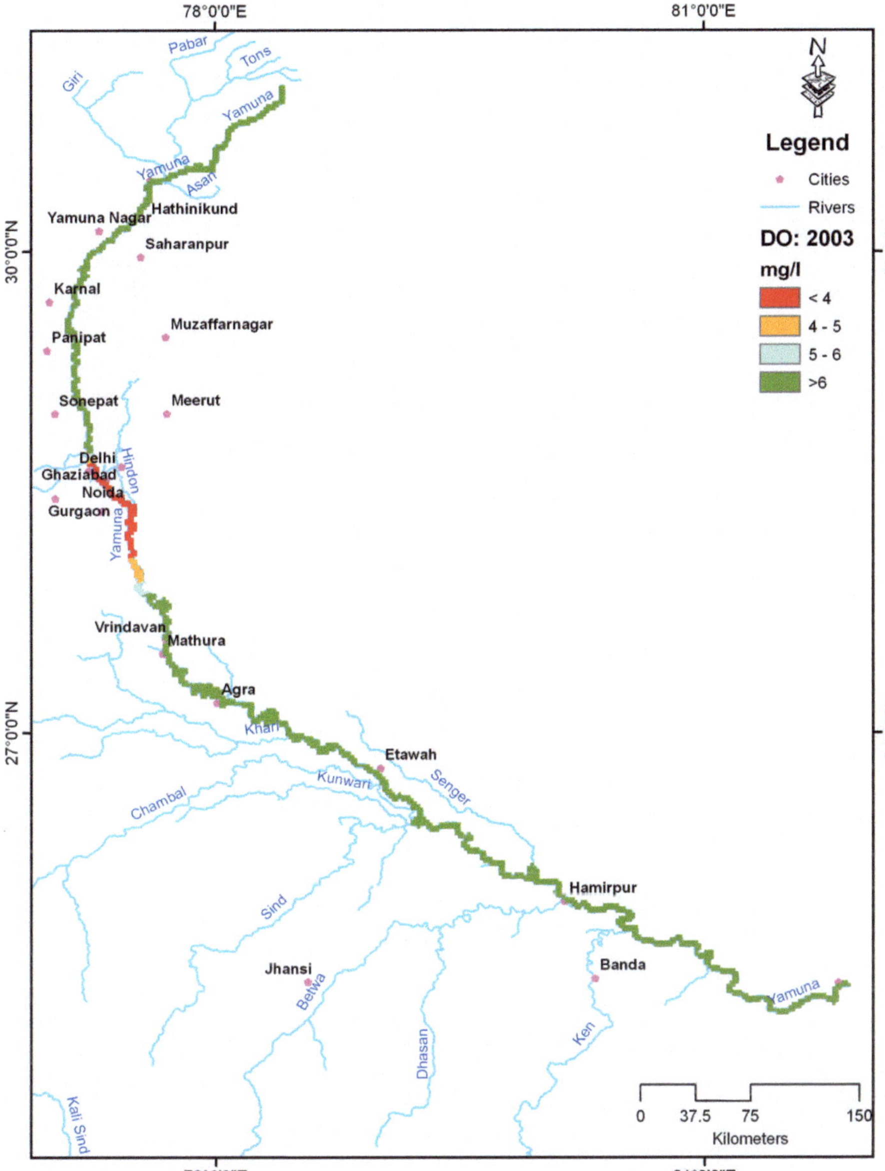

Fig 11.16 Water quality status using the DO mapping for the year 2003

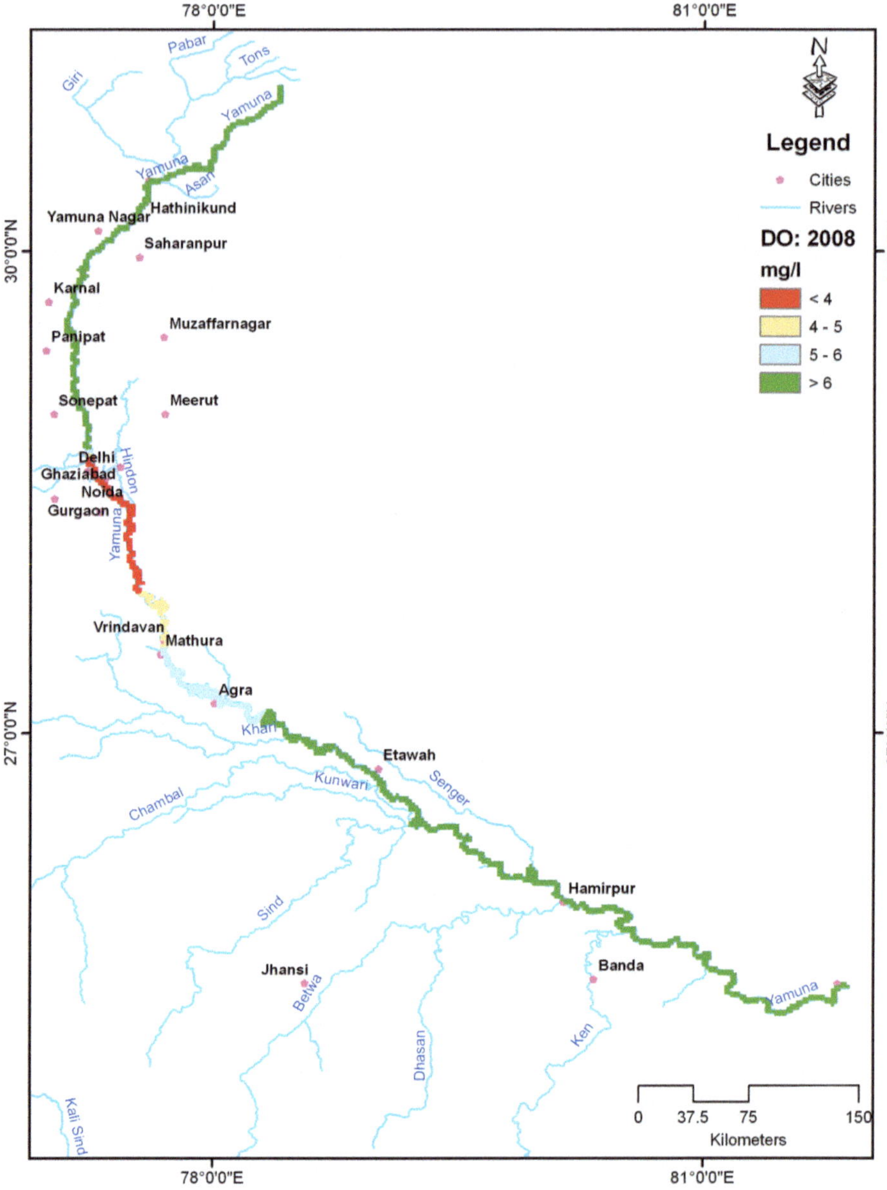

Fig 11.17 Water quality status using the DO mapping for the year 2008

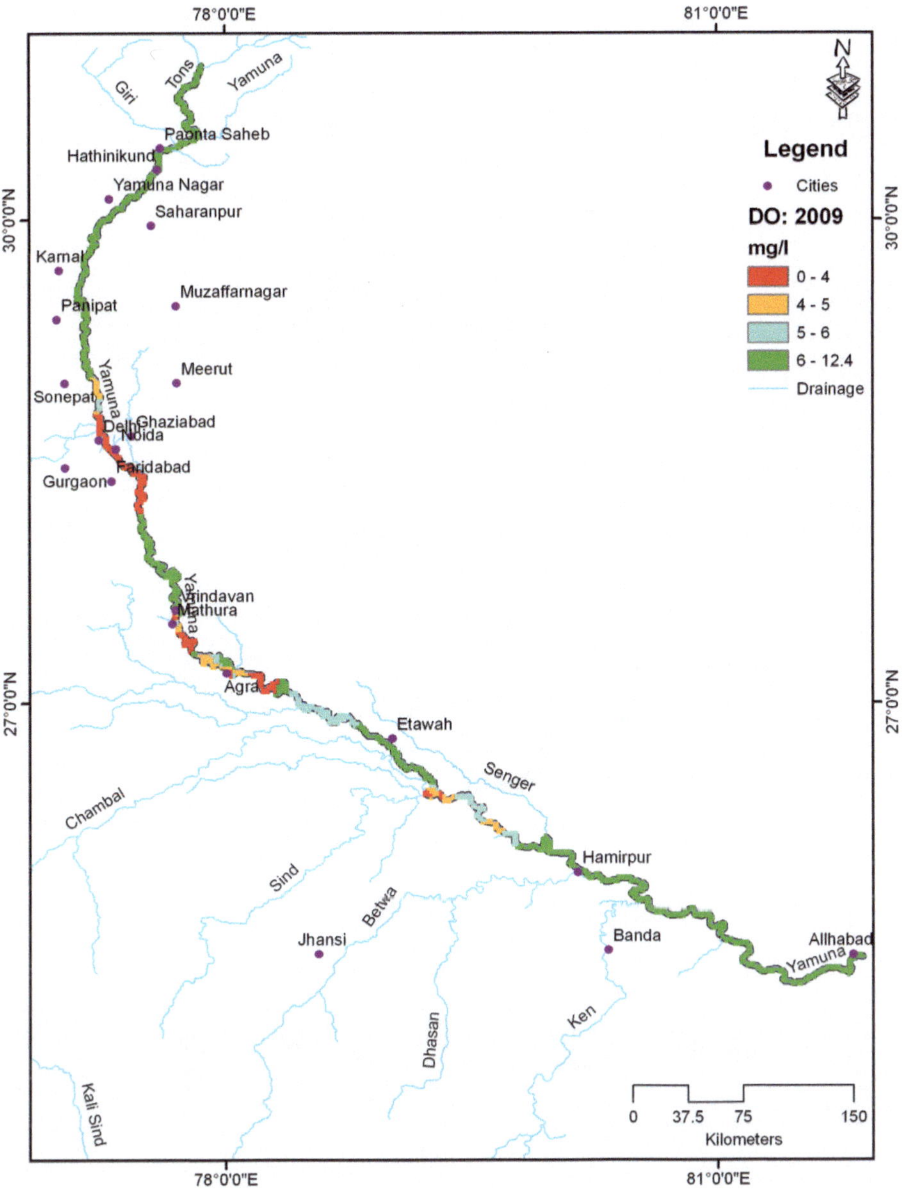

Fig 11.18 Water quality status using the DO mapping for the year 2009

Drain No. 8: Wazirabad barrage gets water through Storm Water Drain (No. 8) via Palla [pH = 7.49, EC = 201μmhos/cm, TDS = 120.6 mg/l, Turbidity = 8.5 NTU, NH_4 = 0 mg/l, BOD = 2.0 mg/l, TC = 500 MPN/100ml, FC = 450, DO = 8.3 mg/l]

Yamuna at Palla [pH = 7.47, EC = 255.4, TDS = 153, Turbidity = 77, NH_4 =0.13, BOD = 4.77, TC = 67,066, FC = 7,033, DO = 5.5, Q = 11 cumec]

Yamuna at Nigambodh Ghat [pH = 7.4, EC = 1,449, TDS = 886, Turbidity = 40.8, NH_4 = 3.47, BOD = 43.46, TC = 3,100,080, FC = 21,00,031, DO = 0.0, Q = 13.72]

Yamuna at Nizamuddin Bridge [pH = 7.37, EC = 1,364, TDS = 823, Turbidity = 20, NH_4 = 8.24, BOD = 25.4, TC = 412,042, FC = 247,624, DO = 0.49, Q = 21.93]

Yamuna at Okhla Barrage [pH = 6.91, EC = 11,02.5, TDS = 673, Turbidity = 17.5, NH_4 = 3.85, BOD = 19.35, TC = 262,775, FC = 262,537, DO = 2.28]

Yamuna at Kosi Shergarh [pH = 7.76, EC = 1,805, TDS = 1119, Turbidity = 11.5, NH_4 = 8, BOD = 10.7, TC = 310, FC = 260, DO = 12.4, Q = 10]

Plates 11.1(a) Photographs showing the status of Yamuna River and its water quality parameters

Yamuna at Vrindavan [pH = 7.76, EC = 1,367, TDS = 844, Turbidity = 10.5, NH_4 = 21.4, BOD = 25.9, TC = 755, FC = 419, DO = 9.3, Q = 13]

Yamuna near Taj Mahal [pH = 7.04, EC = 1,309, TDS = 779, Turbidity = 20.5, NH_4 = 6.3, BOD = 15.75, TC = 170,000, FC = 6,000, DO = 7.55, Q = 11.44]

Yamuna at Bateshwar [pH = 7.81, EC = 1,383, TDS = 826, Turbidity = 30, NH_4 = 5.7, BOD = 16, TC = 900,000, FC = 20,000, DO = 5.7]

Yamuna near Etawah [pH = 7.91, EC = 1,434, TDS = 857, Turbidity = 38, NH_4 = 4.6, BOD = 49, TC = 24,000, FC = 200,00, DO = 6.9]

Yamuna at Hamirpur [pH = 8.7, EC = 1,110, TDS = 710, Turbidity = 29, NH_4= 0.0, BOD = 41.6, TC = 460, FC = 21, DO = 6.7, Q = 45.15]

Yamuna at Allahabad [pH = 8.46, EC = 790, TDS = 505, Turbidity = 20, NH_4 = 0.44, BOD = 31.4, TC = 75, FC = 20, DO = 8.3, Q=62.7 cumec]

Plates 11.1(a) (continued)

Chambal River at Kondal [pH = 8.14, EC = 618, TDS = 367, Turbidity = 27, NH_4 = 2.1, BOD = 12, TC = 300, FC = 300, DO = 12, Q = 8.4 cumec]

Sind River [pH = 7.86, EC = 568, TDS = 343, Turbidity = 97, NH_4 = 4.5, BOD = 21.1, TC = 160,000, FC = 160,00, DO = 3.6, Q = 4.97 cumec]

Betwa River [pH = 8.63, EC = 480, TDS = 307, Turbidity = 32, BOD = 0.0, TC = 150, FC = 21, DO = 8.8, Q = 4.02 cumec]

Ken River [pH = 8.53, EC = 540, TDS = 345, Turbidity = 35, NH_4 = 0.44, BOD = 0.0, TC = 23, FC = 4, DO = 7.2, Q = 2.78 cumec]

Plates 11.1(a) (continued)

Nazafgarh outfall: [pH = 7.21, EC = 1,517, TDS = 922, Turbidity = 53, NH_4 = 61, BOD = 61, COD = 176, Q = 13.03 cumec]

Barapulla Drain [pH = 8.4, EC = 1,421, TDS = 909, Turbidity = 78, BOD = 62.4, COD = 187, Q = 0.5 cumec]

Plates 11.1(b) Photographs showing the status of drains polluting the Yamuna River

Power House Drain [pH = 7.16, EC = 1,239, TDS = 768, Turbidity = 76.2, NH_4 = 8.8, BOD = 54.6, COD = 249, Q = 0.623 cumec]

Sen Nursing Home drain [pH = 7.21, EC = 1553, TDS = 955, Turbidity = 66, NH_4 = 6.84, BOD = 77, COD = 216, Q = 0.516 cumec]

Mantola Drain, Agra [pH = 7.69, EC = 2,687, TDS = 1,657, Turbidity = 71, NH_4 = 7.55, BOD = 77.7, COD = 217.4 (Flow measurement was not performed due to deep sludge)]

Plates 11.1(b) (continued)

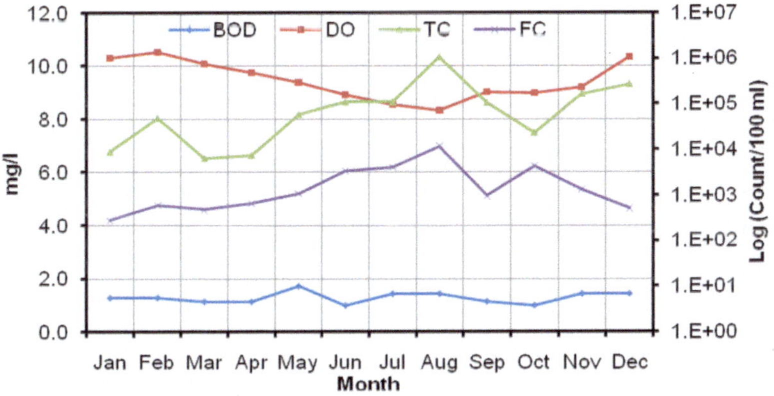

Fig. 11.19 Monthly variation in water quality at Hathinikund

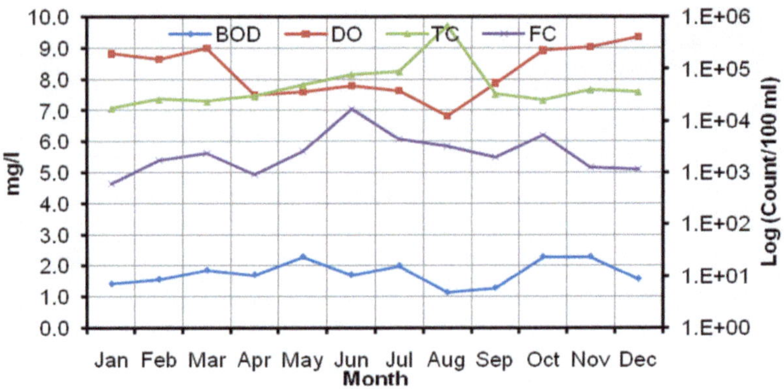

Fig. 11.20 Monthly variation in water quality at Palla

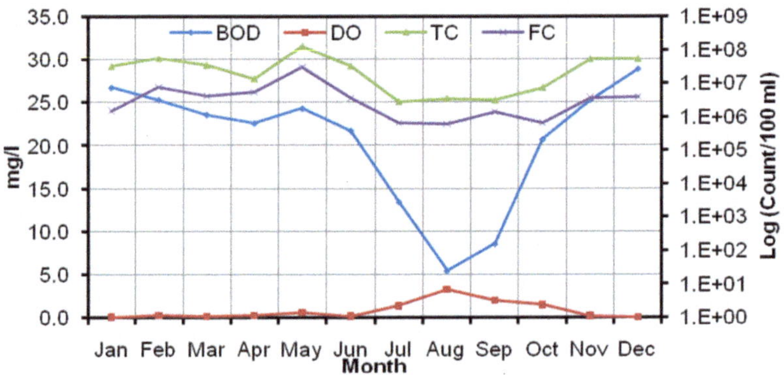

Fig. 11.21 Monthly variation in water quality at Nizamuddin Bridge

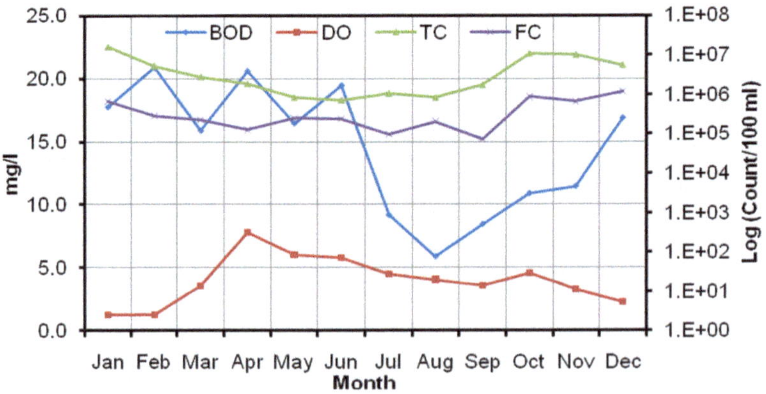

Fig. 11.22 Monthly variation in water quality at Mazawali

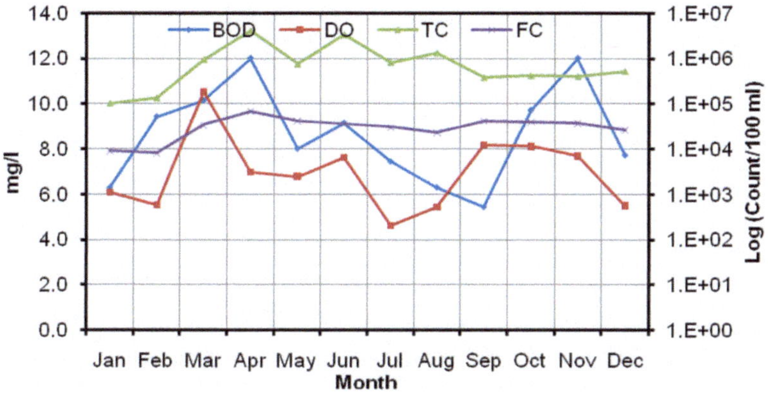

Fig. 11.23 Monthly variation in water quality at Mathura U/S

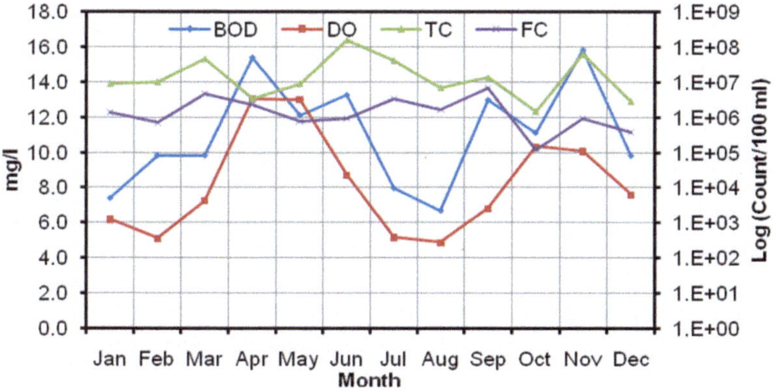

Fig. 11.24 Monthly variation in water quality at Agra D/S

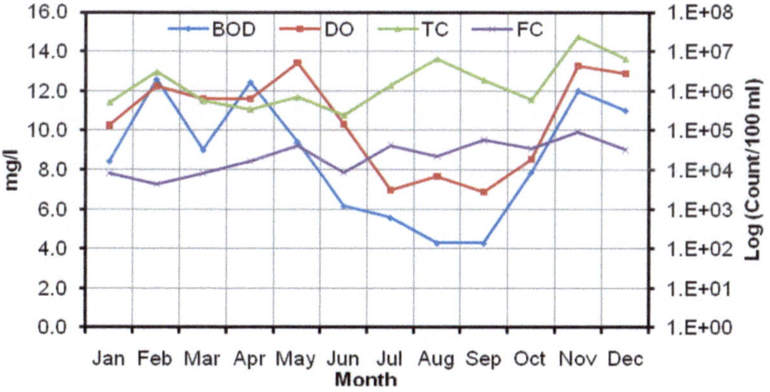

Fig. 11.25 Monthly variation in water quality at Etawah

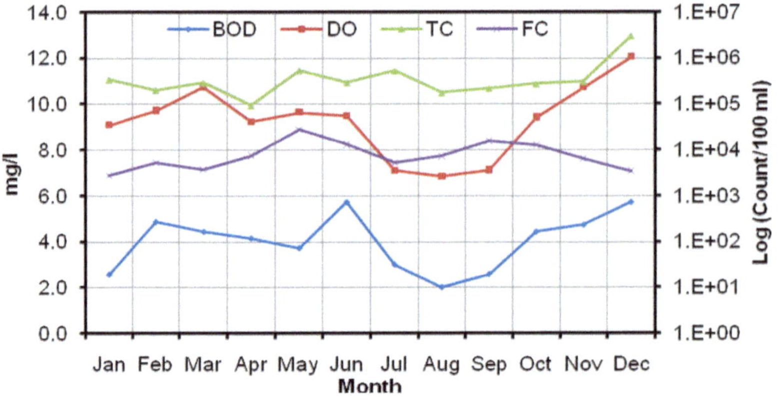

Fig. 11.26 Monthly variation in water quality at Juhika

11.3 Concluding Remarks

Yamuna River maintains reasonable good quality upstream of Delhi and thus it satisfies the designated best use criteria most of the times except sometimes in summer season. Addition of huge amounts of untreated domestic sewage in Delhi and non-availability of dilution water results in degradation of water quality downstream stretch of the river. The river in its 500 km stretch from Delhi to the Chambal confluence is not fit for its designated best use even in the monsoon season when sufficient dilution is available. The pollution is predominantly organic in nature. Therefore, the depletion of oxygen is the major impact on the polluted stretch of the river which disturbs the river ecosystem to a large extent. The biodegradation of this organic pollution results in the release of nutrients which in turn promotes the growth of algae and other aquatic plants in the river. This excessive growth of unwanted plants results in the situation of eutrophication. Due to eutrophic conditions in this segment of the river, the dissolved oxygen during nights is depleted to a large extent resulting in mass killing of fishes and other aquatic life. The conditions are worsened by the addition of untreated domestic sewage from Mathura and Agra. Thus in this entire stretch of 500 km, the oxygen is the key factor which determines the health of the river and distribution of aquatic life in it.

The microbial pollution is prevailing over the entire Yamuna River due to the contribution of human wastes. The Yamuna River is used for almost all the uses identified under the designated best use classification of CPCB. Considering its holy nature people from all over the country gather at several places along the river to take a holy dip in it, especially in places like Vrindavan, Mathura, Bateshwar and Allahabad where lakhs of people gather on some auspicious occasions to take a holy dip. The results of CPCB monitoring indicate that the river is not meeting the designated best use criteria for outdoor bathing. At present there is no study to establish the health impact of taking bath in the river.

Results of micro-pollutant analysis in water and sediment indicate that some of these pollutants especially pesticides like DDT and BHC are presented in quite significant amounts. Other micro-pollutants are generally found either below the detection limit or in low concentration. The micro-pollutants are generally discharged along with huge amounts of organic matter and thus they are adsorbed on the surface of organic particles and settle down in the sediment in the bottom of drains and river bed. This deposited sludge is annually flushed during floods in the monsoon season. Thus, the river and drains are being annually cleaned and are not allowed to build up micro-pollutants in their beds year after year.

References

Ahmad S, David KS, Gerald S (2004) Environmental assessment: an innovation index for evaluation water quality in streams. Environ Manage 34:406–414

Bhargava DS (1987) Nature and the Ganga. Environ Conserv 14:307–318

Boyacioglu H (2007) Development of a water quality index based on a European classification scheme. Water SA 33(1):101–106

Brown RM, McLelland NI, Deininger RA, O'Connor MF (1972) A water quality index-crashing the psychological barrier. In: Advances in Water Pollution Research, Proceedings of the 6th International Conference. Pergamon Press, New York, pp. 787–794

Couillard D, Lefebvre Y (1985) Analysis of water quality indices. J Environ Manage 21:161–179

CPCB (1978) Union Territory of Delhi (Abridged), Control of Urban Pollution Series, CUPS/1/1978-79, Central Pollution Control Board of India

CPCB (1981) Basin Sub Basin Inventory of Water Pollution – The Ganga Basin Part One: The Yamuna sub – basin, Assessment and Development study of River Basin Series, ADSORBS/2/1980-81, Central Pollution Control Board of India

CPCB (1982–1983) Quality and Trend of Yamuna River(1977–1982), Assessment and Development Study of River Basin Series, ADSORBS/10/1982-83, Central Pollution Control Board of India

CPCB (1984–2008) Water Quality Statistics of India, Monitoring of Indian Notional Aquatic Resources Series (MINARS/1 to 30/1984 to 2008), Central Pollution Control Board of India

CPCB (1994) Water Quality Status of Yamuna River, Assessment and Development Study of River Basin Series, ADSORBS/29/1993-94, Central Pollution Control Board of India

CPCB (2006) Status of Sewage Treatment in India, Control of Urban Pollution Series: CUPS/61/2005-2006, Central Pollution Control Board of India

Dinius SH (1987) Design of an index of water quality. Water Resour Bull 23:833–843

EC (European Council) (1991) Consolidated Text Produced by the CONSLEG System of the Office for Directive of 16 June 1975 Concerning the Quality Required of Surface Water Intended for the Abstraction of Drinking Water in the Member States (75/440/EEC). Office for Official Publications of the European Communities. CONSLEG: 1975L0440 31/12/1991

Faisal K, Tahir H, Ashok L (2003) Water quality evaluation and trend analysis in selected watersheds of the Atlantic region of Canada. Environ Monit Assess 88:221–248

Harrison TD, Cooper JAG, Ramm AEL (2000) Water quality and aesthetics of South African estuaries. Department of Environmental Affairs and Tourism. South Africa. www.environment.gov.za/soer/reports/ehi/ehi_ch4.pdf

Horton RK (1965) An index number system for rating water quality. J Water Pollut Contam Fed 37:300–305

Landwehr JM (1979) A statistical view of a class of water quality indices. Water Resour Res 15:460–468

Landwehr JM, Deininger RA (1976) A comparison of several water quality indices. J Water Pollut Cont Fed 48(5):954–958
Nasirian M (2007) A new water quality index for environmental contamination contributed by mineral processing: A case study of Amang (Tin Tailing) processing activity. J App Sci 7(2):2977–2987
Nives SG (1999) Water quality evaluation by index in Dalmatia. Water Res 33:3423–3440
O'Connor MF (1971) The application of multi-attribute scaling procedures to the development of Indices of water quality. Ph.D. dissertation, University of Michigan
Official Gazette (1988) Turkish Water Pollution Control Regulation, No. 19919. $ September 1988, Ankara
Sharifi M (1990) Assessment of surface water quality by an index system in Anzali basin. The Hydrological Basis for Water Resources Management (Proceedings of the Beijing Symposium, October 1990). IAHS Pub. No. 197, pp 163–171
Shiow-Mey L, Shang-Lien L, Shan-Hsien W (2004) A generalized water quality index for Taiwan. Environ Monit Assess 96:35–52.
Swamee PK, Tyagi A (2000) Describing water quality with aggregate index. J Environ Eng 126:451–455
Walski TM, Parker FL (1974) Consumer's water quality index. J Environ Eng 100:593–611

Chapter 12
Environmental Flows

Abstract This chapter presents techniques for the estimation of environmental flows for a river, including the advantages and disadvantages of various methodologies for environmental flow estimation for a river. A component based technique for environmental flows is presented for the river, which are highly intercepted and used for pollutant transport. The technique describes the various water demands of the river, such as water requirements for aquatic habitat (i.e., ecological requirement), water requirements for groundwater conservation (i.e., seepage loss), water requirements for evaporation (i.e., evaporation loss), and water requirements for assimilation of river pollution (i.e., dilution water demand). Comparison between various hydrology-based techniques, such as Tennant method, modified Tennant method, and environmental management class based environmental flow estimator, is also included in the chapter.

Water resources and catchment development activities often intercept natural flows in rivers and streams. The interception leads to the impairment of various day to day water-based activities and ecosystem functions. In order to satisfy environmental flow requirements, a minimum continuous water supply to the streams must be maintained, which helps in the sustenance of aquatic life and ecosystems, without being a burden to the development. Environmental flows also make a critical contribution to the river health, economic development and poverty alleviation. Such flows even facilitate groundwater recharge and ensure the continued availability of the many benefits that healthy river and groundwater systems bring to society (O'Keefe, 2000; Tharme and King, 1998; Tharme, 2003; Smakhtin et al., 2006). Environmental flow, besides being a lifeline of aquatic habitat, also helps in the reduction of water footprint, which is defined as the total volume of freshwater used by any individual or community to produce goods and services consumed by the individual or the community. Multiple services rendered by the environmental flow decline the water footprint through the enhancement of virtual water consumption.

12.1 Definition

The term environmental flows (EFs) are also termed with different names, such as "Environmental Water Requirement"; "Environmental Water Demand"; etc. On the other hand, EF should not be confused with some similar terminologies

R.K. Rai et al., *The Yamuna River Basin*, Water Science and Technology Library 66, DOI 10.1007/978-94-007-2001-5_12, © Springer Science+Business Media B.V. 2012

like “In-stream Flow Requirement (IFR)”, which mostly refers to the flow for fish; “Drought IFR”, which is a reduced flow in dry years to maintain aquatic species without assuring their reproduction; and “Minimum Flow” that merely limits the abstraction of flow in the dry season. However, the “Maintenance IFR” although not synonymous to environmental flow, is similar. It refers to the flow regime required for maintaining all river ecosystem functions. But unlike EF, the maintenance IFR does not assure socio-economic or hydrological benefits.

The terminology “Ecological Demand of Water”, too sometimes creates confusion. Being consistent with term Ecology, the primary focus of ecological demand most frequently remains the water requirement for ecosystem services only, and overlooks other important roles of water, like aquifer recharging, soil moisture maintenance, prevention of salinization in the estuary, the drift of silt material, evapotranspiration requirement, and more (Iyer, 2005).

Based on this discussion, the environmental flow, in general, should at least include the water required for the assimilation of pollutants, evaporation, groundwater conservation, and aquatic-habitat conservation.

12.2 Methods for Assessment of Environmental Flows

Ecologists agree that the major criteria for determining environmental flows should include the maintenance of both spatial and temporal patterns of river flow, i.e., the flow variability which affects the structural and functional diversity of rivers and their flood plains and which in turn influences the species diversity of the river (Richter et al., 1997; Poff et al., 1997; Dyson et al., 2003; Gordon et al., 2004; Ward and Tockner, 2001; Ward et al., 2001; Knights, 2002; Nature Conservancy, 2006). Bunn and Arthington (2002) have formulated four basic principles that emphasize the role of flow regime in structuring aquatic life and show the link between flow and ecosystem changes: (i) Flow is a major determinant of physical habitat in rivers and in turn is the major determinant of biotic composition. Therefore, river flow modifications eventually lead to changes in the composition and diversity of aquatic communities. (ii) Aquatic species have evolved life history strategies primarily in response to natural flow regimes. Therefore, flow regime alterations can lead to the loss of biodiversity of native species. (iii) Maintenance of natural patterns of longitudinal and lateral connectivity in river floodplains determines the ability of many aquatic species to move between the river and flood plain or between the main river and its tributaries. The loss of longitudinal and lateral connectivity can lead to local extinction of species. (iv) The invasion of exotic and introduced species in rivers is facilitated by the alteration of flow regimes. Inter-basin water transfer may represent a significant mechanism for the spread of exotic species.

For the estimation of environmental flows, several methodologies have been developed and used in different rivers. Depending on data requirements for calculation, these methodologies can be categorized into four major groups (Palau, 1994; Jowett, 1997; Alcaźar et al., 2008): (i) Hydrology-based methods: These are based on the study of historical flow regime records (Tennant, 1976; Richter et al., 1997; Hughes and Münster, 2000; Hughes and Hannart, 2003; Smakhtin and Anputhas, 2006; and Kashaigili et al., 2007). (ii) Hydraulic methods: These methods are based on the study of the hydraulic geometry of stream channels (Collings, 1972). (iii) Habitat methods: These are based on the simulation of physical habitats, such as the Instream Flow Incremental Methodology (Bovee, 1982). (iv) A fourth group of methodologies labeled "holistic" have appeared more recently as a response to a changing conception of water resources management arising from a more global vision of the riverine ecosystem (e.g., King and Louw, 1998). A brief description of these methodologies is summarized in Table 12.1.

Table 12.1 Various approaches for estimating environmental flows

S. No.	Methods	Description
1.	Look-up tables	Using the thumb rule on simple indices given in the look-up tables. This method is used when few or no local ecological data are available. In this method Q_{95} (i.e. the flow equaled or exceeded 95% of time) is considered as environmental flow. This method is purely based on river hydrology.
2.	Tennant method	This method was based on the index developed by calibrating the data from hundreds of rivers in the United States of America (USA). Percentages of mean annual flow (MAF) are specified that provide different quality habitats for fish; e.g. (i) 10% of MAF: poor quality (survival), (ii) 30% of MAF: moderate habitats (Satisfactory), (iii) 60% of MAF: excellent habitats.
3.	Range of variability approach	It is a desktop method, which identifies the components of a natural flow regime, indexed by magnitude (of both high and low flows), timing (indexed by monthly statistics), frequency (number of events) and duration (indexed by moving average minima and maxima).
4.	Hydraulic rating method	It is also a desktop method, which uses changes in hydraulic variable such as those in the wetted perimeter, and the area of riverbed submerged, to define environmental flows. These provide simple indices of available habitat in a river at a given discharge. This method provides simple indices of available habitat in a river at a given discharge and perimeter. More appropriate to support scenario based decision-making and water allocation negotiations.
5.	Building block method	It is based on the functional analysis of the system. The basic premise is that riverine species are dependent on basic elements (building blocks) of the flow regime, including low flows and floods that maintain the sediment dynamics and geomorphologic structure of the river. This method is flexible, robust, more focused on the whole ecosystem.
6.	Habitat analysis	This method is replicable and predictive; however, it is expensive to collect hydraulic and ecological data.

Despite wide applications of the hydrology-based methods for estimating environmental flows, such as Tennant method (Tennant, 1976), desktop reserve model (Hughes and Münster, 2000; Kashaigili et al., 2007), range of variability approach (Richter et al, 1997), and environmental management class (EMC) based environmental flow estimation (Smakhtin and Anputhas, 2006; Smakhtin and Eriyagama, 2008), do not account for the different components of environmental flows, and therefore the applicability of these methodologies has become limited when a river is highly intercepted, receives wastewater discharge, and has a lack of historical records. On the other hand, in the fourth category, a few studies have been undertaken (King and Louw, 1998, Yang et al., 2009). In the study by Yang et al. (2009), environmental flows for Yellow River were estimated, based on the classification and regionalization of ecosystems, multiple ecological management objectives, and spatial variability. This approach considered natural water consumption, such as water losses for evaporation and infiltration in the area above the section that fulfills environmental flow requirements of the terrestrial ecosystems. However, the water requirement for assimilating pollutants was not considered, which the major concern is when the river is used for pollutant transport also.

This chapter describes the application of two hydrology-based methods, such as the Tennant method (Tennant, 1976) and Environment Management Class based approach developed by Smakhtin and Anputhas (2006) and Smakhtin and Eriyagama (2008). Looking into the limitations of these methods for Indian River conditions, a comprehensive method has been developed and is presented here.

12.3 The Tennant Method

The Tennant method bases its stream flow requirements on the observation that aquatic-habitat conditions are similar in streams carrying the same proportion of the Mean Annual Flow (MAF). The method establishes stream flow requirements on the basis of a predetermined percentage of MAF (Tennant, 1976), and associates aquatic-habitat conditions with different percentages of MAF (Table 12.2). The Tennant method is less sensitive to summer water withdrawals than methods that use low flow statistics to determine stream flow requirements because the Tennant stream flow requirements are derived from the mean annual flow statistics, which is largely determined by high flows.

Minimum stream flows for small streams during summer were established by the Tennant method as 40, 30, and 10% of MAF, which represent good, fair, and poor habitat conditions, respectively. At the 30% MAF, most of the stream substrate is submerged, but at 10% MAF half or more of the stream substrate can be exposed (Tennant, 1976). The 10% MAF value is often used to determine minimum stream flow requirements in summer. To account for seasonal variability, Tennant (1976) established different stream flow requirements for summer and winter seasons. In the mountainous Western United States, where the tenant method was developed, precipitation patterns and snowmelt runoff typically result in low

Table 12.2 Relations between aquatic-habitat conditions and mean annual flow described by the Tennant method for small streams

Aquatic-habitat condition for small streams	% MAF (April–Sept) [for India: monsoon season (June–Sept)]	% MAF (Oct–March) [for India: non-monsoon season (Oct–May)]
Flushing flows	200	200
Optimum range	60–100	60–100
Outstanding	60	40
Excellent	50	30
Good	40	20
Fair	30	10
Poor	10	10
Severe degradation	<10	<10

stream flows in fall and early winter, and high stream flows in spring and summer. Therefore, the Tennant stream flow recommendations are higher in summer than in winter. Similar criteria can also be evaluated for Indian streams, where during the Monsoon (June to September) high flows occur and for the rest of the year low flow condition is achieved. This condition is almost similar to the mountainous Western United States and therefore can be adopted for Indian rivers. Considering river flow variability, the Tennant method was modified for Indian rivers by distributing the estimated environmental flow throughout the year considering similar monthly flow distribution patterns. For Indian conditions, the monthly distribution of 25% MAF may be considered as the ecological flows except for the highly polluted river conditions like the Wazirabad to Okhla barrage reach of the Yamuna River.

12.3.1 The Modified Tennant Method

Looking into the importance of the flow variability in the river system, the constant allowance for environmental flow based on the mean annual flow (MAF) will be not adequate for the Indian River system which has a large variation in the flow during the monsoon and non-monsoon periods. It leads to the revision in the methodology, in which the allowable estimated environmental flows should be temporally distributed considering the temporal variation of the stream flows.

12.3.2 Application of Modified Tennant Method

Using the Tennant method, environmental flows for different reaches of the Yamuna River have been estimated. The reaches considered for the estimation of environmental flows are:

(i) Origin to the Hathinikund barrage (Paonta),
(ii) Hathinikund barrage to the Wazirabad barrage (Kalanaur site),
(iii) Wazirabad barrage to Okhla barrage (Delhi Railway Bridge site, DRB),
(iv) Okhla barrage to the Agra discharge site (Agra discharge site), and
(v) Agra to the Allahabad at the confluence of Ganga River (Pratappur discharge site).

In the analysis, 10-daily discharge data for the period of 1989–2007 have been used. The first step of this methodology is to estimate the mean annual flow (MAF) of the respective reaches, and is given in Table 12.3 which also includes the allocation of environmental flow volume with respect to the different aquatic-habitat conditions. The second important step for applying the modified Tennant method is to derive the monthly flow distribution of MAF for the reach. This step is carried out by analyzing the monthly discharge data followed by the estimation of mean monthly flow. The estimated monthly distribution of MAF is presented in Table 12.4.

Using the appropriate monthly distribution (Table 12.4) with respect to the site, monthly volume of environmental flows at different aquatic-habitat conditions (using Table 12.3) can be obtained, and is given in Tables 12.5, 12.6, 12.7, 12.8, and 12.9. These tables also give the mean monthly e-flow requirements.

Since the Yamuna River water is highly regulated due to the upstream uses and Delhi water supply from the Wazirabad barrage, therefore the flow is no more virgin. Under such circumstances the first two condition (i.e., poor and fair classes cannot be considered). Therefore, to maintain the environmental flows in the Delhi Stretch (i.e. Wazirabad to Okhla reach), it is highly recommended to use the "Good Class (i.e., 40% MAF)" aquatic-habitat condition. The temporal flow volumes and mean discharges are shown in Figs. 12.1 and 12.2. However, in this computation the assimilation requirement for dilution of pollutants has been

Table 12.3 Estimated environmental flow allocation for different sites of Yamuna River using Tennant method

				Aquatic-habitat condition			
S. No.	Reach	Site name	MAF (MCM)	Poor: 10% MAF	Fair: 25% MAF	Fair: 30% MAF	Good: 40% MAF
1	Origin to Hathinikund	Paonta	4,818.0	481.2	1,203.0	1,443.6	1,924.8
2	Hathinikund to Wazirabad	Kalanaur	4,167.0	416.7	1,041.7	1,250.1	1,666.8
3	Wazirabad to Okhla	DRB	4,547.0	454.7	1,136.7	1,364.1	1,818.8
4	Okhla to Agra	Agra	4,225.0	422.5	1,056.2	1,267.5	1,690.0
5	Agra to Allahabad	Pratappur	51,556.0	5,156.1	12,890.2	15,468.3	20,624.4

Table 12.4 Monthly percent distribution of MAF for different sites/reaches of the Yamuna River

	Discharge sites of Yamuna				
Months	Paonta	Kalanaur	DRB	Agra	Pratappur
June	4.57	2.23	2.16	1.66	1.79
July	16.83	18.50	14.63	9.96	12.61
August	33.15	38.88	37.17	32.07	34.33
September	21.51	24.24	25.84	30.96	31.33
October	5.82	5.76	5.50	9.96	6.06
November	2.87	3.24	3.30	4.97	2.86
December	2.27	0.77	2.11	2.77	2.42
January	2.27	0.77	2.02	1.89	1.90
February	2.83	0.60	1.78	1.66	1.84
March	3.99	2.81	1.65	1.33	1.79
April	1.66	1.10	2.42	1.66	1.79
May	2.24	1.10	1.43	1.11	1.26

Table 12.5 E-flows using modified Tennant method for the origin to the Hathinikund reach

		Poor: 10%MAF		Fair: 25%MAF		Fair: 30%MAF		Good: 40%MAF	
Month	%Dist	MCM	m^3/s	MCM	m^3/s	MCM	m^3/s	MCM	m^3/s
Jun	4.57	22.00	8.49	55.00	21.22	66.00	25.46	88.00	33.95
Jul	16.83	81.00	30.24	202.50	75.60	243.00	90.73	324.00	120.97
Aug	33.15	159.50	59.55	398.75	148.88	478.50	178.65	638.00	238.20
Sep	21.51	103.50	39.93	258.75	99.83	310.50	119.79	414.00	159.72
Oct	5.82	28.00	10.45	70.00	26.14	84.00	31.36	112.00	41.82
Nov	2.87	13.80	5.32	34.50	13.31	41.40	15.97	55.20	21.30
Dec	2.27	10.90	4.07	27.25	10.17	32.70	12.21	43.60	16.28
Jan	2.27	10.90	4.07	27.25	10.17	32.70	12.21	43.60	16.28
Feb	2.83	13.60	5.62	34.00	14.05	40.80	16.87	54.40	22.49
Mar	3.99	19.20	7.17	48.00	17.92	57.60	21.51	76.80	28.67
Apr	1.66	8.00	3.09	20.00	7.72	24.00	9.26	32.00	12.35
May	2.24	10.80	4.03	27.00	10.08	32.40	12.10	43.20	16.13
Total	100.0	481.2		1,203.0		1,443.6		1,924.8	

not considered as this component is not in the scope of this method. The E-flow requirements for other reaches are graphically shown in Figs. 12.2, 12.3, 12.4, and 12.5.

12.4 EMC Based Environmental Flow Calculator

The estimation of environmental flows, based on the Smakhtin and Anputhas (2006) and Smakhtin and Eriyagama (2008), requires establishing the environmental management classes followed by the derivation of environmental flow duration

Table 12.6 E-flows using modified Tennant method for the Hathinikund to Wazirabad reach

		Poor: 10%MAF		Fair: 25%MAF		Fair: 30%MAF		Good: 40%MAF	
Month	%Dist	MCM	m^3/s	MCM	m^3/s	MCM	m^3/s	MCM	m^3/s
Jun	2.23	9.30	3.59	23.25	8.97	27.90	10.76	37.20	14.35
Jul	18.50	77.10	28.79	192.75	71.96	231.30	86.36	308.40	115.14
Aug	38.88	162.00	60.48	405.00	151.21	486.00	181.45	648.00	241.94
Sep	24.24	101.00	38.97	252.50	97.42	303.00	116.90	404.00	155.86
Oct	5.76	24.00	8.96	60.00	22.40	72.00	26.88	96.00	35.84
Nov	3.24	13.50	5.21	33.75	13.02	40.50	15.63	54.00	20.83
Dec	0.77	3.20	1.19	8.00	2.99	9.60	3.58	12.80	4.78
Jan	0.77	3.20	1.19	8.00	2.99	9.60	3.58	12.80	4.78
Feb	0.60	2.50	1.03	6.25	2.58	7.50	3.10	10.00	4.13
Mar	2.81	11.70	4.37	29.25	10.92	35.10	13.10	46.80	17.47
Apr	1.10	4.60	1.77	11.50	4.44	13.80	5.32	18.40	7.10
May	1.10	4.60	1.72	11.50	4.29	13.80	5.15	18.40	6.87
Total	100.0	416.7		1,041.7		1,250.1		1,666.8	

Table 12.7 E-flows using modified Tennant method for the Wazirabad to the Okhla reach

		Poor: 10%MAF		Fair: 30%MAF		Fair: 25%MAF		Good: 40%MAF	
Month	%Dist	MCM	m^3/s	MCM	m^3/s	MCM	m^3/s	MCM	m^3/s
Jun	2.16	9.80	3.78	29.40	11.34	24.50	9.45	39.20	15.12
Jul	14.63	66.50	24.83	199.50	74.48	166.25	62.07	266.00	99.31
Aug	37.17	169.00	63.10	507.00	189.29	422.50	157.74	676.00	252.39
Sep	25.84	117.50	45.33	352.50	136.00	293.75	113.33	470.00	181.33
Oct	5.50	25.00	9.33	75.00	28.00	62.50	23.33	100.00	37.34
Nov	3.30	15.00	5.79	45.00	17.36	37.50	14.47	60.00	23.15
Dec	2.11	9.60	3.58	28.80	10.75	24.00	8.96	38.40	14.34
Jan	2.02	9.20	3.43	27.60	10.30	23.00	8.59	36.80	13.74
Feb	1.78	8.10	3.35	24.30	10.04	20.25	8.37	32.40	13.39
Mar	1.65	7.50	2.80	22.50	8.40	18.75	7.00	30.00	11.20
Apr	2.42	11.00	4.24	33.00	12.73	27.50	10.61	44.00	16.98
May	1.43	6.50	2.43	19.50	7.28	16.25	6.07	26.00	9.71
Total	100	454.7		1,364.1		1,136.75		1,818.8	

curves (EFDC) and finally the environmental flow time series. Environmental flows aim to maintain an ecosystem in, or upgrade it to, some prescribed or negotiated condition, which is also referred to as "desired future state", "environmental management class (EMC)", "ecological management category" or level of environmental protection (DWAF, 1997; Aceman and Dunbar, 2004). The higher the EMC, the more water will need to be allocated for the ecosystem maintenance or conservation and more flow variability will need to be preserved.

Table 12.8 E-flows using modified Tennant method for the Okhla to the Agra site reach

		Poor: 10%MAF		Fair: 25%MAF		Fair: 30%MAF		Good: 40%MAF	
Month	%Dist	MCM	m^3/s	MCM	m^3/s	MCM	m^3/s	MCM	m^3/s
Jun	1.66	7.01	2.70	17.53	6.76	21.03	8.11	28.04	10.82
Jul	9.96	42.10	15.72	105.25	39.30	126.30	47.16	168.40	62.87
Aug	32.07	135.50	50.59	338.75	126.47	406.50	151.77	542.00	202.36
Sep	30.96	130.80	50.46	327.00	126.16	392.40	151.39	523.20	201.85
Oct	9.96	42.10	15.72	105.25	39.30	126.30	47.16	168.40	62.87
Nov	4.97	21.00	8.10	52.50	20.25	63.00	24.31	84.00	32.41
Dec	2.77	11.70	4.37	29.25	10.92	35.10	13.10	46.80	17.47
Jan	1.89	8.00	2.99	20.00	7.47	24.00	8.96	32.00	11.95
Feb	1.66	7.01	2.90	17.53	7.24	21.03	8.69	28.04	11.59
Mar	1.33	5.60	2.09	14.00	5.23	16.80	6.27	22.40	8.36
Apr	1.66	7.01	2.70	17.53	6.76	21.03	8.11	28.04	10.82
May	1.11	4.67	1.74	11.68	4.36	14.01	5.23	18.68	6.97
Total	100.0	4,222.5		1,056.2		1,267.5		1,690.0	

Ideally, these classes should be based on empirical relationships between flow and ecological status/conditions associated with clearly identifiable thresholds. However, at present the evidence for such thresholds is not sufficient and these categories are simply a management concept, which has been developed and used in the world because of the need to make decisions under conditions of limited knowledge. Placing a river into certain EMC is often accomplished by expert judgment using the scoring system (DWAF, 1997; Environmental Agency, 2001). Alternatively, EMCs may be seen as default "scenarios" of environmental protection and the corresponding EF as "scenarios" of environmental water demand.

Smakhtin and Anputhas (2006) defined six EMCs corresponding to the default levels of environmental flows (Table 12.10). A river which falls into classes C to F would normally be present in densely populated areas with multiple men induced impacts. Poor ecosystem conditions (Class E and F) are sometimes not considered acceptable from the management perspective and the management intention is always to "move" such rivers up to the least acceptable class D through rehabilitation measures. It can be noted that the ecosystems in class F are likely to be those which have been modified beyond rehabilitation to anything resembling a natural condition.

In this approach, the environmental flow duration curve is derived for different EMC classes from the reference flow duration curve, and finally the monthly time series spatial interpolation technique is used (Smakhtin and Anputhas, 2006; and Hughes and Smakhtin, 1996). In this manner, the general pattern of flow variability is preserved, although with every shift, part of the variability is lost. The procedure of deriving the environmental flow duration curve followed by the monthly time series of e-flows is presented in the following section.

Table 12.9 E-flows using modified Tennant method for the Agra to the Allahabad reach

Month	% Dist	Poor: 10%MAF		Fair: 25%MAF		Fair: 30%MAF		Good: 40%MAF	
		MCM	m^3/s	MCM	m^3/s	MCM	m^3/s	MCM	m^3/s
Jun	1.79	92.51	35.69	231.27	89.23	277.53	107.07	370.04	142.76
Jul	12.61	650.06	242.71	1,625.16	606.76	1,950.19	728.12	2,600.25	970.82
Aug	34.33	1,770.17	660.91	4,425.43	1,652.27	5,310.51	1,982.72	7,080.69	2,643.63
Sep	31.33	1,615.16	623.13	4,037.89	1,557.83	4,845.47	1,869.39	6,460.63	2,492.53
Oct	6.06	312.53	116.69	781.33	291.71	937.59	350.06	1,250.12	466.74
Nov	2.86	147.51	56.91	368.79	142.28	442.54	170.73	590.06	227.65
Dec	2.42	125.01	46.67	312.53	116.69	375.04	140.02	500.05	186.70
Jan	1.90	98.11	36.63	245.27	91.57	294.33	109.89	392.44	146.52
Feb	1.84	95.01	39.27	237.52	98.18	285.03	117.82	380.04	157.09
Mar	1.79	92.51	34.54	231.27	86.35	277.53	103.62	370.04	138.16
Apr	1.79	92.51	35.69	231.27	89.23	277.53	107.07	370.04	142.76
May	1.26	65.01	24.27	162.52	60.68	195.02	72.81	260.03	97.08
Total	100.0	5,156.1		12,890.2		15,468.3		20,624.4	

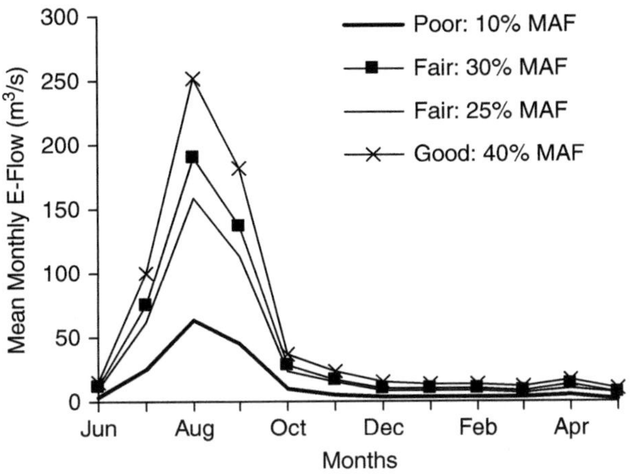

Fig. 12.1 Mean monthly e-flow rates for the Wazirabad-Okhla reach

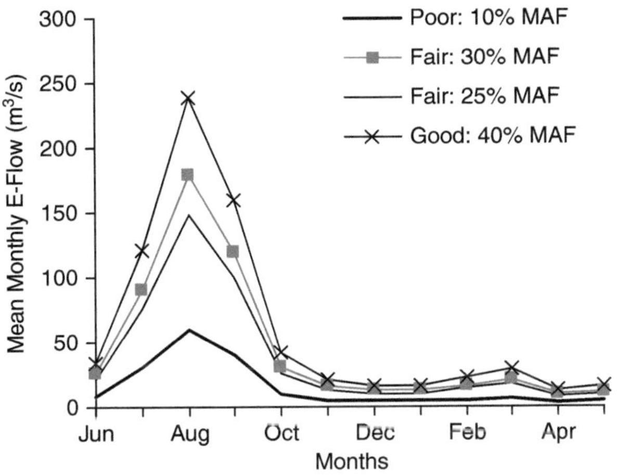

Fig. 12.2 Mean monthly e-flow rates for the Origin-Hathinikund reach

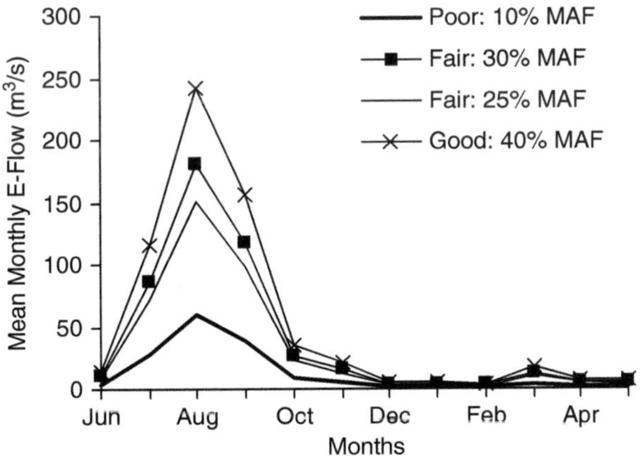

Fig. 12.3 Mean monthly e-flow rates for the Hathinikund-Wazirabad reach

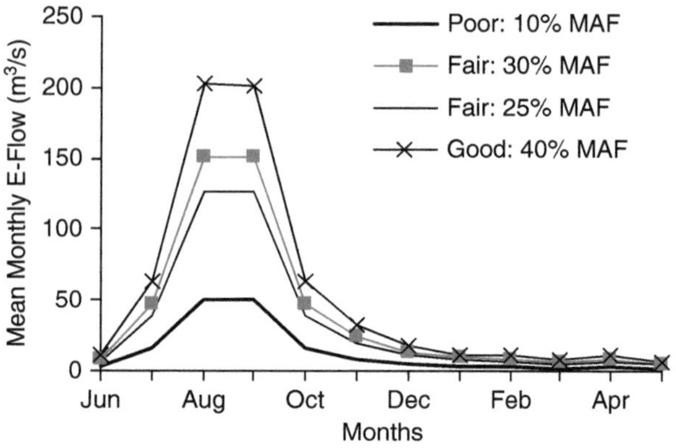

Fig. 12.4 Mean monthly e-flow rates for the Okhla-Agra reach

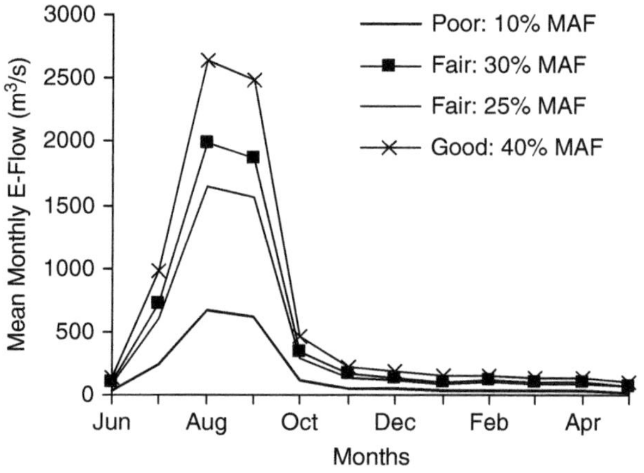

Fig. 12.5 Mean monthly e-flow rates for the Agra-Pratappur reach

12.4.1 Establishing Environmental Flow Duration Curve

A simple approach has been proposed by Smakhtin and Anputhas (2006) to determine the default FDC representing the summary of EF for each EMC. In this approach, the original reference FDC is represented by table of flow values covering the entire range of probabilities of occurrence. For representing the FDC, 17 fixed percentage points: 0.01, 0.1, 1, 5, 10, 20, 30, 40, 50, 60, 70, 80, 90, 95, 99, 99.9, and 99.99% are considered. These points adequately cover the entire range of flow regime. These percentage points are used to determine the environmental flow duration curve (E-FDC) corresponding to the EMC by lateral shifting of the original reference FDC to the left along the probability axis.

An FDC shift by one step means that a flow which is exceeded 99.99% of the time in the original FDC will now be exceeded 99.9% of the time; the flow at 99.9% becomes the flow at 99%; the flow at 99% becomes the flow at 95%; etc. The procedure for deriving the E-FDC in graphical and tabular is shown in Fig. 12.6 and Table 12.11, respectively.

Table 12.10 Environmental Management Classes (EMCs)

EMC	Status	Ecological description	Management perspective	Default FDC shift limit
A	Natural	Pristine condition or minor modification of in-stream and riparian habitat	• Protected river and basins. • Reserves and national parks. • No new water projects (dam, diversions, etc.) allowed	Lateral shift of reference FDC one percentage point to the left along the time axis from the original FDC position
B	Slightly modified	Largely intact biodiversity and habitats despite water resources development and/or basin modifications	Water supply schemes or irrigation development present and/or allowed	Lateral shift of reference FDC one percentage point to the left along the time axis from the position of the FDC for A class
C	Moderately modified	The habitat and dynamics of the biota have been disturbed, but basic ecosystem functions are still intact. Some sensitive species are lost and/or reduced in extent. Alien species present.	Multiple disturbances associated with the need for socio-economic development, e.g., dams, diversion, habitat modification and reduced water supply	Lateral shift of reference FDC one more percentage point to the left along the time axis from the position of the FDC for B class
D	Largely modified	• Large changes in natural habitat, biota and basic ecosystem functions have occurred. • A clearly lower than expected species richness. • Much lowered presence of intolerant species. • Alien species prevail	Significant and clearly visible disturbances associated with basin and water resources development, including dams, diversions, transfers, habitat modification and water quality degradation	Lateral shift of reference FDC one more percentage point to the left along the time axis from the position of the FDC for C class

Table 12.10 (continued)

EMC	Status	Ecological description	Management perspective	Default FDC shift limit
E	Seriously modified	• Habitat diversity and availability have declined. • A strikingly lower than expected species richness. • Only tolerant species remain. • Indigenous species can no longer breed. • Alien species.	High human population density and extensive water resources exploitation	Lateral shift of reference FDC one more percentage point to the left along the time axis from the position of the FDC for D class
F	Critically modified	• Modifications have reached a critical level and ecosystem has been completely modified with almost total loss of natural habitat and biota. • In the worst case, the basic ecosystem functions have been destroyed and the changes are irreversible.	This status is not acceptable from the management perspective. Management interventions are necessary to restore flow pattern, river habitats, etc. (if still possible/feasible) – to "move" a river to a higher management category.	Lateral shift of reference FDC one more percentage point to the left along the time axis from the position of the FDC for E class

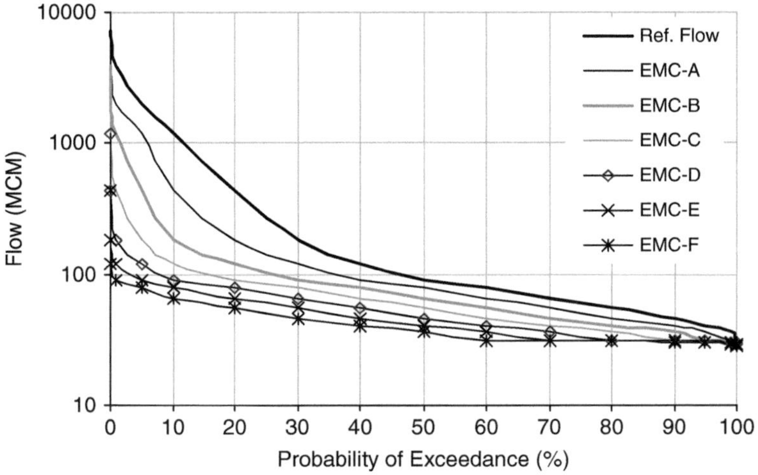

Fig. 12.6 An example E-FDCs for Delhi Railway Bridge site of Yamuna River

Table 12.11 Computational procedure of environmental FDC (E-FDC)

%	Reference FDC	EMC-A	EMC-B	EMC-C	EMC-D	EMC-E	EMC-F
0.01	7,148	7,148	3,910	2,018	1,204	431	180
0.1	7,148	3,910	2,018	1,204	431	180	120
1	3,910	2,018	1,204	431	180	120	89.9
5	2,018	1,204	431	180	120	89.9	77.9
10	1,204	431	180	120	89.9	77.9	64.6
20	431	180	120	89.9	77.9	64.6	55.5
30	180	120	89.9	77.9	64.6	55.5	45.7
40	120	89.9	77.9	64.6	55.5	45.7	40.1
50	89.9	77.9	64.6	55.5	45.7	40.1	36.2
60	77.9	64.6	55.5	45.7	40.1	36.2	31.4
70	64.6	55.5	45.7	40.1	36.2	31.4	31
80	55.5	45.7	40.1	36.2	31.4	31	30.5
90	45.7	40.1	36.2	31.4	31	30.5	30
95	40.1	36.2	31.4	31	30.5	30	29.6
99	36.2	31.4	31	30.5	30	29.6	29.1
99.9	31.4	31	30.5	30	29.6	29.1	28.7
99.99	31	30.5	30	29.6	29.1	28.7	28.3

12.4.2 Simulating Continuous Monthly Time Series of Environmental Flows

An environmental FDC, discussed above, for any EMC gives only a summary of the EF regime acceptable for this EMC. However, one such curve is determined; it is also possible to convert it into actual environmental monthly flow time series.

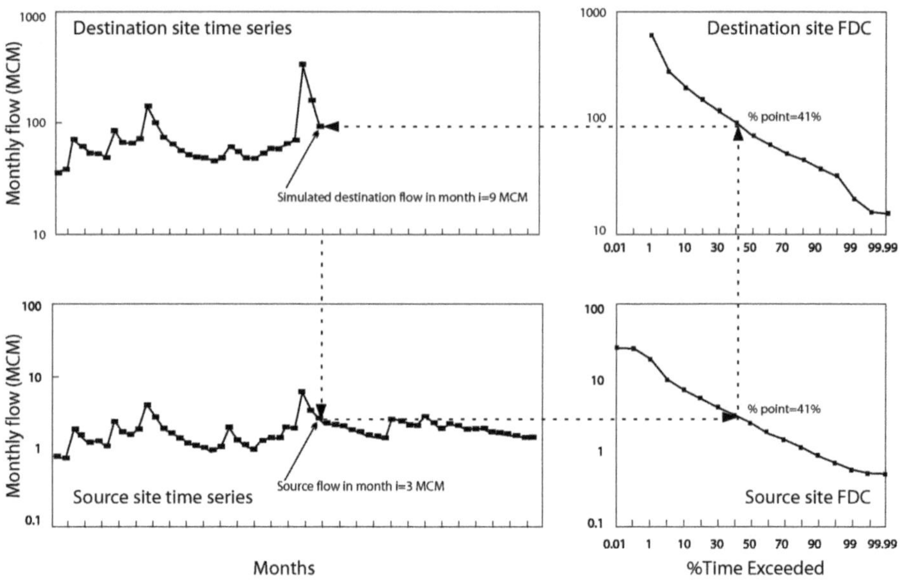

Fig. 12.7 Spatial interpolation technique (Hughes and Smakhtin, 1996)

To accomplish this, a spatial interpolation technique proposed by Hughes and Smakhtin (1996) can be used. The underlying principle in this technique is that flows occurring simultaneously at sites in reasonable close proximity to each other correspond to similar percentage points on their respective FDCs. The site at which stream flow time series is generated is referred as destination site. The site with available time series, which is used for the generation, is called a "source site". In essence, the procedure is to transfer the streamflow time series from the location where the data are available to the destination site. The destination FDC is the one representing the EF sequence to be generated, while the source FDC and time series are those representing the reference natural flow regime at the same site. The procedure is therefore, to generate the modified time series (environmental flow time series) based on the linear transformation through two sets of FDCs (i.e. reference and E-FDC) corresponding to the reference time series. The procedure can be further elaborated through the following steps (Fig. 12.7):

(i) Identify the percentage point position on the reference FDC corresponding to the natural stream flow.
(ii) Using the identified percentage point, estimate the stream flow volume from the environmental FDC using the linear transformation.
(iii) Convert the stream flow volume as monthly stream flow discharge.
(iv) The final discharge value calculated, in the above steps, will be the monthly discharge corresponding to the time of natural stream flow time series. This results in the monthly time series of the environmental flow at particular EMC.

The generated environmental flow time series can be suitably interpreted by different specialists, like aquatic ecologist, hydrologist, etc., for making decisions.

12.4.3 Application of the Methodology

To implement the technique, 10-daily data from various sites at Yamuna River for the period of 1989–2007 are first converted into the monthly flow data, and used to calculate the flow duration curve followed by the environmental flow duration curve and the monthly environmental flow time series. The sample calculation of the E-FDC for the Paonta Saheb is given in Table 12.12. The derived monthly flow time series, so derived following the aforesaid procedure, is analyzed to derive the allocation rule for monthly environmental flows.

Using the discharge data of different sites, the environmental flows is estimated for all the considered reaches of the Yamuna River. Site wise results are summarized in Tables 12.12, 12.13, 12.14, 12.15, 12.16, and 12.17) as well as in the graphical form through Figs. 12.8 through 12.22 for various stretches of the Yamuna River (i.e. Sections 12.4.3.1, 12.4.3.2, 12.4.3.3, 12.4.3.4 and 12.4.3.5).

Table 12.12 Reference and environmental FDCs for different environment management classes (EMCs) for Paonta Saheb

%	Ref Flow	EMC-A	EMC-B	EMC-C	EMC-D	EMC-E	EMC-F
0.01	2,859	2,859	2,418	1,662	1,308	677	387
0.1	2,859	2,418	1,662	1,308	677	387	224
1	2,418	1,662	1,308	677	387	224	144
5	1,662	1,308	677	387	224	144	94.2
10	1,308	677	387	224	144	94.2	61.8
20	677	387	224	144	94.2	61.8	44.1
30	387	224	144	94.2	61.8	44.1	33.1
40	224	144	94.2	61.8	44.1	33.1	28.6
50	144	94.2	61.8	44.1	33.1	28.6	23.7
60	94.2	61.8	44.1	33.1	28.6	23.7	19.6
70	61.8	44.1	33.1	28.6	23.7	19.6	19.2
80	44.1	33.1	28.6	23.7	19.6	19.2	18.8
90	33.1	28.6	23.7	19.6	19.2	18.8	18.4
95	28.6	23.7	19.6	19.2	18.8	18.4	18
99	23.7	19.6	19.2	18.8	18.4	18	17.6
99.9	19.6	19.2	18.8	18.4	18	17.6	17.3
99.99	19.2	18.8	18.4	18	17.6	17.3	16.9

Table 12.13 Monthly distribution of environmental flows for Paonta site of the Yamuna

	Mean Flow		EMC-A		EMC-B		EMC-C		EMC-D	
Month	MCM	m^3/s	MCM	m^3/s	MCM	m^3/s	MCM	m^3/s	MCM	m^3/s
Jun	220.0	84.9	142.0	54.8	82.0	31.6	48.0	18.5	42.0	16.2
Jul	810.0	302.4	505.0	188.5	278.0	103.8	179.0	66.8	112.0	41.8
Aug	1,595.0	595.5	1,100.0	410.7	685.0	255.7	398.0	148.6	230.0	85.9
Sep	1,035.0	399.3	631.0	243.4	392.0	151.2	235.0	90.7	140.0	54.0
Oct	280.0	104.5	160.0	59.7	97.0	36.2	71.0	26.5	50.0	18.7
Nov	138.0	53.2	93.0	35.9	71.0	27.4	41.0	15.8	33.0	12.7
Dec	109.0	40.7	71.0	26.5	56.0	20.9	41.0	15.3	28.0	10.4
Jan	109.0	40.7	71.0	26.5	56.0	20.9	41.0	15.3	28.0	10.4
Feb	136.0	56.2	93.0	38.4	71.0	29.3	41.0	16.9	33.0	13.6
Mar	192.0	71.7	117.0	43.7	83.0	30.9	56.0	20.9	36.0	13.4
Apr	80.0	30.8	71.0	27.4	41.0	15.8	28.0	10.8	28.0	10.8
May	108.0	40.3	71.0	26.5	41.0	15.3	28.0	10.4	28.0	10.4

Table 12.14 Monthly distributions of e-flows for the Hathinikund barrage to Wazirabad barrage reach

	Mean flow		EMC-A		EMC-B		EMC-C		EMC-D	
Month	MCM	m^3/s	MCM	m^3/s	MCM	m^3/s	MCM	m^3/s	MCM	m^3/s
Jun	93	35.88	56.00	21.60	31.00	11.96	19.00	7.33	13.00	5.02
Jul	771	287.86	411.00	153.45	161.00	60.11	75.00	28.00	39.00	14.56
Aug	1,620	604.84	1,028.00	383.81	537.00	200.49	238.00	88.86	96.00	35.84
Sep	1,010	389.66	501.00	193.29	225.00	86.81	95.00	36.65	48.00	18.52
Oct	240	89.61	150.00	56.00	102.00	38.08	49.00	18.29	27.00	10.08
Nov	135	52.08	57.00	21.99	32.00	12.35	20.00	7.72	16.00	6.17
Dec	32	11.95	18.00	6.72	16.00	5.97	12.00	4.48	11.00	4.11
Jan	32	11.95	18.00	6.72	16.00	5.97	12.00	4.48	11.00	4.11
Feb	25	10.33	18.00	7.44	16.00	6.61	12.00	4.96	15.00	6.20
Mar	117	43.68	56.00	20.91	38.00	14.19	22.00	8.21	15.00	5.60
Apr	46	17.75	37.00	14.27	18.00	6.94	12.00	4.63	11.00	4.24
May	46	17.17	37.00	13.81	18.00	6.72	12.00	4.48	11.00	4.11

Table 12.15 Monthly distribution of environmental flows for DRB site of the Yamuna

	Mean Flow		EMC-A		EMC-B		EMC-C		EMC-D	
Month	MCM	m^3/s	MCM	m^3/s	MCM	m^3/s	MCM	m^3/s	MCM	m^3/s
Jun	98	37.81	65	25.08	56	21.60	42	16.20	42	16.20
Jul	665	248.28	316	117.98	131	48.91	92	34.35	68	25.39
Aug	1,690	630.97	976	364.40	486	181.45	210	78.41	118	44.06
Sep	1,175	453.32	560	216.05	299	115.35	160	61.73	100	38.58
Oct	250	93.34	140	52.27	112	41.82	75	28.00	62	23.15
Nov	150	57.87	125	48.23	56	21.60	50	19.29	46	17.75
Dec	96	35.84	65	24.27	56	20.91	50	18.67	41	15.31
Jan	92	34.35	65	24.27	48	17.92	42	15.68	39	14.56
Feb	81	33.48	65	26.87	47	19.43	41	16.95	38	15.71
Mar	75	28.00	63	23.52	56	20.91	54	20.16	49	18.29
Apr	110	42.44	91	35.11	68	26.23	52	20.06	42	16.20
May	65	24.27	62	23.15	52	19.41	50	18.67	42	15.68

Table 12.16 Monthly distribution of environmental flows for the Okhla to Agra reach of the Yamuna

	Mean flow		EMC-A		EMC-B		EMC-C		EMC-D	
Month	MCM	m^3/s	MCM	m^3/s	MCM	m^3/s	MCM	m^3/s	MCM	m^3/s
Jun	70.10	27.04	42.1	16.24	25.2	9.72	15.8	6.10	15.0	5.79
Jul	421.00	157.18	196.0	73.18	109.0	40.70	65.0	24.27	48.0	17.92
Aug	1,355.00	505.90	799.0	298.31	404.0	150.84	200.0	74.67	112.0	41.82
Sep	1,308.00	504.63	743.0	286.65	379.0	146.22	190.0	73.30	110.0	42.44
Oct	421.00	157.18	238.0	88.86	135.0	50.40	82.0	30.62	58.0	21.65
Nov	210.00	81.02	126.0	48.61	75.7	29.21	50.5	19.48	35.0	13.50
Dec	117.00	43.68	70.0	26.14	50.5	18.85	40.0	14.93	26.0	9.71
Jan	80.00	29.87	69.0	25.76	43.1	16.09	32.0	11.95	24.0	8.96
Feb	70.10	28.98	56.0	23.15	42.1	17.40	31.0	12.81	22.0	9.09
Mar	56.00	20.91	46.0	17.17	36.0	13.44	30.0	11.20	20.0	7.47
Apr	70.10	27.04	40.0	15.43	33.4	12.89	24.0	9.26	16.0	6.17
May	46.70	17.44	24.0	8.96	24.0	8.96	16.8	6.27	11.0	4.11

Table 12.17 Monthly distributions of environmental flows for the Agra to Allahabad reach of the Yamuna

	Mean flow		EMC-A		EMC-B		EMC-C		EMC-D	
Month	MCM	m^3/s	MCM	m^3/s	MCM	m^3/s	MCM	m^3/s	MCM	m^3/s
Jun	925	356.87	561	216.44	502	193.67	421	162.42	421	162.42
Jul	6,500	2,426.82	3,095	1,155.54	1,499	559.66	1,121	418.53	883	329.67
Aug	17,700	6,608.42	10,374	3,873.21	5,060	1,889.19	2,243	837.44	1,346	502.54
Sep	16,150	6,230.71	9,857	3,802.85	5,020	1,936.73	2,523	973.38	1,388	535.49
Oct	3,125	1,166.74	1,710	638.44	1,121	418.53	981	366.26	715	266.95
Nov	1,475	569.06	1,121	432.48	940	362.65	694	267.75	648	250.00
Dec	1,250	466.70	841	313.99	748	279.27	561	209.45	470	175.48
Jan	981	366.26	841	313.99	748	279.27	561	209.45	463	172.86
Feb	950	392.69	561	231.89	561	231.89	561	231.89	421	174.02
Mar	925	345.36	561	209.45	561	209.45	421	157.18	421	157.18
Apr	925	356.87	561	216.44	561	216.44	421	162.42	411	158.56
May	650	242.68	561	209.45	460	171.74	379	141.50	379	141.50

12.4.3.1 Paonta Saheb (Origin to Hathinikund reach)

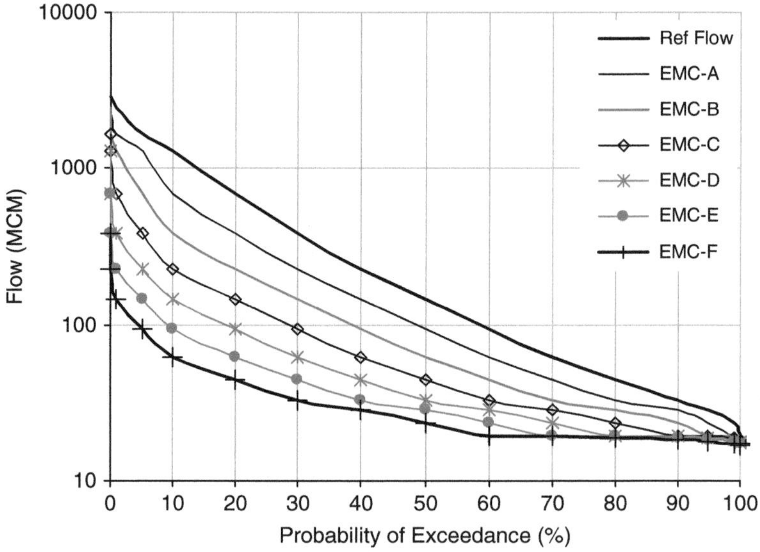

Fig. 12.8 Reference and environmental FDCs for Paonta site at the Yamuna

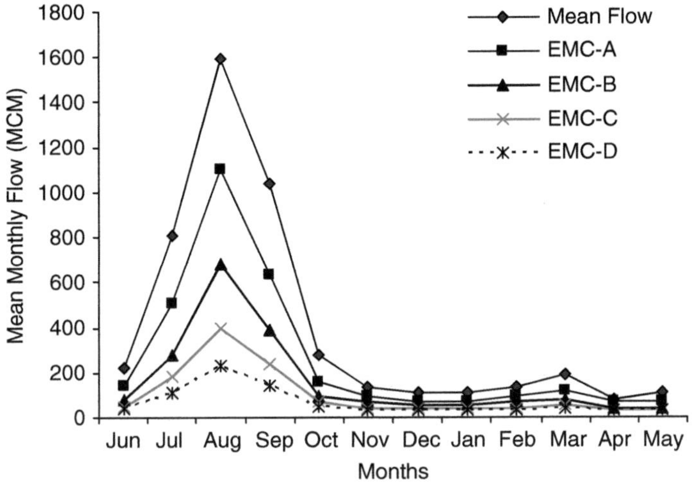

Fig. 12.9 Mean monthly allocation of e-flow volume for different management classes for the origin to Hathinikund barrage reach

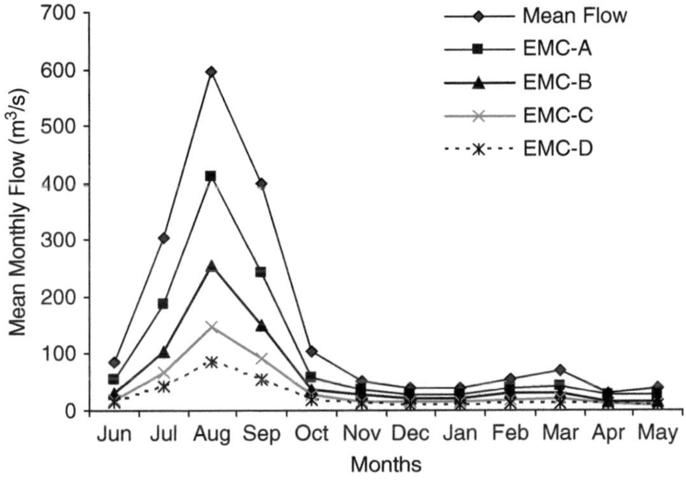

Fig. 12.10 Mean monthly environmental flows for different management classes for the origin to Hathinikund barrage reach

12.4.3.2 Kalanaur Site (Hathinikund Barrage to Wazirabad Barrage Reach)

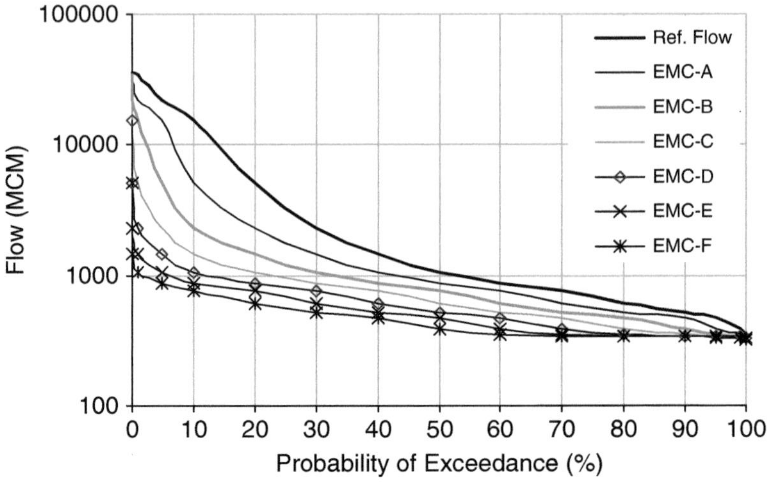

Fig. 12.11 Reference and environmental FDCs for Kalanaur site at the Yamuna

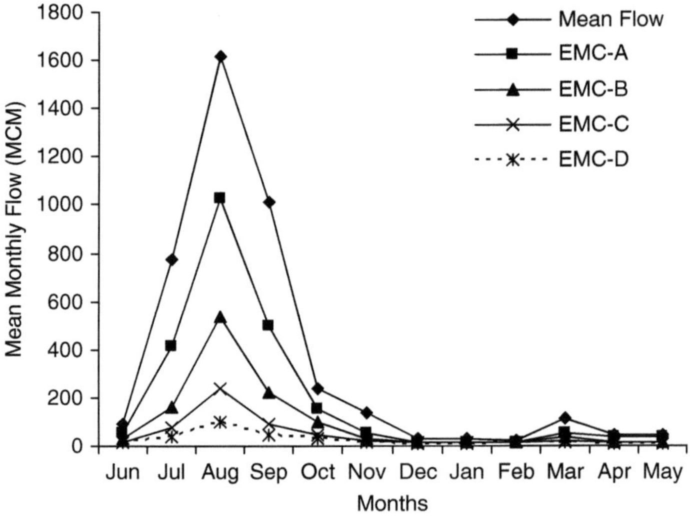

Fig. 12.12 Mean monthly allocation of environmental flow volume for different management classes for the Hathinikund barrage to Wazirabad barrage reach

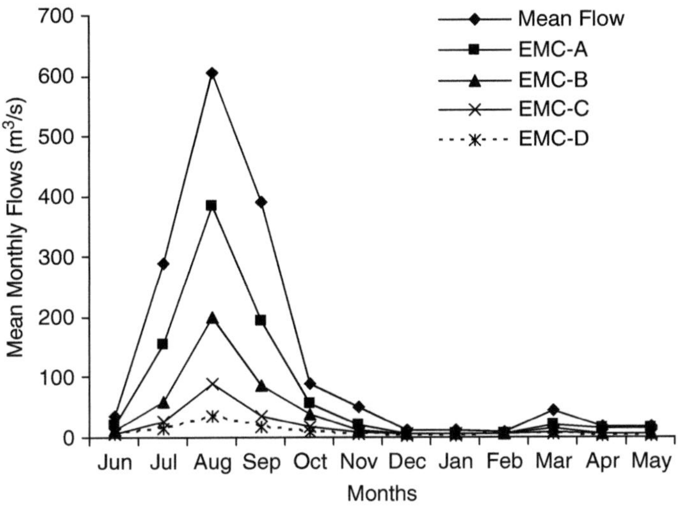

Fig. 12.13 Mean monthly environmental flows for different management classes for the Hathinikund barrage to Wazirabad barrage reach

12.4.3.3 Delhi Railway Bridge, DRB (Wazirabad to Okhla Reach)

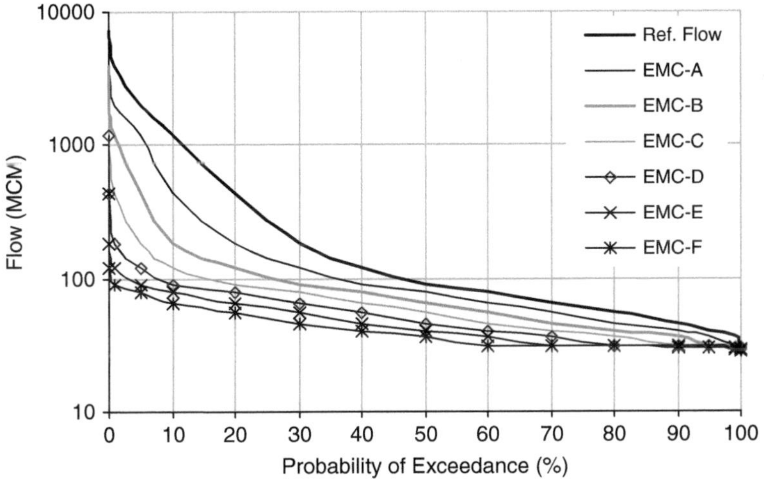

Fig. 12.14 Reference and Environmental FDCs for DRB site of Yamuna

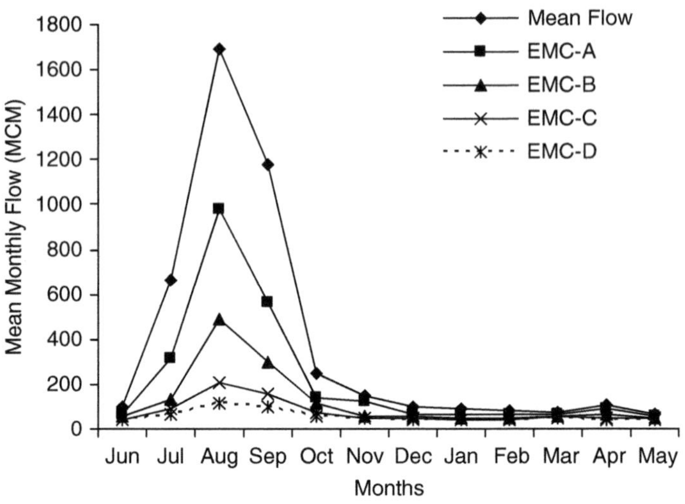

Fig. 12.15 Mean monthly allocation of environmental flow volume for different management classes for the Wazirabad to Okhla reach

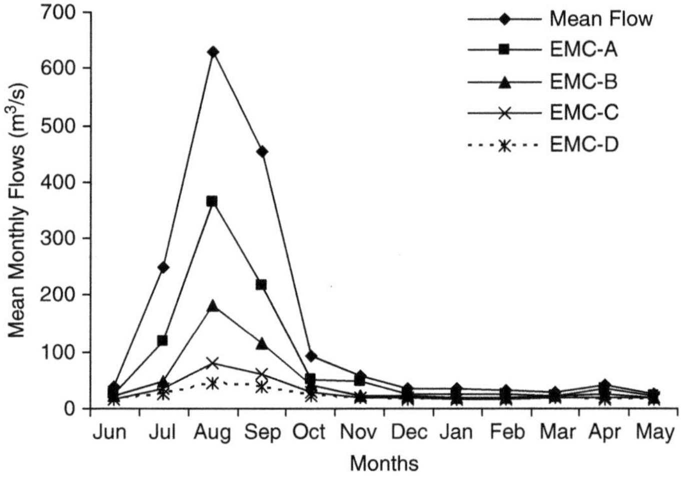

Fig. 12.16 Mean monthly environmental flows for different management classes for the Wazirabad to Okhla reach

12.4.3.4 Agra site (Okhla to Agra Reach)

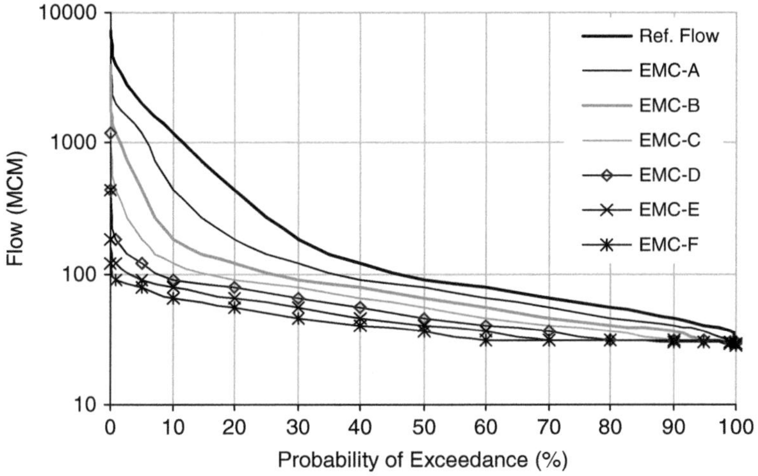

Fig. 12.17 Reference and Environmental FDCs for Agra site at the Yamuna

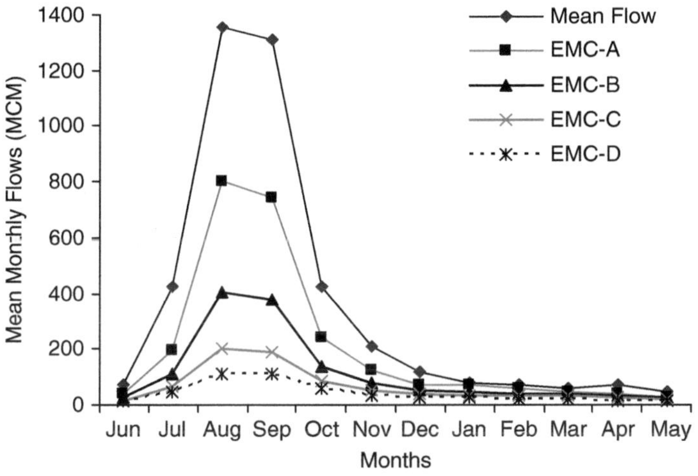

Fig. 12.18 Mean monthly allocation of environmental flow volume for different management classes for the Okhla to Agra reach

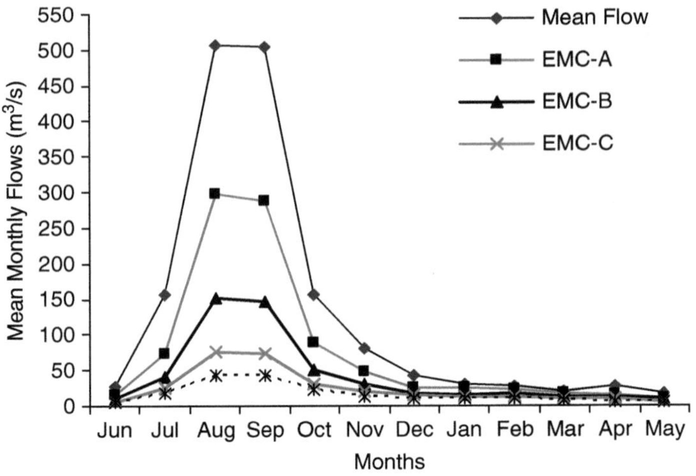

Fig. 12.19 Mean monthly environmental flows for different management classes for the Okhla to Agra reach

12.4.3.5 Pratappur Site (Agra to Allahabad Reach)

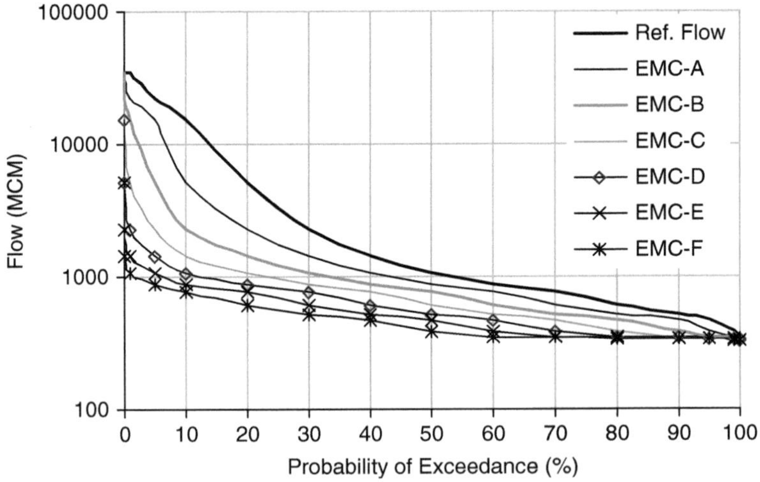

Fig. 12.20 Reference and Environmental FDCs for Pratappur site at the Yamuna

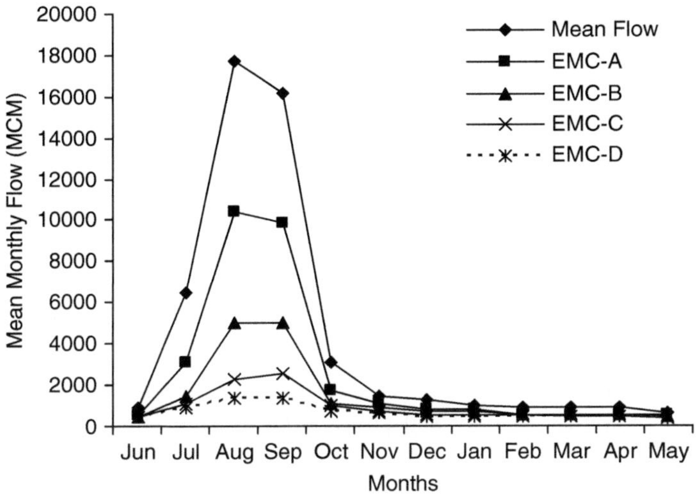

Fig. 12.21 Mean monthly allocation of environmental flow volume for different management classes for the Agra to Allahabad reach

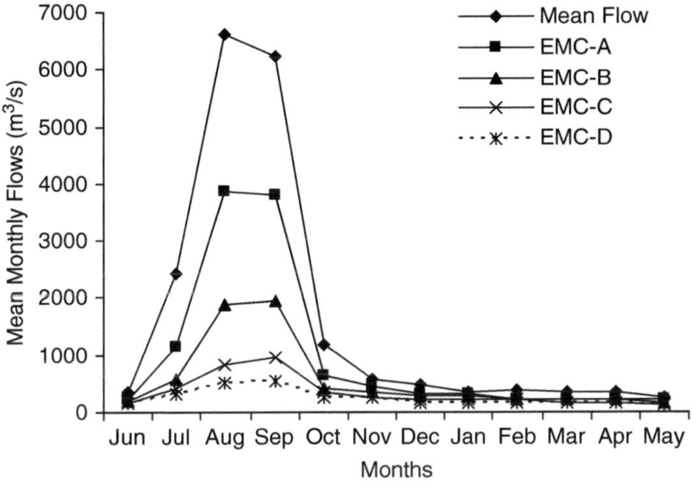

Fig. 12.22 Mean monthly environmental flows for different management classes for the Agra to Allahabad reach

The above two methods has the limitation that they are completely based on the hydrology, and do not consider the pollution, interception of the river, groundwater conservation, etc. Looking into these limitations a comprehensive method has been developed for the Indian Rivers which are highly intercepted and being used for pollution transport. The following section gives the detailed investigation of the methodology.

12.5 Development of Component Based Environmental Flow Estimator

Incorporating water demands for the assimilation of polluted river water, evaporation, conservation of groundwater, and ecological requirements, the proposed technique requires component-wise development and analysis for the river system. The component-wise methodology is now described.

12.5.1 Water Requirement for Assimilation of Polluted River Water

The assimilation capacity of a river is defined as its capacity to digest pollutants by means of biological activity and physical purification, both of which depend on the use of the water body and the quality standards adopted by the management agency (Lee and Wen, 1996). Calculation of the quantity of water needed for assimilation of polluted river water requires calculation of the inverse of assimilative capacity; that is, it represents the minimum quantity of water needed to permit self-purification and dilution of pollutants. A mathematical expression for the quantity of water needed for the assimilation of polluted water in a river can be derived as follows.

Let Q_0^* be the flow rate from the upstream (m^3/s), C_0 be the concentration of pollutants from the upstream (mg/l), q_1 be the flow rate from pollutant sources (m^3/s), C_1 be the concentration of pollutants from pollutant sources (mg/l), Q_1 be the flow rate released from the environmental reservoir (m^3/s), C be the concentration of pollutants from the environmental reservoir (mg/l), C_N be the standard for water quality (mg/l), k be the degradation coefficient (1/day), x_1 be the length of river stretch, and u be the average velocity of flow (m/s) (Fig. 12.23).

The intermixing concentration of pollutants, L_0, can be calculated as:

$$L_0 = \left(\frac{q_1 C_1 + Q_0^* C_0}{q_1 + Q_0^*} \right) \tag{12.1}$$

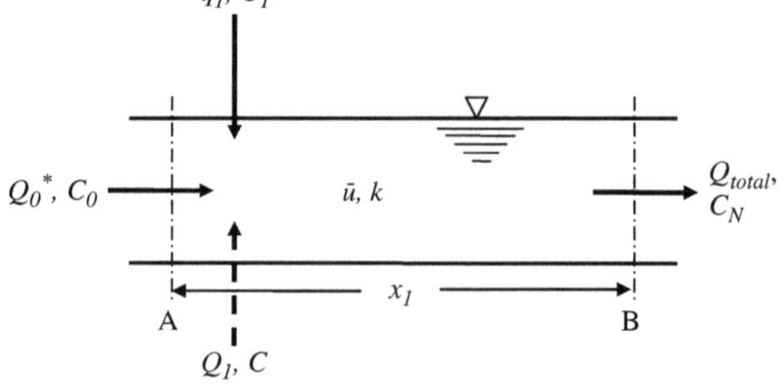

Fig. 12.23 Components of continuity equation in a river stretch under control volume

If k is the degradation coefficient of the pollutant, then the pollutant concentration of environmental reservoir flow and the weighted concentration of intermixed pollutants after time t can be expressed as:

$$C_t = C \exp(-kt) \tag{12.2}$$

and

$$L_t = L_0 \exp(-kt) \tag{12.3}$$

where

$$t = x_1/\bar{u} \tag{12.4}$$

The mass conservation equation after time t and distance x_1 can be expressed as:

$$Q_1 C_t + L_t \left(q_1 + Q_0^*\right) = Q_{\text{total}} C_N \tag{12.5}$$

where

$$Q_{\text{total}} = Q_1 + q_1 + Q_0^* \tag{12.6}$$

Using eqs. (12.2) through (12.6), eq. (12.5) can be recast as:

$$\begin{aligned} &Q_1 C \exp(-kt) + \left(\frac{q_1 C_1 + Q_0^* C_0}{q_1 + Q_0^*}\right)\left(q_1 + Q_0^*\right) \exp(-kt) \\ &= \left(Q_1 + q_1 + Q_0^*\right) C_N \end{aligned} \tag{12.7}$$

Simplification of eq. (12.7) yields

$$Q_1 = \frac{\left(q_1 C_1 + Q_0^* C_0\right) \exp(-kt) - \left(q_1 + Q_0^*\right) C_N}{C_N - C \exp(-kt)} \tag{12.8}$$

Substituting eq. (12.4) into eq. (12.8) yields

$$Q_1 = \frac{\left(q_1 C_1 + Q_0^* C_0\right) \exp(-k \cdot x_1/\bar{u}) - \left(q_1 + Q_0^*\right) C_N}{C_N - C \exp(-kt)} \tag{12.9}$$

Equation (12.9) gives the water required for assimilation of pollutants in the river stretch.

12.5.2 Water Required for Evaporation

Evaporation losses are an important part of a river's environmental water demand, especially during summer. Evaporation decreases the quantity of river water without greatly affecting the quantity of pollutants in the river. The evaporative loss of the river can be estimated using the following formula:

$$Q_2 = \begin{cases} 10^{-3}\, T \cdot L_C \cdot (E - P)\,, & E > P \\ 0, & E \leq P \end{cases} \tag{12.10}$$

where Q_2 = the water demand created by evaporation (m^3), T = the average top width of river, which depends upon the depth of flow (m), L_C = the length of river stretch (m), P = the monthly rainfall (mm), and E = the monthly evaporation (mm). Evaporation from the water surface can be estimated using the Penman (1948) method, whereas the monthly normal values are used for monthly rainfall.

12.5.2.1 Estimation of Evaporation

The Penman method (Penman, 1948), a well-known combination equation (i.e., combination of an energy balance and an aerodynamic formula) can be expressed as:

$$\lambda E = \frac{\{\Delta(R_n - G)\} + (\gamma E_a)}{(\Delta + \gamma)} \tag{12.11}$$

where λE = the evaporative latent heat flux (MJ $m^{-2}d^{-1}$), λ = the latent heat of vaporization (MJ kg^{-1}) = 2.45 MJ kg^{-1}, Δ = the slope of the saturated vapor pressure curve (i.e. $\partial e_s/\partial T$) (kPa °C^{-1}), e_s = the saturated vapor pressure (kPa), T = the temperature (°C^{-1}), R_n = the net radiation flux (MJ $m^{-2}d^{-1}$), G = the sensible heat flux into soil (MJ $m^{-2}d^{-1}$), γ = the psychometric constant (kPa °C^{-1}) = 0.059 kPa °C^{-1}, E_a = the vapor transport flux (MJ $m^{-2}d^{-1}$) = $f\{u_2, (e_s - e_a)\}$, u_2 = the wind speed (m s^{-1}), and e_a = the actual vapor pressure (kPa). Variables used in eq. (12.11) can be estimated from various relationships summarized in Table 12.18.

12.5.3 Water Required for Groundwater Conservation

The amount of water infiltrated from a channel can be expressed as a function of the hydraulic conductivity of channel surroundings and geometric dimensions of the channel. For stream channels, where the depth of water table is large relative to the water depth in the stream, the seepage loss can be mathematically described as (Harr, 1962).

$$q_3 = K(B + \beta h) \tag{12.12}$$

where q_3 is the unit seepage flow rate (m^3 s^{-1} per m of channel length), K is the hydraulic conductivity (m s^{-1}), B is the channel width (m) = $f(h)$, h is the depth of water (m), and β is the coefficient = f (channel geometry). Equation (12.12) can be modified for the total seepage loss in the stretch as:

$$Q_3 = KP_wL_C \tag{12.13}$$

Table 12.18 Auxiliary equations used for the Penman method

Parameter	Relationships
Relative humidity, RH (%)	$\text{RH} = \frac{e_a}{e^0(T)} \times 100\%$ $e^0(T)$ is the saturation vapor pressure at the same temperature (kPa), T is temperature (°C), and e_a is the actual vapor pressure (kPa)
Saturation vapor pressure, e_s (k Pa)	$e^0(T) = 0.6108\ \exp\left\{\frac{17.27\,T}{T + 237.3}\right\}$ $e_s = 0.5\left[e^0(T_{max}) + e^0(T_{min})\right]$ T_{max} and T_{min} are the daily maximum and minimum temperatures (°C)
Actual vapor pressure, e_a (kPa)	$e_a = e^0(T_{dew}) = 0.6108\ \exp\left\{\frac{17.27\,T_{dew}}{T_{dew} + 237.3}\right\}$ $e_a = e_s \times (\text{RH}/100)$
Slope if the saturation vapor pressure, Δ (kPa °C^{-1})	$\Delta = \frac{4098\left[0.6108\ \exp\left(\frac{17.27\,T}{T + 237.3}\right)\right]}{(T + 237.3)^2}$
Extraterrestrial radiation, R_a (MJ m^{-2}d^{-1})	$R_a = \frac{24\,(60)}{\pi} G_{sc} d_r\left[\omega_s\ \sin(\varphi) \cdot \sin(\delta) + \cos(\varphi) \cdot \cos(\delta) \cdot \sin(\omega_s)\right]$ G_{sc} = solar constant (0.0820 MJ m^{-2} min^{-1}) $d_r = 1 + 0.033\ \cos(2\pi J/365)$ J = number of the day in the year between 1 (1st January) and 365 or 366 (31st December) $\omega_s = \arccos[-\tan(\varphi) \cdot \tan(\delta)]$ φ = latitude (radian) [radian = π (decimal degree)/180] $\delta = 0.409\ \sin\left[\frac{2\pi J}{365} - 1.39\right]$
Solar radiation, R_s (MJ m^{-2}d^{-1}) When $n = N$, the solar radiation will becomes the clear sky solar radiation.	$R_s = \left(a_s + b_s\frac{n}{N}\right) \cdot R_a$ n = actual duration of sunshine hours (hours); N = maximum possible daylight hours (hours); n/N = relative sunshine hour (dimensionless); $a_s = 0.25$, and $b_s = 0.50$ $N = 24\,\omega_s/\pi$
Net shortwave radiation, R_{ns} (MJ m^{-2}d^{-1})	$R_{ns} = (1 - \alpha) \cdot R_s$ α = albedo or reflection coefficient. For hypothetical grass reference, $\alpha = 0.23$ For deep water α = 0.04–0.09 For shallow water, α = 0.09–0.12

Table 12.18 (continued)

Parameter	Relationships
Net longwave radiation, R_{nl} (MJ m^{-2}d^{-1})	$R_{nl} = \sigma \left[\frac{T_{max,K^4} + T_{min,K^4}}{2}\right] \cdot (0.34 - 0.14\sqrt{e_a}) \times \left(1.35\frac{R_s}{R_{so}} - 0.35\right)$ σ = Stefan-Boltzman constant (= 4.903×10^{-9} MJ K^{-4} m^{-2} day^{-1}) $T_{max,K}$ =daily maximum temperature (K) [K = °C + 273.16]; $T_{min,K}$ =daily minimum temperature (K); R_s/R_{so} =relative shortwave radiation ($\leq$1.0).
Net radiation, R_n (MJ m^{-2}d^{-1})	$R_n = R_{ns} - R_{nl}$
Soil heat flux, G (MJ m^{-2}d^{-1})	For water surface, $G = 0$

where P_w is the wetted perimeter (m), L_C is the channel length (m), and Q_3 is the seepage loss (m^{3}s^{-1}). For a wide open channel (i.e., B >> 2 h), eq. (12.13) can be simplified as:

$$Q_3 = KBL_C \tag{12.14}$$

Using the wetted perimeter, the length of river stretch and hydraulic conductivity, the water required for groundwater conservation can be estimated using the formula given through eq. (12.13). However, most of the Indian rivers carry sediments with different magnitudes, depending upon the seasonal flow variation. Due to sedimentation processes, sediments get deposited during the low flow season. Deposition of finer particle causes clogging of the porous medium resulting in a 50–75% reduction of hydraulic conductivity. Therefore, in the study, instead of using the hydraulic conductivity, an effective hydraulic conductivity of the river bed was used. Accordingly, eq. (12.13) was modified as:

$$Q_3 = K_{eff}P_wL_C \tag{12.15}$$

where K_{eff} is the effective hydraulic conductivity of the porous material (m/s).

12.5.3.1 Estimation of Wetted Perimeter

For a particular discharge, the wetted perimeter of the stream can be estimated using appropriate Manning's roughness, longitudinal slope and river cross-section (Fig. 12.24). The following steps can be considered for estimating the wetted perimeter based on the discharge value:

Step 1: Derive the rating curve (i.e., stage vs discharge relationship using the cross-section data in the form of x–z) based on estimated water area, wetted perimeter, hydraulic radius, longitudinal slope, and Manning's roughness. For

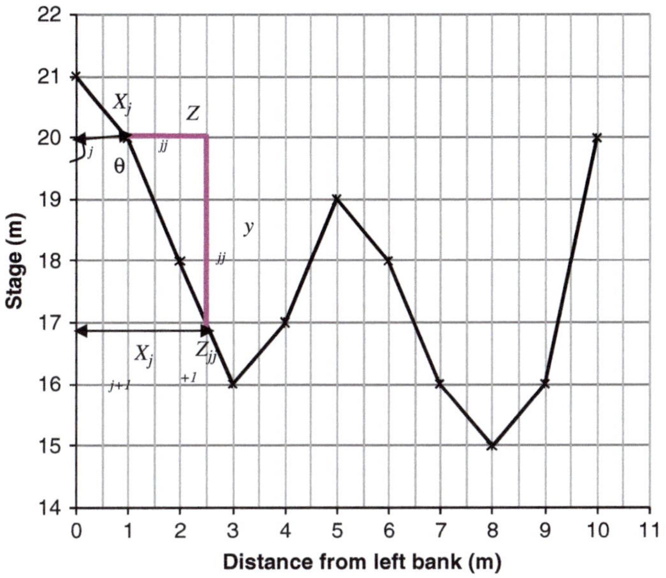

Fig. 12.24 Typical diagram of river cross-section

discharge estimation, Manning's formula is used:

$$Q = A_w R^{2/3} S^{1/2} / n \quad (12.16)$$

where Q is the discharge (m^3/s), n is the Manning roughness [L$^{-1/3}$T], and S is the longitudinal slope (m/m). Once the discharge is estimated corresponding to the depth of flow, a rating curve can be developed.

Step 1a: Estimate the water area with respect to depth, using the cross-section (x–z) data, with the trapezoidal rule, which is:

$$\int_{x_0}^{x_0+nh} f(x)\,dx = \frac{h}{2}(y_0 + y_n) + h(y_1 + y_2 + y_3 + \cdots + y_{n-1}) \quad (12.17)$$

Step 1b: Estimate the wetted perimeter using the following formula:

$$p_{w\,jj} = \left(x_{jj+1} - x_{jj}\right) / \cos\left[\tan^{-1}\left\{\left(z_{jj} - z_{jj+1}\right) / \left(x_{jj+1} - x_{jj}\right)\right\}\right] \quad (12.18)$$

where p_w is the wetted perimeter.

Step 1c: Estimate the hydraulic radius at each depth using the following relationship:

$$R = A_w / p_w \quad (12.19)$$

where R is the hydraulic radius (m), A_w is the water area (m^2), and p_w is the wetted perimeter (m).

Step 2: Based on the desired discharge value, stage, water area and wetted perimeter can be estimated using a linear interpolation technique. The discharge value was considered corresponding to the 25% MAF (Tennant, 1976), so that ecological flow can be maintained.

12.5.4 Water Required for Aquatic-Habitat Conservation

The ecological flow for the river is considered to maintain the aquatic diversity in the river. Based on the methodology presented by Tennant (1976), 25% of the mean annual flow (MAF) can be considered appropriate for the ecological maintenance. He found that rivers with more stable flow regimes had relatively higher flow requirements than rivers with more variable flow regimes. This is because in highly variable flow regimes the biota would have adjusted to the relative scarcity of water, while in more reliably flowing rivers the biota are more sensitive to the reduction in flow (Hughes and Hannart, 2003). Considering this aspect, for Indian conditions and required flow variability, the modified Tennant method at 25% MAF is found more appropriate than the uniform flow allocation.

12.5.5 Application of Methodology

The proposed methodology is applied to estimate environmental flows for different stretches of Yamuna River, a major tributary of Ganga River. Due to large abstractions at the Hathinikund barrage, the downstream part of the river is almost dry, though the 4.5 cumec water is allocated for the downstream release during December to June. However, after traversing a few kilometers, this water disappears and the river becomes dry up to the Palla. After the Wazirabad barrage near Delhi, there is no release during the dry period (i.e., December to June). The water available in the Wazirabad barrage to the Okhla barrage stretch is due to wastewater outfalls of more than 30 cumecs with recorded BOD of more than 30 mg/l. After the Okhla barrage, a similar situation persists and the river has little water all the way up to Etawah where the Chambal River joins. However, due to base flow, the river does not get dry. After Etawah, the river quality gets improved and up to the confluence with Ganga River, it becomes environmentally good. Considering these aspects, Yamuna River is of major interest for the estimation of environmental flow. The stretch-wise important parameters are given in Table 12.19.

12.5.5.1 Data Preparation and Analysis

The data and sources used for different environmental flow components are listed in Table 12.20 which also includes the steps required for deriving data. Using the

Table 12.19 Important parameters for different stretches of the Yamuna River

S. No.	Stretch	Length, L_c (km)	Avg. slope (m/m)	Manning's n	Latitude at centroid (°)	Avg. waste water discharge (m^3/s)	Avg. BOD (mg/l)[a]	Degradation rate coeff., k (per day)
1	Hathinikund to Palla	254.0	0.00055	0.035	29.5249	5.0	20.0	0.40
2	Wazirabad to Okhla	22.0	0.000327	0.040	28.6510	30.0	30.0	0.25
3	Okhla to Agra	310.0	0.000139	0.035	27.4571	10.0	30.0	0.38
4	Agra to Etawah	305.0	0.000136	0.035	26.9548	4.0	30.0	0.40
5	Etawah to Pratappur (Allahabad)	465.0	0.000055	0.035	25.6785	4.0	30.0	0.42

[a] BOD level after treatment; BOD concentration of dilution water was assumed 2 mg/l

dataset and river characteristics (Table 12.19), the seepage loss component of environmental flow is estimated using eqs. (12.15) through (12.19). The water required to assimilate pollutants in the river is estimated using eq. (12.9). The river evaporation loss is estimated using eq. (12.10) with auxiliary equations. The water requirement for the conservation of aquatic-habitat was considered as monthly distribution of 25% MAF.

Table 12.20 List of data required for computation using component based method

Components	Data required	Source/procedure
1. Water requirement for assimilating the pollutants in the river	• the flow rate from upstream (m^3/s), • the concentration of pollutants from upstream (mg/l), • the flow rate from the pollutant sources (m^3/s), • the concentration of pollutants from the pollutant sources (mg/l), • the concentration of pollutants in the environmental reservoir (mg/l), • the standard for water quality (mg/l), • the degradation coefficient (1/day), • the length of river stretch, and • the average velocity of flow (m/s)	• Central Pollution Control Board • Primary measurements

Table 12.20 (continued)

Components	Data required	Source/procedure
2. Water requirement for groundwater conservation	• The effective hydraulic conductivity (m/s) • Manning's roughness • Mean monthly flows at 25% MAF (m^3/s) • River cross section ° Wetted perimeter (m) ° Water area (m^2) ° Rating curve • Length of stretch (m)	• Based on the river bed texture, hydraulic conductivity is fixed from literature and adjusted for effective hydraulic conductivity. • Based on the river condition Manning's roughness is fixed using the available literatures. • Long term mean monthly flows were derived from the available 10-daily discharge data of different cites of Yamuna River. The data were collected from the Central Water Commission. • Using the mean monthly flows and estimated mean monthly flow (MAF) at the site, percentage distribution of MAF is derived. • At 25% of MAF, the corresponding monthly distribution is estimated • Survey was conducted for river cross-section
3. Water requirement for evaporation loss	• Top width of river water (m) derived from cross-section • Length of stretch (m) • Monthly normal rainfall (mm) • Monthly Evaporation (mm) from Penman method ° Maximum, minimum, and mean temperature (°C) ° Dew point temperature (°C) ° Latitude (degree decimal) ° Albedo	• Cross-section and the mean monthly flows is used for deriving the depth of flows. Corresponding to the depth of flow top width of river is computed. • Monthly rainfall normals were collected from the India Meteorological Department. • Data required for the estimation of the evaporation is collected from India Meteorological Department, and National Climatic Data Center (freely available).
4. Water requirement for aquatic-habitat conservation	• Discharge data (m^3/s) ° Mean annual flow (MCM) ° Monthly distribution of mean annual flow (%)	• In component 2, it is covered.

The methodology is individually applied to each stretch, namely, the Hathinikund barrage to Palla, Wazirabad barrage to Okhla barrage, Okhla barrage to Etawah, and Etawah to Pratappur near the confluence with Ganga River. For all these stretches, results are obtained; for economy of space, different components of environmental flow for the Hathinikund to Palla stretch only are presented in Fig. 12.25. Tabular results for all the stretches are given in Tables 12.21, 12.22, 12.23, 12.24, 12.25,

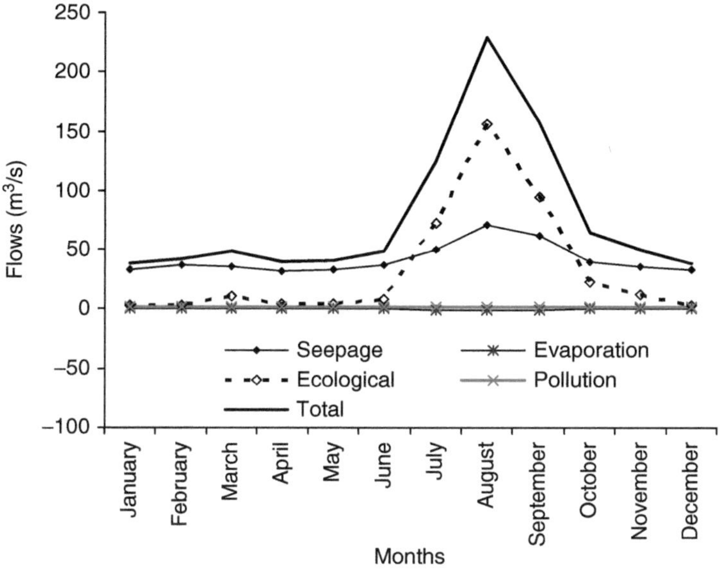

Fig. 12.25 Components of environmental flows for the Hathinikund-Palla stretch

and 12.26. It is observed that the water requirement for groundwater conservation is quite high, followed by dilution water requirement. The dilution component is another important component when the river is being utilized for pollutant transport. It is also observed that the environmental water demand is high during the period of June to October, but is not a major concern for the Indian rivers, because the Monsoon is active during these months. During the rest of the year, attention needs to be paid.

For the Wazirabad to Okhla stretch which is a short stretch, the water required to assimilate pollutants is very high, as compared to other demands (Table 12.22). For this stretch, dilution is not effectively modified by the river water. Therefore, the assimilation water demand is further estimated under different influent conditions (Table 12.23). In Table 12.23, dilution water was estimated for combinations of treatment quality and effluent discharge. The estimated dilution water is very high as compared to the availability. Under such a condition, without diverting the waste discharge for other uses after treatment, it is not possible to maintain the environmental flow in the river for this stretch. Similarly, the stretch from Okhla barrage to Agra is also critical and needs to be taken care of.

The total environmental flow requirement along with the available flow was compared, as shown in Figs. 12.26, 12.27, 12.28, 12.29, and 12.30. These figures reveal that during the non-monsoon period, the water available in the river does not satisfy environmental flows for all the stretches except for the Etawah to Pratappur stretch.

Table 12.21 Environmental flow components for the Hathinikund to Palla stretch of Yamuna River

Month	Mean monthly water requirement (MCM)					Mean monthly discharge required (m^3/s)				
	Seepage	Evapo.	Ecolog. (25%MAF)	Pollution ($5\ m^3/s$ + 30 BOD)	Total EF	Seepage	Evapo.	Ecolog. (25%MAF)	Pollution ($5\ m^3/s$ + 30 BOD)	Total EF
(i)	(ii)	(iii)	(iv)	(v)	(vi)	(vii)	(viii)	(ix)	(x)	(xi)
January	90.06	0.54	8.00	5.68	104.28	33.62	0.20	2.99	2.12	38.93
February	90.72	0.91	6.25	5.13	103.01	37.50	0.38	2.58	2.12	42.58
March	95.24	2.00	29.25	5.68	132.17	35.56	0.75	10.92	2.12	49.35
April	82.13	2.30	11.50	5.50	101.43	31.69	0.89	4.44	2.12	39.13
May	90.63	2.92	11.50	5.68	110.73	33.84	1.09	4.29	2.12	41.34
June	95.10	1.86	23.25	5.50	125.71	36.69	0.72	8.97	2.12	48.50
July	134.65	−0.96	192.75	5.68	332.12	50.27	−0.36	71.96	2.12	124.00
August	185.63	−1.77	405.00	5.50	594.36	71.62	−0.68	156.25	2.12	229.30
September	164.80	−0.53	252.50	5.68	422.45	61.53	−0.20	94.27	2.12	157.72
October	102.07	1.34	60.00	5.50	168.91	39.38	0.52	23.15	2.12	65.16
November	95.16	1.03	33.75	5.68	135.62	35.53	0.38	12.60	2.12	50.63
December	90.05	0.69	8.00	5.68	104.42	33.62	0.26	2.99	2.12	38.99

Table 12.22 Environmental flow components for the Wazirabad to Okhla stretch of Yamuna River

	Mean monthly water requirement (MCM)					Mean monthly discharge required (m^3/s)				
Month	Seepage	Evapo.	Ecolog. (25%MAF)	Pollution (30 m^3/s + 30 BOD)	Total EF	Seepage	Evapo.	Ecolog. (25%MAF)	Pollution (30 m^3/s + 30 BOD)	Total EF
(i)	(ii)	(iii)	(iv)	(v)	(vi)	(vii)	(viii)	(ix)	(x)	(xi)
January	15.76	0.084	23.00	321.78	360.62	5.88	0.03	8.59	120.14	134.64
February	15.52	0.116	20.00	290.64	326.28	6.42	0.05	8.27	120.14	134.87
March	14.05	0.197	18.70	321.78	354.73	5.25	0.07	6.98	120.14	132.44
April	17.93	0.312	27.50	311.40	357.14	6.92	0.12	10.61	120.14	137.79
May	13.06	0.263	16.25	321.78	351.35	4.88	0.10	6.07	120.14	131.18
June	16.71	0.240	24.50	311.40	352.85	6.45	0.09	9.45	120.14	136.13
July	56.73	0.125	166.25	321.78	544.89	21.18	0.05	62.07	120.14	203.44
August	67.31	0.079	422.50	311.40	801.29	25.97	0.03	163.00	120.14	309.14
September	140.59	0.787	293.70	321.78	756.86	52.49	0.29	109.66	120.14	282.58
October	31.73	0.282	62.50	311.40	405.91	12.24	0.11	24.11	120.14	156.60
November	22.32	0.158	37.50	321.78	381.76	8.33	0.06	14.00	120.14	142.53
December	16.16	0.090	24.00	321.78	362.03	6.03	0.03	8.96	120.14	135.17

Table 12.23 Environmental flow requirement for the Wazirabad to Okhla stretch of Yamuna River under different effluent flow conditions

Month	Seepage (m^3/s)	Evapo. (m^3/s)	Ecolog. (25%MAF) (m^3/s)	Water required to assimilate the pollution				Total EF (m^3/s)			
				30 m^3/s + 10 mg/l BOD	30 m^3/s + 5 mg/l BOD	5 m^3/s + 10 mg/l BOD	5 m^3/s + 5 mg/l BOD	30 m^3/s + 10 mg/l BOD	30 m^3/s + 5 mg/l BOD	5 m^3/s + 10 mg/l BOD	5 m^3/s + 5 mg/l BOD
(i)	(ii)	(iii)	(iv)	(v)	(vi)	(vii)	(viii)	(ix)	(x)	(xi)	(xii)
January	5.88	0.03	8.59	120.14	26.29	25.46	6.36	134.64	40.79	39.96	20.86
February	6.42	0.05	8.27	120.14	26.29	25.46	6.36	134.87	41.02	40.19	21.09
March	5.25	0.07	6.98	120.14	26.29	25.46	6.36	132.44	38.59	37.76	18.66
April	6.92	0.12	10.61	120.14	26.29	25.46	6.36	137.79	43.94	43.11	24.01
May	4.88	0.10	6.07	120.14	26.29	25.46	6.36	131.18	37.33	36.50	17.40
June	6.45	0.09	9.45	120.14	26.29	25.46	6.36	136.13	42.28	41.45	22.35
July	21.18	0.05	62.07	120.14	26.29	25.46	6.36	203.44	109.59	108.76	89.66
August	25.97	0.03	163.00	120.14	26.29	25.46	6.36	309.14	215.29	214.46	195.36
September	52.49	0.29	109.66	120.14	26.29	25.46	6.36	282.58	188.73	187.90	168.80
October	12.24	0.11	24.11	120.14	26.29	25.46	6.36	156.60	62.75	61.92	42.82
November	8.33	0.06	14.00	120.14	26.29	25.46	6.36	142.53	48.68	47.85	28.75
December	6.03	0.03	8.96	120.14	26.29	25.46	6.36	135.17	41.32	40.49	21.39

Table 12.24 Environmental flow components for the Okhla to Agra stretch of Yamuna River

Month	Mean monthly water requirement (MCM)					Mean monthly discharge required (m^3/s)				
	Seepage	Evapo.	Ecolog. (25%MAF)	Pollution (10 m^3/s + 30 BOD)	Total EF	Seepage	Evapo.	Ecolog. (25%MAF)	Pollution (10 m^3/s + 30 BOD)	Total EF
(i)	(ii)	(iii)	(iv)	(v)	(vi)	(vii)	(viii)	(ix)	(x)	(xi)
January	92.63	0.98	20.00	26.09	139.70	34.58	0.37	7.47	9.74	52.16
February	91.35	1.29	17.53	23.56	133.73	37.76	0.53	7.25	9.74	55.28
March	82.43	1.96	14.00	26.09	124.48	30.78	0.73	5.23	9.74	46.47
April	104.82	3.04	17.53	25.25	150.64	40.44	1.17	6.76	9.74	58.12
May	76.57	2.53	11.68	26.09	116.87	28.59	0.94	4.36	9.74	43.63
June	97.80	2.68	17.53	25.25	143.26	37.73	1.03	6.76	9.74	55.27
July	207.85	1.13	105.25	25.09	340.32	77.60	0.42	39.30	9.74	127.06
August	234.33	0.04	338.75	25.25	598.37	90.41	0.02	130.69	9.74	230.85
September	233.10	1.09	327.00	26.09	587.28	87.03	0.41	122.09	9.74	219.26
October	190.91	2.84	105.25	25.25	324.25	73.65	1.10	40.61	9.74	125.09
November	131.95	1.61	52.50	26.09	212.15	49.26	0.60	19.60	9.74	79.21
December	95.81	0.98	29.25	26.09	152.13	35.77	0.37	10.92	9.74	56.80

Table 12.25 Environmental flow components for the Agra to Etawah stretch of Yamuna River

	Mean monthly water requirement (MCM)					Mean monthly discharge required (m^3/s)				
Month	Seepage	Evapo.	Ecolog. (250%MAF)	Pollution (4 m^3/s + 30 BOD)	Total EF	Seepage	Evapo.	Ecolog. (25%MAF)	Pollution (4 m^3/s + 30 BOD)	Total EF
(i)	(ii)	(iii)	(iv)	(v)	(vi)	(vii)	(viii)	(ix)	(x)	(xi)
January	121.55	1.32	48.00	15.88	186.75	45.38	0.49	17.92	5.93	69.73
February	120.26	1.73	42.00	14.35	178.34	49.71	0.72	17.36	5.93	73.72
March	111.36	2.68	33.00	15.88	162.92	41.58	1.00	12.32	5.93	60.83
April	134.67	3.91	42.06	15.37	196.01	51.96	1.51	16.23	5.93	75.62
May	105.38	3.50	28.02	15.88	152.78	39.34	1.31	10.46	5.93	57.04
June	127.23	3.40	42.06	15.37	188.06	49.09	1.31	16.23	5.93	72.55
July	273.90	0.54	252.60	15.88	542.92	102.26	0.20	94.31	5.93	202.70
August	317.08	−1.47	813.00	15.37	1,143.98	122.33	−0.57	313.66	5.93	441.35
September	309.90	0.90	784.80	15.88	1,111.48	115.70	0.34	293.01	5.93	414.98
October	217.02	3.24	252.60	15.37	488.23	83.73	1.25	97.45	5.93	188.36
November	159.72	1.96	126.00	15.88	303.56	59.63	0.73	47.04	5.93	113.34
December	123.97	1.30	70.20	15.88	211.35	46.29	0.49	26.21	5.93	78.91

Table 12.26 Environmental flow components for the Etawah to Pratappur (Allahabad) stretch of Yamuna River

	Mean monthly water requirement (MCM)					Mean monthly discharge required (m^3/s)				
Month	Seepage	Evapo.	Ecolog. (25%MAF)	Pollution (4 m^3/s + 30 BOD)	Total EF	Seepage	Evapo.	Ecolog. (25%MAF)	Pollution (4 m^3/s + 30 BOD)	Total EF
(i)	(ii)	(iii)	(iv)	(v)	(vi)	(vii)	(viii)	(ix)	(x)	(xi)
January	536.27	6.02	245.27	3.27	790.83	200.22	2.25	91.57	1.22	295.26
February	539.62	7.70	237.52	2.95	787.79	223.06	3.18	98.18	1.22	325.64
March	522.69	12.65	231.27	3.27	769.88	195.15	4.72	86.35	1.22	287.44
April	535.49	15.65	231.27	3.16	785.57	206.59	6.04	89.22	1.22	303.08
May	434.41	14.38	162.52	3.27	614.58	162.19	5.37	60.68	1.22	229.46
June	534.52	12.91	231.27	3.16	781.86	206.22	4.98	89.22	1.22	301.64
July	911.06	−7.55	1,625.16	3.27	2,531.94	340.15	−2.82	606.77	1.22	945.32
August	1,401.81	−19.23	4,425.43	3.16	5,811.17	540.82	−7.42	1,707.34	1.22	2,241.96
September	1,389.56	−0.91	4,037.89	3.27	5,429.81	518.80	−0.34	1,507.58	1.22	2,027.26
October	956.71	8.91	781.33	3.16	1,750.11	369.10	3.44	301.44	1.22	675.20
November	567.92	7.04	368.79	3.27	947.02	212.04	2.63	137.69	1.22	353.58
December	549.82	6.05	312.53	3.27	871.67	205.28	2.26	116.69	1.22	325.44

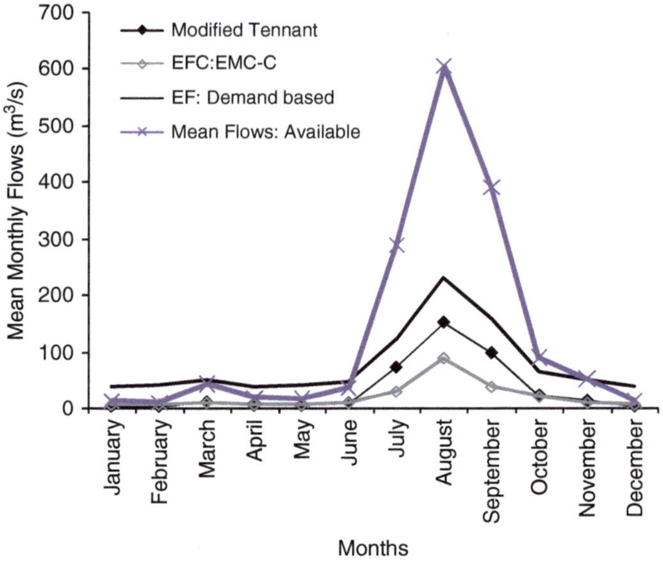

Fig. 12.26 Comparison of different methods of EF estimation for the Hathinikund-Palla stretch ($L_c = 254.0$ km, $Q_{\text{waste}} = 5\ \text{m}^3/\text{s}$, BOD $= 30$ mg/l)

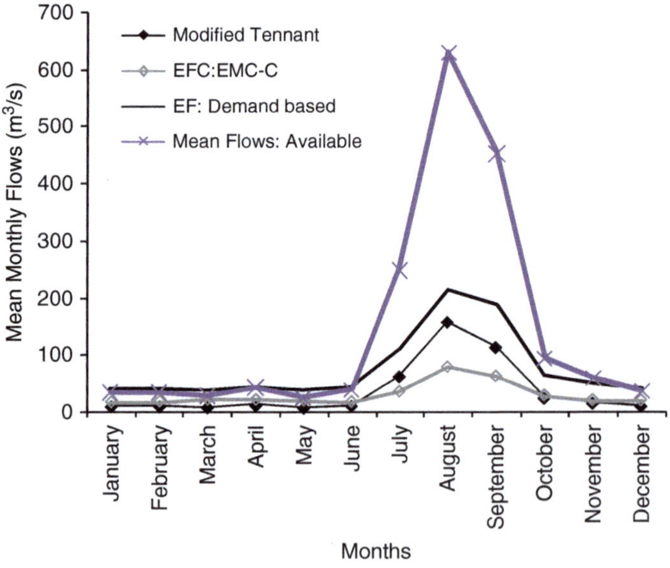

Fig. 12.27 Comparison of different methods of EF estimation for the Wazirabad-Okhla stretch ($L_c = 22.0$ km, $Q_{\text{waste}} = 5\ \text{m}^3/\text{s}$, BOD $= 10$ mg/l)

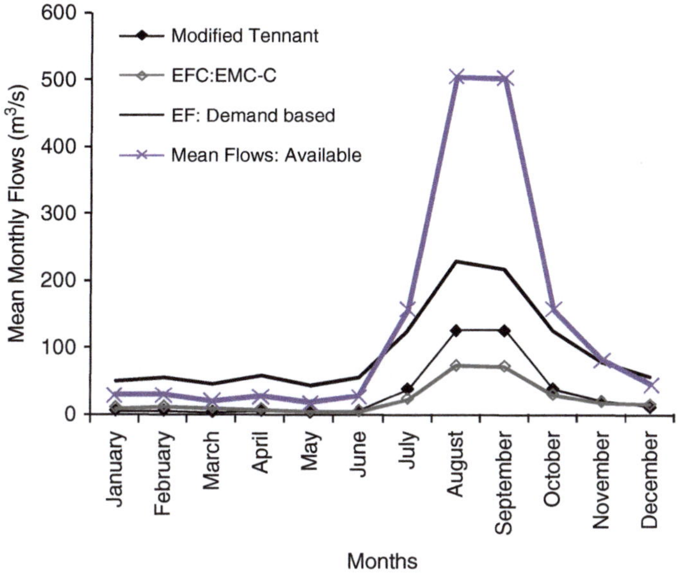

Fig. 12.28 Comparison of different methods of EF estimation for the Okhla-Agra stretch (L_c = 310.0 km, Q_{waste} = 10 m^3/s, BOD = 30 mg/l)

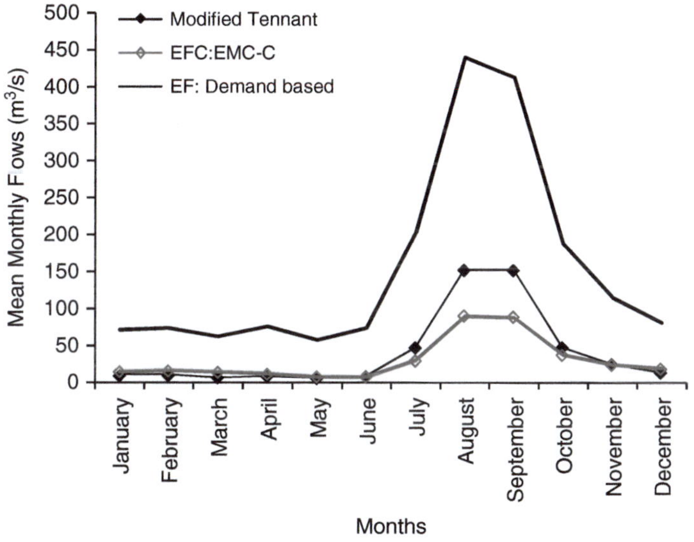

Fig. 12.29 Comparison of different methods of EF estimation for the Agra-Etawah stretch (L_c = 305.0 km, Q_{waste} = 4 m^3/s, BOD = 30 mg/l)

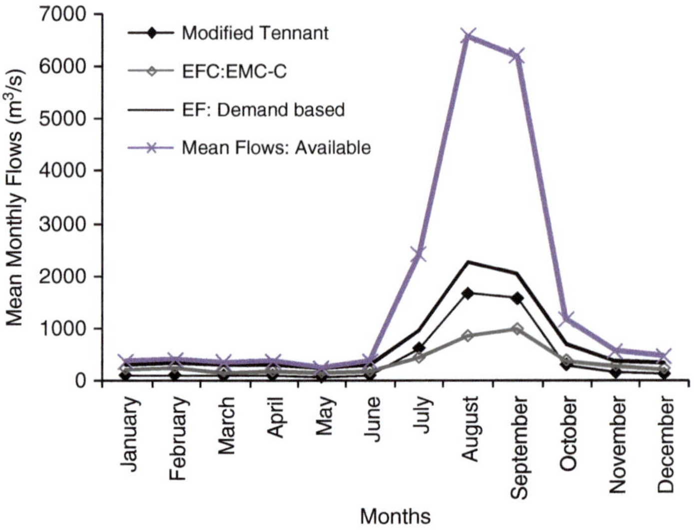

Fig. 12.30 Comparison of different methods of EF estimation for the Etawah-Pratappur stretch ($L_c = 465.0$ km, $Q_{waste} = 4\ m^3/s$, BOD = 30 mg/l)

12.6 Comparison of the Approaches

The proposed approach is compared with two hydrology-based approaches: the Tennant method and EMC based environmental flow calculator. A description of these methods has already been presented in earlier sections.

Results obtained from all the methods for all the stretches are shown in Figs. 12.26, 12.27, 12.28, 12.29, and 12.30, sequentially for the Hathinikund-Palla, Wazirabad-Okhla, Okhla-Agra, Agra-Etawah, and Etawah-Pratappur stretches. Comparison of results shows that the proposed method gives higher values than the other two methods, and seems practical when the river morphology is considered. The most important component of environmental flows is seepage and is very large in the alluvial rivers as the average width of the river is 650 m with the range of 350–1,800 m.

The method "Component based Environmental flow calculator" is comprehensive when river is highly intercepted and being used for the disposal of pollutants either treated/partially treated/untreated than the other methods. The hydrology-based methods, on the other hand, require good historical discharges and hydrologic regime. The length of record should include the entire alteration made in the river. A hydrology-based method has a major disadvantage in that it is a black box method whose accuracy largely depends on the length of record. Further, the hydrology-based method does not account for the water requirement for assimilating pollutants in the river. This means that if the river is highly intercepted and used for pollutant transport then hydrology-based methods do not yield realistic estimates of environmental flows.

Table 12.27 Summary of the environmental flow components for Yamuna River

S. No.	Segment	Length (km)	Ecological (MCM)	Seepage (MCM)	Evaporation (MCM)	Pollution (MCM)	Total E-flows (MCM)
(i)	(ii)	(iii)	(iv)	(v)	(vi)	(vii)	(viii)
1	Hathinikund to Pall	254.0	1,041.75 (42.78)	1,316.24 (54.05)	10.33 (0.42)	66.86 (2.75)	2,435.00
2	Wazirabad to Okhla (Delhi Segment)	22.0	1,136.40 (21.22)	427.87 (7.99)	2.73 (0.05)	3,788.70 (70.74)	5,355.71
3	Okhla to Agra	310.0	1,056.27 (39.94)	1,639.50 (54.23)	20.17 (0.67)	307.16 (10.16)	3,023.15
4	Agra to Etawah	305.0	2,534.30 (52.08)	2,122.00 (43.61)	23.01 (0.47)	187.01 (3.84)	4,866.40
5	Etawah to Pratappur (Allahabad)	465.0	12,890.25 (58.93)	8,879.88 (40.60)	63.6 (0.29)	38.47 (0.18)	21,872.20

Values in the parenthesis are percent demand of total environmental flows

The proposed method does not consider the water required for the sediment transport. It is assumed that if it is considered once in a year then it is automatically possible during the wet season (i.e., August and September).

For aquatic-habitat conservation (i.e., ecological flow) in the river, monthly distribution of 25% MAF is sufficient (Tennant, 1976). However, the question is "how to maintain this ecological flow in the river?". The importance of each component for the river can be seen from Table 12.27 which shows that seepage and dilution components are important. If only ecological component estimated by Tennant (1976) (i.e., 25% MAF criteria) is allocated for the river; all the flow will disappear due to the seepage loss. More clearly, to maintain the ecological demand throughout the river stretch, it is required to satisfy other demands quantitatively, such as seepage and evaporation. Further, if river is receiving pollutants, water quality will become inadequate for aquatic conservation, and therefore additional water needs to be added to assimilate the pollutants within the stretch. This means that if there are no losses and effluent flow is considered then environmental flow estimated by the Tennant method will be appropriate, although it is not practically possible. This is the basic reason for overestimating the environmental flow demand by the proposed method as compared to other methods.

12.7 Concluding Remarks

For environmental flow estimation various hydrology-based techniques are commonly used. These techniques either estimate the ecological flow or environmental flow without considering the pollution and manmade alterations. In this chapter, four important components, i.e., ecological, seepage, evaporation, and dilution of pollutant assimilation are considered for a comprehensive assessment of environmental flows in the river. Of course, the ecological flow is most important for proper ecosystem functioning of the river, but it cannot be maintained if other three components (i.e., seepage, evaporation and dilution) are not satisfied.

It may be affirmed that the seepage component is important for the Yamuna River that largely affects the allocation of ecological flow (Table 12.27). Besides the seepage component, the dilution component is another essential component for maintaining water quality for the survival of aquatic habitat as most of the rivers in India are getting wastewater effluent (Table 12.27).

The water availability in the Yamuna River is not enough to meet environmental demands during the months of December to June except for the Etawah to Allahabad stretch. This stretch has no major industrial and domestic influence. Also, this stretch receives good quality water from four major tributaries, such as Chambal, Sind, Betwa and Ken.

The Delhi stretch is of high concern to the Ministry of Environment and Forests, Government of India, as this stretch receives pollutants from 23 wastewater drains and there is no release of water from the Wazirabad barrage. Therefore, for this stretch, the water demand for assimilating pollutants becomes very high as compared to the available water. Under such circumstances, there is only one solution for Delhi: "Do not use the river for pollutant transport." To be more precise, all the wastewater either treated or untreated should be diverted for other uses, such as irrigation, construction, etc., and should not be discharged into the river.

References

Aceman M, Dunbar MJ (2004) Defining environmental river flow requirements-a review. Hydrol Earth Syst Sci 8:861–876

Alcaźar J, Palau A, Vega-Garcı C (2008) A neural net model for environmental flow estimation at the Ebro River Basin, Spain. J Hydrol 349:44–55

Bovee KD (1982) A guide to stream habitat analysis using the instream flow incremental methodology. Instream Flow Information Paper 12. USDI Fish Wildl. Serv., Office of Biol Serv FWS/OBS-82/26, 248pp

Bunn SE, Arthigton AH (2002) Basic principles and ecological consequences of altered flow regimes for aquatic biodiversity. Environ Manage 30:492–507

Collings M (1972) A methodology for determining instream flow requirements for fish. In: Proceedings of the instream flow methodology workshop, Washington, Department of Ecology (Olympia), pp 72–86

DWAF (1997) While paper on a National Water Policy for South Africa. Department of Water Affairs and Forestry, Pretorio, Soth Africa

Dyson M, Bergkamp G, Scanlon G (eds) (2003) Flow. The essentials of environmental flows. IUCN, Gland, Switzerland and Cambridge, UK, p 118

Environmental Agency (2001) Managing water abstraction: the catchment abstraction management strategy process. Environmental Agency, Bristol, UK

Gordon ND, McMahon TA, Finlayson BL, Gippel CJ, Nathan RJ (2004) Stream hydrology: an introduction for ecologists. Wiley, New York, NY, p 429

Harr ME (1962) Groundwater and seepage. McGraw-Hill, New York

Hughes DA, Münster F (2000) Hydrological Information and Techniques to Support the Determination of the Water Quantity Component of the Ecological Reserve for Rivers. Report to the Water Research Commission by the Institute for Water Research, Rhodes University, WRC Report No. 867/3/2000, Pretoria, South Africa

Hughes DA, Hannart P (2003) A desktop model used to provide an initial estimate of the ecological in-stream flow requirements of rivers in South Africa. J Hydrol 270(3–4):167–181

Hughes DA, Smakhtin V (1996) Daily flow time series patching or extension: a spatial interpolation approach based on flow duration curves. J Hydrol Sci 41(6):851–871

Iyer RR (2005) The notion of environmental flows: a caution. NIE/IWMI Workshop on Environmental Flows, New Delhi, March 23–24, 2005

Jowett IG (1997) Instream flow methods: a comparison of approaches. Regul Riv: Res Manage 13:115–127

Kashaigili JJ, Mccartney M, Mahoo HF (2007) Estimation of environmental flows in the Great Ruaha River Catchment, Tanzania. Phys Chem Earth 32:1007–1014

King JM, Louw MD (1998) Instream flow assessments for regulated rivers in South Africa using the Building Block Methodology. Aquat Ecosyst Health Manage 1:109–124

Knights P (2002) Environmental flows: lessons from an Australian experience. In: Proceedings of international conference: dialog on water, food, and environment, Hanoi, Vietnam, 18pp

Lee CS, Wen CG (1996) Application of multi-objective programming to water quality management in a river basin. J Environ Manage 47:11–26

Nature Conservancy (2006) Conservation by design: a strategic frameworks for mission success. The Nature Conservancy, Virginia, USA

O'Keefe J (2000) Environmental flow assessments: background and assumptions. In: King JM, Tharme RE, De Villiers MS (eds) Environmental flow assessments for rivers: manual for the building block methodology. Water Research Commission, Report No, TT 131/00, Pretoria, South Africa

Palau A (1994) Los mal llamados caudales "ecológicos". Bases para una propuesta de cálculo. Obras Públicas 28(2):84–95

Penman HL (1948) Natural evaporation from open water, bare soil and grass. Proc R Soc Lond (England) 193:120–146

Poff NL, Allan JD, Bain MB, Karr JR, Prestegaard KL, Richter BD, Sparks RE, Stromberg JC (1997) The natural flow regime: a paradigm for river conservations and restoration. BioScience 47(11):769–784

Richter BD, Baumgartner JV, Wigington R, Braun DP (1997) How much water does a river need? Freshwater Biol 37:231–249

Smakhtin V, Anputhas M (2006) An assessment of environmental flow requirements of Indian River basins. RR 107, International Water Management Institute, Colombo, Sri Lanka

Smakhtin V, Eriyagama N (2008) Developing a software package for global desktop assessment of environmental flows. Environ Model Softw 23:1396–1406

Smakhtin VU, Shilpakar RL, Hughes DA (2006) Hydrology-based assessment of environmental flows: an example from Nepal. Hydrol Sci J 51(2):207–222

Tennant DL (1976) Instream flow regimens for fish, wildlife, recreation and related environmental resources. Fishiries 1:6–10

Tharme RE (2003) A global perspective on environmental flow assessment: emerging trends in the development and application of environmental flow methodologies for rivers. River Res Appl 19:397–441

Tharme RE, King JM (1998) Development of the building block methodology for instream flow assessments and supporting research on the effects of different magnitude flows on riverine ecosystems. Water Research Commission Report No 576/1/98, Pretoria, South Africa, p 452

Ward JV, Tockner K (2001) Biodiversity: towards a unifying theme for river ecology. Freshwater Biol 46:807–819

Ward JV, Tockner K, Uehlinger U, Malard F (2001) Understanding natural patterns and processes in river corridors as the basis for effective river restoration. Regul Riv Res Manage 17:311–323.

Yang ZF, Sun T, Cui BS, Chen B, Chen GQ (2009) Environmental flow requirements for integrated water resources allocation in the Yellow River Basin, China. Commun Nonlinear Sci Numerical Simulation 14:2469–2481

Chapter 13
Cropping Plan for Water Resources Management

Abstract In India, nearly 75% of utilizable water resources are used for irrigation. Due to intensive agriculture and changes in climatic patterns, water scarcities are becoming frequent for agriculture, industry, energy and households. Therefore, it is imperative to implement cropping patterns that can lead to reduced agricultural water demand, while preserving farmers' economic returns. Looking into this scenario, the chapter determines economically feasible intercropping patterns for the Yamuna River basin.

In India, approximately 70% of the population depends on agriculture and the agricultural sector is a major part of the Indian economy. For sustained agriculture, water and climate are vital, for they govern agricultural productivity, market value, employment, and the living standard. The other governing factors for agricultural operation, selection of crops, cultivation methods, and irrigation methods are topography and soil. Agriculture, including crop planning, crop diversity, consumptive use of crops, and spatio-temporal availability of water resources for irrigation are highly sensitive to global warming or climate change.

In India, three distinct crop seasons can be identified: *kharif* (July to October), *rabi* (October to March) and *Zaid* (April to June), and agricultural water demands are met by three sources: rainfall, canal irrigation, and groundwater pumping. However, most of the *kharif* crop planning is dependent upon the onset of monsoons whose patterns seem to have changed due to climate change. Rai et al. (2010) reported that in the Yamuna River basin the date of onset of effective monsoons has positively shifted and the number of rainy days has reduced. These two changes are critical for water resources management, because most of the rainfall becomes confined to a shorter duration, resulting in increased flooding and declining aquifer recharge. Declining recharge induces crop water deficit for *rabi* crops and supplementary irrigation stress for *kharif* crops. Under such water deficit conditions, crop water management seems to be the only way for sustainable agriculture. Therefore, the objective of this study is to discuss the suitability of cropping patterns for sustainable agriculture from the perspective of both water demand and economy. Specific objectives include: (i) identification of crops based on climate, soil, and landscapes; (ii) identification of cropping patterns; (iii) estimation of crop water requirements; (iv) identification of cropping patterns based on minimum water use; and (v) economic evaluation of cropping patterns and net savings of irrigation water.

Yamuna River (Fig. 13.1), a holy river and a major tributary of the Ganga River which originates from the Shivalik range of Lesser Himalayas and after traversing

R.K. Rai et al., *The Yamuna River Basin*, Water Science and Technology Library 66, DOI 10.1007/978-94-007-2001-5_13,

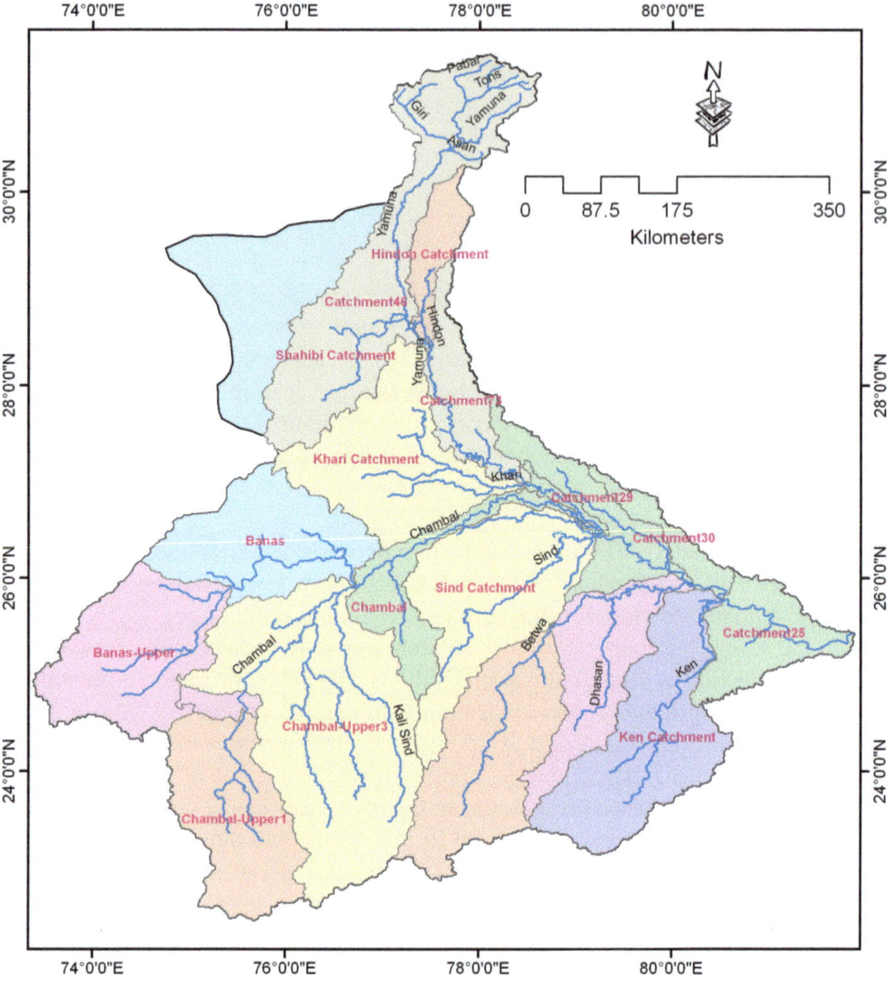

Fig. 13.1 Sub catchments of Yamuna River basin

approximately 1,350 km it joins the Ganga at the right bank near Allahabad (called "Sangam"). The detailed description of Yamuna river basin such as climate, soil, landuse, water resources etc. has been presented in previous chapters. As stated in previous chapter that many tributaries are joining the river Yamuna after Okhla barrage. Most of these tributaries are highly intercepted by various irrigation and multi-purpose projects resulting in little discharge before their confluence. Because of high utilization, the Yamuna River carries very little discharge during the lean season.

A water balance study of Yamuna River basin is presented in Chapter 8, which showed that all the sub-catchments of the Yamuna, excepting the Upper Himalayan and Ken sub-catchments, were under water deficit (see Table 13.1). Under such

Table 13.1 Summary of basin water balance

S. No	Catchments	Area (km^2)	DWR (MCM)	CWR (MCM)	IWR (MCM)	GWR (MCM)	SMS (MCM)	NGW (MCM)	SWR (MCM)	TWR (MCM)	S/D (MCM)
(i)	(ii)	(iii)	(iv)	(v)	(vi)	(vii) = (iv) + (v) + (vi)	(viii)	(ix)	(x)	(xi) = (viii) + (ix) + (x)	(xii) = (xi)−(vii)
1	Himalayan	11,058	54.3	1,332	10.8	1,397.1	1,935	457	3,605	5,997	4,599
2	Chambal	134,967	721	62,771	144.2	63,636.2	23,619	93,82.9	26,736	59,739	−3,897
3	Sind	27,940	166.8	13,786	33.3	13,986.1	4,890	3,441.7	1,888	10,220	−3,766
4	Betwa	43,836	287.5	26,456	57.5	26,801	7,671	5,701	11,803	25,175	−1,626
5	Ken	28,806	137.2	12,803	27.4	12,972.6	5,041	3,113	8,089	16,244	3,271
6	Yamuna	365,870	3,713	200,761	742.6	205,216.6	64,027	45,428.8	61,219	170,675	−34,541

DWR Domestic water requirement, *IWR* Industrial water requirement, *CWR* Crop water requirement, *GWR* Gross water requirement, *SMS* Soil moisture storage, *NGW* Net groundwater availability, *SWR* Surface water resources, *TWR* Total water resources, *S/D* Surplus/deficit circumstances, it becomes essential to manage the water resources of the Yamuna basin so that long term equity and economy can be maintained. It is seen from Table 13.1 that agricultural demand is much higher than domestic and industrial demands; therefore, attention needs to be paid to the sustainability of the agricultural sector and in turn the sustainability of the economy of India.

13.1 Landuse and Agriculture

Major land uses in the basin are typified by agriculture, forests, non-agriculture, barren land, permanent pasture, cultivable wasteland, and cultivated fallow. Of the total basin area (i.e., 342,541 km^2), more than 50% of the area (i.e., net sown area is 187,908.6 km^2) is under agriculture. Catchment wise net sown area and irrigation facility are summarized in Table 13.2 which also includes cropping intensity, net irrigated area and sources of irrigation. It is apparent from the table that only 56% of the net sown area is irrigated for which groundwater resource the major water resource is used for irrigation. Approximately 67% of the net irrigated area is being irrigated by ground water. The rest of the net irrigated area uses canal (27%) and other sources (6%). The rest of the net sown area (i.e., 44%) is completely rainfed. To analyze Table 13.2, it is observed that the catchments belonging to the left bank of the Yamuna (i.e., Hindon, Karwan and Senger) are highly irrigated (percent net irrigated area is 89%) due to the Ganga canal network. These catchments fall within the Ganga-Yamuna *Duab*. These catchments not only have canal networks, but the ground water table is relatively high to support supplementary irrigation.

A majority of agricultural lands is sown more than once in a single crop-calendar year. The average cropping intensity in the basin varies between 105 and 212%, depending upon the water availability and topography with an average of 140%. Combinations of cropping patterns largely depend on the crop diversity which is a good indicator of sustainable agriculture. Higher the crop diversity lesser will be the risk involved. There are various crops that can be cultivated in a year in the basin, depending on climate, soil, and water. Based on the physical environment, the suitability of crops is listed in Table 13.3. Major climatic factors affecting crops are temperature, annual rainfall, wind speed and sunshine hours. Crops grown in different climatic zones have different water demands.

Sub-basin wise cultivated crops during *kharif* and *rabi* are summarized in Tables 13.4 and 13.5, respectively. Based on Table 13.4, the area under soybean is the largest in the basin, followed by paddy, cotton, arhar and maize; however, a small value of the coefficient of spatial variation for paddy (i.e., CV = 0.48) shows its dominance as a most common food crop. In spite of the largest area being sown under soybean, its dominance is reduced due to the large value of the coefficient of spatial variation (CV = 2.0). Similarly among *rabi* crops grown (Table 13.5), wheat, mustard, barley and potato are the major crops; however, a less value of the coefficient of spatial variation for wheat (i.e., 0.65) indicates its dominance as a major food crop in the Yamuna basin.

Table 13.2 Agriculture and irrigation sources in the Yamuna basin

Catchments	Net sown area (km^2)	Area sown more than once (km^2)	Grass area (km^2)	Cropping intensity (%)	Net irrigated area, NIA (km^2)	NIA by canal (%)	NIA by tube well (%)	NIA by other sources (%)
Sahibi	10,914.3	6,257.3	17,171.6	164	7,685.4	21.28	78.66	0.05
Hindon	5,052.2	2,687.9	7,740.1	153	4,519.3	*18.45*	*79.88*	1.67
Khri	13,395.2	6,316.0	19,711.2	148	10,270.6	9.98	89.39	0.63
Chambal	62,349.2	14,239.6	76,588.7	125	27,753.7	15.24	76.15	8.61
Sind	14,918.3	3,283.4	18,201.7	123	7,335.3	33.92	57.41	8.66
Betwa	26,853.1	6,304.5	33,157.6	126	12,129.2	22.65	60.57	16.78
Ken	15,834.4	4,043.9	19,878.2	127	6,640.1	15.14	67.00	17.86
Karwan	3,503.4	2,000.5	5,503.9	157	3,160.8	*18.37*	*80.10*	1.53
Sengar	4,136.4	2,399.0	65,35.4	158	3,607.2	*25.94*	*73.85*	0.21
Yamuna	*187,908.6*	*66,094.7*	*254,003.3*	*141*	*105,178.3*	*27.16*	*66.93*	*5.90*

Table 13.3 Suitability of crops under different physical environments

Crop	Climatic factors					
	Annual rainfall (mm)	Germination temperature (°C)	Optimum temperature (°C)	Maximum temperature (°C)	Climate zone	Non-climatic factor: soil
Paddy	>800	10	20–35	40	Tropical and Sub tropical. Need hot, humid climate and prolong sunshine	Alluvial soil, red soil and black soil are very good for rice cultivation. The pH of soil ranges from 5 to 8.
Sorghum (Jowar)	500–600	12	20–25	32	Semi arid tropical	Light to medium deep black soil
Maize	700–1,000	12	>12 (no frost)	35	Tropical, Sub tropical and temperate	Sandy to heavy clay (pH of soil between 7.5 and 8.5)
Pearl Millet (Bajra)	500–600	10	23–28	30	Semi arid tropical	Wide range of soil from very light soil to red loam soil
Arhar (Pigeon pea)	800–1,000	20	25–30	45	Tropical and sub tropical	Sandy to heavy clay loam
Soybean	700–1,000	20	30	35	Sub humid to humid	Sandy loam to clay soil with pH of 7.0
Cotton	>500	15	21–27	40	Tropical and sub tropical	Well drained deep alluvial soil to black clayey soil
Wheat	300–700	20	22–23	28	Semi arid to humid	Well drained clay loam and loam
Barley	300–700	20	23–25	30	Temperate climate having cold dry winter with low rainfall: Semi arid to sub humid	Well drained fertile loam or light clay soil
Gram	800–2,000	20	25–35	35	Tropical	Sandy, heavy loam, well drained loam soil
Potato	150–200	18–20	20 (no frost)	30	Temperate and cool climate	Well drained sandy loam and fertile loam soils are most suitable for potato cultivation
Groundnut	500–1,250	20	30	35	Tropical region and extends to sub tropical	Sandy loam, loamy soil and in black soil with drainage
Mustard	250–400	20	22–23	28	Tropical as well as temperate zone	Light to heavy loam soil. pH around 7.0

Table 13.4 Common cropping systems for *kharif* season in the Yamuna basin (in km^2)

Catchments	Paddy	Jowar	Maize	Bajra	Arhar	Soybean	Cotton
Sahibi	1,283.6	122.0	16.8	788.8	72.6	0.07	556.8
Hindon	800.4	34.6	285.5	80.5	74.0	0.04	2.6
Khri	429.7	60.5	47.4	475.9	86.9	3.6	152.8
Chambal	671.3	5.0	350.9	277.7	601.3	18,530.0	1,143.0
Sind	584.9	0.8	51.7	2.6	591.6	1,654.51	0.01
Betwa	233.4	0.5	21.2	3.1	1,454.7	5,388.5	0.7
Ken	643.6	0.1	0.2	0.0	533.8	1,674.1	0.01
Karwan	371.3	2.6	336.5	65.5	70.8	0.0	6.8
Sengar	634.2	2.5	246.3	30.7	115.5	0.25	6.0
Yamuna	*13,526.8*	*1,014.6*	*1,901.7*	*24,66.2*	*35,12.3*	*25,790.4*	*6,117.44*
CV	**0.48**	1.65	0.99	1.44	1.15	2.01	1.91

Table 13.5 Common cropping systems for *rabi* season in the Yamuna basin (in km^2)

Catchments	Wheat	Barley	Gram	Potato	Groundnut	Mustard	Sugarcane
Sahibi	4,067.0	294.3	336.9	4.4	252.4	3,174.8	140.0
Hindon	2,472.3	62.4	3.0	68.3	16.4	79.7	1,900.8
Khri	5,089.4	482.3	112.9	324.7	278.3	5,441.1	75.4
Chambal	12,252.9	667.6	3,044.7	404.9	1,248.6	10,347.1	105.5
Sind	4,407.9	104.7	866.9	67.7	426.4	2,052.7	65.9
Betwa	6,923.9	190.0	2,712.0	48.2	676.9	771.0	93.2
Ken	3,571.7	51.6	1,904.4	54.0	75.9	199.9	177.7
Karwan	2,167.5	164.9	19.2	142.7	1.2	379.5	270.6
Sengar	25,54.87	188.4	32.8	231.5	2.2	515.9	67.5
Yamuna	*560,10.4*	*3,967.8*	*8,691.3*	*1,566.6*	*2,394.8*	*25,938.4*	*5,756.6*
CV	**0.65**	0.84	1.22	0.93	1.25	1.34	1.85

13.2 Parameters of Cropping Pattern

The parameters, most influential in choosing an optimum cropping pattern, are: crop water requirement, crop suitability based on the physical environment, cropping system, productivity, water utilization under different conditions, and economic value to the farmer. Among these parameters, the net crop return is perhaps the most important parameter used in the adoption of a particular pattern. The procedure of crop water requirement is now described.

13.2.1 Estimation of Crop Water Requirement

The crop water requirement for a specific crop is estimated using the FAO 56 procedure (Allen et al., 1998), in which the crop coefficient is multiplied with the reference evapotranspiration using:

$$ET_c = K_c \times ET_0 \tag{13.1}$$

where ET_c is the crop water requirement (mm), K_c is the crop coefficient, and ET_0 is the reference evapotranspiration (mm). The value of daily ET_0 depends on the climatic factor and location of the area, and can be estimated using the Penman-Montieth formula as:

$$ET_0 = \frac{0.408\,\Delta\,(R_n - G) + \gamma\,\frac{900}{T+273}\,u_2\,\left(e_z^0 - e_z\right)}{\Delta + \gamma\,(1 + 0.34\,u_2)} \tag{13.2}$$

In eq. (13.2) ET_0 = the grass reference ET (mm/d), R_n = the net radiation (MJ m^{-2} d^{-1}), G = the sensible heat exchange from the surface to the soil or water (MJ m^{-2} d^{-1}), T = the mean daily temperature (°C), Δ = the slope of the saturation vapor pressure versus temperature (kPa °C^{-1}), γ = the psychometric constant (= 0.0642 kPa °C^{-1}), u_2 = the mean 24-h wind speed at 2 m above the ground (ms^{-1}), e_z^0 = the saturation vapor pressure based on measurements at 1.5–2.0 m (kPa), and e_z = the actual vapor pressure (kPa).

When wind speed is measured at a height other than 2 m above the ground surface then the adjusted value of wind speed can be computed as (Jensen et al., 1990):

$$u_2 = u_1\left[\ln\left\{(z_2 - d)/z_{om}\right\} / \ln\left\{(z_1 - d)/z_{om}\right\}\right] \tag{13.3}$$

The wind level adjustments are relatively insensitive to the value of z_{om}, and for water a value of d is generally considered to be zero. Therefore, eq. (13.3) takes on the following reduced form:

$$u_2 = u_1\left[\ln(z_2) / \ln(z_1)\right] \tag{13.4}$$

where z_1 = the wind measurement height (m), and z_1 = 2 m.

The saturation vapor pressure for daily or longer time step can be computed as (Allen et al., 1989; Jensen et al., 1990):

$$e_z^0 = \left[e^0(T_{max}) + e^0(T_{min})\right]/2 \tag{13.5}$$

where T_{max} and T_{min} are the daily maximum and minimum temperature (°C), respectively; and e^0 (in kPa) can be computed using the relationship proposed by Tetens (1930) and Murray (1967):

$$e^0 = \exp\left[\frac{16.78\,T - 116.9}{T + 237.3}\right] \tag{13.6}$$

where T is the daily temperature (°C). Equation (13.6) calculates within 0.1% of values in the Smithsonian Meteorological Tables for temperatures in the range of 0–50°C (Allen et al., 1989).

The slope of the saturation vapor pressure curve Δ (in kPa °C^{-1}) can be computed using the expression given by Allen et al. (1989):

$$\Delta = \frac{4098\, e^0}{(T + 237.3)^2} \tag{13.7}$$

where T is in °C. The actual vapor pressure e (kPa) is computed as:

$$e = \frac{\text{RH}}{100} \times e^0 \tag{13.8}$$

where RH is the mean relative humidity (%).

The net radiation flux R_n can be measured directly using hemispherical net radiometers or estimated from the net short-wave and net long-wave components as:

$$R_n = (1 - \alpha)\, R_s^{\downarrow} - R_b^{\uparrow} \tag{13.9}$$

where R_s is the short-wave radiation received by a water, soil or vegetated surface; α is the short-wave reflectance or albedo; and R_b is the net back, or net outgoing, long-wave radiation.

For vegetated surfaces the value of α can be taken as 0.23 within a range of 0.20–0.25. The short-wave radiation R_s (in MJ m^{-2} d^{-1}) can be estimated as (Doorenbos and Pruitt, 1977):

$$R_s = \left(0.25 + 0.50\,\frac{n}{N}\right) \cdot R_A \tag{13.10}$$

where R_A is the extra-terrestrial radiation (MJ m^{-2} d^{-1}), n/N is the ratio between actual measured bright sunshine hour and maximum possible sunshine hours. The value of R_A can be computed for a location as a function of latitude and day of the year using the following set of mathematical equations given by Duffie and Beckman (1991):

$$R_A = \frac{24\,(60)}{\pi} \cdot G_{sc} \cdot d_r \cdot [\omega_s \sin(\phi) \sin(\delta) + \cos(\phi) \cos(\delta) \sin(\omega_s)] \tag{13.11}$$

where R_A = the daily extra-terrestrial radiation (MJ m^{-2} d^{-1}), and ϕ = the latitude of the station (radians) (negative for southern latitudes). The declination δ (in radians) can be estimated as:

$$\delta = 0.4093\ \sin\left[2\pi\,\frac{(284 + J)}{365}\right] \tag{13.12}$$

where J is the day of the year (January 1st = 1). The term d_r in eq. (13.11), the relative distance of the earth from the sun, was computed as:

$$d_r = 1 + 0.033 \cos\left(\frac{2\pi}{365} J\right) \tag{13.13}$$

The sunset hour angle ω_s (in radians) can be estimated as:

$$\omega_s = \arccos\left[-\tan(\phi)\tan(\delta)\right] \tag{13.14}$$

$$\omega_s = \frac{\pi}{2} - \arctan\left[\frac{-\tan(\phi)\tan(\delta)}{\left[1 - \tan^2(\phi)\tan^2(\delta)\right]^{1/2}}\right] \tag{13.15}$$

G_{sc} in eq. (13.11) is the solar constant (= 0.0820 MJ m^{-2} min^{-1}) as recommended by the International Association of Meteorology and Atmospheric Physics (London and Fröhlich, 1982). The value of maximum possible sunshine hours N was approximately computed as:

$$N = \omega_s (24/\pi) \tag{13.16}$$

For the Yamuna basin the stage wise ET_c was calculated for different crops for different sub-basins. The estimated values of crop coefficients (K_c) corresponding to the stage development, suggested by FAO 56, were used for the estimation of crop water requirements as given in Table 13.6. The estimated average consumptive use of different crops and reported values by ICAR (Indian Council of Agricultural Research) are given in Table 13.7.

The crop water requirement varies with the stage of growth. When water supply is limited, it is necessary to take into account the critical stages of crop growth with respect to moisture. The term critical stage is commonly used to define the growth when plants are most sensitive to the shortage of water causing a drastic reduction in yield. The critical stages of different crops are summarized in Table 13.8.

13.2.2 Suitability of Crops Based on Physical Environment

The physical environment for crops includes climate (i.e., germination temperature, optimum temperature, maximum temperature, rainfall, and climate), soil, and topography. Based on these factors suitable crops for the Yamuna basin is listed in Table 13.3.

13.2.2.1 Cropping Systems

A study of cropping systems is not only useful to understand the overall sustainability of agriculture but also in generating parameters that are useful in the assessment of water resources availability. Cropping systems of a region are also determined by soil and climatic parameters which determine the overall agro-ecological setting for nourishment and of a crop or set of crops for cultivation. The important cropping systems are: (i) sole cropping (monoculture) referring to a single crop grown

Table 13.6 Approximate date of sowing, growth stage and crop coefficients for important crops of the Yamuna basin

Crop	Sowing date (Approx)	Growth stage Initial	Development	Middle	Late	Total duration
Wheat	1st November	20	60	20	50	150
Kc		*0.40*	*1.00*	*1.20*	*0.60*	
Barley	1st November	15	25	50	30	120
Kc		*0.20*	*1.00*	*1.15*	*0.25*	
Gram	20th October	20	30	30	20	100
Kc		*0.35*	*0.75*	*1.05*	*0.50*	
Potato	25th October	25	30	30	30	115
Kc		*0.52*	*0.81*	*1.10*	*0.70*	
Mustard	1st November	25	35	35	40	135
Kc		*0.38*	*0.38*	*1.00*	*0.56*	
Bajra	10th June	20	35	25	30	110
Kc		*0.41*	*1.00*	*1.20*	*0.40*	
Paddy	20th June	30	30	60	30	150
Kc		*1.05*	*1.20*	*1.20*	*0.75*	
Jowar	20th June	20	40	45	20	125
Kc		*0.42*	*1.00*	*1.20*	*0.40*	
Maize	15th May	20	35	40	30	125
Kc		*0.42*	*0.95*	*1.00*	*0.50*	
Arhar	15th June	20	35	40	30	125
Kc		*0.63*	*0.63*	*0.63*	*0.63*	
Soybean	15th May	20	35	60	25	140
Kc		*0.40*	*1.00*	*1.15*	*0.50*	
Groundnut	15th May	35	45	35	25	140
Kc		*0.52*	*0.85*	*1.00*	*0.55*	
Cotton	10th April	30	50	60	55	195
Kc		*0.43*	*1.00*	*1.15*	*0.60*	
Sugarcane	15th November	30	50	180	60	320
Kc		*0.47*	*1.00*	*1.25*	*0.70*	

in a season; (ii) rotational cropping wherein cropping is done in a defined sequence; (iii) mixed cropping referring to a process of growing two or more crops together on the same piece of land, mixed cropping is widely practiced in India, particularly during the kharif season; and (iv) intensive cropping wherein the process of growing a number of crops on the same piece of land during a given period of time. The aim is to increase the income per unit area within a specified period of time. Intensive cropping can be either multiple or intercropping. In multiple cropping systems, two or more crops are grown in succession within a year, for example, "paddy-potato-sunflower" followed by "paddy-wheat-gram"; whereas, in an intercropping system more than one crop is grown in the same field during a growing season.

The existing cropping pattern sub-catchment-wise during *Kharif* and *Rabi* for the Yamuna basin is given in Tables 13.9, 13.10, 13.11, 13.12, 13.13, 13.14, 13.15, and 13.16. Among the *kharif* crops, soybean, arhar and cotton account for the highest share of the total *kharif* cropped area for Chambal, Sind and Betwa catchments; and

Table 13.7 Estimated crop water requirements

Kharif season			Rabi season		
Crop	ET_c (FAO 56) [mm]	ET_c (ICAR) [mm]	Crop	ET_c (FAO 56) [mm]	ET_c (ICAR) [mm]
Paddy	530–834 (784.4, 34.8)	900–1,500	Wheat	207–527 (427.3, 46.5)	450–650
Sorghum (Jowar)	337–522 (487, 23.5)	450–650	Groundnut	370–559 (526, 26.3)	500–700
Maize	422–631 (584, 35.4)	500–800	Potato	140–364 (291, 32.2)	300–700
Cotton	544–940 (891, 56.3)	700–1,300	Mustard	166–432 (347, 39.0)	300–450
Soybean	445–668 (627, 32.4)	450–700	Gram	102–295 (213, 25.0)	100–250
Arhar	273–417 (391, 19.2)	350–450	Sugarcane	1,647–1,822 (1,708, 53.4)	1,500–2,500
Bajra	301–447 (414, 23.7)	400–650	Barley	140–375 (296, 35.0)	250–350

Table 13.8 Critical stages of different crops

Crops	Critical satges
Paddy	Transplanting, Panicle initiation, flowering.
Sorghum (Jowar)	Booting, Blooming, milky and dough stage
Maize	Teaseling, silking stages to early grain formation
Bajra	Panicle initiation, flowering
Cotton	Flowering and boll development
Soybean	Blooming and seed formation
Wheat	Crown root initiation, shooting, earing
Groundnut	Flowering, Peg penetration, Seed development
Pulses	Flowering and pod formation
Potato	Tuber initiation to tuber maturity

for other catchments paddy is the most dominating crop, followed by maize and arhar. Wheat is grown in almost every district occupying the maximum share of the total *rabi* cropped area. Most of the irrigation water is shared among these major crops.

Tables 13.9, 13.10, 13.11, 13.12, 13.13, 13.14, 13.15, and 13.16 also include the sub-catchment and district-wise net sown area along with the area under major *kharif* and *rabi* crops and their percentage distribution. These numbers were further used for the estimation of crop-wise water demand and to finally determine suitable crop combinations based on water use and gross return.

The final selection of cropping patterns and their implementation largely depend on crop yield, productivity, and gross return. For all crops, the Ministry of Agriculture, Government of India (GoI), in concert with the State Governments, fixes the minimum support price for farmers. These prices are fixed on the basis

Table 13.9 Sub-catchment wise crops sown

District	NSA (ha)	Krf (ha)[a]	Rabi (ha)[a]	Kharif Crops							Rabi Crops					
				Padd	Jowa	Maiz	Bajr	Arha	Soyb	Cott	Whea	Barl	Gram	Pota	Grou	Must
Catchment: Chambal Upper 1																
Chittaurgarh	26,751	10,460	16,479	0.10	0.00	5.80	0.00	2.00	89.80	2.30	39.00	2.00	3.40	0.00	18.90	36.30
Mandsaur	426,899	279,192	99,041	0.00	0.00	0.00	0.00	0.30	99.40	0.20	31.50	0.10	8.20	0.00	2.00	58.20
Jhalawar	32,111	20,166	11,528	0.30	0.00	0.20	0.00	2.00	97.50	0.00	33.40	0.10	4.20	0.10	2.00	60.20
Shajapur	44,779	32,599	12,672	0.00	0.00	0.00	0.00	1.60	98.40	0.00	40.40	0.10	52.00	3.30	1.40	2.40
Ratlam	192,940	125,604	39,553	0.00	0.00	0.50	0.00	0.80	84.30	14.40	62.20	0.00	19.70	0.40	0.60	17.10
Ujjain	449,192	374,177	123,079	0.00	0.00	0.00	0.00	0.00	99.90	0.00	51.20	0.00	35.90	1.70	0.40	10.60
Dewas	53,026	43,375	18,294	0.00	0.00	0.00	0.00	0.80	88.70	10.60	57.30	0.00	37.10	4.20	0.50	0.00
Dhar	96,289	66,632	19,739	0.00	0.00	0.60	0.00	0.20	68.30	31.00	85.30	0.00	8.00	0.70	4.50	0.30
Indore	187,346	166,551	61,637	0.00	0.00	0.00	0.00	0.10	99.80	0.00	68.10	0.00	15.10	15.40	0.50	0.90
Catchment: Chambal Upper 2																
Chittaurgarh	27,646	10,810	17,030	0.10	0.00	5.80	0.00	2.00	89.80	2.30	39.00	2.00	3.40	0.00	18.90	36.30
Mandsaur	93,347	61,049	21,657	0.00	0.00	0.00	0.00	0.30	99.40	0.20	31.50	0.10	8.20	0.00	2.00	58.20
Catchment: Banas																
Sikar	14	1	6	0.00	0.00	0.03	91.05	6.48	0.00	2.44	36.45	10.11	14.49	0.00	12.30	26.63
Jaipur	3,597	370	2,575	0.00	0.45	3.71	79.03	15.46	0.04	1.31	33.31	11.12	2.46	0.00	13.13	39.97
Sawai Madhopur	2,487	20	2,487	5.07	0.93	1.91	53.35	2.79	35.94	0.00	16.68	0.29	0.36	0.00	2.11	80.54
Ajmer	16,041	670	2,951	0.00	0.52	14.17	1.97	2.17	0.00	81.18	32.41	12.83	1.39	0.00	3.56	49.76
Tonk	11,213	66	9,094	0.17	1.41	42.11	15.49	20.97	1.66	18.19	15.28	1.25	0.77	0.00	3.63	79.05
Bhilwara	123,916	11,301	53,904	0.92	0.07	51.33	0.00	10.55	4.74	32.38	38.04	7.66	2.06	0.04	10.61	41.49
Bundi	88	21	88	15.41	0.01	7.07	0.36	6.16	70.89	0.09	37.94	0.28	0.53	0.04	0.13	59.77
Udaipur	46,830	1,686	21,307	0.00	0.00	11.85	0.00	60.73	24.65	2.77	63.41	8.81	4.06	0.00	5.75	17.73
Chittaurgarh	122,205	47,782	75,278	0.06	0.00	5.84	0.00	1.97	89.81	2.31	38.97	1.97	3.38	0.02	18.91	36.26
Mandsaur	10,289	6,729	2,387	0.00	0.00	0.00	0.00	0.35	99.44	0.21	31.48	0.14	8.18	0.03	1.99	58.18

[a] Area under major crops *Padd* paddy, *Jowa* Jowar, *Maiz* Maize, *Arha* Arhar, *Soyb* Soybean, *Cott* Cotton, *Whea* Wheat, *Barl* Barley; *Pota* Potato, *Gron* Groundnut, *Must* Mustard

Table 13.10 Sub-catchment wise crops sown

District	NSA (ha)	Krf (ha)	Rabi (ha)	Kharif crops							Rabi crops					
				Padd	Jowa	Maiz	Bajr	Arha	Soyb	Cott	Whea	Barl	Gram	Pota	Grou	Must
Catchment: Chambal Upper 3																
Sawai Madhopur	688,173	5,505	688,173	5.07	0.93	1.91	53.35	2.79	35.94	0.00	16.68	0.29	0.36	0.00	2.11	80.54
Morena	185,102	814	168,813	46.60	0.00	0.09	0.00	35.92	17.39	0.00	29.01	0.67	2.20	0.17	0.07	67.32
Tonk	27,051	160	21,938	0.17	1.41	42.11	15.49	20.97	1.66	18.19	15.28	1.25	0.77	0.00	3.63	79.05
Bhilwara	32,686	2,981	14,219	0.92	0.07	51.33	0.01	10.55	4.74	32.38	38.04	7.66	2.06	0.04	10.61	41.49
Shivpuri	1,913	543	1,248	3.30	0.00	0.00	0.00	7.92	88.77	0.00	38.42	0.48	11.77	0.15	25.02	23.57
Bundi	216,220	52,325	216,220	15.41	0.01	7.07	0.36	6.16	70.89	0.09	37.94	0.28	0.53	0.04	0.13	59.77
Kota	656,579	328,290	535,112	5.29	0.00	0.16	0.00	0.15	94.39	0.00	31.64	0.15	0.57	0.09	0.16	67.38
Chittaurgarh	15,550	6,080	9,579	0.06	0.00	5.84	0.00	1.97	89.81	2.31	38.97	1.97	3.38	0.02	18.91	36.26
Guna	307,946	69,596	46,808	0.18	0.00	0.00	0.00	0.71	99.12	0.00	58.19	0.05	30.43	0.09	0.11	10.39
Mandsaur	96,744	63,270	22,445	0.00	0.00	0.00	0.00	0.35	99.44	0.21	31.48	0.14	8.18	0.03	1.99	58.18
Jhalawar	274,571	172,431	98,571	0.35	0.00	0.16	0.00	1.95	97.53	0.01	33.39	0.14	4.21	0.09	1.97	60.17
Vidisha	24,019	7,374	9,656	0.00	0.00	0.01	0.00	17.60	82.39	0.00	51.79	0.05	47.12	0.06	0.44	0.42
Shajapur	402,798	293,237	113,992	0.00	0.00	0.00	0.00	1.56	98.43	0.00	40.43	0.08	52.01	3.26	1.45	2.44
Rajgarh	426,801	275,287	99,445	0.00	0.00	0.00	0.00	2.01	97.98	0.01	44.57	0.12	52.33	0.79	1.22	0.88
Bhopal	28,471	17,396	15,090	0.00	0.00	0.00	0.00	1.94	98.05	0.01	73.96	0.00	24.76	0.09	0.52	0.21
Ujjain	30,392	25,316	8,327	0.00	0.00	0.00	0.00	0.04	99.91	0.04	51.16	0.03	35.95	1.72	0.38	10.63
Sehore	167,442	126,084	87,237	0.00	0.01	0.00	0.00	1.53	98.41	0.04	68.47	0.01	30.03	0.06	0.22	0.03
Dewas	108,343	88,625	37,378	0.00	0.00	0.00	0.00	0.80	88.65	10.55	57.33	0.03	37.12	4.16	0.46	0.03
Agra	43,497	405	41,105	34.76	6.75	4.54	0.41	44.54	0.00	9.00	50.71	3.83	1.08	11.65	0.00	32.50
Etawah	24,455	9,953	19,687	81.52	0.15	10.64	0.53	7.15	0.00	0.00	75.68	3.02	0.93	5.15	0.11	14.08

Table 13.11 Sub-catchment wise crops sown

District	NSA (ha)	Krf (ha)	Rabi (ha)	Kharif crops							Rabi crops					
				Padd	Jowa	Maiz	Bajr	Arha	Soyb	Cott	Whea	Barl	Gram	Pota	Grou	Must
Catchment: Sind																
Etawah	12,551	5,108	10,104	81.50	0.20	10.60	0.50	7.10	0.00	0.00	75.70	3.00	0.90	5.10	0.10	14.10
Morena	108,548	478	98,996	46.60	0.00	0.10	0.00	35.90	17.40	0.00	29.00	0.70	2.20	0.20	0.10	67.30
Bhind	307,811	9,696	220,085	39.90	0.10	0.00	0.00	58.20	1.80	0.00	33.40	1.80	0.90	0.30	0.00	63.50
Jalaun	47,775	2,771	18,871	4.60	0.00	0.00	0.00	90.60	4.80	0.00	84.20	3.90	3.20	0.20	0.10	6.90
Gwalior	153,562	153,562	91,523	88.30	0.00	0.00	0.00	11.70	0.00	0.00	89.20	0.50	6.10	1.00	0.00	0.00
Datia	180,424	32,296	114,028	3.20	0.00	0.00	0.00	94.50	2.30	0.00	64.10	2.00	11.60	0.20	8.10	10.10
Jhansi	74,498	8,865	371,75	2.00	0.00	0.00	0.00	94.30	3.70	0.00	69.00	1.60	9.50	0.20	15.50	4.10
Shivpuri	321,525	91,313	209,634	3.30	0.00	0.00	0.00	7.90	88.80	0.00	38.40	0.50	11.80	0.20	25.00	23.60
Guna	141,106	31,890	21,448	0.20	0.00	0.00	0.00	0.70	99.10	0.00	58.20	0.00	30.40	0.10	0.10	10.40
Vidisha	39,290	12,062	15,795	0.00	0.00	0.00	0.00	17.60	82.40	0.00	51.80	0.00	47.10	0.10	0.40	0.40
Catchment: Ken																
Fatehpur	744	231	440	97.10	0.00	0.10	0.00	2.80	0.00	0.00	82.90	0.10	0.10	3.40	0.20	7.80
Hamirpur	404,992	239,755	80,593	0.20	0.00	0.00	0.00	99.10	0.70	0.00	64.30	0.60	13.90	0.20	7.70	9.30
Banda	43,013	7,699	7,355	99.70	0.00	0.00	0.00	0.10	0.20	0.00	92.80	0.10	1.40	0.20	1.30	3.00
Chhatapur	2,764	406	1,440	0.00	0.00	0.00	0.00	58.50	41.50	0.00	58.40	4.40	26.10	0.60	3.50	6.80
Satna	12,230	673	4,623	7.80	0.00	0.00	0.00	8.20	84.10	0.00	65.60	1.50	29.40	1.30	0.00	2.00
Panna	233,270	7,931	76,513	33.40	0.00	0.00	0.00	57.80	8.80	0.00	48.20	1.20	45.50	0.80	0.20	3.80
Sagar	285,030	120,283	112,017	0.00	0.00	0.00	0.00	11.70	88.30	0.00	52.50	0.20	44.80	0.90	0.80	0.40
Damoh	412,228	68,018	147,165	5.80	0.00	0.00	0.00	4.30	90.00	0.00	42.70	0.00	55.60	0.60	0.50	0.60
Jabalpur	25,359	837	9,459	10.40	0.10	0.10	0.00	81.00	8.40	0.00	60.90	0.00	34.00	0.30	0.10	3.80
Raisen	16,673	3,168	7,386	1.60	0.00	0.00	0.00	2.90	95.50	0.00	61.60	0.00	37.20	0.10	0.10	0.10
Narsimhapur	8,981	2,173	4,787	0.10	0.00	0.00	0.00	16.50	83.40	0.00	33.80	0.00	49.60	0.40	0.00	0.20

Table 13.12 Sub-catchment wise crops sown

District	NSA (ha)	Krf (ha)	Rabi (ha)	Kharif crops							Rabi crops					
				Padd	Jowa	Maiz	Bajr	Arha	Soyb	Cott	Whea	Barl	Gram	Pota	Grou	Must
Catchment: Betwa Upper																
Jhansi	893	106	446	2.00	0.00	0.00	0.00	94.30	3.70	0.00	69.00	1.60	9.50	0.20	15.50	4.10
Shivpuri	641,67	18,223	41,837	3.30	0.00	0.00	0.00	7.90	88.80	0.00	38.40	0.50	11.80	0.20	25.00	23.60
Tikamgarh	18	4	11	0.00	0.00	0.00	0.00	40.40	59.50	0.00	58.00	6.30	15.10	0.90	10.80	8.60
Lalitpur	508,587	127,147	254,294	0.00	0.00	0.00	0.00	90.60	9.40	0.00	65.30	4.20	22.40	0.30	5.40	2.30
Guna	242,493	54,803	36,859	0.20	0.00	0.00	0.00	0.70	99.10	0.00	58.20	0.00	30.40	0.10	0.10	10.40
Sagar	195,916	82,677	76,995	0.00	0.00	0.00	0.00	11.70	88.30	0.00	52.50	0.20	44.80	0.90	0.80	0.40
Vidisha	113,474	34,837	45,617	0.00	0.00	0.00	0.00	17.60	82.40	0.00	51.80	0.00	47.10	0.10	0.40	0.40
Bhopal	30,731	18,777	16,287	0.00	0.00	0.00	0.00	1.90	98.00	0.00	74.00	0.00	24.80	0.10	0.50	0.20
Raisen	161,344	30,655	71,475	1.60	0.00	0.00	0.00	2.90	95.50	0.00	61.60	0.00	37.20	0.10	0.10	0.10
Sehore	4,299	3,237	2,240	0.00	0.00	0.00	0.00	1.50	98.40	0.00	68.50	0.00	30.00	0.10	0.20	0.00
Catchment: Betwa Lower																
Kanpur Dehat	40,957	9,297	26,049	77.20	0.10	8.20	1.20	13.00	0.20	0.00	75.10	2.30	1.50	2.00	0.00	16.50
Jalaun	62,683	3,636	24,760	4.60	0.00	0.00	0.00	90.60	4.80	0.00	84.20	3.90	3.20	0.20	0.10	6.90
Fatehpur	96,153	29,807	56,923	97.10	0.00	0.10	0.00	2.80	0.00	0.00	82.90	0.10	0.10	3.40	0.20	7.80
Hamirpur	225,502	133,497	44,875	0.20	0.00	0.00	0.00	99.10	0.70	0.00	64.30	0.60	13.90	0.20	7.70	9.30
Jhansi	204,732	24,363	102,161	2.00	0.00	0.00	0.00	94.30	3.70	0.00	69.00	1.60	9.50	0.20	15.50	4.10
Tikamgarh	180,308	367,83	115,037	0.00	0.00	0.00	0.00	40.40	59.50	0.00	58.00	6.30	15.10	0.90	10.80	8.60
Chhatapur	481,599	70,795	250,913	0.00	0.00	0.00	0.00	58.50	41.50	0.00	58.40	4.40	26.10	0.60	3.50	6.80
Lalitpur	142,820	35,705	71,410	0.00	0.00	0.00	0.00	90.60	9.40	0.00	65.30	4.20	22.40	0.30	5.40	2.30
Sagar	190,453	80,371	74,848	0.00	0.00	0.00	0.00	11.70	88.30	0.00	52.50	0.20	44.80	0.90	0.80	0.40
Raisen	8,894	1,690	3,940	1.60	0.00	0.00	0.00	2.90	95.50	0.00	61.60	0.00	37.20	0.10	0.10	0.10

Table 13.13 Sub-catchment wise crops sown

				Kharif crops							Rabi crops					
District	NSA (ha)	Krf (ha)	Rabi (ha)	Padd	Jowa	Maiz	Bajr	Arha	Soyb	Cott	Whea	Barl	Gram	Pota	Grou	Must
Catchment: 73																
Muzaffarnagar	3,444	510	3,444	85.30	0.00	1.70	0.10	12.70	0.00	0.20	38.00	0.40	0.10	0.70	0.00	1.00
Meerut	41,081	5,587	41,081	77.40	0.10	12.10	0.30	10.00	0.00	0.20	39.70	0.10	0.10	2.80	0.00	1.70
Ghaziabad	51,786	12,480	41,481	54.30	0.50	24.70	17.60	2.90	0.00	0.00	65.40	2.70	0.00	2.50	0.00	1.50
Bulandshahr	113,074	59,477	100,410	34.10	0.30	55.00	6.70	3.70	0.00	0.30	71.50	4.60	0.20	2.80	0.00	3.30
Faridabad	32,838	8,801	23,873	61.50	16.70	0.10	13.40	7.80	0.00	0.50	89.80	1.00	0.00	0.20	0.00	4.50
Aligarh				38.60	0.00	44.20	5.10	10.70	0.00	1.40	76.00	8.50	0.40	3.10	0.00	8.90
Etah	1,518	427	1,169	31.80	0.00	41.70	7.60	18.80	0.00	0.10	78.20	6.50	1.90	3.90	0.20	6.30
Mathura	204,706	31,934	204,706	91.50	1.00	1.30	0.90	1.90	0.00	3.50	71.60	5.00	0.10	2.90	0.00	16.80
Firozabad	33,398	4,743	31,795	63.90	0.10	19.00	0.80	16.20	0.00	0.00	63.40	8.90	1.10	17.20	0.00	9.20
Agra	79,426	739	75,058	34.80	6.70	4.50	0.40	44.50	0.00	9.00	50.70	3.80	1.10	11.60	0.00	32.50
Catchment: 46																
Kinnaur	47	10	10	0.00	0.00	15.30	0.00	84.70	0.00	0.00	8.10	76.00	0.00	15.90	0.00	0.00
Shimla	51,568	1,753	5,286	65.70	0.00	30.30	0.00	3.90	0.00	0.00	16.10	3.90	0.00	80.00	0.00	0.00
Uttarkashi	9,948	1,622	2,139	95.10	0.00	0.00	0.00	3.00	1.90	0.00	52.60	1.20	0.00	31.50	0.00	14.60
Dehradun	30,337	6,401	10,921	99.00	0.00	0.40	0.00	0.60	0.10	0.00	64.30	0.20	0.00	3.70	1.30	6.80
Sirmaur	44,399	12,299	12,920	38.40	0.00	51.90	0.00	9.70	0.00	0.00	75.30	4.70	0.40	12.00	0.00	0.00
Tehri Garhwal	4,281	608	702	92.40	0.00	0.00	0.00	0.50	7.10	0.00	72.00	0.70	0.00	13.90	0.00	13.40
Yamunanagar	69,649	33,780	62,336	96.80	0.00	2.00	0.00	1.20	0.00	0.00	61.90	0.00	0.00	1.10	0.10	1.90
Saharanpur	124,812	34,823	108,961	93.00	0.00	0.30	0.00	6.60	0.00	0.10	46.90	0.00	0.00	0.30	1.70	0.70
Hardwar	172	25	156	96.40	0.00	0.60	0.00	2.90	0.00	0.10	35.20	0.00	0.00	0.40	1.20	0.70
Kurukshetra	17,152	12,864	14,734	98.80	0.00	0.20	0.10	1.00	0.00	0.00	85.00	0.00	0.10	3.80	0.00	0.50
Karnal	100,152	90,537	98,550	98.80	0.00	0.20	0.70	0.30	0.00	0.00	92.60	0.10	0.10	0.50	0.00	0.40
Jind	4,312	2,949	3,820	56.50	0.10	0.00	14.20	0.00	0.00	29.20	92.30	0.30	0.10	0.00	0.00	5.70

Table 13.14 Sub-catchment wise crops sown

District	NSA (ha)	Krf (ha)	Rabi (ha)	Kharif crops							Rabi crops					
				Padd	Jowa	Maiz	Bajr	Arha	Soyb	Cott	Whea	Barl	Gram	Pota	Grou	Must
Catchment: 46 (Continued. . .)																
Muzaffarnagar	50,833	7,523	50,833	85.30	0.00	1.70	0.10	12.70	0.00	0.20	38.00	0.40	0.10	0.70	0.00	1.00
Panipat	81,037	62,885	79,416	98.90	0.00	0.10	0.90	0.10	0.00	0.10	91.80	0.00	0.10	0.20	0.00	0.80
Meerut	37,100	5,046	37,100	77.40	0.10	12.10	0.30	10.00	0.00	0.20	39.70	0.10	0.10	2.80	0.00	1.70
Rohtak	362,119	166,937	327,356	27.60	30.50	0.10	8.10	13.60	0.00	20.00	68.80	0.60	0.30	0.00	0.00	21.10
Sonipat	100,451	50,125	92,515	76.00	15.10	0.50	5.80	0.40	0.00	2.20	88.10	0.40	0.00	0.20	0.00	3.40
Bhiwani	111,159	31,569	89,038	11.00	2.10	0.10	21.60	1.00	0.00	64.30	37.00	1.50	3.90	0.00	0.00	57.10
Ghaziabad	10,928	2,634	8,753	54.30	0.50	24.70	17.60	2.90	0.00	0.00	65.40	2.70	0.00	2.50	0.00	1.50
Delhi	31,352	10,346	16,836	66.80	32.50	0.30	0.30	0.10	0.00	0.00	99.70	0.20	0.10	0.00	0.00	0.00
Bulandshahr	2,361	1,242	2,097	34.10	0.30	55.00	6.70	3.70	0.00	0.30	71.50	4.60	0.20	2.80	0.00	3.30
Gurgaon	75,735	6,968	45,517	12.60	5.30	0.00	80.60	0.40	0.00	1.10	42.10	1.60	0.00	0.00	0.30	56.00
Jhunjhunun	42,361	3,516	24,019	0.00	0.00	0.00	91.50	4.60	0.00	3.90	28.80	3.50	19.90	0.00	0.70	47.10
Faridabad	10,208	2,736	7,421	61.50	16.70	0.10	13.40	7.80	0.00	0.50	89.80	1.00	0.00	0.20	0.00	4.50
Mahendragarh	147,890	28,691	137,833	0.00	0.10	0.00	53.40	0.30	0.00	46.30	27.00	0.40	2.50	0.00	0.30	69.80
Rewari	109,986	30,136	101,627	1.80	3.70	0.00	71.30	0.00	0.00	23.10	36.10	0.70	0.10	0.00	0.30	62.90
Alwar	202,952	27,804	188,542	0.20	0.60	3.80	52.90	0.10	0.00	42.30	33.70	2.80	1.60	0.00	0.20	61.70
Sikar	64,141	5,644	29,120	0.00	0.00	0.00	91.10	6.50	0.00	2.40	36.40	10.10	14.50	0.00	12.30	26.60
Jaipur	636,48	6,556	45,572	0.00	0.40	3.70	79.00	15.50	0.00	1.30	33.30	11.10	2.50	0.00	13.10	40.00

Table 13.15 Sub-catchment wise crops sown

District	NSA (ha)	Krf (ha)	Rabi (ha)	Kharif crops							Rabi crops					
				Padd	Jowa	Maiz	Bajr	Arha	Soyb	Cott	Whea	Barl	Gram	Pota	Grou	Must
Catchment: Kali-Hindon																
Saharanpur	46,696	13,028	40,766	93.00	0.00	0.30	0.00	6.60	0.00	0.10	46.90	0.00	0.00	0.30	1.70	0.70
Hardwar	15,869	2,269	14,425	96.40	0.00	0.60	0.00	2.90	0.00	0.10	35.20	0.00	0.00	0.40	1.20	0.70
Muzaffarnagar	61,823	9,150	61,823	85.30	0.00	1.70	0.10	12.70	0.00	0.20	38.00	0.40	0.10	0.70	0.00	1.00
Meerut	2,671	363	2,671	77.40	0.10	12.10	0.30	10.00	0.00	0.20	39.70	0.10	0.10	2.80	0.00	1.70
Catchment: Hindon Lower																
Saharanpur	18,574	5,182	16,215	93.00	0.00	0.30	0.00	6.60	0.00	0.10	46.90	0.00	0.00	0.30	1.70	0.70
Muzaffarnagar	57,634	8,530	57,634	85.30	0.00	1.70	0.10	12.70	0.00	0.20	38.00	0.40	0.10	0.70	0.00	1.00
Meerut	77,738	10,572	35,837	77.40	0.10	12.10	0.30	10.00	0.00	0.20	39.70	0.10	0.10	2.80	0.00	1.70
Ghaziabad	57,736	13,914	46,247	54.30	0.50	24.70	17.60	2.90	0.00	0.00	65.40	2.70	0.00	2.50	0.00	1.50
Delhi	525	173	282	66.80	32.50	0.30	0.30	0.10	0.00	0.00	99.70	0.20	0.10	0.00	0.00	0.00
Bulandshahr	52,35	2,754	4,649	34.10	0.30	55.00	6.70	3.70	0.00	0.30	71.50	4.60	0.20	2.80	0.00	3.30
Faridabad	70	19	51	61.50	16.70	0.10	13.40	7.80	0.00	0.50	89.80	1.00	0.00	0.20	0.00	4.50
Catchment: 29																
Aligarh	47,547	13,931	43,933	38.60	0.00	44.20	5.10	10.70	0.00	1.40	76.00	8.50	0.40	3.10	0.00	8.90
Etah	49,776	13,987	38,328	31.80	0.00	41.70	7.60	18.80	0.00	0.10	78.20	6.50	1.90	3.90	0.20	6.30
Mathura	154	24	154	91.50	1.00	1.30	0.90	1.90	0.00	3.50	71.60	5.00	0.10	2.90	0.00	16.80
Firozabad	93,055	13,214	88,588	63.90	0.10	19.00	0.80	16.20	0.00	0.00	63.40	8.90	1.10	17.20	0.00	9.20
Mainpuri	92,365	47,568	75,832	63.20	0.30	24.20	2.10	10.10	0.00	0.00	83.40	2.50	0.80	7.90	0.10	5.00
Agra	7,213	67	6,816	34.80	6.70	4.50	0.40	44.50	0.00	9.00	50.70	3.80	1.10	11.60	0.00	32.50
Etawah	290,743	118,332	234,048	81.50	0.20	10.60	0.50	7.10	0.00	0.00	75.70	3.00	0.90	5.10	0.10	14.10
Kanpur Dehat	83,836	19,031	53,320	77.20	0.10	8.20	1.20	13.00	0.20	0.00	75.10	2.30	1.50	2.00	0.00	16.50
Jalaun	256,284	14,864	101,232	4.60	0.00	0.00	0.00	90.60	4.80	0.00	84.20	3.90	3.20	0.20	0.10	6.90
Fatehpur	19	6	11	97.10	0.00	0.10	0.00	2.80	0.00	0.00	82.90	0.10	0.10	3.40	0.20	7.80
Hamirpur	13,227	7,830	2,632	0.20	0.00	0.00	0.00	99.10	0.70	0.00	64.30	0.60	13.90	0.20	7.70	9.30
Jhansi	14,321	1,704	7,146	2.00	0.00	0.00	0.00	94.30	3.70	0.00	69.00	1.60	9.50	0.20	15.50	4.10

Table 13.16 Sub-catchment wise crops sown

				Kharif crops							Rabi crops					
District	NSA (ha)	Krf (ha)	Rabi (ha)	Padd	Jowa	Maiz	Bajr	Arha	Soyb	Cott	Whea	Barl	Gram	Pota	Grou	Must
Catchment: 25																
Fatehpur	184,573	57,218	109,267	97.10	0.00	0.10	0.00	2.80	0.00	0.00	82.90	0.10	0.10	3.40	0.20	7.80
Banda	474,949	85,016	81,216	99.70	0.00	0.00	0.00	0.10	0.20	0.00	92.80	0.10	1.40	0.20	1.30	3.00
Chhatapur	254	37	132	0.00	0.00	0.00	0.00	58.50	41.50	0.00	58.40	4.40	26.10	0.60	3.50	6.80
Satna	43,334	2,383	16,380	7.80	0.00	0.00	0.00	8.20	84.10	0.00	65.60	1.50	29.40	1.30	0.00	2.00
Panna	31,590	1,074	10,362	33.40	0.00	0.00	0.00	57.80	8.80	0.00	48.20	1.20	45.50	0.80	0.20	3.80
Catchment: 30																
Mainpuri	20,191	10,398	16,577	63.20	0.30	24.20	2.10	10.10	0.00	0.00	83.40	2.50	0.80	7.90	0.10	5.00
Etawah	101,061	41,132	81,354	81.50	0.20	10.60	0.50	7.10	0.00	0.00	75.70	3.00	0.90	5.10	0.10	14.10
Kanpur Dehat	82,913	18,821	52,733	77.20	0.10	8.20	1.20	13.00	0.20	0.00	75.10	2.30	1.50	2.00	0.00	16.50
Fatehpur	72,225	22,390	42,757	97.10	0.00	0.10	0.00	2.80	0.00	0.00	82.90	0.10	0.10	3.40	0.20	7.80
Hamirpur	2,517	1,490	501	0.20	0.00	0.00	0.00	99.10	0.70	0.00	64.30	0.60	13.90	0.20	7.70	9.30
Banda	3,548	635	607	99.70	0.00	0.00	0.00	0.10	0.20	0.00	92.80	0.10	1.40	0.20	1.30	3.00
Catchment: 41																
Firozabad	34,524	4,902	32,867	63.90	0.10	19.00	0.80	16.20	0.00	0.00	63.40	8.90	1.10	17.20	0.00	9.20
Agra	24,054	224	10,824	34.80	6.70	4.50	0.40	44.50	0.00	9.00	50.70	3.80	1.10	11.60	0.00	32.50
Etawah	80,527	32,774	64,824	81.50	0.20	10.60	0.50	7.10	0.00	0.00	75.70	3.00	0.90	5.10	0.10	14.10
Bhind	350	11	250	39.90	0.10	0.00	0.00	58.20	1.80	0.00	33.40	1.80	0.90	0.30	0.00	63.50

of probable crop production and estimated cultivation cost. The probable crop production is estimated based on the climatic support to crops. Catchment-wise estimated average yield, and minimum support price by GoI are summarized in Table 13.17. However, while estimating the actual production, district-wise crop yield is used. Considering water requirement, minimum support price, and average yield (Table 13.17), it is found that during the *kharif* season cultivation of jowar, soybean, bajra and arhar are more economic than the cultivation of other crops; whereas during the rabi season, mustard, wheat, potato lead to good economic returns with less water requirement and high yield. Barley and gram are of equally good economic value and can be taken as *rabi* crop.

13.3 Identification of Suitable Crop Rotation

Rotational cropping is one of the major and widely used agricultural practices in India. A well-planned rotation not only reduces diseases, pests and weeds but also offers other advantages, such as increasing the soil fertility and utilization of cropping land. Based on the physical environment, soil, historical crop rotation, and minimum support price, twelve combinations of crop rotations are identified: (i) cotton-wheat, (ii) paddy-potato, (iii) paddy-barley, (iv) paddy-barley-maize, (v) paddy-barley-bajra, (vi) paddy-wheat, (vii) groundnut-jowar, (viii) groundnut-barley, (ix) soybean-wheat-maize, (x) arhar-wheat, (xi) paddy-gram, and (xii) paddy-mustard. From the estimated reference evapotranspiration and crop coefficients, theoretical crop water demands are estimated for each district in the sub-catchments. The total crop water demand and gross income for each combination and each sub-catchment are summarized in Tables 13.18 and 13.19, respectively. From Table 13.18 it is observed that the rotational combination of crops "soybean-wheat-maize", "paddy-wheat", and "arhar-wheat" are the major crop rotations used for income. Based on gross returns (Table 13.19), different crop rotations for catchments are marked in "bold."

From Tables 13.9, 13.10, 13.11, 13.12, 13.13, 13.14, 13.15, 13.16, 13.18, and 13.19, it can be remarked that the arhar-wheat rotation can be practiced in all the districts in the Yamuna basin, as it can be grown in the prevailing climatic conditions and the water consumption for this combination is less as compared to others.

In the Chambal basin, most frequently practiced crop rotations are cotton-wheat, soybean-wheat, arhar-wheat, and groundnut-barley. In the Sind and Ken catchments, paddy is the most dominating crop among all the cereals; that is why all the rotations followed by paddy are practiced throughout. Sorghum and millets must take over a good proportion of the kharif land as water requirements (Tables 13.7 and 13.18) for these crops are less than for paddy (for example, paddy-wheat practice can be shifted to sorghum-wheat and paddy-gram can be shifted to bajra-gram to save water).

In the Betwa catchment, the cultivated area under paddy is almost negligible (Table 13.12), except for Kanpur Dehat and Fatehpur that fall within the Ganga-Yamuna *Duab*. Most of the seasonal practice for *kharif-rabi* combination can be possible with only *kharif* crops like Arhar or soyabean. *Kharif*

Table 13.17 Catchment wise estimated average crop yields and minimum support prices

	Kharif season							Rabi season						
Catchment/Crops	Padd	Jow	Maiz	Bajr	Arha	Soyb	Cott	Whea	Barl	Gram	Pota	Grou	Must	Suga
MSP: 2008–2009 (Rs/qt)	850	840	840	840	2,000	1,350	2,500	1,080	**680**	**1,730**	**500**	2,100	**1,830**	89
MSP: 2009–1010 (Rs/qt)	950	**840**	840	**840**	**2,300**	**1,350**	2,500	**1,080**						139
ET_c (mm): Average	784	**487**	584	**414**	**391**	**627**	891	**427**	**296**	**213**	**291**	526	**347**	1,708
ET_c (mm): Range (lo)	530	337	422	301	273	445	544	207	140	102	140	370	166	1,647
ET_c (mm): Range (up)	834	522	631	447	417	668	940	527	375	295	364	559	432	1,822
Sub-catchment wise average yield (qt/ha)														
Chambal up 1	5.7	1.0	3.3	7.8	–	20.7	11.2	24.1	18.2	6.1	1,333.2	11.3	157.7	201.9
Chambal up 2	11.5	0.0	4.1	6.8	–	22.4	9.6	27.0	22.3	5.4	3,987.5	10.6	150.0	250.0
Chambal up 3	9.5	3.5	9.1	9.0	7.9	18.5	11.5	22.2	16.9	6.1	2,277.4	11.1	138.3	306.3
Banas	1.3	0.0	7.5	7.2	7.2	18.4	9.6	25.0	22.5	13.4	–	10.6	–	–
Sind	8.4	14.8	9.3	11.7	9.9	18.2	9.2	24.2	20.9	9.5	21,053.9	11.4	148.6	268.8
Ken	–	9.2	3.9	9.4	5.8	–	9.2	17.2	13.2	9.3	1,997.3	10.1	140.2	316.8
Betwa Upper	–	6.7	2.2	10.2	7.3	–	11.0	18.5	17.7	7.3	17,737.1	10.8	155.3	296.7
Betwa Lower	0.8	10.3	2.5	11.5	7.6	11.3	8.1	21.9	17.4	10.8	42,305.0	10.5	167.4	163.6
Catch 73	12.8	22.0	7.1	11.9	11.5	14.1	5.2	34.6	27.4	9.6	68,077.6	4.0	205.8	35.0
Catch-46	15.5	21.5	3.1	7.6	12.4	11.8	4.6	32.7	21.7	11.0	13,145.4	0.9	162.7	45.0
Kali-Hindon	14.3	23.0	2.9	10.5	9.5	12.3	5.7	33.8	24.5	6.9	1,398.6	3.3	186.1	55.0
Hindon Lower	14.1	23.8	4.7	10.3	10.5	12.4	5.4	35.3	26.7	8.3	6,441.9	2.0	177.6	55.0
Catch 29	8.5	17.6	7.1	13.2	10.7	14.5	6.3	29.7	23.6	12.4	66,990.0	4.7	193.2	50.0
Catch 25	10.0	8.8	2.7	9.4	5.3	11.9	5.7	17.0	11.7	9.3	1,135.7	6.3	115.1	55.0
Catch 30	7.8	17.1	10.1	13.6	9.5	15.0	6.6	26.2	19.8	15.4	17,095.9	5.4	148.3	50.0
Catch 41	10.6	21.5	10.4	13.9	12.5	18.5	8.4	28.8	26.6	12.1	38,998.4	2.9	148.9	45.0
Average Yield (qt/ha)	10.5	**18.4**	5.6	**11.3**	**9.9**	**13.5**	6.2	**28.9**	***22.2***	***10.6***	**28,398.7**	4.4	**167.2**	61.5

Table 13.18 Catchment-wise water requirements under different crop rotations

	Crop-Rotation Combination and Estimated Water Requirement (ha m)											
	Cotton-wheat	Paddy-potato	Paddy-barley	Paddy-barley-maize	Paddy-barley-bajra	Paddy-wheat	Groundnut-jowar	Groundnut-barley	Soyabean-wheat-maize	Arhar-wheat	Paddy-gram	Paddy-mustard
Catchment	1	2	3	4	5	6	7	8	9	10	11	12
Chambal up 1	140,251	120	177	873	110	16,972	2,121	3,775	535,456	103,388	2,218	7,693
Chambal up 2	3,649	6	114	464	114	110	1,612	1,947	6,228	3,484	138	2,318
Chambal up 3	174,233	21,609	23,160	26,843	9,252	237,824	10,815	15,741	1,198,192	324,005	279,53	472,126
Chambal	**318,133**	**21,735**	**23,450**	**28,179**	**9,477**	**254,906**	**14,547**	**21,463**	**1,739,876**	**430,877**	**30,308**	**482,137**
Banas	34,164	115	1,897	6,726	1,907	23,292	10,422	13,546	60,267	31,392	955	21,605
Sind	5,743	115,159	120,671	8,769	7,923	291,041	17	38,466	162,658	206,965	130,273	105,558
Ken	250,99	12,354	12,037	240	0	117,597	6	4,849	2,866	208,894	53,811	16,030
Betwa Upper	37,968	1,444	4,386	539	0	133,384	3	16,489	*45,487*	*200,931*	31,225	8,224
Betwa Lower	13,923	30,849	33,728	32,928	6,437	181,638	80	31,308	*113,604*	*318,661*	48,383	40,417
Betwa	**51,891**	**32,293**	**38,114**	**33,467**	**6,437**	**315,022**	**82**	**47,797**	**159,091**	**519,592**	**79,608**	**48,641**
Kali-Hindon	18,001	18,040	17,928	18,087	16,192	35,831	0	550	10,689	16,929	16,133	16,399
Hindon Lower	29,164	23,661	23,500	27,407	24,650	52,270	5	271	8,216	30,425	23,062	23,699
Hindon	**47,165**	**41,701**	**41,428**	**45,494**	**40,842**	**88,101**	**5**	**821**	**18,906**	**47,354**	**39,195**	**40,097**
Catch 29	51,252	58,923	133,990	157,117	135,526	335,068	449	9,164	155,100	227,294	127,879	149,674
Catch 25	0	111,294	110,229	43,995	0	189,660	139	894	39,553	80,630	111,295	114,183
Catch 30	0	35,818	61,711	58,937	38,801	114,263	148	1,182	61,785	62,043	60,891	68,948
Catch 41	10,940	4,571	24,716	27,023	24,808	54,960	76	1,555	22,529	33,082	23,424	28,586
Catch 73	138,610	63,326	62,386	85,698	65,721	193,638	852	2,946	975	139,602	56,634	78,646
Catch-46	399,831	239,007	287,991	293,569	310,904	592,449	7,461	12,240	21,051	340,380	252,952	385,205
Total (ha-m)	**1,082,828**	**736,295**	**918,618**	**789,215**	**642,345**	**2,569,997**	**34,205**	**154,923**	**2,444,655**	**2,328,105**	**967,227**	**1,539,309**

Table 13.19 Catchment-wise gross-returns under different crop rotations

Catchment	Gross return (million Rs)						
	Cott-whet	Pad-Pot	Pad-Barl	Pad-Barl-Maiz	Pad-Barl-Pl Mlt	Pad-Wht	Gn-Jowr
Chambal Up 1	**6,409**	632	8	40	8	4,970	87
Chambal Up 2	**393**				0	391	8
Chambal Up 3	**19,233**	639	276	394	323	**19,154**	97
Banas	**1,422**					**1,422**	1
Sind	**10,658**	3,164	3,179	3,188	3,179	**13,693**	1,471
Ken	**3,368**	204	111	111		**3,464**	33
Betwa up	**6,241**	90	131	131		**6,252**	533
Betwa lower	**10,989**	823	873	878	873	**11,574**	639
Catch-73	0	3,045	1,732	2,342	1,776	**14,033**	7
Catch-46	**31,858**	6,355	6,235	6,467	7,558	**33,799**	153
Kali-Hindon	0	419	388	391	388	**1,906**	12
Hindon Lwr	0	720	602	695	632	**3,495**	6
Catch 29	0	4,672	3,449	4,033	3,476	**19,757**	26
Catch 25	0	1,907	1,711	1,712	1,711	**5,952**	11
Catch 30	0	1,596	1,428	1,551	1,433	**6,525**	4
Catch 41	0	1,056	691	764	693	**3,146**	2
Total Dr Rtn (Rs)	90,571	25,321	20,817	22,697	22,050	**149,535**	3,090
Total (ha-m)	1,082,828	736,295	918,618	789,215	642,345	**2,569,997**	34,205

Table 13.19 (continued)

Catchment	Gross return (million Rs)				
	Gn-Barl	Soy-Wht-Maiz	Arhr-Wht	Pad-Gram	Pad-Must
Chambal Up 1	95	**20,385**	**5,009**	1,145	
Chambal Up 2	14	**1,267**	**396**		
Chambal Up 3	201	**33,341**	**19,312**	3,368	577
Banas	29	**1,649**	**1,442**		
Sind	1,614	**12,330**	**11,392**	4,328	6,123
Ken	48	**5,725**	**3,591**	2,567	199
Betwa up	654	**9,419**	**9,119**	2,658	96
Betwa lower	927	**12,720**	**13,077**	3,309	1,008
Catch-73	417	**13,327**	**12,826**	1,351	2,979
Catch-46	302	**28,055**	**28,520**	6,270	14,414
Kali-Hindon	16	**1,524**	**1,547**	385	401
Hindon Lwr	36	**3,015**	**2,967**	574	607
Catch 29	570	**17,450**	**17,827**	3,116	4,452
Catch 25	15	**4,263**	**4,323**	1,825	1,832
Catch 30	73	**5,292**	**5,377**	1,399	1,905
Catch 41	108	**2,634**	**2,651**	612	955
Total Dr Rtn (Rs)	5,116	**172,396**	**139,377**	32,907	35,549
Total (ha-m)	154,923	**2,444,655**	**2,328,105**	967,227	1,539,309

intercropping can be suggested as the best solution for the sustainable use of the net sown area and also improving the productivity. In catchment 73, all the districts have an average area of more than 50% used for paddy cultivation during *kharif*. Based on the estimated water demand (Tables 13.7 and 13.18), the combination with paddy would be the appropriate rotation. The water requirement in the paddy-gram and paddy-barley rotation is quite less than paddy-wheat. In the Kali-Hindon catchment, paddy cultivation area in the *kharif* season is more than 80% (Table 13.15). Sequentially the rotation with paddy is advisable with gram, mustard, barley and potato, according to the estimated water requirement.

13.4 Identification of Suitable Cropping Patterns

Increasing crop productivity and saving irrigation water are two interrelated issues raising a lot of concern these days in India. Legume or cereal intercropping pattern is generally more productive than the sole crop and can be a way of saving irrigation water. Considering the land use pattern and major crops grown during *kharif* and *rabi*, intercropping for the Yamuna basin is categorized into two groups. The possible intercropping patterns taken for this study area are given in Table 13.20.

During the kharif season, paddy, arhar and soybean are the major crops (see Tables 13.9, 13.10, 13.11, 13.12, 13.13, 13.14, 13.15, and 13.16). Specific intercropping patterns (the row spacing for the particular type crop combination) considered for the study are chosen from various documents (ICAR, 2006). Then, district-wise water requirements for intercropping patterns are estimated for all the sub-catchments. The sub-catchment wise water and gross-income summary are given in Tables 13.21, 13.22, 13.23, and 13.24. From Table 13.20, it may be

Table 13.20 Possible intercropping combinations for the Yamuna basin

Crop Season	Combination	Area ratio
Kharif	Soybean+Paddy	1:2
	Soybean+Cotton	1:1
	Soybean+Arhar	1:1
	Soybean+Arhar	4:2
	Soybean+Jowar	2:2
	Bajra+Soyabean	4:2
	Bajra+Soyabean	1:1
	Arhar+ Sorghum	1:2
	Arhar+ Bajra	1:2
	Arhar+Maize	1:2
	Arhar+ Paddy	1:3
Rabi	Groundnut +Wheat	3:1
	Wheat+Gram	4:2
	Mustard+Potato	1:3
	Mustard+Wheat	1:9
	Mustard+Gram	1:2

Table 13.21 Sub-catchment wise estimated water requirements for inter cropping patterns for Rabi

	Rabi				
	Grout +Whet	Wht+Gram	Must+Pot	Must+Wht	Must+Gram
Catchment	12	13	14	15	16
Chambal up 1	200,869	132,213	143,772	195,503	120,172
Chambal up 2	19,081	12,015	13,033	17,685	10,920
Chambal up 3	684,125	678,140	733,759	1,001,204	615,040
Chambal	***904,075***	***822,368***	***890,564***	***1,214,393***	***746,132***
Banas	83,751	51,305	52,054	75,087	46,645
Sind	426,760	240,167	256,858	357,375	217,145
Ken	221,382	136,368	147,658	202,820	123,554
Betwa Upper	270,068	165,998	179,915	247,148	150,423
Betwa Lower	385,724	227,361	245,622	338,181	205,763
Betwa	***655,793***	***393,359***	***425,537***	***585,329***	***356,186***
Kali-Hindon	60,113	27,524	30,511	42,864	23,320
Hindon Lower	80,687	39,990	43,081	59,223	36,158
Hindon	***140,800***	***67,514***	***73,592***	***102,087***	***59,478***
Catch 73	266,305	138,066	148,526	204,864	124,730
Catch-46	812,720	402,713	437,071	602,399	358,280
Catch 29	336,415	181,341	144,684	269,887	163,680
Catch 25	109,480	75,067	75,323	96,534	79,030
Catch 30	100,185	54,960	41,610	81,744	49,616
Catch 41	56,300	30,139	18,479	44,841	27,203
Total (ha m)	**4,113,966**	**2,593,367**	**2,711,955**	**3,837,359**	**2,351,680**

remarked that the estimated water requirement for the soybean intercropping is less than for paddy. Among the combinations, the soybean-arhar combination requires less water than others. The soybean-sorghum intercropping not only requires less water but also provides better crop rotation with millets. It is apparent from Tables 13.21 and 13.22 that bajra (millets) with soybean and arhar can save a lot of water. The arhar-paddy intercropping consumes less water than soybean with paddy or cotton intercropping.

It is observed from Table 13.23 that among the various cropping patterns suitable for the Yamuna River basin, four inter-cropping patterns for the Kharif season are identified based on economic returns and water requirements. These inter-crops are maize-arhar (1:1), soybean-arhar (1:2), soybean-arhar (1:1), and paddy-arhar (1:1).

Catchment-wise inter-cropping patterns can be identified from Tables 13.17 and 13.18. Among the various patterns for the rabi season, it is observed from Table 13.24 that the mustard-potato (1:1) combination is the most economic and water conservative, followed by the wheat-gram (1:1) and mustard-wheat (1:1) combinations. However, catchment-wise rabi inter-crops can be selected based on Tables 13.21 and 13.24.

Table 13.22 Sub-catchment wise estimated water requirements for inter cropping patterns for Kharif

	Kharif										
	Soy+Padd (1:2)	Soy+Cott (1:1)	Soy+Arhr (1:1)	Soy+Arhr (4:2)	Soy+Jowr (2:2)	Bajr+Soy (4:2)	Bajr+Soy (1:1)	Arhr+Jowr (1:2)	Arhr+Bajr	Arhr+Maiz	Arhr+Padd
Catchment	1	2	3	4	5	6	7	8	9	10	11
Chambal up 1	797,281	816,460	542,491	583,154	596,582	511,972	550,099	492,625	430,647	552,976	752,888
Chambal up 2	51,239	52,616	35,034	37,672	38,479	33,101	35,563	31,713	27,825	35,720	48,318
Chambal up 3	1,108,944	1,143,573	760,291	817,886	833,793	718,394	772,064	685,509	603,205	775,258	1,044,536
Chambal	**1,957,464**	**2,012,648**	**1,337,816**	**1,438,712**	**1,468,854**	**1,263,467**	**1,357,726**	**1,209,847**	**1,061,676**	**1,363,955**	**1,845,741**
Banas	48,902	50,112	33,461	35,975	36,770	31,662	33,997	30,331	26,634	34,123	46,118
Sind	142,332	266,577	177,922	191,638	194,493	169,150	181,630	158,869	141,719	182,109	240,274
Ken	323,802	337,548	223,981	241,272	244,943	212,272	228,167	200,057	177,689	229,689	303,858
Betwa Upper	261,907	272,074	180,414	194,225	197,348	170,433	183,287	161,558	142,811	184,562	246,198
Betwa Lower	307,846	321,050	213,240	229,723	233,097	202,228	217,343	190,265	169,261	218,743	288,765
Betwa	**569,754**	**593,124**	**393,654**	**423,948**	**430,445**	**372,661**	**400,630**	**351,824**	**312,071**	**403,305**	**534,963**
Kali-Hindon	18,613	19,502	12,658	13,852	14,348	12,642	13,541	11,331	10,254	13,247	17,119
Hindon Lower	30,786	32,132	21,636	23,333	23,661	20,802	22,283	19,246	17,409	22,312	28,749
Hindon	**49,399**	**51,634**	**34,294**	**37,184**	**38,009**	**33,444**	**35,824**	**30,576**	**27,663**	**35,559**	**45,868**
Catch 73	87,056	97,917	65,859	70,984	71,996	63,125	67,652	58,667	52,876	67,796	87,938
Catch-46	489,772	510,738	342,461	369,957	376,460	330,827	354,357	305,306	275,835	353,640	456,632
Catch 29	186,138	194,218	132,199	142,393	144,315	126,106	135,274	117,773	105,719	126,708	173,767
Catch 25	103,818	111,126	89,006	93,668	80,694	70,247	74,646	75,121	68,837	86,015	107,048
Catch 30	61,224	70,901	49,700	53,540	54,262	47,382	50,841	21,424	39,703	47,902	65,494
Catch 41	27,857	30,041	20,083	21,619	21,913	19,166	20,548	18,472	16,093	18,097	25,948
Total (ha m)	**4,047,518**	**4,326,583**	**2,900,437**	**3,120,890**	**3,163,152**	**2,739,508**	**2,941,293**	**2,578,267**	**2,306,516**	**2,948,899**	**3,933,650**

Table 13.23 Sub-catchment wise estimated gross returns for inter cropping patterns for Kharif

Catchment	Soya+Pad (1:2)	Soya+Cot (1:1)	Soy+Arhr (1:1)	Soy+Arhr (1:2)	Soy+Jowr (1:1)	Bajr+Soy (4:2)	Bajr+Soy (1:1)	Jowr+Arhr (1:1)	Bajr+Arhr (1:1)	Maiz+Arhr (1:1)	Pad+Arhr (1:1)
Chambal Up 1	5, 892	16, 233	13, 449	12,548	9,111	5,451	8,126	4,962	3,649	16,782	**3,258**
Chambal Up 2	290	912	799	775	453	359	487	268	312	1,175	**182**
Chambal Up 3	5,426	15,829	15,882	16,290	10,054	5,387	7,705	9,332	6,201	22,092	**4,882**
Banas	32	101	405	509	163	415	335	391	621	1,009	**179**
Sind	5,882	3,857	5,597	5,157	5,009	4,061	4,775	3,493	3,181	6,600	**5,093**
Ken	3,118	1,402	2,158	1,970	1,367	908	1,362	538	531	4,078	**2,885**
Betwa Up	3,017	3,622	5,175	5,198	2,580	1,702	2,554	1,784	1,748	3,590	**2,789**
Betwa lwr	3,348	1,774	4,307	4,574	1,933	1,178	1,760	1,942	1,712	4,146	**3,728**
Catch 73	1,593	6	1,109	1,475	436	468	354	1,308	1,199	1,916	**2,339**
Catch 46	6,562	6,695	6,311	8,380	469	6,173	4,655	4,731	10,313	5,644	**10,474**
Kali-Hindon	328	62	198	223	63	44	64	93	94	238	**392**
Hindon Lwr	582	58	375	462	155	234	205	342	408	551	**770**
Catch 29	3,062	352	3,464	4,384	1,228	1,391	1,219	3,243	3,231	4,546	**4,737**
Catch 25	1,181	15	1,286	1,705	272	331	256	1,190	1,169	1,836	**1,953**
Catch 30	1,199	136	1,628	2,080	524	587	508	1,513	1,491	1,820	**1,993**
Catch 41	492	0	549	732	144	188	141	558	554	785	**828**
Total Rtn (M Rs)	42,003	51,053	**62,694**	**66,462**	33,962	28,878	34,506	35,688	36,413	**76,808**	**46,481**
Total WR (ha m)	4,047,518	4,326,583	**2,900,437**	**3,120,890**	3,163,152	2,739,508	2,941,293	2,578,267	2,306,516	**2,948,899**	**3,933,650**

Table 13.24 Sub-catchment wise estimated gross returns for inter cropping patterns for Rabi

Catchment	Gn+Wht (3:1)	Wht+Gram (1:1)	Must+Pot (1:1)	Must+Wht (1:1)	Gram +Must (1:1)
Chambal Up 1	9,230	6,682	13,760	8,816	3,417
Chambal Up 2	605	664	731	999	294
Chambal Up 3	25,617	46,441	14,153	53,472	27,275
Banas	1,008	2,546	81	3,498	1,250
Sind	16,971	18,970	29,580	19,839	14,371
Ken	6,640	5,573	14,062	5,996	3,779
Betwa Up	11,109	9,686	20,376	9,751	7,525
Betwa lwr	11,193	15,257	22,994	15,736	11,676
Catch 73	6,774	14,069	27,414	18,559	11,479
Catch 46	18,000	29,333	24,098	47,344	22,331
Kali-Hindon	2,295	2,512	4,631	3,521	1,914
Hindon Lwr	2,576	4,476	7,067	5,734	3,354
Catch 29	13,227	17,665	24,170	21,363	14,854
Catch 25	2,967	3,441	5,307	4,743	2,482
Catch 30	4,097	5,277	6,195	6,412	4,415
Catch 41	2,295	3,011	2,872	3,573	2,606
Total Rtn (M Rs)	134,603	**185,602**	**217,491**	**229,356**	133,023
Total WR (ha m)	4,113,966	**2,593,367**	**2,711,955**	**3,837,359**	2,351,680

13.5 Water Conservation Under Different Cropping Patterns

Analyzing the gross income and water requirements for inter-cropping patterns for all the catchments (Tables 13.21, 13.22, 13.23, and 13.24), the best inter-cropping patterns are identified, as given in Table 13.25 which also includes theoretical water requirements for combinations of crop in *kharif* and *rabi* seasons. Based on dual crop seasons, the total water requirements are estimated on an annual basis, considering the major crop area only.

When analyzing cropping patterns there is the possibility of skipping a few more crops and their cultivated areas. Therefore, the estimated theoretical crop water demands are projected considering the *kharif* ratio and *rabi* Ratio to cover the entire Net Sown Area (NSA) of the catchments (Table 13.26). The estimated crop water requirements under different combinations are further compared with the crop water requirements reported in the baseline (Table 13.1) and water saving statistics are computed, as shown in Table 13.27. From these comparative statistics, it is apparent that the best inter-cropping pattern can save significant amounts of water.

The range of water saving for the Chambal, Sind, Betwa, and Ken are 2.5–10.1, 16.5–30.0, 4.4–11.4, and 5.7–8.5%, respectively. However, the annual combinations, II+I, II+II, and III+I for the Ken catchment are not appropriate,

Table 13.25 Catchment-wise inter-cropping patterns and water requirements considering the Kharif and Rabi areas under major crops

	Kharif			Rabi		Total CWR (MCM) with combinations				
Catchment	I	II	III	I	II	I + I	I + II	II + I	II + II	III + I
Chambal	Maiz+Arhr (1:1)	Soy+Arhr (1:1)	Soy+Arhr (1:2)	Wht+Grm (1:1)	Grm+Must (1:1)					
CWR (MCM)	13,981	13,713	14,747	8,737	7,928	22,718	21,909	22,449	21,641	23,484
Sind	Soyb+Pad (1:1)	Maiz+Arhr (1:1)	Soyb+Arhr (1:1)	Must+Pot (1:1)	Wht+Grm (1:1)					
CWR (MCM)	1,423	1,821	1,779	2,569	2,402	3,992	3,825	4,390	4,223	4,348
Betwa	Soyb+Arhr (1:1)	Soy+Arhr (1:2)	Maiz+Arhr (1:1)	Must+Pot (1:1)	Wht+Grm (1:1)					
CWR (MCM)	3,937	4,240	4,033	4,255	3,934	8,192	7,870	8,495	8,173	8,288
Ken	Maiz+Arhr (1:1)	Pad+Arhr (1:1)	Soyb+Pad (1:1)	Must+Pot (1:1)	Wht+Grm (1:1)					
CWR (MCM)	2,297	3,038	3,238	1,477	1,364	3,774	3,661	4,515	4,402	4,715
Yamuna	Maiz+Arhr (1:1)	Soy+Arhr (1:2)	Soy+Arhr (1:1)	Must+Pot (1:1)	Wht+Grm (1:1)					
CWR (MCM)	29,489	31,209	29,004	27,120	25,934	56,609	55,423	58,328	57,143	56,124

Table 13.26 Catchment-wise inter-cropping patterns and water requirements considering the reported Kharif and Rabi areas and their projection up to the Net Sown Area (NSA)

			Kharif			Rabi		Total CWR (MCM) with combinations				
Catchment	Kharif ratio	Rabi ratio	I	II	III	I	II	I + I	I + II	II + I	II + II	III + I
Chambal	2.59	2.63	13,981	13,713	14,747	8,737	7,928	59,237	57,108	58,542	56,413	61,223
Sind	3.99	1.66	1,423	1,821	1,779	2,569	2,402	9,926	9,649	11,511	11,235	11,344
Betwa	3.71	2.24	3,937	4,240	4,033	4,255	3,934	24,162	23,440	25,287	24,565	24,520
Ken	3.20	3.20	2,297	3,038	3,238	1,477	1,364	12,082	11,720	14,456	14,094	15,096
Yamuna	2.86	1.82	29,489	31,209	29,004	27,120	25,934	133,670	131,511	138,586	136,426	132,285

Kharif Ratio = NSA/Kharif Area Under Major Crops; Rabi Ratio = NSA/Rabi Area Under Major Crops

Table 13.27 Catchment-wise water savings under different combinations of inter-cropping patterns

Catch	CWR: baseline (Table 13.1)	Total CWR (MCM) with combinations					% Water savings				
		I + I	I + II	II + I	II + II	III + I	I + I	I + II	II + I	II + II	III + I
Chambal	62,771	59,237	57,108	58,542	56,413	61,223	5.6	9.0	6.7	10.1	2.5
Sind	13,786	9,926	9,649	11,511	11,235	11,344	28.0	30.0	16.5	18.5	17.7
Betwa	26,456	24,162	23,440	25,287	24,565	24,520	8.7	11.4	4.4	7.1	7.3
Ken	12,808	12,082	11,720	14,456	14,094	15,096	5.7	8.5	12.9	10.0	−17.9
Yamuna	200,761	133,670	131,511	138,586	136,426	132,285	33.4	34.5	31.0	32.0	34.1

as more water is required under these combinations. Beside this, for the entire Yamuna River basin 31.0–34.5% water can be saved by using these combinations.

13.6 Impact of Change in Cropping Patterns on Water Allocation

The above 34.5% saving in irrigation water does not only satisfy the deficit in the Yamuna basin but can also create additional water of approximately 34.7 BCM at Pratappur near the confluence of River Ganga. The monthly additional water availability in the basin at Pratappur is shown in Fig. 13.2, whereas the additional water in the river in terms of flow is given in Fig. 13.3. The distribution of the flow is based on the actual long term monthly distribution of annual flows in the river at different sites. Similarly, additional water availability and flows for the Agra and Delhi Railway Bridge (DRB) sites are shown in Figs. 13.4, 13.5, 13.6, and 13.7, respectively.

Further, the enhanced flow condition in the river at the Delhi Railway Bridge site is compared with environmental flows (Rai et al., 2010) for the Wazirbad-Okhla stretch, as shown in Fig. 13.8. If crop planning is achieved up to 75% in the catchment corresponding to the Delhi Railway Bridge site (i.e., 30,793 km^2), environmental flows in the river can be achieved if the pollution discharge is regulated up to 10 mg/l BOD with 5 cumecs.

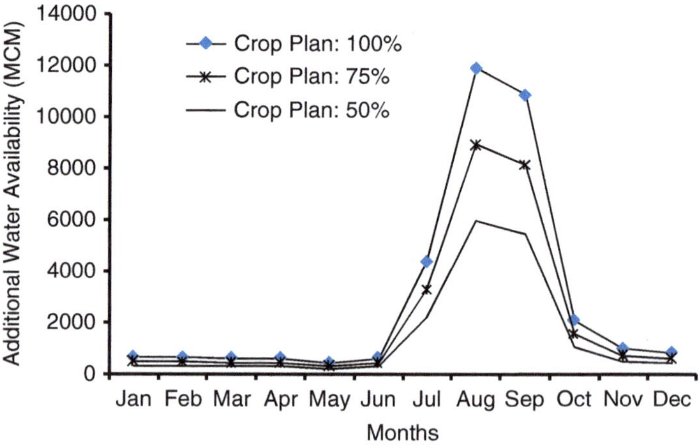

Fig. 13.2 Additional water availability in the Yamuna basin at Pratappur when crop planning is considered

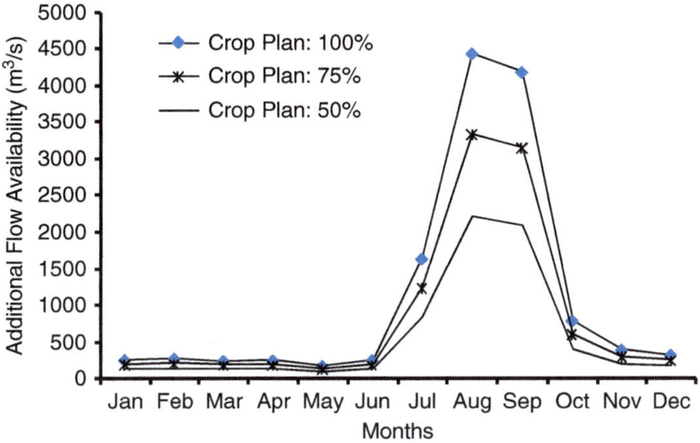

Fig. 13.3 Additional flow availability in the Yamuna basin at Pratappur when crop planning is considered

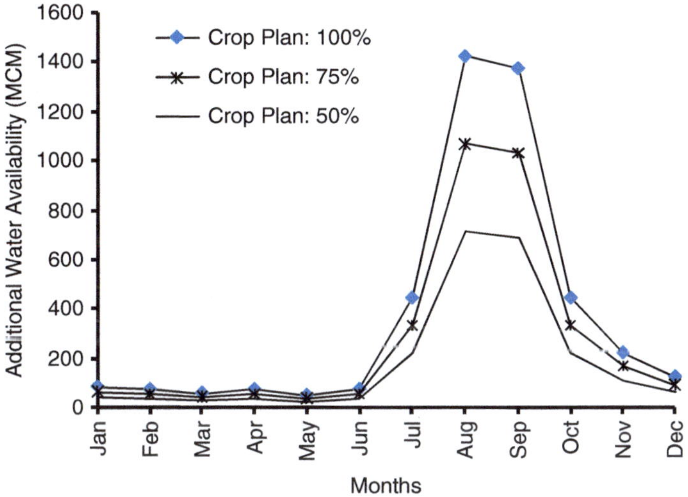

Fig. 13.4 Additional water availability in the Yamuna basin at Agra when crop plan is considered

13.7 Concluding Remarks

The following conclusions can be drawn from this study:

(i) Considering water requirements, minimum support price, and average yield, it is found that during the *kharif* season cultivation of jowar, soybean, bajra and arhar crops are more economic than other crops; whereas during the *rabi* season, mustard, wheat, and potato produce high yields and good economic

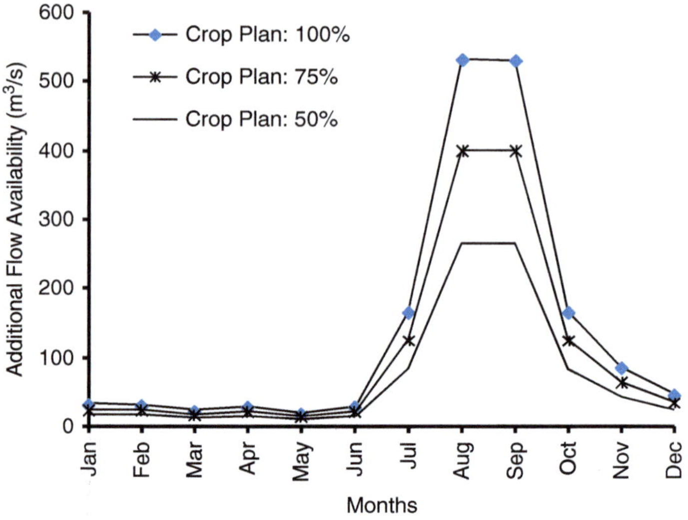

Fig. 13.5 Additional flow availability in the Yamuna basin at Agra when crop planning is considered

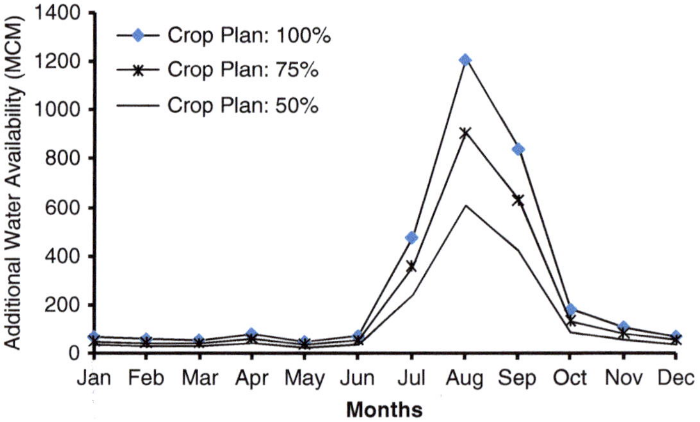

Fig. 13.6 Additional water availability in the Yamuna basin at DRB when crop planning is considered

returns and require less water. Barley and gram are of equally good economic value and can be implemented as *rabi* crops.

(ii) Based on water requirements and gross income, major rotational combinations of crops in the basin are soybean-wheat-maize, paddy-wheat, and arhar-wheat.

(iii) For paddy growing areas, the sequential rotation with gram, mustard, barley and potato is advisable, according to the estimated water requirement.

(iv) Amongst all identified combinations, the soybean-arhar combination requires less water than others. The soybean-jowar intercropping not only requires less

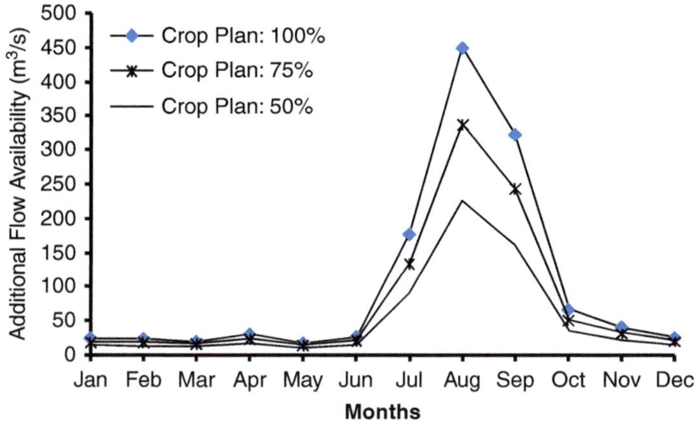

Fig. 13.7 Additional flow availability in the Yamuna basin at DRB when crop planning is considered

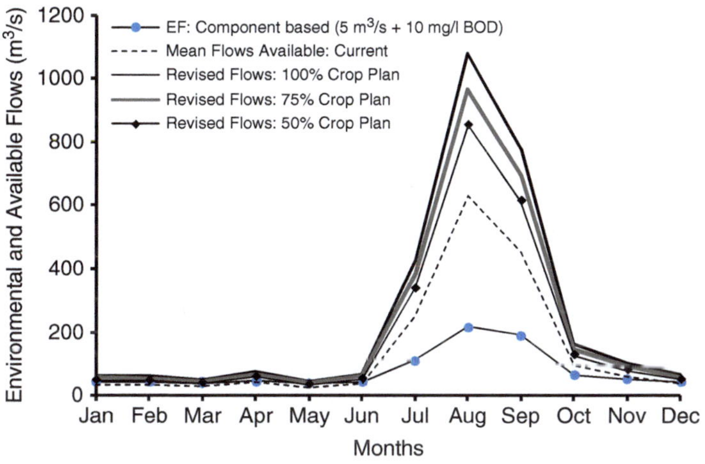

Fig. 13.8 Environmental flows and available flows with crop planning for the Wazirabad to Okhla stretch

water but also provides better crop rotation with bajra. It is also found that bajra (millets) with soybean and arhar can save significant amounts of water. The arhar-paddy intercropping consumes less water than the soybean-paddy or cotton intercropping.

(v) Amongst the various cropping patterns suitable for the Yamuna basin, four inter-cropping patterns for the kharif season are identified based on economic returns and water requirements. These patterns include: maize-arhar (1:1), soybean-arhar (1:2), soybean-arhar (1:1), and paddy-arhar (1:1). Similarly for the *rabi* season the mustard-potato (1:1) pattern is found most economic and water conservative, followed by wheat-gram (1:1) and mustard-wheat (1:1) combinations.

(vi) Using the best inter-cropping pattern, 31.0–34.5% irrigation water can be saved in the entire Yamuna basin, which helps reduce ground water exploitation.

(vii) If crop planning is achieved up to 75% of the catchment corresponding to the Delhi Railway Bridge site (i.e., 30,793 km^2), environmental flows for the Wazirabad-Okhla stretch can be achieved, with the proviso that the pollution discharge is regulated up to 10 mg/l BOD with 5 m^3/s.

References

Allen RG, Jensen ME, Wright JL, Burman RD (1989) Operational estimates of reference evapotranspiration. Agron J 81:650–662

Allen RG, Pereira LS, Raes D, Smith M (1998) Crop evapotranpiration: guildlines for computing crop water requirements. FAO Irrigation and Drainage Paper No 56. Food and Agriculture Organisation, Land and Water. Rome, Italy

Doorenbos J, Pruitt WO (1977) Guidelines for predicting crop water requirements. Irrigation and Drainage Paper 24, Food and Agriculture Organization of the United Nations, Rome, Italy, 179pp

Duffie JA, Beckman WA (1991) Solar engineering of thermal processes, 2nd edn. Wiley, New York, NY, 994p. Harrison LP (1963) Fundamentals concepts and definitions relating to humidity. In: Wexler A (ed) Humidity and moisture, vol 3. Reinhold Publishing Co, New York

ICAR (2006) Handbook of agriculture, 6th edn. Indian Council of Agricultural Research, New Delhi

Jensen ME, Burman RD, Allen RG (ed) (1990) Evapotranspiration and Irrigation Water Requirements. ASCE Manuals and Reports on Engineering Practices No. 70, Am Soc Civil Engrs, NY, 360p

London J, Fröhlich C (1982) Extended abstracts presented at the symposium of the solar constants and the spectral distribution of solar irradiance. In IAMAP third scientific assembly, August 17–28, 1981, Hamburg. Federal Republic of Germany.

Murray FW (1967) On the computation of saturation vapor pressure. J Appl Meteorol 6:203–204

Rai RK, Upadhyay Alka, Ojha CSP (2010) Temporal variability of climatic parameters of Yamuna River Basin: spatial analysis of persistence, trend and periodicity. Open Hydrol J 4:184–210

Tetens O (1930) Ober einige meteorologische Begriffe. Z Geophys 6:297–309

Appendix A
Salient Feature of Dams in the Yamuna Basin

DOI 10.1007/978-94-007-2001-5,

Name of dam	River/ tributary	District	State	Sub-basin	Basin	Year of completion	Catch area (km^2)	Submergence ('000 km^2)	Gross capacity ('000 cu m)	Live capacity ('000 cu m)	Spillway capacity ('000 cumec)	Type of dam	HT of dam (m)	Length of dam (m)	Vol of dam ('000 cu m)	Purpose
Baisly	Kaliasote	Bhopal	MP	Betwa	Betwa	1988	365.2	5,295	35,887	34,410	1,355	TE	67.08	1,080	–	IRR
Balwantpura	Local	Bhopal	MP	Betwa	Betwa	1994	29.53	3,560	6,130	6,100	430	TE	16.1	93	–	IRR
Banchor	Kerwan	Bhopal	MP	Betwa	Betwa	1976	64.5	4,824	25,026	22,600	453	TE	22.6	396.5	168	IRR
BandaBedra	Local	Bhopal	MP	Betwa	Betwa	1986	15.54	1,904	7,179	5,415	52.29	TE	17.14	1,954.5	200	IRR
Bandha	Ajnal	Bhopal	MP	Betwa	Betwa	1962	36.9	3,130	16,277	15,216	133.1	TE	17.1	1,581	–	I/Ind
Baniganj	Local	Bhopal	MP	Betwa	Betwa	1984	4.14	387	1,315	1,222	56.37	TE	11.26	253.44	87	IRR
Baniganj	Ghodp'r	Bhopal	MP	Betwa	Betwa	1986	34.6	2,059	12,410	11,570	121.44	TE	23.42	1,841	568	IRR
Bansichikhli	Local	Bhopal	MP	Betwa	Betwa	1991	51.52	5,340	14,822	12,680	201	TE	14.9	2,438.42	448	IRR
Barana Bamori	–	Bhopal	MP	Betwa	Betwa	2002	–	–	1,960	1,764	53.37	TE	12.1	750	–	IRR
Barbatpur	–	Bhopal	MP	Betwa	Betwa	2004	–	–	2,370	2,133	63.45	TE	18.8	1,005	–	IRR
Bardha		Chhatarpur	MP	Betwa	Betwa	1963	3,496	8,666	11,250	11,006	170	TE	21.34	1,341		IRR.
Barkheda		Chhatarpur	MP	Betwa	Betwa	1966	16.88	1,741	7,950	7,470	60	TE	15.85	442		IRR.
Barkhedi		Chhatarpur	MP	Betwa	Betwa	1970	2.9		1,054	979	38	TE	11.79	1,440		IRR.
Barkhedi		Chhatarpur	MP	Betwa	Betwa	1972	4.14	523	1,105	975	56	TE	10.98	178		IRR.
Barmandal		Chhatarpur	MP	Betwa	Betwa	1986	7.32	613	963	1,849	86	TE	20.48	386	113	IRR.
Barnoo		Chhatarpur	MP	Betwa	Betwa	1986	5.5	319	1,529	1,435	65	TE	19.44	579	92	IRR.
Barod		Chhatarpur	MP	Betwa	Betwa	1986	3.2	477	881	813	47	TE	10.56	915	43	IRR.
Bhuhrla Tank	Kaliasot	Dewas	MP	Betwa	Betwa	1988	365.2	5,295	35,380	34,410	1,355	TE	67.08	1,080		IRR
BhutiaNallaT.	Local	Dewas	MP	Betwa	Betwa	1994	29.53	3,560	6,130	6,100	430	TE	16.1	93		IRR
Bila	Kerwan	Dewas	MP	Betwa	Betwa	1976	64.5	4,824	25,026	22,600	453	TE	22.6	396.5	168	IRR
Birmawal	Local	Dewas	MP	Betwa	Betwa	1986	15.54	1,904	7,179	5,415	52	TE	17.14	1,954.5	200	IRR
Bisonia	Local	Dewas	MP	Betwa	Betwa	1984	4.14	387	1,316	1,222	56	TE	11.26	253.44	87	IRR
Bithlee	Ghorap'r	Dewas	MP	Betwa	Betwa	1986	34.6	2,059	12,410	11,570	121	TE	23.42	1,841	568	IRR
BodiT1	Local	Dewas	MP	Betwa	Betwa	1992	11.65	1,156	3,807	3,502	43	TE	13.71	108	1,820	IRR
Daroli	Local	Guna	MP	Betwa	Betwa	1973	124.1	5,380	21,650	18,810	72,920	TE	29.87	3,169.93		IRR.
Datoda	Local	Guna	MP	Betwa	Betwa	1956	46.54	2,150	13,050	11,700	396	TE	25	667.57		IRR.
Deopur	Local	Guna	MP	Betwa	Betwa	1917	53.41	3,000	11,270	10,660	207	TE	14.63	2,700		I/WS
Devikheda	Local	Guna	MP	Betwa	Betwa	1976	9	1,181	1,593	1,206	97	TE	11.8	911.35		IRR.
Dhamdhusar	Local	Guna	MP	Betwa	Betwa	1916	7.15	108	2,410	2,369	122	TE	10.82	1,432		IRR.
Imlikheda	Arialkha	Raisen	MP	Betwa	Betwa	1972	14.76	1,635	7,644	7,219	47	TE	19.65	310	400	IRR
IndiraGandhiT	Local	Raisen	MP	Betwa	Betwa	U/C	9	815	4,736	4,408	62	TE	18.1	716		IRR
IndiraGandhiT	Khand	Raisen	MP	Betwa	Betwa	1997	25.9	1,760	17,082	8,316	235	TE	20.7	210	216	IRR

Name of dam	River/ tributary	District	State	Sub-basin	Basin	Year of completion	Catch area (km^2)	Submergence ('000 km^2)	Gross capacity ('000 cu m)	Live capacity ('000 cu m)	Spillway capacity ('000 cumec)	Type of dam	HT of dam (m)	Length of dam (m)	Vol of dam ('000 cu m)	Purpose
IndraDaman	Local	Raisen	MP	Betwa	Betwa	1978	6.21	1,466	2,650	2,480	64	TE	16.46	894	38.485	IRR
Jabera	Local	Raisen	MP	Betwa	Betwa	1936	85.45	2,130	13,872	12,796	430	TE	15.24	990		IRR
Jaglatank	Bhoghvi	Raisen	MP	Betwa	Betwa	1958	52	6,040	19,364	17,665	533	TE	13.73	780		IRR
Jaguwa	kuhynala	Raisen	MP	Betwa	Betwa	1958	8.68	657	1,646	1,520	97	TE	11.58	445		IRR
Jaguwa	Local	Raisen	MP	Betwa	Betwa	1957	2.59	370	1,031	963	40	TE	10.08	800		IRR
Jajone	Charua	Raisen	MP	Betwa	Betwa	1958	8.29	1,108	2,577	2,435	82	TE	10.05	1,100		IRR
Jalheridhana	Local	Raisen	MP	Betwa	Betwa	1968	2.58	273	1,000	939	25	TE	10.05	296	29.2	IRR
Jamakheri	Local	Raisen	MP	Betwa	Betwa	1998	4.76	347	1,920	1,800	61.62	TE	17.04	418	143	IRR
Jamakheri	Local	Raisen	MP	Betwa	Betwa		2.72	513	1,353	1,286	15.55	TE	11.65	270	442	IRR
Jamber	Local	Raisen	MP	Betwa	Betwa		2.75	265	920	854	30	TE	15.41	273	64.4	IRR
Jamonia	Local	Raisen	MP	Betwa	Betwa	2000	6.48	830	2,730	2,576	37	TE	11.95	927	119	IRR
Jamonia	Local	Raisen	MP	Betwa	Betwa	2000	3.5	296	1,430	1,309	48	TE	11.11	619	94.647	IRR
JamTalai Tank	Local	Raisen	MP	Betwa	Betwa		3.24	276	942	863	42	TE	12.19	720	67	IRR
JamTalai Tank	Barna	Raisen	MP	Betwa	Betwa	1975	3.5	770	5,288	4,558	6,825	TE	16.06	286.5		IRR
JamunBai	Palakmati	Raisen	MP	Betwa	Betwa		50.29	1,943	11,039	10,438	177	TE	22.71	210	169	IRR
Jhamchar	Local	Rajgarh	MP	Betwa	Betwa	1957	5.18	717	2,454	2,429	77	TE	12.2	731		IRR.
Jhandatola	Local	Rajgarh	MP	Betwa	Betwa	1975	7.51	607	2,094	1,921	88	TE	12.19	549		IRR.
Karjuni		Sagar	MP	Betwa	Betwa	1982	7.12	380	3,270	3,250	35	TE	18.68	384		IRR.
Kerwan		Sagar	MP	Betwa	Betwa	1984	4.2	542	1,588	1,488		TE	15.68	762.19		IRR.
Kuda	Local	Sehore	MP	Betwa	Betwa	1984	1.94	232	736	685	29	TE	12.5	675		IRR
KumediTank	Local	Sehore	MP	Betwa	Betwa	1986	4.22	535	1,510	1,326	15	TE	14.96	1,215		IRR
Machhariya	Local	Shivpuri	MP	Betwa	Betwa	1980	83.5	343,530	18,980	15,710	551	TE	27.6	240		IRR.
Madai	Local	Shivpuri	MP	Betwa	Betwa	1995	56.97	2,550	13,840	126	660	TE	33.8	1,980		I/WS
Madwajhari	Local	Shivpuri	MP	Betwa	Betwa	1979	6.37	1,494	9,240	49,490	58	TE	13.51	130		IRR.
Makhani	Local	Shivpuri	MP	Betwa	Betwa	1981	6.87	460	1,000	9,929	69	TE	12.19	732		IRR.
Mala	Local	Shivpuri	MP	Betwa	Betwa	1911	47.16	3,580	3,580	3,191	396	TE	10.98	210		IRR.
Malavar	Local	Shivpuri	MP	Betwa	Betwa	1913	37.81	1,730	3,864	3,793	312	TE	13.5	540		IRR.
Mehgaontola		Tikamgarh	MP	Betwa	Betwa	1968	29.2	2,816	10,248	6,688	96	TE	20.73	137		IRR.
Mohari		Tikamgarh	MP	Betwa	Betwa	1965	11.65		1,911	1,341	41	TE	14.38	137		IRR.
Molipura		Tikamgarh	MP	Betwa	Betwa	1969	11.52	1,074	2,666	2,398	123	TE	17.1	330		IRR.
Moliyakhedi		Tikamgarh	MP	Betwa	Betwa	1976	7.88		2,840	2,590	28	TE	12.5	1,189		IRR.
Mornanpura		Tikamgarh	MP	Betwa	Betwa		5.53		1,560	410	70	TE	12.2	400		IRR.

Name of dam	River/ tributary	District	State	Sub-basin	Basin	Year of completion	Catch area (km^2)	Submergence ('000 km^2)	Gross capacity ('000 cu m)	Live capacity ('000 cu m)	Spillway capacity ('000 cumec)	Type of dam	HT of dam (m)	Length of dam (m)	Vol of dam ('000 cu m)	Purpose
Narayanpura	Naren	Vidisha	MP	Betwa	Betwa	1981	61.44	8,217	20,550	18,328	430	TE	29.85	3,567	479	IRR
Narayanpura	Kethan	Vidisha	MP	Betwa	Betwa	1975	64.75	3,690	19,157	17,574	226	TE	25.36	900	280	IRR
Narola	Local	Vidisha	MP	Betwa	Betwa	1968	17.35	–	6,880	6,331	108.67	TE	18.6	480	104	IRR
Nawadpura T	Jawari	Vidisha	MP	Betwa	Betwa	1968	2.88	357	1,160	1,159	55	TE	12	450	57	IRR
Noonpani	Halali	Vidisha	MP	Betwa	Betwa	1975	699	52,590	25,270	22,700	642	TE	29.56	945	678.27	IRR
Pabra	Local	Mungawali	MP	Betwa	Betwa	1916	7.15	108	2,410	2,369	122	TE	10.82	1,432		IRR.
Panagar	Local	Sehore	MP	Betwa	Betwa	1984	1.94	232	736	685	29	TE	12.5	675		IRR
Panagar	Local	Sehore	MP	Betwa	Betwa	1986	4.22	535	1,510	1,326	15	TE	14.96	1,215		IRR
Pillowa	Local	AshokNagar	MP	Betwa	Betwa	1917	53.41	3,000	11,270	10,660	207	TE	14.63	2,700		I/WS
Raipura	Local	Hoshangabad	MP	Betwa	Betwa	1984	6.48	200	2,820	2,450	86	TE	11.95	1,485		IRR
Rampurakhurd	Local	Raisen	MP	Betwa	Betwa	1978	6.21	1,466	2,650	2,480	64	TE	16.46	894	38	IRR
Ranjeet	kuhynala	Raisen	MP	Betwa	Betwa	1958	8.68	6,573	1,646	1,520	97	TE	11.58	445		IRR
Rechhai			MP	Betwa	Betwa	1963	3,496	8,666	11,250	11,006	170	TE	21.34	1,341		IRR.
VeerSagar	Local	Pichor	MP	Betwa	Betwa	1911	47.16	3,580	3,580	3,191	396	TE	10.98	210		IRR.
VindhyaWasini	Local	Pichor	MP	Betwa	Betwa	1913	37.81	1,730	3,864	3,793	312	TE	13.5	540		IRR.
Barwa Sagar	Barwa Nala	Jhansi	UP	Betwa	Betwa	1694	52		10,200	9,180	6,877	TE	21.03	1,067		I
Magar Pur	Local Drain	Jhansi	UP	Betwa	Betwa	1694	18.13		2,460	2,214	173	TE	10.2	739.1		I
Pachwara Lake	Local	Jhansi	UP	Betwa	Betwa	1694	21.093		6,150	5,850	41.04	TE/PG	13.72	208		I
Parichha	Betwa	Jhansi	UP	Betwa	Betwa	1886	80.2		78,760	77,170	21,547	PG	16.77	1,174.59		I
Dhekwan	Betwa	Jhansi	UP	Betwa	Betwa	1909	194.3		57,800	52,020	15,177	PG	18.97	2,970		I
Pahuj	Pahuj	Jhansi	UP	Betwa	Betwa	1909	402		18,250	18,210	1,528.2	PG	10.67	2,040	1,824	I
Lachura	Sunhari	Hamirpur	UP	Betwa	Betwa	1910	218.5		20,640	10,560	14,444	PG	14.94	542.3		I
Siori Lake	Lakheri	Hamirpur	UP	Betwa	Betwa	1911	19.9		7,820	7,600	452.4	TE	13.94	2,306		I
Majhgawan	Gunchi nala	Hamirpur	UP	Betwa	Betwa	1917			25,460	22,914	169.8	TE	19.43	1,402	37,800	I
Barwar	Bora	Jhansi	UP	Betwa	Betwa	1923	100.64		33,780	31,650	2,831.3	TE	20.4	2,233	2,933.8	I
Saprar	Saprar	Jhansi	UP	Betwa	Betwa	1952	200.5		76,200	65,700	1,143.32	TE	16.76	3,000		I
Govind Sagar	Shahjad	Lalit Pur	UP	Betwa	Betwa	1953	247.88		96,884	50,110	1,445	TE	18.29	3,606	13,406	I
Matatila	Betwa	Jhansi	UP	Betwa	Betwa	1964	138.85		1,132,580	1,019,400	15,350	TE	45.72	6,315.15	5,012.05	I/H
Keolari	Keolari	Hamirpur	UP	Betwa	Betwa	1966	25.667		7,570	7,230	254.06	TE	11.73	23,000		I

Name of dam	River/ tributary	District	State	Sub-basin	Basin	Year of completion	Catch area (km^2)	Submergence ('000 km^2)	Gross capacity ('000 cu m)	Live capacity ('000 cu m)	Spillway capacity ('000 cumec)	Type of dam	HT of dam (m)	Length of dam (m)	Vol of dam ('000 cu m)	Purpose
Chandrawal	Chandrawal	Hamirpur	UP	Betwa	Betwa	1973	119.2		34,731	30,665	1,252	TE	18.2	5,724	52,900	I
Jamni	Jamni	Lalit Pur	UP	Betwa	Betwa	1973	247.235		92,490	84,020	2,521.25	TE	26.22	5,400		I
Rohini	Rohini	Lalit Pur	UP	Betwa	Betwa	1983			12,110	10,899	290	TE	17.82	650		I
Khaprar	Khaprar	Jhansi	UP	Betwa	Betwa	1985	15		3,500	2,980	267.3	TE	11.2	1,100		I
Dongari	Pahuj	Jhansi	UP	Betwa	Betwa	1986	19.2		9,920	8,320	618	TE	13.72	2,760		I
Sajnam	Sajnam	Jhansi	UP	Betwa	Betwa	1990	237.5		83,500	74,250	2,000	TE	22.34	5,147		I
Shahjad	Shahjad	Lalit Pur	UP	Betwa	Betwa	1992	299.3		130,000	96,060	4,000	TE	29	4,160	2,624	I
Pathrai	Pathrai	Jhansi	UP	Betwa	Betwa	U/C	59.4		123,770	78,500	1,263	TE	15.1	3,800		I
Sijar	Sijar	Jhansi	UP	Betwa	Betwa	UC	22.5		55,850	54,800	392.62	TE	13.8	1,900		I
Rajghat	Betwa	Lalitpur	UP	Betwa	Betwa		2,453		2,172,000	1,945,000	33,893	TE/ PG	43.8	10,790		I/H
Pahari	Dhasan	Not known	UP	Dhasan	Betwa	1912	80.3		47,800	46,000	12,776	PG	10	580.95		I
Storage Capacity of Betwa Sub-basin									**4,922,912**	**4,373,265**						
Akhajhiri	Local	Dhar	MP	Chambal	Chambal	1988	2.36	316	416	375	59	TE	22.8	265		IRR.
Amahi	Local	Dhar	MP	Chambal	Chambal	1976	8.18	558	2,410	1,700	94	TE	14.89	450		IRR.
Amiya Bahadurpur	Local	Dhar	MP	Chambal	Chambal	1978	8	630	2,480	2,120	76	TE	15.04	508		IRR.
BandiaNalla	Local	Bhopal	MP	Chambal	Chambal	1985	10.1	934	2,680	2,448	55	TE	11.33	900	86	IRR
Banskheda	Local	Bhopal	MP	Chambal	Chambal	1992	11.65	1,156	3,808	3,502	42.76	TE	13.71	108	1,820	IRR
Barmandal		Chhatarpur	MP	Chambal	Chambal	1964	10.25	400	1,070	911	79	TE	10	2130		IRR.
Bishanda	Local	Dewas	MP	Chambal	Chambal	1985	10.1	934	2,680	2,440	55	TE	11.33	900	86	IRR
BodiTal	Local	Dewas	MP	Chambal	Chambal	1991	51.052	5,340	1,487	12,680	201	TE	14.9	2,438.42	448	IRR
Borena	Local	Dhar	MP	Chambal	Chambal	1988	2.36	316	416	375	59	TE	22.8	265		IRR.
Bori	Local	Dhar	MP	Chambal	Chambal	1976	8.18	558	2,410	1,700	94	TE	14.89	450		IRR.
Chaidipura	Local	Dhar	MP	Chambal	Chambal	1978	8	630	2,480	2,120	76	TE	15.04	508		IRR.
Chillar	Local	Dhar	MP	Chambal	Chambal	1980	13.96	578	1,827	1,497	130	TE	11	530		IRR.
Dahod	Local	Guna	MP	Chambal	Chambal	1985	124.3	4,870	37,500	35,151	1,556	TE	40.85	630		IRR.
Daroli	Local	Guna	MP	Chambal	Chambal	1985	249.12	117	85,010	78,106	2,605	TE	40.15	670		WS
Datla	Local	Guna	MP	Chambal	Chambal	1980	174.1	8,470	46,593	40,983	598	TE	30.5	825.39		IRR.
Dedla	Local	Guna	MP	Chambal	Chambal	1917	89.03	4,337	15,038	14,184	1,450	TE	24.38	1,646		IRR.
DeepSagar	Local	Guna	MP	Chambal	Chambal	1995	36.24	3,104	14,760	13,430	406	TE	25.55	1,285		IRR.
Deori	Local	Guna	MP	Chambal	Chambal	1907	15.54	1,064	4,270	3,850	113	TE	14.45	720		I/WS

Name of dam	River/ tributary	District	State	Sub-basin	Basin	Year of completion	Catch area (km^2)	Submergence ('000 km^2)	Gross capacity ('000 cu m)	Live capacity ('000 cu m)	Spillway capacity ('000 cumec)	Type of dam	HT of dam (m)	Length of dam (m)	Vol of dam ('000 cu m)	Purpose
Depalpur	Local	Guna	MP	Chambal	Chambal	1986	6.22	554	1,557	1,481	76	TE	16.83	630		IRR.
Deri	Local	Guna	MP	Chambal	Chambal	1992	2.95	243	797	760	43	TE	16.43	526		IRR.
DhablaDewan	Local	Guna	MP	Chambal	Chambal	1986	6.86	397	1,530	1,448	81	TE	15.93	680		IRR.
DhablaDewan	Local	Guna	MP	Chambal	Chambal	1984	11.35	1,034	2,582	2,422	120	TE	15.61	450		IRR.
Dhamnod	Local	Guna	MP	Chambal	Chambal	1915	2.59	220	703	635	40	TE	10.28	690		IRR.
Dhamnod	Local	Guna	MP	Chambal	Chambal	1917	30.02	2,140	6,090	5,476	140	TE	11.88	1,380		IRR.
Dhandibada	Sukait	Guna	MP	Chambal	Chambal	2002	43.08	2,810	8,860	8,250	466,650	TE	13.89	1,390	281	IRR.
Dholka	Koil	Gwalior	MP	Chambal	Chambal	1953	54.37	279	7,381	7,206	37	PG	11.89	780		Feeder
DongerBodi	Local	Indore	MP	Chambal	Chambal	1976	14	1,150	2,493	2,158	117	TE	13.3	950		IRR.
Doraha	Local	Indore	MP	Chambal	Chambal	1980	35	465	567	534	50	TE	12.46	1,966.5		IRR.
Doraha	Local	Indore	MP	Chambal	Chambal	1975	2.51	220	623	583	39	TE	10.67	810		IRR.
Dunguna	Local	Indore	MP	Chambal	Chambal	1931	103.56	8,340	22,460	13,390	624	TE	10.05	3,240		IRR.
Gambhir (PHE)	Gambhir	Indore	MP	Chambal	Chambal	1939	485	486	15,963	146	707	TE	18	1,850	14,650	WS
Gryaspura	Local	Mandsaur	MP	Chambal	Chambal	1960	23,025	60,000	73,220	67,974	21,238	PG	62.17	514		I/Hydl
GuradiaRoopchand	Local	Mandsaur	MP	Chambal	Chambal	1988	6.45	343	1,035	880	41	TE	15.79	412.5		IRR.
GuradiaRoopchand	Local	Mandsaur	MP	Chambal	Chambal	1988	8.83	1,164	2,963	1,812	29	TE	15.85	1,410		IRR.
Gurari	Local	Mandsaur	MP	Chambal	Chambal	1956	9.32	107	2,197	2,061	104	TE	19.26	751		IRR.
GwalSagar	Local	Mandsaur	MP	Chambal	Chambal	1954	24.58	2,007	5,657	5,269	216	TE	18.9	792.48		IRR.
Harridih	Local	Mandsaur	MP	Chambal	Chambal	1979	8.8	321	1,200	1,028	99	TE	16.55	457.32		IRR.
Harsi	Local	Mandsaur	MP	Chambal	Chambal	1982	14.75	891	2,660	2,300	146	TE	16.23	1,250		IRR.
Harsi	Local	Mandsaur	MP	Chambal	Chambal	2000	14.75	998	3,554	2,421	146	TE	12.75	960		IRR.
Hasduamudar	Local	Mandsaur	MP	Chambal	Chambal	2003	9.14	418	1,580	1,350	105	TE	12.01	815		IRR.
Jamunia	Local	Raisen	MP	Chambal	Chambal	2002	2.98	876	972	900	1,555	TE	19.63	369	862	IRR
Jamunia	Chhapi	Rajgarh	MP	Chambal	Chambal	1972	86.8	4,825	15,740	13,540	1,305	TE	17.08	1,829		IRR.
JamuniyaRao	Dudhi	Rajgarh	MP	Chambal	Chambal	2001	119.41	5,700	30,770	20,810	1,310	TE/CS	16.25	3,570		IRR.
Jawahargarh	Puchi	Rajgarh	MP	Chambal	Chambal	1992	20.73	1,865	7,264	6,520	189	TE	16.1	915		IRR.
Jawari	Local	Rajgarh	MP	Chambal	Chambal	1982	1.91	277	1,226	1,179	33	TE	15.32	555		IRR.
Jhaloni	Local	Rajgarh	MP	Chambal	Chambal	1991	3.11	336	1,062	1,030	46	TE	13.41	604		IRR.
Jhaloni	Local	Rajgarh	MP	Chambal	Chambal	1979	5.47	862	2,134	2,060	80	TE	13.8	953		IRR.
Jhalpipli	Local	Rajgarh	MP	Chambal	Chambal	1975	11.65	238	6,223	5,827	37	TE	12.19	2,506		IRR.
Jhapdi	Local	Rajgarh	MP	Chambal	Chambal	1975	3.37	258	808	730	48	TE	17.1	640		IRR.
Jhapdi	Local	Rajgarh	MP	Chambal	Chambal	1968	3.18	372	999	932	67	TE	11.89	1,051.56		IRR.

Name of dam	River/ tributary	District	State	Sub-basin	Basin	Year of comple-tion	Catch area (km^2)	Sub-mergence ('000 km^2)	Gross capacity ('000 cu m)	Live capacity ('000 cu m)	Spillway capacity ('000 cumec)	Type of dam	HT of dam (m)	Length of dam (m)	Vol of dam ('000 cu m)	Purpose
Jhikri	Local	Rajgarh	MP	Chambal	Chambal	1991	8.54	855	2,301	2,108	78	TE	11.5	1,676		IRR.
Jojharpur	Local	Rajgarh	MP	Chambal	Chambal	1982	2.85	290	1,081	1,006	43	TE	11.58	762		IRR.
Julwania	Local	Rajgarh	MP	Chambal	Chambal	1982	5.18	355	1,109	986	59	TE	14.97	556		IRR.
Julwania	Local	Rajgarh	MP	Chambal	Chambal	1981	2.7	281	920	817	54	TE	15.31	603		IRR.
Jumnera	Local	Rajgarh	MP	Chambal	Chambal	1972	2.72	300	776	749	20	TE	10.05	1,007.28		IRR.
Jumnera	Local	Rajgarh	MP	Chambal	Chambal	2001	4.35	415	1,330	1,225	80	TE	13.21	845		IRR.
Junapani	Local	Rajgarh	MP	Chambal	Chambal	2002	2.59	260	789	728	57	TE	13.21	655		IRR.
Kabadia	Local	Rajgarh	MP	Chambal	Chambal	2002	14.28	967	3,484	3,071	176	TE	19.75	915	86	IRR.
KacheriFeeder	Gunjari	Rajgarh	MP	Chambal	Chambal	2002	26.8	1,200	4,824	4,174	327	TE	20.38	975	176	IRR.
Kadodia	Local	Rajgarh	MP	Chambal	Chambal	2002	5.9	620	1,407	1,256	88	TE	18.79	718	129	IRR.
Kakarhai	Local	Rajgarh	MP	Chambal	Chambal	2003	8.58	730	3,881	3,358	275	TE	20.3	435	209	IRR.
Kaketo	Local	Rajgarh	MP	Chambal	Chambal	2003	14.68	719	2,148	1,787	208	TE	16.4	536	77	IRR.
Kaketo	Local	Ratlam	MP	Chambal	Chambal	1984	183.88	5,972	54,268	49,916	1,473	TE	36.86	1,360		IRR./WS
Kalrewa	Local	Ratlam	MP	Chambal	Chambal	1981	7.76	490	1,520	1,341	90	TE	12.21	540		IRR.
Kalyanpura	Local	Ratlam	MP	Chambal	Chambal	1997	90.88	396	13,570	11,305	324	TE	18.3	820	1,900	IRR.
KheriaPunawali	Utawali	Sehore	MP	Chambal	Chambal	1983	49.21	4,959	18,060	15,580	386	TE	25	875		IRR
Khor	Local	Sehore	MP	Chambal	Chambal	1938	24.59	3,770	12,496	11,327	216	TE	14.12	2,160		IRR
Khor	Local	Sehore	MP	Chambal	Chambal	1980	6.8	1,160	3,794	3,544	36	TE	14.1	743.9		IRR
Kohru	Local	Sehore	MP	Chambal	Chambal	1976	4.1	347	1,250	1,158	56	TE	18.18	792.68		IRR
Koka	Local	Sehore	MP	Chambal	Chambal	1979	7.35	788	3,164	3,114	1,645	TE	12.8	1,127.76		IRR
Kolwakhedi	Utawali	Sehore	MP	Chambal	Chambal	1988	22.5	1,602	4,259	4,015	144	TE	12.8	2,040		IRR
Konajhir	Local	Sehore	MP	Chambal	Chambal	1985	9.31	233	3,070	2,830	104	TE	12.31	870		IRR
KoopTank	Local	Sehore	MP	Chambal	Chambal	1984	5.18	631	1,725	1,608	55	TE	10.97	1,050		IRR
Kosduna	Local	Sehore	MP	Chambal	Chambal	1984	5.18	649	1,725	1,589	67	TE	10.94	1,127.76		IRR
Kuari	Local	Sehore	MP	Chambal	Chambal		3.5	1,140	1,035	1,005	26	TE	11.07	1,200		IRR
Kulwa	Local	Sehore	MP	Chambal	Chambal	1981	5.7	650	2,090	1,840	32	TE	14.47	1,890		IRR
Kumharwara	Local	Sehore	MP	Chambal	Chambal	1958	3.89		1,307	1,245	38	TE	10.3	1,415.38		IRR
Kumharwara	Local	Sehore	MP	Chambal	Chambal	UC	4.29	420	1,327	1,225	47	TE	11.43	782		IRR
KundaNala	Local	Sehore	MP	Chambal	Chambal	2002	3.36	365	1,038	950	29	TE	11.43	892		IRR
KundaNala	Local	Sehore	MP	Chambal	Chambal	1985	124.3	4,870	37,500	35,160	1,556	TE	40.85	630	–	IRR
Kundujeta	Local	Sehore	MP	Chambal	Chambal	1985	249.12		85,010	78,106	2,605	TE	40.15	670		I/WS
Kundujeta	Local	Shajapur	MP	Chambal	Chambal		16.18	1,705	2,550	2,311	47	TE	10.06	1,860		IRR

Name of dam	River/ tributary	District	State	Sub-basin	Basin	Year of completion	Catch area (km^2)	Submergence ('000 km^2)	Gross capacity ('000 cu m)	Live capacity ('000 cu m)	Spillway capacity ('000 cumec)	Type of dam	HT of dam (m)	Length of dam (m)	Vol of dam ('000 cu m)	Purpose
Kusmeli	Local	Shajapur	MP	Chambal	Chambal	1916	33.28	1,354	3,240	3,087	116	TE	10.02	2,148	182	IRR
Kusmeli	Local	Shajapur	MP	Chambal	Chambal	1980	2.33	300	565	499	34	TE	12.25	930	75	IRR
Kutarinalla	Local	Shajapur	MP	Chambal	Chambal	1982	2.06	142	441	389	34	TE	16.6	175.5	33	IRR
Kutdhar	Local	Shajapur	MP	Chambal	Chambal		3.89	448	1,186	1,086	54	TE	10.99	507		IRR
Laduna	Local	Shajapur	MP	Chambal	Chambal	2000	4.92	546	1,519	1,394	55	TE	11.7	1,020	77	IRR
Lakhundar	Local	Shajapur	MP	Chambal	Chambal	1916	35	2,250	4,280	3,850	280	TE	10.87	2,225	189	IRR
Larpur	Tiller	Shajapur	MP	Chambal	Chambal	1987	174	9,752	52,197	31,120	1,430	TE	26.54	2,220	683	IRR
Larpur	Chiller	Shajapur	MP	Chambal	Chambal	1972	98.42	8,460	34,790	31,110	687	TE	30.48	2,866	683	IRR
Lasudiakangar	Local	Shajapur	MP	Chambal	Chambal	1974	11.48	673	2,860	2,426	127	TE	20.72	821.6	88	IRR
LasurTank	Local	Shajapur	MP	Chambal	Chambal	2000	82.95	7,800	32,150	30,650	632	TE	23.3	2,385	830	IRR
Laxmikheda	Local	Shajapur	MP	Chambal	Chambal	1980	45.34	2,632	11,210	10,120	340	TE	25.87	894	298	IRR
LilgiBundBela	Local	Shajapur	MP	Chambal	Chambal	1989	21.77	1,540	4,080	3,560	198	TE	22.29	1,230	20	IRR
LilgiBundBela	Local	Shajapur	MP	Chambal	Chambal	1983	15.4	1,102	3,742	3,372	123	TE	21.65	1,410	18	IRR
LilgiReservoir	Local	Shajapur	MP	Chambal	Chambal		11.39		1,473	1,473	68	TE	22.18	808		IRR
Limbi	Local	Shajapur	MP	Chambal	Chambal	1980	5.18	277	1,017	911	67	TE	17.84	289.16	4	IRR
LokpalSagar	Local	Shajapur	MP	Chambal	Chambal	1981	6.91	506	1,370	1,212	104	TE	14.24	510	40	IRR
LokpalSagar	Local	Shajapur	MP	Chambal	Chambal	1980	7.77	1,424	1,694	1,497	72	TE	11.4	1,425	134	IRR
Lolki	Local	Shajapur	MP	Chambal	Chambal	1992	3.88	380	1,147	1,054	51	TE	15.63	703	80	IRR
Longerpura	Local	Shajapur	MP	Chambal	Chambal	1983	4.4	371	1,226	1,112	59	TE	14	1,050	73	IRR
Longerpura	Local	Shajapur	MP	Chambal	Chambal	1989	3.36	350	976	894	49	TE	14.59	720	80	IRR
LowerPalakmati	Local	Shajapur	MP	Chambal	Chambal	1989	6.3	817	1,900	1,740	55	TE	14.39	1,545	91	IRR
LowerPalakmati	Local	Shajapur	MP	Chambal	Chambal	1985	6.14	564	1,350	1,186	76	TE	10.1	576	66	IRR
Makroda	Local	Shivpuri	MP	Chambal	Chambal	1982	14.98	10,290	2,330	19,730	148	TE	11.68	763		IRR.
Manikheda	Local	Shivpuri	MP	Chambal	Chambal	1914	11	640	2,270	2,260	96	TE	10.37	640		IRR.
Manikheda	Local	Shivpuri	MP	Chambal	Chambal	1911	26.92	1,880	7,440	7,193	148	TE	12.2	1,280		IRR.
MudiaKheri	Local	Tikamgarh	MP	Chambal	Chambal							TE	12	1,210		IRR
Mundri	Local	Ujjain	MP	Chambal	Chambal	1978	31.07	6,854	7,547	6,862	241	TE	22.78	1,419		IRR.
Mutwan	Gandhi	Ujjain	MP	Chambal	Chambal	1980	82.65	8,146	25,230	22,350	533	TE	18.9	1,859		IRR.
Nagdagajora	Local	Ujjain	MP	Chambal	Chambal	1982	34.95	1,874	6,450	5,550	130	TE	21.33	701		IRR.
Nagdagajora	Kalisindh	Ujjain	MP	Chambal	Chambal	1979	69.4	2,510	7,560	5,110	200	TE	17.72	1,275		IRR.
NagdaNalla	Local	Ujjain	MP	Chambal	Chambal	1981	60.52	3,980	12,460	10,330	200	TE	18.6	3,450		IRR.
Nagri	Local	Ujjain	MP	Chambal	Chambal	1985	6.47	478	1,555	1,113	56	TE	10.44	1,320		IRR.

Name of dam	River/ tributary	District	State	Sub-basin	Basin	Year of completion	Catch area (km^2)	Sub-mergence ('000 km^2)	Gross capacity ('000 cu m)	Live capacity ('000 cu m)	Spillway capacity ('000 cumec)	Type of dam	HT of dam (m)	Length of dam (m)	Vol of dam ('000 cu m)	Purpose
Naharkheda	Local	Ujjain	MP	Chambal	Chambal	1980	20.48	951	2,675	2,360	189	TE	13.1	960		IRR.
Nakheri	Local	Ujjain	MP	Chambal	Chambal	1987	3.52	234	830	742	51	TE	12.19	570		IRR.
Naktora	Local	Ujjain	MP	Chambal	Chambal	1908	39.08	2,160	5,830	4,943	256	TE	11.89	1,560		IRR.
Nalajhiri	Local	Ujjain	MP	Chambal	Chambal	1984	4.76	269	850	776	62	TE	11.8	840		IRR.
NalTank	Local	Ujjain	MP	Chambal	Chambal	1957	5.7	337	1,105	1,065	72	TE	10.06	840		IRR.
Nandanwara	Local	Ujjain	MP	Chambal	Chambal	2001	9.25	828	1,884	1,760	103	TE	12.19	870		IRR.
Nandkho	Local	Ujjain	MP	Chambal	Chambal	2001	20.72	1,023	3,500	2,950	189	TE	13.88	990		IRR.
Padari	Local	Chachoda	MP	Chambal	Chambal	1915	2.59	220	703	635	40	TE	10.28	690		IRR.
Padaria	Utawali	Sehore	MP	Chambal	Chambal	1983	49.21	4,959	18,060	15,580	386	TE	25	875		IRR
Padhar	Local	Sehore	MP	Chambal	Chambal	1938	24.59	3,770	12,496	11,327	216	TE	14.12	2,160		IRR
Pagara	Local	Sehore	MP	Chambal	Chambal	1984	5.18	631	1,725	1,608	55	TE	10.97	1,050		IRR
Panpura	Local	Shajapur	MP	Chambal	Chambal	1980	5.18	277	1,017	911	67	TE	17.84	289.16	4	IRR
Patharehta	Local	Indore	MP	Chambal	Chambal	1980	35	465	567	534	50	TE	12.46	1,966.5		IRR.
Patia	Local	Mandsaur	MP	Chambal	Chambal	1956	9.32	107	2,197	2,061	104	TE	19.26	751		IRR.
Pdamasagar	Local	Mandsaur	MP	Chambal	Chambal	1982	14.75	891	2,660	2,300	146	TE	16.23	1,250		IRR.
Phugawadi	Local	Shajapur	MP	Chambal	Chambal		16.18	1,705	2,550	2,311	47	TE	10.06	1,860		IRR
Phugawadi	Local	Shajapur	MP	Chambal	Chambal	1916	33.28	1,354	3,240	3,087	116	TE	10.02	2,148	182	IRR
PipariyaJagroj	Local	Raghogarh	MP	Chambal	Chambal	1907	15.54	1,064	4,270	3,850	113	TE	14.45	720		I/WS
Pipliyapada	Local	Sehore	MP	Chambal	Chambal	1976	4.1	347	1,250	1,158	56	TE	18.18	792.68		IRR
Pipliyapada	Local	Sehore	MP	Chambal	Chambal	1985	9.31	233	3,070	2,830	104	TE	12.31	870		IRR
Piploda	Local	Sehore	MP	Chambal	Chambal		3.5	1,140	1,035	1,005	26	TE	11.07	1,200		IRR
Piploda	Local	Sehore	MP	Chambal	Chambal	1958	3.89		1,307	1,245	38	TE	10.3	1,415.38		IRR
Piplyakala	Local	Dewas	MP	Chambal	Chambal	1977	8.8	836	2,476	2,343	100	TE	11.5	1,500		IRR
PipraniNo.2T	Local	Shajapur	MP	Chambal	Chambal	1992	3.88	380	1,147	1,054	51	TE	15.63	703	80	IRR
PipraniNo.2T	Local	Shajapur	MP	Chambal	Chambal	1983	4.4	371	1,226	1,112	59	TE	14	1,050	73	IRR
Poiadonger	Local	Dhar	MP	Chambal	Chambal	1983	2.32		530	482		TE	12.4	540		IRR.
Ramova	Local	Karera	MP	Chambal	Chambal	1911	26.92	1,880	7,440	7,193	148	TE	12.2	1,280		IRR.
Rampur	Koil	Gwalior	MP	Chambal	Chambal	1953	54.37	279	7,381	7,206	37	PG	11.89	780		Feeder
Rampur	Local	Shajapur	MP	Chambal	Chambal	1916	35	2,250	4,280	3,850	280	TE	10.87	2,225	189	IRR
Rampura	Chhapi	Rajgarh	MP	Chambal	Chambal	1972	86.8	4,825	15,740	13,540	1,305	TE	17.08	1,829		IRR.
Rampura	Local	Rajgarh	MP	Chambal	Chambal	1982	2.85	290	1,081	1,006	43	TE	11.58	762		IRR.
Rechhai	Local	Dewas	MP	Chambal	Chambal	1997	10.1	532	3,180	2,911	110	TE	16.58	425		IRR

Name of dam	River/ tributary	District	State	Sub-basin	Basin	Year of completion	Catch area (km^2)	Submergence ('000 km^2)	Gross capacity ('000 cu m)	Live capacity ('000 cu m)	Spillway capacity ('000 cumec)	Type of dam	HT of dam (m)	Length of dam (m)	Vol of dam ('000 cu m)	Purpose
Renjhari			MP	Chambal	Chambal	1964	10.25	400	1,070	911	79	TE	10	2,130		IRR.
Rjagarh			MP	Chambal	Chambal	1921	19.42	1,470	4,810	4,507	53	TE	13.26	1,368.5		IRR.
Sendpa	Local	Sheopur	MP	Chambal	Chambal	1934	236	8,185	55,700	40,050	1,246	TE	16.7	1,158		IRR.
Sidhkhad	Local	Guna	MP	Chambal	Chambal	1917	89.03	4,337	15,038	14,184	1,450	TE	24.38	1,646		IRR.
Sikuwa	Local	Dewas	MP	Chambal	Chambal	1916	19.2	173	4,230	4,100	181	TE	11	1,470		IRR
SurajmalJain	Local	Depalpur	MP	Chambal	Chambal	1931	103.56	8,340	22,460	13,390	624	TE	10.05	3,240		IRR.
Tons Barrage	Local	Sabalgarh	MP	Chambal	Chambal	1908	60.63	1,792	5,469	5,462	423	TE	13.39	2,255		IRR.
Umariya	Local	Sabalgarh	MP	Chambal	Chambal	1899	106	2,183	3,823	3,433	642	TE/PG	10.67	1,052		IRR.
Virhana	Local	Dhana	MP	Chambal	Chambal	1914	11	640	2,270	2,260	96	TE	10.37	640		IRR.
Govindpura		Katni	MP	Chambal	Chambal	1917	5.18	1,000	1,393	1,236	72	TE	11.93	1,073		IRR.
Rajsamand	Gomti Banas	Rajsamand	RJ	Banas	Chambal	1671	18.1		107,100	98,650	1,621 22.351	TE	39.2	5,585.05		I
Jai Samand	Gomti	Udaipur	RJ	Banas	Chambal	1730	56		566,000	163,000	4,706	TE/PG	41.1	457.4		I
Swaroop Sagar	Berach	Girwa	RJ	Banas	Chambal	1795	4.18		13,670	9,000	564	TE	22	660		I
Hingonia	Banas	Phagi	RJ	Banas	Chambal	1862			7,523	6,771		TE	15.5	–	NA	–
Chandrana	Banganga	Dausa	RJ	Khri	Chambal	1871			4,930	4,437	1,203	TE	11	–	4,810	I
Kharad	Banganga	JPR – Ramgarh	RJ	Khri	Chambal	1877			11,805	10,625		TE	29.5	–	–	–
Kalakh Sagar	Local nallah	Jaipur	RJ	Banas	Chambal	1883			20,000	16,500		TE	13			I
Madho Sagar	Banganga	Sikrai Ramgarh	RJ	Khri	Chambal	1887			22,603	20,343		TE	14.63	2,440	NA	–
Fateh Sagar	Berach	Girwa	RJ	Banas	Chambal	1889	2.42		12,080	6,990	135	TE	21			I/S
Buchra	Sabi	Kot-putali	RJ	Khri	Chambal	1889			16,360	14,724	731	T	16.31	366	NA	I
Chaparwara	Banas	Dudu	RJ	Banas	Chambal	1894			35,000	35,000	800	TE	13	200	NA	I
Sainthal Sagar	Banganga	Dausa	RJ	Khri	Chambal	1898			13,705	12,335		TE	17	4,180	12,800	–
Sheel Ki Dungri	Banas	Jaipur	RJ	Banas	Chambal	1900			4,160	3,744		TE	19.8	1,981	–	–
Ram Garh	Banqanga	Jamuwa JPR	RJ	Khri	Chambal	1901			58,970	58,970	2,100	TE	26			I
Mandal	Local Nalla	Bhilwara	RJ	Banas	Chambal	1903	0.214		13,877	11,895	1,300	TE	26	1,825		I/S
Dheel	Morel	Bonli	RJ	Banas	Chambal	1911			25,230	25,230		TE	27.62	655		I

Name of dam	River/ tributary	District	State	Sub-basin	Basin	Year of completion	Catch area (km^2)	Sub-mergence ('000 km^2)	Gross capacity ('000 cu m)	Live capacity ('000 cu m)	Spillway capacity ('000 cumec)	Type of dam	HT of dam (m)	Length of dam (m)	Vol of dam ('000 cu m)	Purpose
Bhopal Sagar	Berach	Kapasan	RJ	Banas	Chambal	1936	8.19		18,540	18,400	645	TE	28	287		I
Chittoli	Sabi	Virat Nagar	RJ	Khri	Chambal	1950			24,536	22,082		TE	19.5	–	NA	–
Govta	Manali	Mandal Garh	RJ	Banas	Chambal	1955			6,850	6,850	894	TE	18	168		I
Patan (Deosagar)	Local Nalla	Patan	RJ	Banas	Chambal	1956			6,930	6,237	457.05	TE	13.4	1,652		I
Surwal	Gambhir	S.Modhopur	RJ	Chambal	Chambal	1956	9.85		22,880	22,280	609	TE	32	183		I
Bagolia	Berach	Mavli	RJ	Banas	Chambal	1956	3.5		19,400	13,350	827	TE	12.7	3,079		I
Kalisil	Kalisil Banas	Sapotru	RJ	Banas	Chambal	1956			41,700	37,250	889	TE	25	2,560		I
Man Sarowar	Chambal	Khandar	RJ	Chambal	Chambal	1957	4.65		13,650	12,260	407	TE	26	378.65		I
Paibala Pura	Mej River	Nainwa	RJ	Chambal	Chambal	1957	9.4		10,200	10,000	632	TE	14.63	3,644.52		I
Juggar	Juggar Banas	Hindaun	RJ	Khri	Chambal	1957			39,460	34,850	769	TE	21	239.26		I
Ora	Khari	Sheoganj	RJ	Banas	Chambal	1957	7.46		22,650	22,280	836	TE	20.12	5,618.87		I
Bundika Gothra	Mej River	Hindoli	RJ	Chambal	Chambal	1957	6.7		28,400	28,400	857	TE/PG	20.14	2,613.94		I
Sareri	Khari	Bhilwara	RJ	Banas	Chambal	1957	52.2		55,800	55,010	1,604	TE/PG	42	3,630		I
Khari	Khari	Asind	RJ	Banas	Chambal	1957	8.84		38,940	33,280	1,911	TE/PG	17.67	2,391		I
Arwar	Khari	Shahpura	RJ	Banas	Chambal	1957	18.8		47,910	47,770	2,761	TE/PG	15.8	1,112.6		I
Pachki Baori	Began	Bundi	RJ	Chambal	Chambal	1957			5,600	5,600		TE	19	954		I
Kala Bhala	Khari	Deogarh	RJ	Chambal	Chambal	1958			4,240	3,870	351	TE	14.06	979		I
Bhimlat	Mangli	Bundi	RJ	Chambal	Chambal	1958			10,800	10,700	442	TE/PG	17	274		I
Nandsamand	Banas	Nathwara	RJ	Banas	Chambal	1958	4.4		21,200	20,600	3,240	TE	23	1,733.89		I
Meja	Kothari	Mandal	RJ	Banas	Chambal	1958	26.5		96,220	91,720	5,781	TB/PG	19.2	1,270		I
Gudha	Mejchambal	Hindoli	RJ	Chambal	Chambal	1958			95,650	93,700	5,883	TE	25	2,760		I
Matrakundia	Kotharia	Bhilwara	RJ	Banas	Chambal	1958			49,986	44,987	8,240	TE/PG	16.67			I
Gambhiri	Berach	Nimbahera	RJ	Banas	Chambal	1958	20.57		76,400	65,100	10,800	TE	21	611.12		I
Nagdi	Nagdi	Jahazpur	RJ	Banas	Chambal	1959			5,150	5,090	199	TE	15	805		I
Galwa	Galwa	Uniara	RJ	Banas	Chambal	1960			52,790	52,440	3,240	TE	16.08	6,068		I
Mashi	Bandi	Newai	RJ	Banas	Chambal	1960	10.62		48,100	35,117	11,954	TE	12.88	2,103		I
Kota Barrage	Chambal	Kota	RJ	Chambal	Chambal	1960	4.84		108,060	245,500	21,225	TE	51.9	551.68		I

Name of dam	River/ tributary	District	State	Sub-basin	Basin	Year of completion	Catch area (km^2)	Sub-mergence ('000 km^2)	Gross capacity ('000 cu m)	Live capacity ('000 cu m)	Spillway capacity ('000 cumec)	Type of dam	HT of dam (m)	Length of dam (m)	Vol of dam ('000 cu m)	Purpose
Atawara	Banas	Gangapur	RJ	Banas	Chambal	1961	0.51		5,235	5,094	402	TE	10.36	2,671	NA	I
Banakiya	Local nallah	Chittorgarh	RJ	Banas	Chambal	1962	5.5		9,439	9,269	849.33	TE	12.92	2,555	NA	I
Alnia	Alnia Hukud	Ladpura	RJ	Chambal	Chambal	1962	12.8		43,700	43,710	1,325	PG	14	1,023.2		I
Bithan	Banas	Jaswantpur	RJ	Banas	Chambal	1963			6,390	6,000	354	TE	14	706		I
Parbati Dam	Chambal	Bacera	RJ	Chambal	Chambal	1963	23		115,200	102,800	1,763	TE/PG	28.11	2,318		I
Kaladeh	Local nallah	Chittorgarh	RJ	Banas	Chambal	1964	0.55		1,711	1,550	224	TE	14.75	526	151	I
Berach	Berach	Villabhnagar	RJ	Banas	Chambal	1965			30,470	28,770	2,775	TE	19	5,822		I
Mandawara	Kundal	Swaroopganj	RJ	Sahibi	Chambal	1967	0.65		2,460	2,160	170	TE	15.25	457		I
Ranapratap Sagar	Chambol	Kota	RJ	Chambal	Chambal	1970	198.29		2,898,690	1,566,520	18,408	PG	53.8	1,143		I/H/S
Badgaon	Banas	Vallabh Nagar	RJ	Banas	Chambal	1973	9.79		31,490	30,150	2,874	TE	21.12	457.2		I
Jawahar Sagar	Chambal	Kota	RJ	Chambal	Chambal	1973			52,000	13,200	21,240	PG	36	393		H
Luneri	Local Nalla	Swaroopganj	RJ	Sahibi	Chambal	1976	0.86		2,440	2,450	174	TE	11.4	1,493		I
Panchana	Gambhiri	Karauli	RJ	Khri	Chambal	1977	9.24		59,450	52,650	5,238.6	TE	33.19	1,207.8		I
Morasagar	Banas	Bamenwas	RJ	Banas	Chambal	1978	5.85		12,930	12,870	59	TE	25	94.5		I
Abhaya Pura	Bhimlat	Bundi	RJ	Chmbal	Chambal	1978			7,997	7,270	531	TE	15.64	1,645		I
Jetpura	Eru	Mandalgarh	RJ	Chambal	Chambal	1978	4.3		168OO	16,300	1,064	TE	21.8	585		I
Gopalpura	Chambal	Kishanganj	RJ	Chambal	Chambal	1979	6.29		32,600	31,900	1,175	ER/PG	14.43	4,388.49		I
Kanyakheri	Local Nalla	Bhilwara	RJ	Banas	Chambal	1980			4,300	3,870	171	TE	10.5	3,656		I
Modia Mahadev	Local Nalla	Chittorgarh	RJ	Banas	Chambal	1980			3,885	3,648	595	TE	16.53	625		I
Jhadol	Banas	Kotri	RJ	Banas	Chambal	1980			10,335	9,287	21,160	TE	15	870		I
Titora	Local nallah	Jahazpur	RJ	Banas	Chambal	1981	0.145		1,358	1,231	144.89	TE	12	749	NA	I
Anwasa	Local nallah	Bhilwara	RJ	Banas	Chambal	1981			2,820	2,538	494.2	TE	12.2		–	I
Devliya	Local nallah	Chittorgarh	RJ	Banas	Chambal	1982	2.4		2,380	2,380	230	TE	16.1	250	40	I
Mandol	Local Nalla	Bhilwara	RJ	Banas	Chambal	1982	0.651		6,800	6,263	5,720	TE	12.65	–		I/S
Domti Kokra	Moral	Mandalgarh	RJ	Banas	Chambal	1983			4,720	4,160	303	TE	27	61		I
Saran Kheri	Local Nalla	Jhalra Patan	RJ	Kali Sindh	Chambal	1983			1,230	1,107	330	TE	16	960	–	I
Panghatiya	Local nallah	Mandalgarh	RJ	Banas	Chambal	1984	1.13		1,274	1,133	143	TE	13.9	210	NA	I
Luharia	Local nallah	Chittorgarh	RJ	Banas	Chambal	1984	1.547		2,917	2,832	299	TE	10.2	2,973.6	NA	I
Kothari Stage 1	Banas	Kohri	RJ	Banas	Chambal	1984	7.6		26,000	21,500	6,005	TE/PG	25.4	4,326		I

Name of dam	River/ tributary	District	State	Sub-basin	Basin	Year of completion	Catch area (km^2)	Sub-mergence ('000 km^2)	Gross capacity ('000 cu m)	Live capacity ('000 cu m)	Spillway capacity ('000 cumec)	Type of dam	HT of dam (m)	Length of dam (m)	Vol of dam ('000 cu m)	Purpose
Barkheri	Local nallah	Chittorgarh	RJ	Banas	Chambal	1985	0.52044		1,598	1,473	135.93	TE	15.58	889	NA	I
Bassi	Corail Berach	Gangrar	RJ	Banas	Chambal	1985	2.8		23,200	20,240	2,401	TE	36	1,044		I
Pachanpura	Local Nalla	Bhilwara	RJ	Banas	Chambal	1985	0.306		9,480	9,280	3,549	TE	14.7	–		I
Bhimsagar	Kalisindh	Jhalawar	RJ	Kali Sindh	Chambal	1985	18.02		79,750	72,500		PG	26			I
Sawan Bhado	Ara	Sangod	RJ	Chambal	Chambal	1986	4.4		30,000	27,850	1,346	TE	29.8	4,692		I
Patiyal	Local nallah	Mandalgarh	RJ	Banas	Chambal	1991	0.38		1,708	1,529	5.3	TE	13.9	765	NA	I
Sushil Sagar	Local nallah	Mandalgarh	RJ	Banas	Chambal	1991	0.69		2,202	1,982	216	TE	12.84	430	NA	I
Wageri-Tank	Local nallah	Rajasmand	RJ	Banas	Chambal	1992			2,390	2,160	5,657	PG/TE	22.2	430.25	–	I
Shiv-Sagar	Local Nalla	Bhilwara	RJ	Banas	Chambal	1993	0.312		2,094	2,050	218.28	TE	22.7	247	–	I
Dorai	Brahmani	Begun	RJ	Banas	Chambal	1995			8,500	7,650	657	TE	14.83	1,050		I
Bilas	Bilas	Kishanganj	RJ	Chambal	Chambal	1996	6.5		28,800	26,700	12,148	TE	19.17	4,650		I
Pitha Ka Mund	Local Nalla	Mandalgarh	RJ	Banas	Chambal	1997			1,890	1,701	144.98	TE	11.7	871		I
Sanwaria Sarover	Local Nalla	Chittorgarh	RJ	Banas	Chambal	1997	3.024		5,408	5,180	512	TE	11.08	380		I
Jude Ka Naka	Local Nalla	Bhilwara	RJ	Banas	Chambal	1998			1,780	1,720	257.05	TE	18.7	135	–	
Motipura	Local nallah	Bhilwara	RJ	Banas	Chambal	1999			1,170	1,053	164.82	PG	14.65	180	–	I
Bari Mansarowar	Local nallah	Chittorgarh	RJ	Banas	Chambal	1999	3.74		9,482	8,419	661	TE	14.75	363.25	NA	I
Bisalpur	Banas	Deoli	RJ	Banas	Chambal	1999	218.36		1,095,840	1,040,950	29,046	PG	39.5	574		I/S
Navratan Sagar	Local nallah	Jahazpur	RJ	Banas	Chambal	2000	6.382		2,260	2,040	389	TE	12.2	1,200	NA	I
Kantaliya	Nekhadi	Asind	RJ	Banas	Chambal	2000			1,020	918	546.34	TE	11.35	305		I
Ruparel	Ruparel	Begun	RJ	Banas	Chambal	2004			9,700	8,730	654	TE	16.42	955		I
Chauli	Chauli	Jhalawar	RJ	Kali Sindh	Chambal	2005	12.27		53,500	48,220	1,284	TE	27.65	2,225		I/S
Chhapi	Chhapi	Aklera	RJ	Chambal	Chambal	2006	9.7		7,850C	59,580	4,370	PG	34	310		I
Lassaria	Dai	Kekri	RJ	Banas	Chambal	1981–1982	2.64		11,500	10,650	2,320	TE	20.15	4,325		I
Udai Sagar	Berach	Girwa	RJ	Banas	Chambal	S.T			31,130	27,590	1,419	PG/TE	19.35	315		I
Uncha	Local nallah	Chittorgarh	RJ	Banas	Chambal	ST	1.175		2,605	2,515	65.03	TE	14	1,800	NA	I
Mundliya Kheri	Local Nalla	Jhalawar	RJ	Kali Sindh	Chambal	ST			4,420	4,370	395.26	TE	12.5	1,890	–	I

Name of dam	River/ tributary	District	State	Sub-basin	Basin	Year of completion	Catch area (km^2)	Submergence ('000 km^2)	Gross capacity ('000 cu m)	Live capacity ('000 cu m)	Spillway capacity ('000 cumec)	Type of dam	HT of dam (m)	Length of dam (m)	Vol of dam ('000 cu m)	Purpose
Jaisamand	Banganga	Alwar	RJ	Khri	Chambal	ST			26,942	24,248	410	PG	10	1,671	NA	–
Soniyana	Local nallah	Chittorgarh	RJ	Banas	Chambal	ST	2.92		6,800	6,226	725	TE	11.7	3,060	105	I
Nahar Sagar	Khari	Shahpura	RJ	Banas	Chambal	ST			20,000	20,000	1,414	TE	12	1,560		I
Kukas	Local nallah	Jaipur	RJ	Banas	Chambal	ST			1,925	1,925		TE	13	487		I
Sankal Khera	Local nallah	Chittorgarh	RJ	Banas	Chambal	U/C	0.616		3,170	3,080	315	TE	20.9	1,700	NA	I
Kalikhar	Kalikhar	Manohar Thana	RJ	Kali Sindh	Chambal	U/C			11,328	10,830	1,299	TE	18			I
Lhasi	Lhasi	Baran	RJ	Chambal	Chambal	U/C			30,800	28,300	1,428	TE	17.58			I
Piplad	Piplad	Jhalawar	RJ	Kali Sindh	Chambal	U/C			23,078		1,463	TE	15.77			I
Takli	Takli	Kota	RJ	Chambal	Chambal	U/C			33,740	29,900	2,294	TE	16.68			I/S
Gagrin	Ahu	Jhalawar	RJ	Kali Sindh	Chambal	U/C			52,600	47,100	4,447	TE	25.48	2,535		I
Benthali	Benthali	Baran	RJ	Banas	Chambal	U/C						TE	21.6			
Chakan	Chakan	Bundi	RJ	Chambal	Chambal	U/C						TE	15			
Gararda	Gararda	Bundi	RJ	Chambal	Chambal	U/C						TE	31.76			
Gulandi	Gulandi	Jhalawar	RJ	Kali Sindh	Chambal	U/C						TE	17.14			
Hindlot	Local nallah	Baran	RJ	Banas	Chambal	U/C						TE	19.7			
Bherunda	Moral	S.Madhopur	RJ	Banas	Chambal				2,100	1,900	131	TE	17	36		I
Deopura	Nahari	S.Madhopur	RJ	Banas	Chambal				8,180	7,600	302	TE	23	1,127		I
Silibari	Roop Rail	Alwar	RJ	Sahibi	Chambal				5,600	5,400	346	TE	15	99		I
Reriya	Banganga	Baswa	RJ	Khri	Chambal				4,580	3,840	628	TE	16	100		I
Baretha Bund	Yamuna	Bayana	RJ	Khri	Chambal		13.56		51,510	49,530	968	TE	16.31			I
Basundni	Khari	Kekri	RJ	Banas	Chambal				7,780	7,210	1,241	TE	20	427		I
Gosunda	Banas	Chittaurgerh	RJ	Banas	Chambal		23.39		75,470	67,050	20,643	TE/PG	23	639.5		I
Bishan Samand	Gambhir	Todabhim	RJ	Khri	Chambal				11,550	11,550		TE	11.08	61		I
Madhar	Berach	Girwa	RJ	Banas	Chambal				2,377	2,260		PG	22			I
Gandhi Sagar	Chambal	Mandasur	MP	Chambal	Chambal	1960	660		7,320,000	6,800,000	21,238	PG	62.17	514		I/H
Bisalpur	Banas	Deoli	RJ	Banas	Chambal	1999	212.3		1,100,000	890,000	29,046	PG	39.5	574	1,096,000	I/S
Ranapratap Sagar	Chambal	Kota	RJ	Chambal	Chambal	1969	198.29		2,900,000	1,570,000	18,400	PG	53.59	1,143	567	I/H/S
Storage Capacity of Chambal Sub-basin									**19,612,066**	**15,581,173**						
Bardha	Local	Chhatarpur	MP	Ken	Ken	1995			1,350	1,215	0	TE	12.05	396	0	IRR
Bargone		Chhatarpur	MP	Ken	Ken	1974	76.85	7,620	27,580	26,220	244	TE	22.3	3,000	340	IRR.

Name of dam	River/ tributary	District	State	Sub-basin	Basin	Year of completion	Catch area (km^2)	Submergence ('000 km^2)	Gross capacity ('000 cu m)	Live capacity ('000 cu m)	Spillway capacity ('000 cumec)	Type of dam	HT of dam (m)	Length of dam (m)	Vol of dam ('000 cu m)	Purpose
Barkheda		Chhatarpur	MP	Ken	Ken	1969	47.9	4,391	10,849	10,377	179	TE	15.85	1,410	146	IRR.
BarkhediSiloki		Chhatarpur	MP	Ken	Ken	1975	7.38	971	2,323	2,150	56	TE	10.82	1,220		IRR.
BarothaTank		Damoh	MP	Ken	Ken	1929	161.3		19,960	16,870	887	TE	16.76	2,518		IRR.
Barrach		Damoh	MP	Ken	Ken	1962	37.54	171	5,160	4,900	226	TE	22.8	1,341		IRR.
Basinkhar		Damoh	MP	Ken	Ken	1959	16.36	1,821	4,050	3,390	148	TE	20.03	1,587		IRR.
Basni		Damoh	MP	Ken	Ken	1981	6.67	1,620	4,210	3,920	61	TE	14.6	1,740		IRR.
Bhaisawar		Damoh	MP	Ken	Ken	1913	11.52	460	2,550	2,517	132	TE	11.95	1,575		IRR.
Bhirata		Damoh	MP	Ken	Ken	1994	3.38	443	1,280	1,224	26	TE	16.16	568		IRR.
Hashalpur		Panna	MP	Ken	Ken	1977	13.83	736	4,760	4,528	346	TE	17.38	540		IRR.
Hashalpur		Panna	MP	Ken	Ken	1961		440	1,588	1,481	152	TE	12.8	265		IRR.
Hathaikheda		Panna	MP	Ken	Ken	1975	5.5		1,163	1,073	70	TE	21.12	705		IRR.
Hathaikheda		Panna	MP	Ken	Ken	1989	5.12		1,627	1,498		TE	12.2	630		IRR.
Karmodia		Sagar	MP	Ken	Ken	1985	2.59	277	1,267	1,207	46	TE	18	465		IRR.
Kasturipura		Sagar	MP	Ken	Ken	1969	1.84		886	781	30	TE	16.46	767		IRR.
Katkiya		Sagar	MP	Ken	Ken	1926	79.98		6,000	5,790	225	TE	23	310		IRR.
Kazikhedi		Sagar	MP	Ken	Ken	1978	30.72	1,800	13,280	12,160	256	TE	24.58	945		I/W
Mongra		Tikamgarh	MP	Ken	Ken		9.32		3,499	3,123	95	TE	10.97	854		IRR.
MotiNalla		Tikamgarh	MP	Ken	Ken		7.8		1,732	1,559	41	TE	12.5	681		IRR.
Naren	Karnari	Vidisha	MP	Ken	Ken	1956	15.54	1,041	3,580	3,481	91	TE	14.6	1,560	468	IRR
Narola	Local	Vidisha	MP	Ken	Ken	1964	3.37	708	1,320	1,260	40	TE	11.2	330	65	IRR
Panpura			MP	Ken	Ken	1974	76.85	7,620	27,580	26,220	244	TE	22.3	3,000	340	IRR.
Paronch			MP	Ken	Ken	1926	79.98		6,000	5,790	225	TE	23	310		IRR.
Roomal			MP	Ken	Ken	1962	37.54	171	5,160	4,900	226	TE	22.8	1,341		IRR.
S.AshokSagar			MP	Ken	Ken	1913	11.52	460	2,550	2,517	132	TE	11.95	1,575		IRR.
Sagonikala			MP	Ken	Ken	1961		440	1,588	1,481	152	TE	12.8	265		IRR.
Siloda			MP	Ken	Ken	1912	3		675	665	44		10.94	546		IRR
SimariyaNo.1			MP	Ken	Ken	1919	7.12		3,482	3,134	38	TE	11.62	1,097		IRR.
Sudra			MP	Ken	Ken	1917	5.18	1,000	1,393	1,236	72	TE	11.93	1,073		IRR.
Ghorapachhar		Katni	MP	Ken	Ken	1912	14.89		1,250	1,052		TE	11.74	1,280		IRR.
Ghorapachhar		Katni	MP	Ken	Ken	1966	4.27		1,610	680	39	TE	10.66	1,037		IRR.
Gidwasan		Katni	MP	Ken	Ken	1912	3		675	665	44		10.94	546		IRR
Gangau	Ken	Chhatar Pur	UP	Ken	Ken	1915	149.9		58,472	52,625	13,723	PG	16.15	826		I

Name of dam	River/ tributary	District	State	Sub-basin	Basin	Year of completion	Catch area (km^2)	Submergence ('000 km^2)	Gross capacity ('000 cu m)	Live capacity ('000 cu m)	Spillway capacity ('000 cumec)	Type of dam	HT of dam (m)	Length of dam (m)	Vol of dam ('000 cu m)	Purpose
Raipura	Arjun nala	Mahoba	UP	Ken	Ken	1930			12,453	6,289	283	TE	13	3,509		I
Kabrai	Magaria	Mahoba	UP	Ken	Ken	1955	50.5		14,617	11,944	387	TE	18.24	2,440	425,000	I
Arjun	Arjun	Mahoba	UP	Ken	Ken	1957	180		69,206	63,826	678	TE	25.88	4,206	1,823	I
Rangawan	Bannel Nala	Chhatar Pur	UP	Ken	Ken	1957	25.9		164,238	155,177	2,407.93	TE	27.4	2,073		I
Salar Pur	Kardia	Mahoba	UP	Ken	Ken	1960			4,020	3483	158	TE	11	2,975	13,406	I
Urmil	Urmil	Mahoba	UP	Ken	Ken	1994	250.3		116,600	60,310	5,664	TE	25.56	4,700		I
Maudaha	Birma	Mahoba	UP	Ken	Ken	2003	54.39		200,000	179,000	111,328	TE	22	3,480	1,554	I
Virat Sagar	Dhasan	Mahoba	UP	Ken	Ken	U/C							180.5	1,550		
Storage Capacity of Ken Sub-basin									**811,583**	**691,717**						
Deorikheda	Local	Guna	MP	Sind	Sind	1915	31.03	1,870	4,530	4,361	256	TE	12.19	2,790		IRR.
Depalpur	Local	Guna	MP	Sind	Sind	1916	12.3	932	2,070	1,816	125	TE	12.19	1,676		IRR.
DevendraSagar	Local	Guna	MP	Sind	Sind	1916	16.77	1,094	2,670	2,363	113	TE	12.2	1,221		IRR.
Dhablamata	Local	Guna	MP	Sind	Sind	1995	8.62	867	1,910	1,690	119	TE	17.68	1,800		IRR.
Dhantoli	Parwati	Gwalior	MP	Sind	Sind	1917	1,960	25,230	192,670	179,020	4,758	TE	29.28	2,138		IRR.
Dhapora	Sank	Gwalior	MP	Sind	Sind	1917	414.24	19,319	130,800	134,240	4,067	TE/PG	24.08	1,524		I/WS
Dhapora	Local	Gwalior	MP	Sind	Sind	1931	75.11	8,533	12,960	12,827	708	TE	21.33	384		IRR.
DharamSagar	Local	Gwalior	MP	Sind	Sind	1895	64.75	2,780	8,290	8,440	425	TE	16.47	975		I/WS
Dharwara	Local	Gwalior	MP	Sind	Sind	1976	25.9	9,710	4,308	4,150	250	TE	14.17	2,610		IRR.
Dholawad	Local	Gwalior	MP	Sind	Sind	1913	26	744	2,384	2,146	532	TE	12.19	660		IRR.
Dhumdhuma	Mower	Gwalior	MP	Sind	Sind	1984	83		44,250	16,920	292	TE	24.05	2,150		I/WS
LowerSakarwara	Sindh	Shivpuri	MP	Sind	Sind	1989	5,944	16,243	108	910	265,400	TE/PG	35.63	2,568.7		IRR.
LowerSakarwara		Shivpuri	MP	Sind	Sind	2003	58.28		324	292	324	TE	16.11	690		IRR
MadanSagar	Local	Shivpuri	MP	Sind	Sind	1913	31.06	2,030	6,234	6,154	198	TE	12.5	1,688		
MadanSagar	Local	Shivpuri	MP	Sind	Sind	1966	30.41	1,050	4,758	4,320	253	TE	21.03	580		IRR.
MadarkhaBandhi	Local	Shivpuri	MP	Sind	Sind	1984	22.01	300	2,350	1,830	184	TE	14.08	206		IRR.
MadarkhaBandhi	Local	Shivpuri	MP	Sind	Sind	1918	72.53	2,130	7,780	7,775	482	PG	13.81	2,145		I/WS
Madhar	Local	Shivpuri	MP	Sind	Sind	1969	9.83	1,404	5,870	5,636	147	TE	13.71	810		IRR.
Maglipada	Local	Shivpuri	MP	Sind	Sind	1977	8.1	940	1,630	1,427	97	TE	18.58	652		IRR.
Maglipada	Local	Shivpuri	MP	Sind	Sind	1996	32.68		6,450	5,700	180	TE	11.33	1,950		IRR.
Mahendra Sagar	Local	Shivpuri	MP	Sind	Sind	1991	14.33	920	2,013	1,668	144	TE	15.51	465		IRR.
Majhgawan	Local	Shivpuri	MP	Sind	Sind	1907	51.2		2,050	131	13,200	TE	12.19	3,109		IRR.

Name of dam	River/ tributary	District	State	Sub-basin	Basin	Year of comple-tion	Catch area (km^2)	Sub-mergence ('000 km^2)	Gross capacity ('000 cu m)	Live capacity ('000 cu m)	Spillway capacity ('000 cumec)	Type of dam	HT of dam (m)	Length of dam (m)	Vol of dam ('000 cu m)	Purpose
Majhgawan	Local	Shivpuri	MP	Sind	Sind	1973	77.7	4,310	11,710	9,870	269	TE	15.1	2,055		IRR.
Mala	Local	Shivpuri	MP	Sind	Sind	1978	17.78	1,000	1,800	1,620	170	TE	10.82	1,067		IRR.
Malavar	Local	Shivpuri	MP	Sind	Sind	1987	7.12	656	1,450	1,276	42	TE	15.24	1,050		IRR.
Manjikhedi	Parwati	Shivpuri	MP	Sind	Sind	1934	1,035.6	8,907	80,510	79,180	1,811	PG	37.64	1,047		IRR.
Moorumnalla		Tikamgarh	MP	Sind	Sind	1986	8.02		1,825	1,643	50	TE	12.42	236		IRR.
Pabra	Local	Guna	MP	Sind	Sind	1916	16.77	1,094	2,670	2,363	113	TE	12.2	1,221		IRR.
Phaliamao	Local	Shivpuri	MP	Sind	Sind	1969	9.83	1,404	5,870	5,636	147	TE	13.71	810		IRR.
Piplai	Local	AshokNagar	MP	Sind	Sind	1915	31.03	1,870	4,530	4,361	256	TE	12.19	2,790		IRR.
Pipliyakumar	Local	AshokNagar	MP	Sind	Sind	1916	12.3	932	2,070	1,816	125	TE	12.19	1,676		IRR.
Ramova	Local	Shivpuri	MP	Sind	Sind	1913	31.06	2,030	6,234	6,154	198	TE	12.5	1,688		
ShivgarthBedli	Parwati	Dabra	MP	Sind	Sind	1934	1,035.6	8,907	80,510	79,180	1,811	PG	37.64	1,047		IRR.
Tonkara	Local	Gohad	MP	Sind	Sind	1919	1,140	10,700	15,056	13,550	1,572	TE	12.19	2,134		IRR.
Tora	Local	Sabalgarh	MP	Sind	Sind	1910	28.5	1,852	4,493	4,106	125	TE	12.34	914		IRR.
Udiyapura	Asan	Joura	MP	Sind	Sind	1927	520	13,730	120,530	108,477	1,330	TE/PG	27	1,683		IRR.
Umariya	Asan	Morena	MP	Sind	Sind	1914	1,036	18,390	91,550	90,382	4,918	TE	19.98	1,158		SR
Umariya	Sank	Morena	MP	Sind	Sind	1914	251.42	6,712	16,315	9,645	1,377	TE/PG	13.79	658		IRR.
Upper Chhirpuri	Local	Shivpuri	MP	Sind	Sind	1918	72.53	2,130	7,780	7,775	482	PG	13.81	2,145		I/WS
Upper Palakmati	Local	Karera	MP	Sind	Sind	1907	51.2		2,050	131	13,200	TE	12.19	3,109		IRR.
Yashvant Sagar	Parwati	Dabra	MP	Sind	Sind	1917	1,960	25,230	192,670	179,020	4,758	TE	29.28	2,138		IRR.
YashwantNagar	Sank	Gwalior	MP	Sind	Sind	1917	414.24	19,319	130,800	134,240	4,067	TE/PG	24.08	1,524		I/WS
YashwantNagar	Local	Gwalior	MP	Sind	Sind	1931	75.11	8,533	12,960	12,827	708	TE	21.33	384		IRR.
ZotKheti	Local	Gwalior	MP	Sind	Sind	1895	64.75	2,780	8,290	8,440	425	TE	16.47	975		I/WS
Storage Capacity of Sind Sub-basin									**1,248,082**	**1,165,507**						
Kishau	Tons	Dehradun	UK	Tons	Yamuna	U/C	21.7		1,810,000	133,000	23,019	PG	236	680	9,500	I/H
Lakhwar	Yamuna	Dehradun	UK	Yamuna	Yamuna	U/C	29.5		580,000	330,000	8,000	PG	204	451.78	2,800	I/H
Renuka	Giri	Shimla	HP	Giri	Yamuna	U/C				49,800,000						
Vyasi	Yamuna	Dehradun	UK	Yamuna	Yamuna	U/C	9.65		1,160,000	1,044,000	8,000	PG	88	203	2,250	I/H
Storage Capacity of Yamuna Upper Catchment									**3,550,000**	**51,307,000**			148	844.5		
Storage Capacity of Yamuna Basin									**30,144,643**	**73,118,663**						

Appendix B
District-Wise Domestic Water Demand and Groundwater Resources

S. No.	District	State	Area (km^2)	Total population	Urban population	Rural population	Domestic water demand (MCM)	Total recharge (MCM)	GW flow (MCM)	Net GW availability (MCM)	Total draft (MCM)	Stage GW develop-ment (%)
1	Kinnaur	HP	72.9	874	0	874	0.01	1.16	0.12	1.05	0.28	26.42
2	Shimla	HP	3,660.4	622,266	143,121	479,145	14.05	33.25	3.33	29.93	6.71	22.42
3	Uttarkashi	UK	2,580.2	95,466	7,637	87,829	1.66	0.00	0.00	0.00	0.00	0
4	Dehradun	UK	2,135.3	886,158	469,664	416,494	29.22	373.87	18.69	355.18	20.99	5.91
5	Sirmaur	HP	2,375.9	484,675	48,468	436,207	8.76	81.16	8.12	73.04	12.87	17.62
6	Tehri Garhwal	UK	379.2	51,948	5,195	46,753	0.94	0.00	0.00	0.00	0.00	0
7	Yamunanagar	HR	966.1	570,006	216,602	353,404	15.83	310.62	31.06	279.56	282.42	101.02
8	Saharanpur	UP	3,414.8	2,680,579	696,951	1,983,628	63.30	1,491.12	109.43	1,381.69	1,265.67	91.6
9	Hardwar	UK	329.8	202,175	62,674	139,501	5.12	136.21	13.62	122.59	118.18	96.4
10	Kurukshetra	HR	197.2	101,759	26,457	75,302	2.40	53.32	3.45	49.88	82.60	165.62
11	Kaithal	HR	837.3	331,581	63,000	268,581	7.03	219.23	21.92	197.30	352.65	178.73
12	Ganganagar	RJ	1,331.5	114,505	85,879	28,626	4.65	22.22	2.22	19.99	9.07	45.34
13	Sirsa	HR	1,189.2	310,376	80,698	229,678	7.33	171.94	8.60	163.35	186.73	114.32
14	Karnal	HR	2,091.8	1,370,155	369,942	1,000,213	32.83	1,005.69	60.48	945.21	1,296.29	137.14
15	Jind	HR	2,683.4	1,183,366	236,673	946,693	25.48	750.22	54.26	695.96	485.20	69.72
16	Hissar	HR	5,048.5	1,181,347	307,150	874,197	27.90	467.98	38.47	429.51	289.97	67.51
17	Muzaffarnagar	UP	2,697.3	2,384,413	619,947	1,764,466	56.31	1,268.43	109.39	1,159.04	952.31	82.16
18	Panipat	HR	1,585.4	875,135	358,805	516,330	25.22	326.57	28.82	297.75	464.88	156.13
19	Meerut	UP	2,195.4	2,566,376	1,231,860	1,334,516	80.18	1,112.90	79.71	1,033.18	670.89	64.93
20	Rohtak	HR	4,735.6	1,003,954	351,384	652,570	26.84	293.75	23.82	269.93	179.04	66.33
21	Sonipat	HR	1,221.9	1,158,323	289,581	868,742	26.95	428.65	21.43	407.22	462.94	113.68
22	Bhiwani	HR	4,883.7	1,509,070	286,723	1,222,347	31.97	656.95	44.40	612.56	636.27	103.87
23	Churu	RJ	1,380.8	158,790	114,329	44,461	6.28	11.73	1.12	10.61	7.95	74.89
24	Ghaziabad	UP	1,700.6	2,945,491	1,472,746	1,472,745	94.07	1,119.37	83.03	1,036.35	664.03	61.6
25	Delhi	DL	1,471.0	13,844,958	12,875,811	969,147	836.59	296.98	15.53	281.45	479.25	170.28
26	Bulandshahr	UP	1,555.9	1,048,670	241,194	807,476	23.67	587.21	54.38	532.83	349.30	65.56
27	Gurgaon	HR	2,823.5	1,705,388	375,185	1,330,203	37.91	489.25	46.50	442.76	641.51	144.89

S. No.	District	State	Area (km^2)	Total population	Urban population	Rural population	Domestic water demand (MCM)	Total recharge (MCM)	GW flow (MCM)	Net GW availability (MCM)	Total draft (MCM)	Stage GW development (%)
28	Jhunjhunun	RJ	4,254.7	1,378,536	289,493	1,089,043	30.16	186.66	17.31	169.35	338.77	200.04
29	Faridabad	HR	2,013.7	1,933,104	1,082,538	850,566	65.76	510.70	25.54	485.16	245.02	50.5
30	Mahendragarh	HR	1,989.9	833,785	108,392	725,393	15.93	207.65	20.77	186.88	200.37	107.22
31	Rewari	HR	1,347.8	665,803	119,845	545,958	13.88	245.32	18.38	226.94	273.31	120.43
32	Alwar	RJ	8,168.9	2,801,946	420,292	2,381,654	55.48	811.66	71.24	740.42	1,071.63	144.73
33	Sikar	RJ	1,884.0	552,003	115,921	436,082	12.08	83.45	8.27	75.19	100.56	133.75
34	Aligarh	UP	2,923.3	2,364,942	685,833	1,679,109	58.31	827.82	67.15	760.67	615.00	80.85
35	Etah	UP	700.0	439,591	74,730	364,861	9.01	197.47	19.75	177.72	139.22	78.34
36	Mathura	UP	3,803.1	2,369,319	1,705,910	663,409	93.74	1,236.65	123.67	1,112.98	914.67	82.18
37	Jaipur	RJ	11,041.5	4,206,812	2,061,338	2,145,474	132.90	536.32	48.67	487.65	909.94	186.6
38	Bharatpur	RJ	4,777.6	1,992,263	378,530	1,613,733	42.21	469.31	39.65	429.66	429.21	99.89
39	Firozabad	UP	2,093.8	1,819,530	545,859	1,273,671	45.49	688.86	63.59	625.27	503.27	80.49
40	Mainpuri	UP	1,567.1	910,503	136,575	773,928	18.03	528.91	43.18	485.73	397.44	81.82
41	Agra	UP	4,383.0	3,940,326	1,694,340	2,245,986	116.28	1,166.67	116.67	1,050.01	931.24	88.69
42	Sawai Madhopur	RJ	10,651.7	1,129,080	914,555	214,525	48.20	409.13	38.48	370.65	418.71	112.97
43	Etawah	UP	4,240.5	4,388,959	833,902	3,555,057	92.99	2,729.75	223.16	2,506.59	1,248.21	49.8
44	Ajmer	RJ	5,514.8	1,417,301	566,920	850,381	40.35	229.06	21.26	207.80	255.15	122.79
45	Dholpur	RJ	2,879.7	941,659	169,499	772,160	19.63	236.97	21.41	215.56	229.70	106.56
46	Kanpur (D)	UP	3,770.0	1,157,375	81,016	1,076,359	19.71	710.87	52.18	658.69	325.27	49.38
47	Morena	MP	11,428.8	1,565,746	344,464	1,221,282	34.80	912.83	45.64	867.19	234.30	27.02
48	Bhind	MP	4,689.1	1,505,185	361,244	1,143,941	34.50	830.35	41.52	788.83	197.49	25.04
49	Kanpur (N)	UP	186.1	745,801	499,687	246,114	28.22	180.22	14.43	165.79	111.63	67.33
50	Tonk	RJ	7,213.7	1,204,695	252,986	951,709	26.36	421.73	32.48	389.25	375.20	96.39
51	Jalaun	UP	4,761.8	1,519,008	349,372	1,169,636	34.29	1,006.57	82.65	923.92	300.25	32.5
52	Gwalior	M	5,068.6	1,561,136	936,682	624,454	55.27	662.07	33.10	628.97	184.78	29.38
53	Datia	MP	1,896.9	618,396	136,047	482,349	13.75	433.64	21.68	411.96	182.45	44.29
54	Fatehpur	UP	3,642.4	2,025,191	202,519	1,822,672	36.59	1,218.65	83.67	1,134.98	800.43	70.52

S. No.	District	State	Area (km^2)	Total population	Urban population	Rural population	Domestic water demand (MCM)	Total recharge (MCM)	GW flow (MCM)	Net GW availability (MCM)	Total draft (MCM)	Stage GW develop-ment (%)
55	Hamirpur	UP	7,020.7	3,004,860	570,923	2,433,937	63.67	1,344.57	94.78	1,249.78	716.44	57.33
56	Udaipur	RJ	6,997.3	1,147,559	218,036	929,523	24.31	129.19	12.92	116.27	132.31	113.79
57	Bhilwara	RJ	10,451.4	2,027,572	425,790	1,601,782	44.37	435.98	43.55	392.43	455.27	116.01
58	Jhansi	UP	5,364.2	1,861,384	763,167	1,098,217	53.64	757.16	43.68	713.48	305.55	42.83
59	Shivpuri	MP	10,331.8	1,673,752	284,538	1,389,214	34.30	1,221.22	61.06	1,160.16	791.43	68.22
60	Banda	UP	6,996.8	1,406,361	225,018	1,181,343	28.34	806.58	69.51	737.06	270.57	36.71
61	Bundi	RJ	5,482.1	910,027	172,905	737,122	19.28	318.23	72.81	245.42	266.00	108.39
62	Pali	RJ	26.8	5,684	1,193	4,491	0.12	0.97	0.09	0.88	1.01	114.4
63	Kota	RJ	12,513.7	1,601,754	848,930	752,824	52.82	462.21	45.61	416.60	482.49	115.82
64	Allahabad	UP	2,160.8	1,469,371	352,649	1,116,722	33.68	356.28	30.41	325.88	225.25	69.12
65	Tikamgarh	MP	5,076.6	1,335,154	240,328	1,094,826	27.83	895.01	44.75	850.26	436.96	51.39
66	Chhatapur	MP	8,620.6	1,474,128	324,308	1,149,820	32.77	998.84	49.94	948.90	536.34	56.52
67	Lalitpur	UP	5,038.4	977,442	146,616	830,826	19.35	678.05	55.38	622.67	321.92	51.7
68	Chittaurgarh	RJ	8,252.9	1,444,261	231,082	1,213,179	29.10	349.62	34.31	315.31	449.13	142.44
69	Rewa	MP	0.3	84	13	71	0.00	0.02	0.00	0.02	0.01	41.82
70	Satna	MP	1,164.4	292,259	61,374	230,885	6.40	101.30	5.06	96.23	66.12	68.71
71	Guna	MP	11,253.9	1,710,593	359,225	1,351,368	37.43	662.37	33.12	629.25	342.25	54.39

S. No.	District	State	Area (km^2)	Total population	Urban population	Rural population	Domestic water demand (MCM)	Total recharge (MCM)	GW flow (MCM)	Net GW availability (MCM)	Total draft (MCM)	Stage GW develop-ment (%)
72	Panna	MP	6,658.1	952,104	123,774	828,330	18.19	570.21	28.51	541.70	132.46	24.45
73	Mandsaur	MP	10,101.6	1,333,411	253,348	1,080,063	28.25	839.76	41.99	797.77	873.26	109.46
74	Jhalawar	RJ	6,271.4	1,179,023	165,063	1,013,960	22.94	456.28	26.83	429.44	451.57	105.15
75	Sagar	MP	9,774.3	2,375,157	688,796	1,686,361	58.56	1,428.39	71.42	1,356.97	632.93	46.64
76	Damoh	MP	7,011.7	1,444,408	274,438	1,169,970	30.60	588.88	29.44	559.44	292.96	52.37
77	Vidisha	MP	7,358.6	1,228,881	258,065	970,816	26.89	701.28	35.06	666.22	313.07	46.99
78	Shajapur	MP	6,222.5	1,306,731	248,279	1,058,452	27.69	559.34	27.97	531.38	601.88	113.27
79	Rajgarh	MP	6,365.4	1,311,264	222,915	1,088,349	26.87	979.42	48.97	930.45	723.13	77.72
80	Jabalpur	MP	840.5	194,991	111,145	83,846	6.70	48.71	2.44	46.28	19.43	41.99
81	Ratlam	MP	2,882.0	720,510	216,153	504,357	18.01	377.30	18.87	358.44	420.51	117.32
82	Bhopal	MP	2,820.4	1,923,506	1,538,805	384,701	81.44	357.73	17.89	339.84	242.39	71.32
83	Ujjain	MP	6,086.2	1,716,306	669,359	1,046,947	48.27	836.91	41.85	795.06	866.80	109.02
84	Raisen	MP	3,598.1	546,905	98,443	448,462	11.40	674.28	33.71	640.56	220.90	34.48
85	Sehore	MP	3,099.3	536,181	96,513	439,668	11.17	392.58	19.63	372.95	232.82	62.43
86	Dewas	MP	2,993.7	628,666	169,740	458,926	15.06	445.25	22.26	422.99	280.82	66.39
87	Narsingpur	MP	135.1	28,502	4,560	23,942	0.57	36.03	1.80	34.23	21.39	62.5
88	Dhar	MP	1,563.8	331,530	56,360	275,170	6.79	210.24	10.51	199.73	198.99	99.63
89	Indore	MP	2,896.7	2,047,931	1,433,552	614,379	79.61	497.58	24.88	472.70	493.68	104.44

Appendix C
Area to Point Rainfall Ratios (%) for Deriving the Hourly Distribution of Daily Rainfall

Area	Duration (hours)												
(sq miles)	0.50	1.00	2.00	3.00	4.00	5.00	6.00	9.00	12.00	15.00	18.00	21.00	24.00
10.0	68.33	76.39	82.66	85.60	87.40	88.65	89.59	91.41	92.52	93.28	93.85	94.29	94.65
20.0	61.89	71.23	78.67	82.21	84.40	85.92	87.06	89.31	90.67	91.61	92.31	92.86	93.31
30.0	57.74	67.81	75.98	79.91	82.35	84.05	85.34	87.86	89.39	90.46	91.25	91.87	92.38
40.0	54.63	65.21	73.91	78.13	80.76	82.60	83.99	86.72	88.39	89.55	90.41	91.09	91.64
50.0	52.14	63.10	72.21	76.65	79.43	81.39	82.86	85.77	87.55	88.79	89.71	90.44	91.03
60.0	50.05	61.30	70.75	75.39	78.30	80.34	81.89	84.95	86.83	88.13	89.11	89.87	90.49
70.0	48.26	59.74	69.47	74.27	77.29	79.42	81.03	84.22	86.18	87.55	88.57	89.37	90.02
80.0	46.69	58.36	68.33	73.27	76.39	78.59	80.26	83.57	85.60	87.02	88.08	88.91	89.59
90.0	45.28	57.11	67.29	72.37	75.57	77.84	79.56	82.97	85.07	86.53	87.63	88.49	89.19
100.0	44.02	55.98	66.35	71.54	74.82	77.15	78.91	82.42	84.58	86.09	87.22	88.11	88.83
110.0	42.87	54.94	65.47	70.77	74.12	76.50	78.31	81.90	84.12	85.67	86.83	87.75	88.49
120.0	41.81	53.98	64.66	70.05	73.47	75.90	77.75	81.42	83.70	85.28	86.47	87.41	88.17
130.0	40.84	53.09	63.91	69.38	72.86	75.34	77.22	80.97	83.29	84.92	86.14	87.09	87.87
140.0	39.94	52.25	63.19	68.75	72.29	74.81	76.72	80.55	82.91	84.57	85.81	86.76	87.59
150.0	39.09	51.47	62.52	68.15	71.74	74.30	76.25	80.14	82.55	84.24	85.51	86.51	87.32

Area	Duration (hours)												
(sq miles)	0.50	1.00	2.00	3.00	4.00	5.00	6.00	9.00	12.00	15.00	18.00	21.00	24.00
160.0	38.30	50.73	61.89	67.58	71.23	73.82	75.80	79.76	82.21	83.93	85.22	86.24	87.06
170.0	37.56	50.03	61.28	67.05	70.74	73.37	75.38	79.39	81.88	83.63	84.94	85.98	86.82
180.0	36.86	49.37	60.71	66.53	70.27	72.93	74.97	79.04	81.57	83.34	84.67	85.73	86.58
190.0	36.19	48.74	60.16	66.04	69.82	72.52	74.57	78.70	81.27	83.07	84.42	85.49	86.36
200.0	35.57	48.14	59.64	65.57	69.38	72.11	74.20	78.37	80.98	82.80	84.17	85.26	86.14
210.0	34.97	47.57	59.13	65.12	68.97	71.73	73.84	78.06	80.70	82.54	83.94	85.03	85.93
220.0	34.40	47.02	58.65	64.68	68.57	71.36	73.49	77.76	80.43	82.30	83.71	84.82	85.72
230.0	33.85	46.49	58.18	64.26	68.19	71.00	73.15	77.47	80.16	82.06	83.48	84.61	85.53
240.0	33.33	45.99	57.74	63.86	67.81	70.65	72.82	77.19	79.91	81.83	83.27	84.41	85.34
250.0	32.84	45.50	57.30	63.47	67.45	70.32	72.51	76.91	79.67	81.60	83.06	84.21	85.15
260.0	32.36	45.03	56.88	63.09	67.11	69.99	82.20	76.65	79.43	81.38	82.86	84.02	84.97
270.0	31.90	44.58	56.48	62.72	66.77	69.68	71.90	76.39	79.20	81.17	82.66	83.84	84.80
280.0	31.46	44.14	56.09	62.37	66.44	69.37	71.62	76.14	78.97	80.97	82.47	83.66	84.63
290.0	31.03	43.72	55.71	62.02	66.12	69.07	71.34	75.90	78.75	80.76	82.28	83.48	84.46
300.0	30.62	43.31	55.34	61.69	65.81	68.78	71.06	75.66	78.54	80.57	82.10	83.31	84.30

Sources: CWC (1973). Estimation of design flood peak: A method based on unit hydrograph principle, Rep. No. 1/73, Hydrology for Small Catchments Directorate, Central Water Commission, Government of India, New Delhi

Index

MIX
Papier aus verantwortungsvollen Quellen
Paper from responsible sources
FSC® C105338

If you have any concerns about our products,
you can contact us on
ProductSafety@springernature.com

In case Publisher is established outside the EU,
the EU authorized representative is:
Springer Nature Customer Service Center GmbH
Europaplatz 3, 69115 Heidelberg, Germany

Printed by Libri Plureos GmbH
in Hamburg, Germany